Hydrogen Fuel Cell Technology for Stationary Applications

Gheorghe Badea
Technical University of Cluj-Napoca, Romania

Raluca-Andreea Felseghi
Ştefan cel Mare University of Suceava, Romania

Ioan Aşchilean
Technical University of Cluj-Napoca, Romania

IGI Global
PUBLISHER of TIMELY KNOWLEDGE

A volume in the Advances in Computer and Electrical Engineering (ACEE) Book Series

Published in the United States of America by
IGI Global
Engineering Science Reference (an imprint of IGI Global)
701 E. Chocolate Avenue
Hershey PA, USA 17033
Tel: 717-533-8845
Fax: 717-533-8661
E-mail: cust@igi-global.com
Web site: http://www.igi-global.com

Library of Congress Cataloging-in-Publication Data

Names: Badea, Gheorghe, editor. | Felseghi, Raluca Andreea, 1978- editor. | Aschilean, Ioan, 1966- editor.
Title: Hydrogen fuel cell technology for stationary applications / Gheorghe Badea, Raluca Andreea Felseghi ; Ioan Aschilean, editors.
Description: Hershey, PA : Engineering Science Reference, 2020. | Includes bibliographical references and index. | Summary: "This book focuses on the advantages of hydrogen as a primary energy center and addresses its use in the sustainable future of stationary applications. While highlighting a broad range of topics including cost expectations, production methods, and social impact, this publication explores all aspects of the implementation and dissemination of fuel cell technology in the hope of establishing a sustainable marketplace for it"-- Provided by publisher.
Identifiers: LCCN 2020009145 (print) | LCCN 2020009146 (ebook) | ISBN 9781799849452 (h/c) | ISBN 9781799851301 (s/c) | ISBN 9781799849469 (eISBN)
Subjects: LCSH: Fuel cells. | Fuel cell industry.
Classification: LCC TK2931 .H957 2020 (print) | LCC TK2931 (ebook) | DDC 338.4/7621312429--dc23
LC record available at https://lccn.loc.gov/2020009145
LC ebook record available at https://lccn.loc.gov/2020009146

This book is published in the IGI Global book series Advances in Computer and Electrical Engineering (ACEE) (ISSN: 2327-039X; eISSN: 2327-0403)

British Cataloguing in Publication Data
A Cataloguing in Publication record for this book is available from the British Library.

All work contributed to this book is new, previously-unpublished material. The views expressed in this book are those of the authors, but not necessarily of the publisher.

For electronic access to this publication, please contact: eresources@igi-global.com.

Advances in Computer and Electrical Engineering (ACEE) Book Series

Srikanta Patnaik
SOA University, India

ISSN:2327-039X
EISSN:2327-0403

Mission

The fields of computer engineering and electrical engineering encompass a broad range of interdisciplinary topics allowing for expansive research developments across multiple fields. Research in these areas continues to develop and become increasingly important as computer and electrical systems have become an integral part of everyday life.

The **Advances in Computer and Electrical Engineering (ACEE) Book Series** aims to publish research on diverse topics pertaining to computer engineering and electrical engineering. **ACEE** encourages scholarly discourse on the latest applications, tools, and methodologies being implemented in the field for the design and development of computer and electrical systems.

Coverage

- Electrical Power Conversion
- Chip Design
- Sensor Technologies
- Applied Electromagnetics
- Computer Hardware
- VLSI Fabrication
- Programming
- Computer Architecture
- Circuit Analysis
- Algorithms

IGI Global is currently accepting manuscripts for publication within this series. To submit a proposal for a volume in this series, please contact our Acquisition Editors at Acquisitions@igi-global.com or visit: http://www.igi-global.com/publish/.

The Advances in Computer and Electrical Engineering (ACEE) Book Series (ISSN 2327-039X) is published by IGI Global, 701 E. Chocolate Avenue, Hershey, PA 17033-1240, USA, www.igi-global.com. This series is composed of titles available for purchase individually; each title is edited to be contextually exclusive from any other title within the series. For pricing and ordering information please visit http://www.igi-global.com/book-series/advances-computer-electrical-engineering/73675. Postmaster: Send all address changes to above address.

Titles in this Series

For a list of additional titles in this series, please visit:
http://www.igi-global.com/book-series/advances-computer-electrical-engineering/73675

Computational Methodologies for Electrical and Electronics Engineers
Rajiv Singh (G.B. Pant University of Agriculture and Technology, Pantnagar, India) Ashutosh Kumar Singh (Indian Institute of Information Technology, Allahabad, India) Ajay Kumar Dwivedi (SIET, India & Dr. A.P.J. Abdul Kalam Technical University, Lucknow, India) and P. Nagabhushan (Indian Institute of Information Technology, Allahaba, India)
Engineering Science Reference • © 2021 • 281pp • H/C (ISBN: 9781799833277) • US $225.00

Innovations in the Industrial Internet of Things (IIoT) and Smart Factory
Sam Goundar (British University Vietnam, Vietnam) J. Avanija (Sree Vidyanikethan Engineering College, India) Gurram Sunitha (Sree Vidyanikethan Engineering College, India) K. Reddy Madhavi (Sree Vidyanikethan Engineering College, India) and S. Bharath Bhushan (Sree Vidyanikethan Engineering College, Tirupati, India)
Engineering Science Reference • © 2021 • 311pp • H/C (ISBN: 9781799833758) • US $225.00

Innovative Applications of Nanowires for Circuit Design
Balwinder Raj (National Institute of Technical Teachers Training and Research, Chandigarh, India)
Engineering Science Reference • © 2021 • 263pp • H/C (ISBN: 9781799864677) • US $225.00

Optimizing and Measuring Smart Grid Operation and Control
Abdelmadjid Recioui (Université M'hamed Bougara de Boumerdes, Algeria) and Hamid Bentarzi (Université M'hamed Bougara de Boumerdes, Algeria)
Engineering Science Reference • © 2021 • 400pp • H/C (ISBN: 9781799840275) • US $225.00

New Methods and Paradigms for Modeling Dynamic Processes Based on Cellular Automata
Stepan Mykolayovych Bilan (State University of Infrastructure and Technology, Ukraine) Mykola Mykolayovych Bilan (Mayakskaya Secondary School, Moldova) and Ruslan Leonidovich Motornyuk (Main Information and Computing Center, Ukraine)
Engineering Science Reference • © 2021 • 326pp • H/C (ISBN: 9781799826491) • US $195.00

Examining Quantum Algorithms for Quantum Image Processing
HaiSheng Li (Guangxi Normal University, China)
Engineering Science Reference • © 2021 • 336pp • H/C (ISBN: 9781799837992) • US $195.00

Design and Investment of High Voltage NanoDielectrics
Ahmed Thabet Mohamed (Aswan University, Egypt & Qassim University, KSA)
Engineering Science Reference • © 2021 • 363pp • H/C (ISBN: 9781799838296) • US $195.00

IGI Global
PUBLISHER of TIMELY KNOWLEDGE

701 East Chocolate Avenue, Hershey, PA 17033, USA
Tel: 717-533-8845 x100 • Fax: 717-533-8661
E-Mail: cust@igi-global.com • www.igi-global.com

Editorial Advisory Board

Table of Contents

Detailed Table of Contents

Today, hydrogen is recognized as a non-polluting energy carrier because it does not contribute to global warming if it is produced from renewable energy resources. Hydrogen is the only secondary energy carrier that is suitable for wide application. At the center of attention is the fact that hydrogen can be obtained from a wide range of primary energies. It can be used advantageously for a wide range of applications. Hydrogen can be used in decentralized systems without emitting CO2. Hydrogen is already a part of today's chemical industry, but as an energy resource, its rare benefits can only be achieved through fuel cell technology. The next generations of energy systems for stationary applications based on hydrogen fuel cell have the potential of using and implementing clean energy in the residential buildings sector, as well as in the tertiary and industrial sector, thus having a significant impact on greenhouse gas emissions decreasing, specific characteristics of hydrogen technology having an important role in the decarbonization of energy production systems.

Increasing population and rapid urbanization lead to degradation of the natural environment while waste generation and energy crisis are major challenges in the most developing country. Hydrogen is considered one of the most promising energy carriers and capable to replace fossil fuels and meet the world's energy demand and concomitantly reduce toxic emissions. Currently, the world produces around 50 million tonnes/year from the process (i.e., electrolysis of water, steam reforming of hydrocarbons, and auto-thermal processes), but these processes are not sustainable and economical due to energy requirements and waste/pollutants generation. These challenges required growing interest in renewable energy resources such as hydrogen as an energy carrier. Hydrogen production from renewable sources attracted recent research attention because of its potential for sustainability and diversity. Hydrogen can be produced by various thermal, chemical, and biological technologies that include steam reforming, electrolysis, biomass conversion, solar conversion, and biological conversion.

High efficacy and industrial applicability on electrocatalytic hydrogen production is achieved by proper furnishing of components in the reaction cell. The idea on the basic mechanism of hydrogen evolution reaction (HER), efficient modification of available systems, and recent trends in the development strategies of suitable materials are very important to be explored to design novel systems for large-scale production. This review chapter discusses the scientific details on electrocatalytic HER and plausible materials used for catalyzing the reaction. And it outlines the trends in design and development of transition metal-based catalytic coating systems with a special focus on Ni-P alloy coating and scientific aspects of the methods and the materials used for the HER. On the whole, the discussion on HER and its catalytic systems provides an insight of their potential to be explored for enhanced energy production in hydrogen fuel cell technology for stationary and industrial applications.

The renewable energy for the production of syngas (synthetic gas) by gasification of biomass has specific features and objectives. The present chapter aims to describe the specific features of renewable energy for the production of syngas by gasification of biomass. The specific objective refers to the study of the equipment for the production of synthetic gaseous fuels (syngas) by gasification of biomass. The "biomass" term can be defined generically as the biodegradable part of agricultural products, residues and waste, gathering both animal and plant substances, wood industry, and also the biodegradable component of urban and industrial waste. Biomass sources are the most found renewable resources on Earth, also including all the organic components resulted from the organism's metabolic process, being the first resource of energy man has used, since the discovery of fire.

Energy storage is a vital component in the chain of production-distribution-consumption of energy, even more so if the energy comes from a source that is intermittent and/or is not controllable as is the case with for example solar energy and wind energy. For many people, the term energy storage is the storage of electricity in batteries, as it is the most commonly found way of storing energy. In addition to classic batteries, there are other energy storage alternatives from a primary source for later use. The most valuable forms of energy storage are the ones that can both take over and release the energy on demand, in the form of electricity, such that, in the end, the electrical energy is transformed into thermal or mechanical energy. In stationary applications, energy can be stored in various forms such as batteries, ultracapacitors, or tanks of hydrogen, water, and different types of materials. This chapter will evaluate each form of energy storage.

Future energy systems will be determined by the increasing relevance of renewable energy resources due to global warming, energy crisis, and pollution. Hydrogen is considered one of the promising alternative fuels to replace oil, but its storage remains a significant challenge. The main hydrogen storage technologies can be broadly classified as physical, chemical, and hybrid methods. The physical methods rely on compression and liquefaction of hydrogen, and currently, compressed hydrogen storage is the most mature technology that is commercially available. The chemical methods utilize materials to store hydrogen, and hydrogen can be extracted by on-board regenerable or off-board regenerable chemical reactions depending on the type of material. Hybrid methods take advantage of both physical and chemical storage methods. The most prominent hybrid method is the cryo-adsorption hydrogen storage which utilizes physisorption-based porous materials. The chapter describes these technologies and discusses alternative/novel hydrogen storage material technologies.

Science has shown that there are two sustainable alternatives to providing energy needs: renewable energy resources and fuel cells-hydrogen-based energy, which will play a complementary role in securing global energy resources. By promoting the use of hydrogen-based energy technologies, as clean energy technologies for stationary applications, at the level of local communities, industrial and commercial communities, research topics in this field will help the practical development of sustainable and clean energy systems. This chapter provides an overview of fuel cells highlighting aspects related to fuel cell short history, the main components and operating principles of fuel cells, the main constructive fuel cell types, and the main ways of powering stationary applications through the hydrogen fuel cell technologies.

The purpose of this work is to perform a three-dimensional and stationary numerical study of the heat transfer phenomenon in the planar anode-supported solid oxide fuel cells operating at intermediate temperature (IT-P-AS-SOFC). With particular interest to evaluate and localize the maximum and minimum temperatures in a single cell during their stable operation according to two geometrical configuration types, repetition, and symmetry of the cell stages to determine the best configuration that minimizes and produces more homogeneous thermal stresses and logically improves their lifetime and performance. The considered heat sources are mainly due to electrical overpotentials (Ohm, activation, and concentration). The results are obtained according to a FORTRAN code based on the proposed model that is numerically modeled using the finite difference method. From the obtained result analysis, the achieved temperature values by IT-P-AS-SOFC with cell stages repetition are greater than obtained by IT-P-AS-SOFC with cell stages symmetry.

This chapter presents aspects of hydrogen use in comfort systems and thermal machines in order to improve performance and to reduce pollution emissions. Hydrogen energy is clean, and its use leads to a reduction of CO2 emissions into the atmosphere. It represents an alternative to traditional fuels, oil, coal, and gas. The use of hydrogen preserves traditional fuel resources. In the field of comfort,

Foreword

I will start with a simple statement: this book is about a long-awaited revolution! It is not an exaggeration! Hydrogen is certainly "a buzzy topic in clean energy circles," and it is now unanimous agreed that after a long period of successions of inflated expectations followed by disillusionments, finally finds the way out! So, what change now? After all, hydrogen fuel cell where used in the NASA' spatial programs in the '60s and hydrogen powered the first combustion engines 200 years ago! Well, it seems that the discussion is about integration, and this is the raison I used "revolution" word. This hydrogen hype is about changing almost every side of the human life in a manner that it will modify the way we are powering our cars, heating our homes, supply sustainable and clean electricity and decarbonize every industrial processes.

If we are looking at any of the trickiest problems when it comes to lowering CO_2 emissions – replacing 'process carbon' in metals and chemicals production; powering large vehicles on land, sea, and in the air; generating high-temperature zero-carbon heat; storing the energy from renewable electricity for longer periods; and even reducing emissions from agriculture – and hydrogen is arguably the most promising solution on the horizon.

In spite of long waited hydrogen based technologies chain, finally the momentum of hydrogen arrived and it is maybe not a coincidence that the release of this book is in 2021. The potential replicable example of renewables falling prices and the scientific advances in the electrochemical conversion systems were pushed hydrogen in front of the global "Grand Transition". The idea of a "hydrogen economy" is now enjoying a new wave of enthusiasm — but obviously it is not a silver bullet. Only talking about transportation, hydrogen power needed a new infrastructure, whereas rival battery cars could be charged off the near-ubiquitous electricity grid. Moreover, high-powered batteries were already made well-advanced steps forward and the prime advocate of it, Elon Musk says the process is wasteful because it involves turning electricity into a gas, then back into electricity. "Fool cells", he calls them. But it is not a competition! The future is obvious hybrid.

Companies around the world are making a "trillion-dollar bet" on a hydrogen-fueled future. A series of planned investments in hydrogen power from Europe, Japan, Australia, and United States are foreseen to make the first important steps in the implementation of this "revolution" and eliminate "roughly a quarter of the world's carbon dioxide emissions" in the first stage. The dramatic leap, though, depends on improvement in the technology for producing "green" hydrogen by splitting water molecules. So maybe hydrogen lost the head-on battle for the small type of motor car. But now it's back in the frame for the sort of transport, industry and heating tasks that batteries are struggling to fulfil.

Therefore, the momentum should be emphasized. The challenge is now, to create a new ecosystem to harness the power of hydrogen for lasting change. Only hydrogen can help ensure that the rapid growth of the renewable energy sector continues – and with it the decrease of greenhouse gas emissions. In fact, the declining costs for renewable energy and the increasing share of renewables in our energy mix are rapidly increasing awareness for hydrogen's potential in storing solar and wind power. Many countries in Europe and Asia have taken the opportunity of their post-pandemic economic recovery plans to roll out ambitious hydrogen strategies. Europe become conscious that Covid -19 has dramatically exposed the fragility and endemic risk in all the national economies. Supply chains are stationary, oil prices have gone negative, and fiscal rules have been rewritten. The risk from climate change is seen as many times worse. European investors and analysts change their vision and see now hydrogen as an opportunity for long-term sustainable investment, and will reward company boards for pursuing it. Stimulus packages are giving hydrogen a key role in recovery plans, especially to decarbonise sectors such as chemicals, steel, and heavy transport, where electrification – anticipated as the main route to decarbonisation – may not be an option. It will also create jobs. The less optimistic estimations foresee that new industries such as hydrogen could generate between six and eight jobs for every $1 million invested.

People could ask: Why hydrogen? Simply, it can do things electrons cannot. As a molecule, it can be used as a convertible 'currency' between renewable energy, chemical feedstock, transport fuel, residential and industrial heat. It is an excellent clean energy store for long periods and large energy needs. It sits alongside and in partnership with electricity as a consummate carbon-free energy vector. Hydrogen is rarely an easy solution to any single problem. A lot must be done before it makes a meaningful energy - or climate - or economic - contribution.

However, the road is not straight! In order to become commercial readiness all the pieces of the hydrogen technology chain there are still lot of barriers which should be solved for a large deployment of the hydrogen economy. From lowering the costs until public acceptability, lot of works should be developed to achieve the required advances. And this is the task of research& development! Huge R&D investment are done all over the world to optimize all the pieces of the hydrogen economy system. Each scientific project is pushing a little forward the hydrogen energy technology knowledge. Consequently, tens of thousands of scientific papers and books were published in the last ten years. This book is a piece of this immense puzzle desiring to reveal in a specific manner, several ideas and strategies related to the implementation of diverse components of hydrogen "life" from a technical point of view. Taking into account that the whole "picture" is essentially interdisciplinary and multidisciplinary the book is a kind of hands-on resource for technical staff, addressing also questions and answers for a large categories of public which are interested in the short-term future of sustainable energy.

Mihai Varlam
National Research and Development Institute for Cryogenic and Isotopic Technologies – INC-DTCI-ICSI Râmnicu Vâlcea, Romania

Preface

The book entitled *Hydrogen Fuel Cell Technology for Stationary Applications* was developed based on editors' motivation to provide an overview of the topic in the context of global interest increasing for the use of hydrogen energy as a vector, carrier of clean energy in a sustainable energy system.

The European Council endorsed the goal of achieving a climate-neutral EU by 2050 and, in this context, took note of the European Commission's Communication on the European Green Pact, which aims to make Europe the first climate-neutral continent by 2050, addressing the issue of biodiversity loss and pollution and mitigating the inefficient use of resources by moving to a more circular economy. The European Council's policy emphasizes the importance of implementing a coherent and complementary strategy and promotes actions leading to a 'hydrogen market for Europe' throughout the hydrogen value chain.

Globally, increasing energy consumption - regardless of its type - is making it increasingly difficult to secure the primary resources needed to meet the demand of the energy sector. This fact, corroborated with the environmental pollution factor, respectively the phenomenon of global warming, turns this problem into at least a worrying one. In the current context, the construction of more and more residential and non-residential buildings, offices, factories, as well as the intensification of the circulation of an increasing number of cars and trucks around the world, all this together, its have led to an increase in carbon dioxide emissions into the atmosphere, producing undesirable effects on the environment and human health. Environmental protection for future generations is a major challenge for researchers and scientists, and solutions must be found for the entire planet to be green.

The energy strategy is a complex activity involving the initiation, development and correlation of political, economic, technical and environmental actions in order to ensure global energy security, to diversify sources of internal and external supplies, to improve and diversify forms of energy production, energy efficiency and production, improving the polluting impact of energy production and consumption.

Unconventional energy sources have gained and will continue to gain an increasing share of energy systems around the world. Officially recognized renewable energies originate from the sun's rays, the internal temperature of the earth, or the gravitational interactions of the sun and the moon with the oceans, wind movement. In addition, today, in the context of a circular economy, hydrogen is recognized as a clean energy carrier because it does not contribute to global warming, especially if it is produced from green renewable resources. Hydrogen is the only secondary energy vector that lends itself to a wider application on the market, focusing on the fact that hydrogen can be obtained from a wide range of primary energies. It can be used advantageously for a wide range of applications. Hydrogen is already part of today's chemical industry, but as an energy source, its rare advantages can only be obtained with the help of technologies. Against this background, alternatives are currently being sought to obtain energy through the use of technologies that offer maximum efficiency, high reliability and minimum pollution. Such technology, which is currently considered the cleanest, whereby sustainable energy can be obtained, is based on the fuel cell. With the development of fuel cells, hydrogen-based energy generation becomes a reality, with hydrogen becoming one of the energy alternatives for worldwide. Since hydrogen can be produced from a wide range of primary energies and can be consumed in an increasing number of applications, it will become an energy center, just as electricity is today. The advantage of hydrogen to electricity is that it can be stored from the long term. So, we are on the brink of a new era, characterized by advanced technologies and new fuels. We will find new and completely different ways of producing and using energy, and hydrogen can be considered as a secondary carrier, synthetic energy in a future era, following the fossil fuel economy, within a sustainable energy system.

The Hydrogen Economy Miami Energy Conference, USA March 18-20, 1974 was the first international debate where hydrogen was presented as a permanent solution to conventional fuels depletion and global environmental problems, event that laid the foundations of the International Association for Hydrogen Energy (IAHE). Despite the fact that conventional energy has been permanently subsidized by governments, even if it harms human health, the well-being and integrity of our planet, and the fact that equivalent subsidies have not been extended to the hydrogen energy sector, in the quarter of a century since the hydrogen movement began, however, it has made significant progress in several directions in the energy sector, due to its unmatched, high-level properties and its characteristics as a carrier, a secondary energy vector.

The context of environmental issues, the need to switch from conventional fuels currently used at capitalizing on alternative energies for energy generation and current energy policies create the right framework for the implementation and development of new technologies. The hydrogen-based economy and, implicitly, fuel cell technology are of particular interest in global energy strategies. Recently, a special interest for hydrogen and fuel cells is the potential of this technology regarding energy storage in the context due to the unprecedented development of renewable resources sectors.

Today, there are more opportunities in the world to address the challenges of sustainable development in the field of energy consumption and primary or secondary energy resources. Energy and resource efficiency is a key objective, a challenge for the global research community, with a focus on resolving the phenomena that have unfavorable consequences, taking action in this regard in order to achieve the goals of global policies in the field of energy efficiency, motivated by the multiple benefits of efficient consumption that will contribute favorably to economic and social development, conservation of natural resources, reduction of greenhouse gas emissions and contributes significantly to reducing the impact of economic activities globally.

The extensive use of fossil fuels, the threatening increases in the prices of these energies and last but not least, global climate change require the rapid imagination of concrete solutions to meet our contemporary demands, but to operate under the aegis of important environmental challenges. Energy efficiency and the decarbonisation of the domestic and commercial sector attract special attention because residential and tertiary habitat is solely responsible for 43% of final energy consumption and 25% of greenhouse gas production, especially carbon dioxide emissions.

In this context, hydrogen, energy vector, or ecological synthetic fuel, together with the fuel cell, its conversion technology, can play an important role in streamlining and decarbonizing power generation systems in the field of stationary applications, in particular in the constructions sector. As a result, the topic addressed in the book entitled *'Hydrogen Fuel Cell Technology for Stationary Applications'* falls into the current international context, the importance of the issue being relevant both scientifically, technologically, but also from a socio-economic or cultural point of view.

The next generations of energy systems and technologies for power supply of stationary applications, respectively systems for generating electricity with fuel cells, have significant potential to reduce energy consumption in the residential buildings sector, as well as in the tertiary and industrial buildings sector, implicitly have a significant impact on reducing pollutants and greenhouse gas emissions, being demonstrated the specific characteristics of hydrogen technology with a special role in the efficiency and decarbonization of energy generation systems.

However, in order to harness at maximum potential, state-of-the-art technologies must be accepted as energy-efficient technologies, but also cost-effective, to be widely adopted. Various entities, professional associations and research communities are directly involved in the process of demonstrating and developing these energy efficient systems, having a very important role in the widespread adoption of these technologies.

TARGET AUDIENCE

This book targets advanced-level students, researchers, specialists and practitioners in areas of interest: alternative energy resources, architecture and urbanism, chemistry, electronics, mechanics, physics, building services engineering, civil engineering, electrical engineering, environmental engineering, materials engineering and anyone interested in new clean energy technologies or green power generation systems. Other stakeholders, who are interested in energy efficiency and associated specific technologies will also find this book very useful in their research.

OBJECTIVE OF THE BOOK

This book's key objective is to inventory and disseminate the relevant knowledge regarding hydrogen economy dedicated to the stationary applications, through reviews or overviews on the main components of the hydrogen economy and the fuel cell typology, through theoretically and empirically based studies or through research approaches that highlight the practical applicability of hydrogen fuel cell in stationary applications. The significance of the book resides in its capacity to enlarge our knowledge by emphasizing the complexity of a hydrogen-based economy by presenting in detail and analysing the main aspects related to the production, transport, distribution, storage and use of hydrogen as an energy vector in practical stationary applications. Consequently, the book's main goal is to provide an expansive and nuanced understanding that will deepen the ontology of the subject of 'Hydrogen Fuel Cell Technology for Stationary Applications'.

ORGANIZATION OF THE BOOK

Welcome to *Hydrogen Fuel Cell Technology for Stationary Applications.* There are 10 chapters in the book, each focusing on addressing current issues relevant to the main objective of the proposed topic, namely the use of hydrogen energy and fuel cell technology in stationary applications. The chapters are organized in such a way as to outline a logical structure and provide a clear perspective on hydrogen economy and the possibilities of using fuel cell technology in stationary applications, each chapter being approached to stand alone, advancing a broad perspective on a particular topic. A brief description of each chapter is as follows:

Chapter 1 serves as an introduction to the book entitled 'Hydrogen Fuel Cell Technology for Stationary Applications', providing an overview of the main properties of hydrogen which recommends it as a clean sustainable energy carrier compared to conventional fossil fuels. Also, the chapter briefly presents the main components of hydrogen economy, namely: hydrogen production methods (thermal methods, electrochemical methods and biological methods), insisting on renewable energy-based hydrogen production technologies; hydrogen transport and distribution and hydrogen storage. The authors conclude that the future hydrogen-based economy presents hydrogen as an energy carrier in a safe and sustainable energy system.

Chapter 2 shows the deepening aspects related to hydrogen production from the waste and renewable resources. The authors review most of the hydrogen production process, such as: water electrolysis; thermochemical water splitting, photo-electrolysis; biomass conversion through gasification, pyrolysis and copyrolysis; biological hydrogen production technologies which refers to photolysis direct and indirect photolysis, dark fermentation, photo fermentation, sequential two-stage or multi-stage integrated biological system, microbial electrolysis cells (MECs). In conclusion, the authors point out that most of the hydrogen production methods presented are not yet widely traded and need in-depth research, international collaboration and trade strategies need to be explored in order for these hydrogen production methods to be economically viable.

In Chapter 3, Ameen Sha M and Meenu P C address the idea of the basic mechanism of Hydrogen Evolution Reaction (HER). This chapter discusses the scientific details on electrocatalytic HER and plausible materials used for catalyzing the reaction. And it outlines the trends in the design and development of transition metal-based catalytic coating systems with a special focus on Ni-P alloy coating and scientific aspects of the methods and the materials used for the HER. On the whole, the discussion on HER and its catalytic systems provides an insight into their potential to be explored for enhanced energy production in Hydrogen Fuel Cell Technology for stationary and industrial applications.

Chapter 4 gives an 'Overview Regarding Synthetic Gas Production by Biomass Gasification'. The authors present a general aspect regarding the different types of biomass, including a classification of it, followed by details on the biomass calorific value and an estimated analysis of biomass production at EU level. Consistent with the subject of the book, the authors present the main technologies for biomass capitalizing and its conversion into biofuels, insisting on developing topical on syngas production through gasification process. The main conclusion of the authors of this chapter highlights the fact that gasification and production of syngas represent the ideal solution, from an ecological point of view, of biomass energy recovery by increasing its calorific capacity, limiting the polluting emissions.

Chapter 5 highlights that energy storage is a vital component in the chain of production - distribution - consumption of energy, even more so if the energy comes from a source that is intermittent and/or is not controllable as is the case with for example solar energy and wind energy. The authors address a number of energy storage systems - from classical ones to energy storage in the form of hydrogen energy vector. The types of energy storage technologies presented in this chapter refer to: mechanical storage, electrochemical storage, chemical storage and fuel cells, electromagnetic energy storage and thermal storage.

Chapter 6 details the main hydrogen storage technologies, which can be broadly classified as physical, chemical and hybrid methods. The physical methods rely on compression and liquefaction of hydrogen, and currently, compressed hydrogen storage is the most mature technology that is commercially available. The chemical methods utilize materials to store hydrogen, and hydrogen can be extracted by on-board regenerable or off-board regenerable chemical reactions depending on the type of material. Hybrid methods take advantage of both physical and chemical storage methods. The most prominent hybrid method is the cryo-adsorption hydrogen storage which utilizes physisorption-based porous materials. The present chapter describes these technologies and discusses alternative/novel hydrogen storage material technologies.

Chapter 7 gives an overview of the main types of fuel cells currently developed, highlighting the main components and operating principle of these fuel cells, as follows: Alkaline Fuel Cell (AFC), Direct Methanol Fuel Cell (DMFC), Molten Carbonate Fuel Cell (MCFC), Phosphoric Acid Fuel Cell (PAFC), Proton-Exchange Membrane Fuel Cell (PEMFC), Solid Oxide Fuel Cell (SOFC). Also in this chapter the applicability in the stationary field of fuel cell technologies is approached, being briefly discussed the following directions: Micro – CHP with fuel cell in the residential buildings sector, hydrogen fuel cell technology as back-up power systems by harnessing renewable energy sources and high capacity power plants based on hydrogen fuel cell technology. The main conclusion of this chapter highlights that as practical applications, the fuel cells integrated into the energy generation systems can play various roles, either in micro-/macro-cogeneration of electricity and heat, or back-up power supply by harnessing on renewable sources and waste, or energy generation for remote locations and different capacities power plants for distributed power generation.

Chapter 8 is dedicated to the work presentation that aims to perform a three-dimensional and stationary numerical study of the heat transfer phenomenon in the planar anode-supported solid oxide fuel cells operating at intermediate temperature (IT-P-AS-SOFC). With particular interest to evaluate and localize the maximum and minimum temperatures in a single cell during their stable operation according to 2 geometrical configuration types, repetition and symmetry of the cell stages, to determine the best configuration that minimizes and produces more homogeneous thermal stresses and logically improves their lifetime and performance. The considered heat sources are mainly due to electrical overpotentials (Ohm, activation and concentration). The results are obtained according to a FORTRAN code based on the proposed model that is numerically modeled using the finite difference method. From the obtained result analysis, the achieved temperature values by IT-P-AS-SOFC with cell stages repetition are greater than obtained by IT-P-AS-SOFC with cell stages symmetry.

Chapter 9 presents aspects of hydrogen use in comfort systems and thermal machines in order to improve performance and to reduce pollution emissions. Hydrogen energy is clean, and its use leads to a reduction of CO2 emissions into the atmosphere. It represents an alternative to traditional fuels, oil, coal and gas. The use of hydrogen preserves traditional fuel resources. In the field of comfort, the energy obtained from hydrogen, it is applicable in the heating and air conditioning of spaces, in the production of electricity necessary to create light comfort. The use of hydrogen in boilers leads to the reduction of pollutant emissions. Hydrogen fuelling systems were designed for different experimental thermal machines, designed by authors from spark-ignition engine and compression ignition engine. The characteristic combustion parameters, like maximum pressure, the maximum rate of pressure rise, efficiency and pollutant emissions for hydrogen fuelling are presented and analysed.

Chapter 10 presents considerations regarding the use of fuel cells in combined heat and power for stationary applications, the author insisting on the practical applicability of SOFC in the field of energy cogeneration. Along with the technical aspects related to CHP Performance Indicators, the particularities of CHP through SOFC and the integration of SOFC in CHP systems, the author performs a comparative analysis of CHP technologies having as main comparative criteria: electric efficiency, overall efficiency, typical capacities, electric / thermal yield, investment cost, operational cost, benefits and disadvantages. It is highlighted that fuel cell technology is based on the principle of direct electrochemical energy conversion. In other words, the fuel cells are not similar to the conventional cogeneration plants because the combustion of the fuel does not take place, but the chemical energy of the fuel is converted directly into electricity, without any intermediate stage of mechanical energy production.

The topics addressed in the chapters cover the main aspects related to a hydrogen economy - production, storage, transport, distribution of hydrogen, reviewing the state-of-the-art fuel cell technologies developed, and present the main directions to integrate hydrogen fuel cell in sustainable stationary applications. The topics addressed in the chapters cover the main aspects related to a hydrogen economy - production, storage, transport, distribution of hydrogen, reviewing the state-of-the-art fuel cell technologies developed, and present the main directions to integrate hydrogen fuel cell in sustainable stationary applications. The chapters examined the main technical, environmental and economic characteristics, but also the key challenges in adopting energy systems based on clean hydrogen and evaluated various opportunities to implement hydrogen fuel cell technologies in practical stationary applications with the aim to stimulate the development of energy generation systems under environmental protection.

The technical information, discussions, studies, analyzes and research results presented during this book, but also the problems identified, the technical limitations encountered allow the identification and establishment of new future approaches in the field of 'Hydrogen Fuel Cell Technology for Stationary Applications', each chapter representing a starting point in the elaboration of elaborated, exhaustive studies specific to the addressed subject that to further explore ways to reduce the gaps in the studies and the specialized technical literature.

Gheorghe Badea
Technical University of Cluj-Napoca, Romania

Raluca-Andreea Felseghi
Ştefan cel Mare University of Suceava, Romania

Ioan Aşchilean
Technical University of Cluj-Napoca, Romania

Introduction

It's no secret that the energy problem is already one of the most acute global problems facing humanity. At present, at the planetary level, there is talk everywhere of the blue planet and more specifically how to succeed in having an increasingly blue planet taking into account the alarming depreciation made over time by various industrial technologies and more.

The mission assumed to elaborate the Introduction section for this book entitled 'Hydrogen Fuel Cell Technology for Stationary Applications', put me in the position of asking myself various questions, some of which were recorded as:

What is the reason for writing this book?

How important is this book to the future reader?

What is the reason for writing and talking about hydrogen energy and fuel cell technology?

Why would anyone want to read this book? and obviously the questions can go on.

The answers can be direct with concrete reference, but it is appropriate to evaluate in more depth the problem of hydrogen as an energy vector and alternative in the complex process of increasing the importance of global-planetary energy with unsuspected positive effects.

Currently, in almost all areas of our planet, there is a dizzying increase in energy consumption in its various forms and that ensuring primary energy sources will be increasingly difficult but also with disastrous effects on the environment.

In this context, hydrogen, the energy vector or in other words, the ecological synthetic fuel, in tandem with fuel cell may be capable technologically to play a very important role in decarbonizing power generation systems in a variety of stationary applications, particularly in the buildings domain.

The inventory and dissemination of relevant information and knowledge on the hydrogen energy pathway, made available to the reader, can be considered a mission accomplished as a starting point in the long road of hydrogen use through technologies such as hydrogen fuel cells.

The theme of the book and the need for its elaboration can be supported by the inclusion in a complex process of the hydrogen economy in full development, highlighting the main aspects related to production - storage - transport - distribution and use.

With the conviction that the publication of this book addresses real needs in the current international context of hydrogen use through fuel cell technology, we expect the materialization of the intentions will be reflected in significant achievements in the green energy domain.

Gheorghe Badea
Technical University of Cluj-Napoca, Romania

Chapter 1
Hydrogen–Energy Vector Within a Sustainable Energy System for Stationary Applications

Gheorghe Badea
Technical University of Cluj-Napoca, Romania

Raluca-Andreea Felseghi
https://orcid.org/0000-0001-6310-5932
Ştefan cel Mare University of Suceava, Romania

Ioan Aşchilean
https://orcid.org/0000-0003-3632-5600
Faculty of Civil Engineering, Technical University of Cluj-Napoca, Romania

ABSTRACT

Today, hydrogen is recognized as a non-polluting energy carrier because it does not contribute to global warming if it is produced from renewable energy resources. Hydrogen is the only secondary energy carrier that is suitable for wide application. At the center of attention is the fact that hydrogen can be obtained from a wide range of primary energies. It can be used advantageously for a wide range of applications. Hydrogen can be used in decentralized systems without emitting CO^2. Hydrogen is already a part of today's chemical industry, but as an energy resource, its rare benefits can only be achieved through fuel cell technology. The next generations of energy systems for stationary applications based on hydrogen fuel cell have the potential of using and implementing clean energy in the residential buildings sector, as well as in the tertiary and industrial sector, thus having a significant impact on greenhouse gas emissions decreasing, specific characteristics of hydrogen technology having an important role in the decarbonization of energy production systems.

DOI: 10.4018/978-1-7998-4945-2.ch001

The context of environmental issues, the need to switch from conventional fuels currently used to the use of alternative energies for energy generation and current energy policies create the right environment for the implementation and development of new technologies. The hydrogen-based economy and, implicitly, fuel cell technology are of particular interest in global energy strategies. Recently, a special interest for hydrogen and fuel cells is the potential of this technology regarding energy storage in the context due to the unprecedented development of renewable resources sectors. *(Barreto, L., et.al., 2003).*

The Hydrogen Economy Miami Energy Conference, USA March 18-20, 1974 was the first international debate where hydrogen was presented as a permanent solution to conventional fuels depletion and global environmental problems, event that laid the foundations of the International Association for Hydrogen Energy *(Midilli, A., et.al., 2005; Veziroğlu, T. N., 2000; Veziroğlu, T. N., 2004).*

Despite the fact that conventional energy has been permanently subsidized by governments, even if it harms human health, the well-being and integrity of our planet, and the fact that equivalent subsidies have not been extended to the hydrogen energy sector, in the quarter of a century since the hydrogen movement began, however, it has made significant progress in several directions in the energy sector, due to its unmatched, high-level properties and its characteristics as a carrier, a secondary energy vector. *(Veziroğlu, T. N., 2000; Veziroğlu, T. N., 2004).*

GENERAL ASPECTS

Unconventional energy sources have gained and will continue to gain an increasing share in energy systems around the world, both due to the research effort and political will involved in their development and due to the rising price of energy obtained through the traditional methods. Primary energy sources, generally called renewable, are those sources in the natural environment, available in virtually unlimited quantities or that are regenerated by natural processes, at a faster rate than the one in which they are consumed. *(Badea, G., et.al., 2015; Felseghi, R.A., et.al., 2019, a).* Officially recognized renewable energies originate from the sun's rays, the earth's internal temperature, or the gravitational interactions of the sun and moon with the oceans *(Felseghi, R.A., et.al., 2019 a; Li, Y., et.al.,2020; Filote, C., et.al.,2020).*

Processes and procedures for the production or capture of these types of alternative energies under development, high investment costs and low returns on conversion processes have made renewable energy sources a small part of the needs on a global scale. Optimistic forecasts estimate renewable energy production to account for 30-50% of the total energy market around 2050, but this depends on reducing production costs and finding opportunities for massive storage of electricity. In addition, all these forms of energy cannot provide fuels in satisfactory quantities for various industrial uses. *(Felseghi, R.A., et.al., 2019, a; Colbertaldo, P., et.al., 2019)*

In this context, alternatives are currently being sought to obtain electricity by using technologies that provide maximum efficiency, high reliability and minimum pollution. Such a technology, currently considered the cleanest, through which sustainable energy can be obtained, is based on fuel cells. With the development of fuel cells, the production of hydrogen-based energy is becoming a reality, hydrogen being one of the safest energy alternatives for mankind. The future hydrogen-based economy presents hydrogen as an energy carrier in a safe and sustainable energy system. *(Felseghi, R.A., et.al., 2019, b; Ren, X., et.al., 2020; Cloete, S., Ruhnau, O., & Hirth, L., 2021)*

Humanity is on the threshold of a new era characterized by advanced technologies and new fuels. We will witness new and completely different ways of producing and using energy. Electricity could be generated by sources with virtually no pollution. Hydrogen can be considered as a synthetic fuel, a secondary energy carrier in a future era, after the fossil fuel economy. *(Felseghi, R.A., et.al., 2019, b; Sazali, N., 2020; Hosseini, S. E., & Wahid, M. A., 2020; Pareek, A., 2020).*

HYDROGEN - ENERGY VECTOR IN A SUSTAINABLE ENERGY SYSTEM

Today hydrogen is recognized as a clean energy carrier because it does not contribute to global warming if it is produced from renewable sources. In addition, hydrogen is the only secondary energy carrier suitable for wide applications on the market. The focus is on the fact that hydrogen can be obtained from a wide range of primary energies. It can be used advantageously for a wide range of applications, from transport and portable to stationary. In addition, hydrogen can also be used in decentralized systems without emitting carbon dioxide. Hydrogen is already a component of today's chemical industry, but as an energy source, its rare advantages can only be achieved with the help of technologies such as fuel cells. *(Felseghi, R.A., et.al., 2019, a; Weeda, M., & Segers, R., 2020).*

As hydrogen can be produced from a wide range of primary energies and can be consumed in an even greater number of applications, it will become an energy center, just as electricity is today. The advantage of hydrogen over electricity is that it can be stored for long periods of time and transported away from the place of production. Consequently, an energy carrier helps to increase the stabilization of energy security and price, giving rise to competition between different energy sources. *(Felseghi, R.A., et.al., 2019, a; Zhiznin, S. Z., Vassilev, S., & Gusev, A. L., 2019; Midilli, A., Ay, M., Dincer, I., & Rosen, M. A., 2005).*

T. N. Veziroglu, editor of the journal specializing in hydrogen technology and energy, International Journal of Hydrogen Energy briefly presents some features that recommend the use of hydrogen as an energy secondary vector produced on the basis of unconventional technologies:

- hydrogen concentrates primary energy sources, which it makes available to the consumer in a convenient form;
- the production cost is relatively low as a result of technological developments and improvements;
- offers the possibility of transformation into various other forms of energy through conversion processes characterized by high efficiency;
- it is an inexhaustible source if it is obtained electrolytically from water; hydrogen production and consumption is a closed cycle, the source of production - water remains constant and is a classic cycle of recirculation of this type of raw material;
- is the lightest and cleanest fuel; hydrogen combustion is almost entirely free of pollutant emissions;
- has a much higher gravimetric energy density compared to other fuels;
- hydrogen can be stored in various ways, such as: gas at normal or high pressure, in the form of liquid hydrogen or in the form of solid hydrides;
- can be transported over long distances stored in the form or in one of the ways presented above;
- the equipment for converting hydrogen into electricity, of fuel cell type, has an efficiency of over 60%. *(Midilli, A., et.al., 2005; Veziroğlu, T. N., 2000; Veziroğlu, T. N., 2004).*

Hydrogen is considered by more and more specialists to be a true fuel of the future. Hydrogen condenses at -252.77 °C, and the specific gravity of liquefied hydrogen is 71 g/L, which gives it the highest energy density per unit mass of all fuels and energy carriers: 1 kg of hydrogen it contains as much energy as 2.1 kg of natural gas or 2.8 kg of oil. This feature has made hydrogen the fuel used in the propulsion and power supply of spacecraft. Unlike other fuels, such as oil, natural gas and coal, hydrogen is renewable and non-toxic when used in fuel cells. Hydrogen has great potential as an environmentally friendly fuel and in reducing the import of energy resources. Even if the use of hydrogen now seems unprofitable, following the improvement of technologies, we could witness in the not too distant future the development of a hydrogen-based economy. *(Molnarne, M., & Schroeder, V.,2019; Felseghi, R.A., et.al., 2019, a)*

Hydrogen - Carrier of Energy and Synthetic Fuel

Etymologically, the word hydrogen is a combination of two Greek words, meaning "to make water". *(Iordache, I., & Stefănescu, I., 2010)*. Produced from non-fossil sources and raw materials, by using various forms of alternative energy (solar, wind, hydroelectric, geothermal, biomass, etc.), hydrogen is considered to be a prime fuel in the supply of so-called "green energy". Thus, hydrogen-powered systems can be considered one of the best solutions for accelerating and ensuring global stability (*Dincer, I., & Zamfirescu, C., 2012)*. Hydrogen is expected to play an important role in the world's future energy scenarios, the most important factor that will determine the specific role of hydrogen will probably be energy demand. At the same time, hydrogen can, to some extent, replace fossil fuels and become the preferred clean, non-toxic energy carrier in the near future. (*McCarty, R. D., Hord, J., & Roder, H. M., 1981; Felseghi, R.A., et.al., 2019, a)*

The main properties of hydrogen, presented in Table 1, recommend it as an alternative fuel to the classic ones.

Table 1. Properties of hydrogen.

Characteristics	Measure Unit	Value
Density	kg/m^3	0,0838
HVV/ LH_2	MJ/kg (liquid)	141,90 - 119,90
HVV/ CGH_2	MJ/m^3 (volumetric)	11,89 - 10,05
Boiling point	K	20,41
Freezing point	K	13,97
Density (liquid)	kg/m^3	70,8
Air diffusion coefficient	cm^2/s	0,61
Specific heat	kJ/kg K	14,89
Ignition limits in air	% (volume)	4 - 75
Ignition energy in air	Millijoule	0,02
Ignition temperature	K	585,00
Flame temperature in air	K	2318,00
Energy exploding	kJ/g TNT	58,823
Flame emissivity	%	17 - 25
Stoichiometric mixture in air	%	29,53
Stoichiometric air/fuel	kg/kg	34,30/1
Burning speed	cm/s	2,75
Energy reserve factor	-	1,00

Source: (Felseghi, R.A., et.al., 2019, a)

In order to highlight the advantages of hydrogen compared to other fuels, Table 2 summarizes the main properties of the various fuels currently used.

Table 2. Comparison between the main properties of hydrogen and other fuels.

Fuel Type	Energy/ Mass Unit (J/kg)	Energy/ Volume Unit (J/m³)	Energy Reserve Factor	Specific Carbon Emission (kgC/kg eq.)
Liquid hydrogen	141,90	10,10	1,00	0,00
Gaseous hydrogen	141,90	0,013	1,00	0,00
Fuel oil	45,50	38,65	0,78	0,84
Benzine	47,40	34,85	0,76	0,86
Jet fuel	46,50	35,30	0,75	-
GPL	48,80	24,40	0,62	-
GNL	50,00	23,00	0,61	-
Methanol	22,30	18,10	0,23	0,50
Ethanol	29,90	23,60	0,37	0,50
Biodiesel	37,00	33,00	-	0,50
Natural gases	50,00	0,04	0,75	0,46
Coal	30,00	-	-	0,50

Source: (Felseghi, R.A., et.al., 2019, a)

Analyzing the tabular information, it can be concluded that the main arguments in favor of using hydrogen as a synthetic fuel, obtained from renewable sources, are the following: it has the highest energy/ unit of mass of all types of fuel; it is ecological, from its combustion resulting water vapors, observing that for hydrogen the amount of carbon emissions is zero; it has the highest energy reserve factor, respectively the highest conversion factor into electricity, being considered for this reason the best of the presented fuels, and the energy efficiency is very high. Hydrogen is expected to play an important role in future global energy scenarios. (*Mekhilef, S., Saidur, R., & Safari, A., 2012; Tabak, J.,2009)*

The advantages that promote it as an energy vector compared to other forms of energy are as follows:

- hydrogen can be transported remotely through pipelines, in safe conditions;
- hydrogen is a non-toxic energy transport operator, with a high specific energy per mass unit (for example, the energy obtained from 9.5 kg of hydrogen is equivalent to that of 25 kg of gasoline);
- hydrogen can be generated from various energy sources, including renewable ones;
- compared to electricity, hydrogen can be stored for relatively long periods of time;
- hydrogen can be used advantageously in all sectors of the economy (as a raw material in the industry, as a fuel for automobiles and as an energy carrier in sustainable energy systems to generate electricity through fuel cells). *(Felseghi, R.A., et.al., 2019, a)*

Barriers to be overcome concern issues related to:

- hydrogen burns in the presence of air, and this can cause safety problems in operation;
- storage of hydrogen in liquid form is difficult because very low temperatures are needed to liquefy hydrogen;
- high costs of hydrogen technologies and processes;
- high costs of technologies for converting hydrogen into electricity through fuel cells;
- the viability/cost ratio of hydrogen and fuel cell production technologies is relatively low;
- lack of logistics, hydrogen transmission and distribution infrastructure to final consumers involves costly investments. *(Felseghi, R.A., et.al., 2019, a)*

Given the research and development programs supported in this field, the deficient technical problems regarding the production, storage and distribution of hydrogen, at the same time with the decrease of costs and the increase of the life of the equipment used in hydrogen-based energy generation, will be solved shortly. and hydrogen will become a possible solution for providing fuel, while also being an alternative energy resource to the traditional one. *(Bădea, G., et.al., 2017)*

HYDROGEN ECONOMY

The hydrogen economy represents the totality of energy infrastructures at the national and international level that have hydrogen as their energy resource *(Koyuncu, I., et.al., 2020)*. Building an infrastructure that allows, easily and efficiently, the transport and delivery of hydrogen energy is a critical step towards a future hydrogen-based economy *(Cloete, S., Ruhnau, O., & Hirth, L., 2021)*. Hydrogen has great potential as a clean fuel and is also a means of supporting the reduction of energy imports in some cases. Before hydrogen can play an important energy role and become an alternative to the use of oil, and infrastructure must be built that contains both centralized or decentralized production facilities and distribution systems and networks *(Linssen, J., & Hake, J. F., 2016)*. Some facilities will be needed for the production of hydrogen, for its storage and transportation, but also for the construction of economical fuel cells, and consumers will need the education necessary to use it safely. *(Eiler, J. M., 2020; Nuttall, W. J., & Bakenne, A. T., 2020)*.

In order to emphasize the complexity of a hydrogen-based economy, it is necessary to analyze the main aspects related to the production, transport, distribution, storage and use of hydrogen.

Hydrogen Production Processes

Hydrogen cannot be found in nature in its pure state, so it cannot be mined in the same way as oil or coal. Hydrogen must be extracted from chemical compounds. Regardless of the resource from which hydrogen is extracted, a production process is needed and it involves energy consumption. The big advantage, however, is that for the generation of hydrogen it is not strictly necessary to use energy from fossil fuels, you can use renewable, hydraulic, solar, wind, geothermal energy. *(Liu, W., et.al., 2020)*

The main processes currently used for hydrogen manufacture are summarized in Table 3.

Hydrogen production is not new. Every year, approx. 500 billion m^3 of hydrogen is produced, stored, transported and used, mainly in the chemical and petrochemical industry *(Acar, C., & Dincer, I., 2014)*. However, most of it is obtained as a by-product from the processing of fossil fuels (oil and natural gas).

There are already established technologies in the manufacture of hydrogen, used in the chemical industry, as well as new processes developed especially for fuel cells. Each of the existing technologies has its advantages and disadvantages, so research is needed to optimize them.

All processes for obtaining hydrogen are based on separating it from the products that contain it. The nature of the raw material dictates the choice of the separation method. Today, two main methods of hydrogen separation are used: thermo-chemical and electrochemical. The third method, the biological one, is in the research-development phase. *(Iordache, I., & Stefănescu, I., 2010; Dawood, F., Anda, M., & Shafiullah, G. M., 2020, Kannah, R. Y., et.al., 2020)*.

Table 3. Processes used for hydrogen manufacture.

Basic Method	Process	Raw Material	Energy Type	Pollutant Emissions
Thermal method	Steam reforming	Natural gases	High temperature steam	Some emissions, carbon sequestration may decrease their effect.
	Thermochemical splitting of water	Water	High temperature obtained from the cooling of nuclear reactors	No emissions.
	Gasification	Coals, Biomass	Steam and oxygen at high temperature and pressure	Some emissions, carbon sequestration may decrease their effect.
	Pyrolysis	Biomass	Moderate temperature steam	Some emissions, carbon sequestration may decrease their effect.
Electrochemical method	Electrolysis	Water	Wind, solar, hydro and nuclear electricity	No emissions.
	Electrolysis	Water	Electricity produced from coal and natural gas	Some emissions from electricity production.
	Photochemical	Water	Direct sunlight	No emissions.
Biological method	Photobiological	Water and algae species	Direct sunlight	No emissions.
	Anaerobic fermentation	Biomass	High temperature	No emissions.
	Fermentative microorganisms	Biomass	High temperature	No emissions.

Source: (Iordache, I., & Stefănescu, I., 2010)

Renewable Energy-Based Hydrogen Production Technologies

Numerous technologies and equipment have currently been developed that offer the possibility of obtaining hydrogen through mechanisms of water splitting under the action of renewable energy, especially solar and wind energy. In this part of the chapter are briefly presented only the technologies that use water as raw material, and the resulting products are hydrogen and oxygen, although in practice there are other types of raw materials used in the production of organic hydrogen, and as objectives of this presentation aims to analyze and compare the most important water-splitting technologies under the action of various components of renewable energy.

a. Thermal and Thermochemical Technologies for Hydrogen Production

In the case of using the thermal component related to renewable energy, the water molecules can be cleaved directly under the action of thermal energy or indirectly with the help of intermediate auxiliary chemicals. In principle, when the temperature reaches higher than 2000 °C, the direct decomposition of water begins without the use of any other auxiliary chemicals. However, the temperature of direct thermolysis is too high, requiring special construction of refractory equipment *(Cobîrzan, N., et.al., 2018)*. In addition, the thermolysis product is a mixture of hydrogen and oxygen gases, which indicates a considerable risk of explosion. *(Wang, Z. et al.,2012; Safari, F.& Dincer, I., 2020)*.

Therefore, in order to reduce the temperature required for the dissociation of the water molecule, some auxiliary chemicals can be introduced. This type of water decomposition is often called the thermochemical cycle because water enters the process along with the chemical compounds that will be recovered and reused, and the results of the process are hydrogen and oxygen. If a small part of the cycle's input energy is electricity, then we are talking about a type of hybrid thermochemical cycle, rather than a pure thermochemical cycle. Currently, more than 300 thermochemical cycles have been indicated, either purely thermal or hybrid. *(Wang, Z. et al.,2012; Safari, F.& Dincer, I., 2020).*

Currently, the most studied thermochemical cycle is the sulfur - iodine cycle, which consists of three chemical reactions in which sulfur and iodine compounds decompose water at high temperatures into oxygen and hydrogen. Sulfur and iodine compounds are recycled, the process requiring a heat source. In the first stage of the process, iodine, sulfur dioxide and water vapor react to produce a mixture of sulfuric and iodine acid. The iodine acid is then separated by distillation. The reaction that requires the highest temperature, 800-900 °C, is the decomposition of H_2SO_4. Water, sulfur dioxide and residual acid are separated from the oxygen produced by condensation. The third reaction is the breakdown of iodine acid into iodine and hydrogen, the reaction takes place at about 450 °C. Iodine, possible water residues or sulfur dioxide are separated by condensation, hydrogen being the only gaseous product of the separation process. *(Ying, Z., et al.,2020; Mandal, S. & Jana, A. K., 2020).*

The coldest cycle is copper - chlorine-based, where the maximum required temperature does not exceed 550 °C. The Cu-Cl thermochemical cycle is represented by a series of chemical reactions that lead to the cleavage of the water molecule into hydrogen and oxygen, using interim compounds based on copper and chlorine that will be recovered and regenerated. Basically, the process consumes thermal energy and water, and the copper and chlorine compounds undergo transformations during the cycle without being consumed. *(Mehrpooya, M. & Habibi, R.,2020; Kiani, R., et al., 2020).*

As mentioned there is a very wide variety of thermochemical cycles from a theoretical point of view, but practically only a limited number of such cycles have technological potential. In addition to those mentioned above, the thermochemical cycles that use metal oxides (cerium, iron, zinc, etc.) may be of interest from certain perspectives. (*Carrillo, R. J., & Scheffe, J. R., 2017; Ishaq, H., & Dincer, I., 2019).*

Different cycles require different inlet temperatures, which also determine the availability of the energy source. At present, a high temperature of up to 3500 °C can be obtained in solar concentration furnaces only in the pilot phase laboratory. By comparison, the maximum temperature of future generation planned IV reactors is up to 1000 °C. This suggests that solar thermal energy has a more feasible potential to power a thermochemical cycle to obtain hydrogen. *(Wang, Z. et al.,2012; Oruc, O.& Dincer, I., 2021).*

b. Water Electrolysis

Electrolysis comprises all the processes that take place when the direct electric current passes through the solution or melt of an electrolyte, in other words, it is the process of orienting and separating the ions of an electrolyte with the help of the direct electric current. In the electrolysis process, the positive ions are directed towards the negative pole, and the negative ions towards the positive pole, where they lose their charge and deposit or enter a chemical reaction, at the anode there is an oxidation process, while at the cathode a reduction one. *(Yuzer, B., et.al., 2020; Kakoulaki, G., et.al., 2021)*

In water electrolysis, it is separated by direct current to produce hydrogen. Direct current passes through two electrodes immersed in water, anode and cathode, and hydrogen is obtained on the surface of the cathode when the electric potential is high enough. In this regard, in order to avoid confusion with other terminologies, such as photoelectrolysis or other photoelectrochemical methods, it is suggested that the electrolysis terminology will be used only when electricity is fully received from an external power source (eg energy electricity generated by photovoltaic panels). Therefore, electrolyzers do not receive sunlight to dissociate water. *(Wang, Z. et al.,2012; Pareek, A., et al.,2020; Kannah, R. Y., et al.,2020)*

Conventional electrolysis is a mature technology, and the efficiency of a commercially available electrolyzer is 50-80%. Regarding the electrolytic production of hydrogen, there are three types of technologies and implicitly three types of distinct electrolyzers: alkaline electrolyzers, electrolyzers with ion exchange membranes and electrolyzers with solid oxide electrolyte. Currently, the efficiency of electricity production, either with photovoltaic panels or other types of renewable energy conversion equipment, is about 10-20%, therefore, the efficiency of hydrogen production by electrolysis is below 16%. It is anticipated that the efficiency of hydrogen production based on renewable energies may be increased in the future, through the use of new types of materials and equipments. *(Wang, Z. et al.,2012; Wang, H., et.al., 2020; Yadav, D., & Banerjee, R., 2018).*

In addition to technological maturity, another major advantage of water electrolysis is that hydrogen production can provide the necessary energy during the night or on days with adverse weather conditions, by using renewable energy stored in hydrogen. *(Wang, Z. et al.,2012; Maggio, G., et.al., 2019).*

c. Photoelectrolysis and Photoelectrochemical Technologies

The terminology suggests that these technologies are adopted when the electrode uses at least one absorption light and electricity from external power sources to carry out the reduction or oxidation reactions. *(Van de Krol, R., & Grätzel, M., 2012; Wang, Z. et al.,2012)*

The hydrogen production unit in this case may be an electrochemical electrolyzer or photoelectrolyzer, which consists of an electrode that absorbs sunlight and a counter electrode immersed in the electrolyte. The sunlight absorption electrode must be arranged so as to be incident with the sunlight window to capture sufficient solar radiation in the process of generating electric potential, the electricity being generated by this electrode which is made of materials that can create electric potential. due to its exposure to the sun. *(Wang, Z. et al.,2012; Grimm, A., et.al., 2020; Bai, S., et.al., 2021)*

d. Photolysis and Photochemical Technologies

These technologies suggest the use of sunlight in the interaction with a semiconductor-electrolyte surface for the generation of hydrogen from water or solvents, the appearance of which is observed at the semiconductor-electrolyte interface in the phenomenon of photocorrosion. (*Valente, A., et.al, 2017).*

The decomposition of water into hydrogen and oxygen by photochemical mechanisms involves a chain of endergonic or endoergic reactions of the "up-hill" type, which are accompanied by a positive variation of Gibbs free energy *(Punekar, N. S., 2018).*Through these reactions, the energy of the photon is converted to chemical energy, identical to the reactions that take place in the process of green plants photosynthesis.

The basic components of a photochemical cell for hydrogen production are: the solar window, the sensitizer and the catalyst, which are immersed and evenly distributed in the water. Briefly, the photocatalytic decomposition process can be summarized in three simple steps: photons with energy greater than 1.23 eV ($\lambda = 1000$ nm) are absorbed by the catalyst, photogenerated pairs of electron-gaps migrate separately to the catalyst surface, species absorbed on the surface of the catalyst (water) are reduced and oxidized by electrons and positive voids. *(Dong, Y., et.al., 2021; Cornwell, O., et.al, 2018)*

e. Hybrid Technologies

Two or more of the technologies presented in the previous sections can be combined together for hydrogen production, which would mean a hybrid production technology with a higher energy supply potential and thus an improved efficiency of hydrogen production. *(Wang, Z. et al.,2012; Dawood, F., Anda, M., & Shafiullah, G. M., 2020).*

A typical hybrid technology, which uses both thermal and electrical energy, is high temperature-electrolysis (HTE). In HTE, electricity could be the external power supply, and the overall hydrogen production efficiency can reach 40%. *(Wang, Z. et al.,2012; Ayers, K., et.al., 2019)*

Other examples of hybrid hydrogen generation technologies using solar energy are: electricity and photonic energy - used by water photoelectrolysis technology, photonic energy and biochemical energy - used by methods such as water biophotolysis or artificial photosynthesis. *(Abdel-Basset, M., et.al., 2021)*

Assuming that hydrogen can be produced from a wide variety of raw materials available almost anywhere in nature, using a wide variety of technological processes it can be concluded that each region of the world may be able to produce more of its own energy with minimal impact on the environment *(Holladay, J. D., 2009)*, while increasing energy efficiency through primary energy savings, reducing grid losses, reducing the price and cost of electricity for consumers, all together contributing to the security and sustainability of the regional, national and global energy system *(Felseghi, R. A., & Papp, A., 2014).*

Hydrogen production technologies using renewable energy were presented, innovative technologies that have recently made significant progress in the development of alternative hydrogen production systems. A simplified comparison of the main characteristics and particularities of the types of technologies *(Midilli, A., et.al., 2005, a)* developed and currently used in obtaining hydrogen using renewable energy is presented in Table 4.

Following a critical analysis of the current state of development of hydrogen production technologies based on renewable energy, it can be stated that the electrolysis of water fed by energy generated by these sources is the most mature of the technologies, and thermochemical cycles offer the possibility of obtaining hydrogen on a scale. high production while minimizing energy losses caused by high-temperature requirements *(Wang, Z. et al.,2012; Midilli, A., et.al., 2005, b; Wieczorek, A. K., et.al., 2006).* Hydrogen production technologies based on photochemical and photoelectrolytic methods are limited, in terms of efficiency, significantly by the quality of materials, photocatalysts and photoelectrolytes that are part of the constructive composition of systems.

Sustained research and development programs in the field of hydrogen production systems and technologies will eliminate technical shortcomings while reducing costs, and hydrogen will compete with traditional energy resources. Any form of renewable energy, in order to be a convincing replacement for fossil fuels, must allow for storage and transport. Hydrogen is often proposed as an attractive option to supplement the use of renewable energy, while hydrogen can prove to be the most ubiquitous fuel available. *(Holladay, J. D., 2009; Felseghi, R.A., et.al., 2019, a)*

Table 4. Particularities in obtaining hydrogen based on renewable energy

Method	Thermochemical	Electrolysis	Photoelectrolysis	Photochemical
Power supply	External	External	Intern	Intern
Basic system components	more than 3 thermal reactors	2 electrodes + electrolyte	2 electrodes, electrolyte + solar lighting window	at least 1 sensitizer + 1 catalyst and sunlight window
Overall efficiency of H_2 production	45%	30%	16%	10%
Energy form at the input	Thermal	Electric	Electric	Photonics
Reaction mechanism	Thermal splitting of water	Electric potential	Electric potential	Photons activate electrons
Production scale	Wide scale	Medium-scale or local fuel stations	Local fuel stations	Local fuel stations
getting H_2 / O_2	Separated	Separated	Mixed	Mixed
Distribution network	Needed	It depends on the production scale	It's not necessary	It's not necessary
prod. H_2 constructive type & solar energy capture	No	No	Yes	Yes

Source: (Iordache, I., & Stefănescu, I., 2010)

Hydrogen Transport and Distribution

The detailed analyzes that can be found in the literature on the distance/cost ratio of hydrogen transport from the place of production to the place of consumption can be summarized as follows: for small consumers with low demand it is preferable to supply hydrogen in the form of gas tablet (CGH_2), for moderate long-distance consumption hydrogen is delivered in liquid form (LH_2), and for crowded urban centers with high demand hydrogen will be delivered through distribution networks. *(Wulf, C., et.al., 2018; Armaroli, N., & Balzani, V., 2011; Iulianelli, A., & Basile, A., 2014; EHFCTP, 2005).*

Land or naval means of transport can be used to transport hydrogen from the manufacturer to distribution stations, in the form of compressed gas, cryogenic liquid or stored in special compounds, carriers that store hydrogen in the form of reversible chemical bonds (metal hydrides, carbon nanofibers).

Worldwide, more than 820 approved hydrogen stations were registered worldwide for 2019, intended to power cars that use fuel cells to generate electricity. Hydrogen is delivered through distribution stations in the form of liquid hydrogen at -253 °C and hydrogen gas at 300-350 bar and 700 bar, and as sources of hydrogen are: pipelines from plants that generate secondary hydrogen, plants local electrolysis or catalytic reforming of methane gas. *(E4tech, 2019)*

The use of hydrogen transport pipelines is not a novelty, but given the specificity of hydrogen transport, for example, the hydrogenation of steel leads to a deterioration of its quality becoming breakable, the construction of such pipelines requires very large initial investments. An advanced solution to accelerate the transition to a hydrogen economy would be to use the current gas network to transport a mix of 80% natural gas and 20% hydrogen, or a liquid compound (ethanol) from which hydrogen can be easily extracted (by reforming). *(Cerniauskas, S., et.al., 2020; Abd, A. A., et.al., 2020).*

Scientific studies analyze the possibility of creating a super-network for the simultaneous transport of liquid hydrogen and electricity, the same pipe/cable in which liquid hydrogen would have the role of coolant necessary to obtain the state of superconductivity and would be used at the exit of the cable as and fuel in hydrogen filling stations. (*Al-Breiki, M., & Bicer, Y., 2021).*

Hydrogen Storage

A key issue in ensuring the viability of a hydrogen-based economy is the storage of manufactured hydrogen. This problem arose with the development of hydrogen production processes using renewable energy, unpredictable, weather-dependent energy. With classical technologies, hydrogen is difficult to store due to its physicochemical properties. Thus, in the presence of air, an explosive mixture occurs, because hydrogen, mixed with oxygen in a proportion of 5–85%, is flammable. Due to the extremely small size of its molecules, hydrogen diffuses through a large number of materials, the diffusion process being favored by high pressure and heat. In prolonged contact with steel, its hydrogenation occurs, which makes it brittle. Due to its very low density, the volume energy power of hydrogen is three times lower than that of methane gas, which means that, at the same amount stored, a tank with a pressure three times higher than for methane gas. *(Zhou, L., 2005; Eberle, U., Felderhoff, M., & Schueth, F., 2009).*

All these problems are currently being solved by developing new materials and storage technologies *(Ren, J., et.al. 2017).* Thus, by using new materials, diffusion losses were reduced and special seals were made, which allowed the storage of hydrogen in the form of gas under pressure (CGH_2). Liquid hydrogen tanks (LH_2) are currently used for large quantities. This reduces the reactivity of hydrogen and increases its density by approx. 800 times the gaseous state at normal temperature, but heat loss on evaporation occurs. High-cost issues involve the thermal insulation of tanks and pipes, which is necessary to avoid the occurrence of boiling, but also the formation of ice around them. *(Chapelle, D., et.al., 2021)*

To reduce the pressure of molecular hydrogen, research is being done to dissolve hydrogen in other materials that are either liquid or solid. The principle consists in the storage of molecular hydrogen in the voids of the network of the respective materials. These hydrogenated materials are called hydrides. The problems raised by the storage of hydrogen in hydrides are the temperature and pressure of absorption and release of hydrogen, toxicity in contact with the skin and eyes, but also the weight of the tanks. Among the hydrides possible to be used for hydrogen storage are magnesium, sodium, lithium or calcium hydrides and aluminum or boron. (*Dwivedi, S. K., & Vishwakarma, M., 2018)*

Another means of storing hydrogen may be ammonia (NH_3), from which it can be released by a catalytic reforming process. Ammonia has the advantage of having a high density of hydrogen storage in liquid form and without special pressure and temperature problems. When mixed with water, it can be stored at normal ambient temperature and pressure. Also, there are no problems with the production, transport and distribution of ammonia, there is already a well-developed infrastructure in this regard, and by catalytic reforming, hydrogen is obtained without toxic waste. The disadvantage, however, is that the manufacture of ammonia is energy-consuming. (*Makepeace, J. W., et.al., 2019; Palys, M. J., & Daoutidis, P., 2020; Ishaq, H., & Dincer, I., 2020)*

In recent years, however, a new storage technique has emerged, using graphite nanotubes, called fullerenes. Hydrogen is stored between layers of graphite nanotubes with diameters of 5–10 nm. Thus, about 30 liters of hydrogen are stored for every gram of graphite. The disadvantage is that the charging time of a 25-liter tank is 4–24 hours and can only be done 4-5 times. Also, the theory of storage mode has not yet been fully elucidated. Ongoing research is also on the storage of hydrogen in metal-organic networks using a series of synthetic porous materials, polymers with attached titanium atoms, glass microspheres of 10-300 m. (*Lyu, J., Kudiiarov, V., & Lider, A., 2020; Ahamed, M. I., et.al., 2021)*

It is worth mentioning the initiative taken by the European Commission, which financed through the public-private partnership FCH JU (Fuel Cells and Hydrogen Joint Undertaking) within the HyUnder project (Assessment of the potential, the actors and relevant business cases for large scale and seasonal storage of renewable electricity by hydrogen underground storage in Europe). The idea behind the project is to establish a European initiative to implement hydrogen-based energy technologies generated by increasing the percentage of capitalization of renewable resources. This research project focused on large-scale hydrogen storage in underground caverns, related to the energy market and existing storage technologies, and identified and analyzed areas of applicability, possible stakeholders, safety rules, regulatory framework and the societal impact on public acceptance. Within this project, case studies were carried out in several areas of Europe, including Romania. Each case study analyzed the competitiveness of hydrogen storage compared to other types of energy storage, the geological zonal potential of hydrogen storage and how this hypothesis of hydrogen storage can be implemented in the energy market. (*Felseghi, R.A., et.al., 2019, a; Filote, C., et.al, 2020)*

FUTURE RESEARCH DIRECTIONS

In the light of the new strategies of global policies on climate neutrality of energy systems that propose plans for the energy system of the future and place clean hydrogen in the spotlight, it is necessary to address as soon as possible the following study directions:

- carrying out a study to outline the socio-economic perception, viability and public acceptance on the use of hydrogen as an energy alternative and the regional transition to sustainable and ecological hydrogen-based energy generation systems;
- identifying and inventorying the availability of hydrogen economy implementation, estimating the potential for hydrogen production, storage, transport, distribution and use at the regional level;
- in the context of the often discussed topic of the circular economy, a series of studies can be developed on how to obtain hydrogen from various wastes, but also from wastewater by using renewable energies in production processes;
- conducting feasibility studies on the implementation of energy systems that integrate hydrogen and fuel cells in the power supply of individual consumers and at the level of local communities.

CONCLUSION

Hydrogen, called the energy vector, is currently designated as the fuel of the future for transportation, for industry, for heating buildings and more. Its most important aspect is that it does not emit CO_2 and does not pollute the atmosphere when used. Currently, in the large world, the amount of hydrogen is produced mainly from fossil fuels with harmful effects on the environment. Hydrogen production is currently based on a variety of processes, but it is known that almost all of them are associated with a wide range of emissions with harmful implications for the environment, but also with a different footprint in terms of production costs.

For an overview, an important part of the processes and methods of hydrogen production can be highlighted, keeping in mind:

- hydrogen based on electricity, when it is produced by water electrolysis - in which case the greenhouse gas emissions belong entirely to the electricity production cycle;
- hydrogen based on fossil fuels - in which case a variety of processes are considered that use natural gas as a raw material by reforming or coal by gasification. It can be appreciated that currently, this method is the most used, and with the hydrogen production is created the most important amount of harmful greenhouse gases;
- hydrogen based fossil fuels with the use - carbon capture process - a situation in which the effect of greenhouse gas emissions is greatly diminished.

After the review it can be easily highlighted that hydrogen obtained from renewable sources is the most suitable because it is the "cleanest", which is why it is the only one compatible with the objectives of each country and implicitly of the U.E. on climate neutrality. For the future, the differentiation between the hydrogen types depending on the technologies used for its production will be the basis of the strategies adopted, the main objective being to obtain clean hydrogen.

ACKNOWLEDGMENT

This work was supported by a grant from the Romanian Ministery of Research and Innovation, CCCDI-UEFISCDI, project number PN-III-P1-1.2-PCCDI-2017-0776/No. 36 PCCDI/15.03.2018, within PNCDI III.

REFERENCES

E4tech. (n.d.). *The Fuel Cell Industry Review 2018*. Available online: https://www.californiahydrogen.org/wpcontent/uploads/2019/01/TheFuelCellIndustryReview2018.pdf

Abd, A. A., Naji, S. Z., Thian, T. C., & Othman, M. R. (2021). Evaluation of hydrogen concentration effect on the natural gas properties and flow performance. *International Journal of Hydrogen Energy*, *46*(1), 974–983. doi:10.1016/j.ijhydene.2020.09.141

Abdel-Basset, M., Gamal, A., Chakrabortty, R. K., & Ryan, M. J. (2021). Evaluation of sustainable hydrogen production options using an advanced hybrid MCDM approach: A case study. *International Journal of Hydrogen Energy*, *46*(5), 4567–4591. doi:10.1016/j.ijhydene.2020.10.232

Ahamed, M. I., Shakeel, N., Anwar, N., Khan, A., Asiri, A. M., & Dzudzevic-Cancar, H. (2021). Graphene-based nanocomposite for hydrogen storage application. In *Nanomaterials for Hydrogen Storage Applications* (pp. 57–78). Elsevier. doi:10.1016/B978-0-12-819476-8.00006-2

Al-Breiki, M., & Bicer, Y. (2021). Comparative life cycle assessment of sustainable energy carriers including production, storage, overseas transport and utilization. *Journal of Cleaner Production*, *279*, 123481. doi:10.1016/j.jclepro.2020.123481

Armaroli, N., & Balzani, V. (2011). The hydrogen issue. *ChemSusChem*, *4*(1), 21–36. doi:10.1002/cssc.201000182 PMID:21226208

Ayers, K., Danilovic, N., Ouimet, R., Carmo, M., Pivovar, B., & Bornstein, M. (2019). Perspectives on low-temperature electrolysis and potential for renewable hydrogen at scale. *Annual Review of Chemical and Biomolecular Engineering*, *10*(1), 219–239. doi:10.1146/annurev-chembioeng-060718-030241 PMID:31173524

Badea, G., Felseghi, R., Aschilean, I., Bolboaca, A., Muresan, D., Soimosan, T., ... Raboaca, S. (2015, August). Energen system for power supply of passive house: case study. In *2015 Second International Conference on Mathematics and Computers in Sciences and in Industry (MCSI)* (pp. 24-31). IEEE. 10.1109/MCSI.2015.31

Badea, G., Felseghi, R. A., Aşchilean, I., Răboacă, S. M., & Şoimoşan, T. (2017, December). The role of hydrogen as a future solution to energetic and environmental problems for residential buildings. In AIP Conference Proceedings (Vol. 1917, No. 1, p. 030004). AIP Publishing LLC. doi:10.1063/1.5018277

Bai, S., Tian, K., Meng, J. C., Zhao, Y., Sun, J., Zhang, K., Feng, Y., Luo, R., Li, D., & Chen, A. (2021). Reduced graphene oxide decorated SnO2/BiVO4 photoanode for photoelectrochemical water splitting. *Journal of Alloys and Compounds*, *855*, 156780. doi:10.1016/j.jallcom.2020.156780

Barreto, L., Makihira, A., & Riahi, K. (2003). The hydrogen economy in the 21st century: A sustainable development scenario. *International Journal of Hydrogen Energy*, *28*(3), 267–284. doi:10.1016/S0360-3199(02)00074-5

Carrillo, R. J., & Scheffe, J. R. (2017). Advances and trends in redox materials for solar thermochemical fuel production. *Solar Energy*, *156*, 3–20. doi:10.1016/j.solener.2017.05.032

Cerniauskas, S., Junco, A. J. C., Grube, T., Robinius, M., & Stolten, D. (2020). Options of natural gas pipeline reassignment for hydrogen: Cost assessment for a Germany case study. *International Journal of Hydrogen Energy*, *45*(21), 12095–12107. doi:10.1016/j.ijhydene.2020.02.121

Chapelle, D., Maynadier, A., Bebon, L., & Thiébaud, F. (2021). Hydrogen Storage: Different Technologies, Challenges and Stakes. Focus on TiFe Hydrides. In *Advances in Renewable Hydrogen and Other Sustainable Energy Carriers* (pp. 1–10). Springer. doi:10.1007/978-981-15-6595-3_1

Cloete, S., Ruhnau, O., & Hirth, L. (2021). On capital utilization in the hydrogen economy: The quest to minimize idle capacity in renewables-rich energy systems. *International Journal of Hydrogen Energy*, *46*(1), 169–188. doi:10.1016/j.ijhydene.2020.09.197

Cloete, S., Ruhnau, O., & Hirth, L. (2021). On capital utilization in the hydrogen economy: The quest to minimize idle capacity in renewables-rich energy systems. *International Journal of Hydrogen Energy*, *46*(1), 169–188. doi:10.1016/j.ijhydene.2020.09.197

Cobîrzan, N., Thalmaier, G., Balog, A. A., Constantinescu, H., Timiş, I., & Streza, M. (2018). Thermophysical properties of fired clay bricks with waste ceramics and paper pulp as pore-forming agent. *Journal of Thermal Analysis and Calorimetry*, *134*(1), 843–851. doi:10.100710973-018-7092-3

Colbertaldo, P., Agustin, S. B., Campanari, S., & Brouwer, J. (2019). Impact of hydrogen energy storage on California electric power system: Towards 100% renewable electricity. *International Journal of Hydrogen Energy*, *44*(19), 9558–9576. doi:10.1016/j.ijhydene.2018.11.062

Cornwell, O., Radford, S. E., Ashcroft, A. E., & Ault, J. R. (2018). Comparing hydrogen deuterium exchange and fast photochemical oxidation of proteins: A structural characterisation of wild-type and ΔN6 β2-microglobulin. *Journal of the American Society for Mass Spectrometry*, *29*(12), 2413–2426. doi:10.100713361-018-2067-y PMID:30267362

Dawood, F., Anda, M., & Shafiullah, G. M. (2020). Hydrogen production for energy: An overview. *International Journal of Hydrogen Energy*, *45*(7), 3847–3869. doi:10.1016/j.ijhydene.2019.12.059

Dincer, I., & Zamfirescu, C. (2012). Sustainable hydrogen production options and the role of IAHE. *International Journal of Hydrogen Energy*, *37*(21), 16266–16286. doi:10.1016/j.ijhydene.2012.02.133

Dong, Y., Peng, W., Liu, Y., & Wang, Z. (2021). Photochemical origin of reactive radicals and halogenated organic substances in natural waters: A review. *Journal of Hazardous Materials*, *401*, 123884. doi:10.1016/j.jhazmat.2020.123884 PMID:33113752

Dwivedi, S. K., & Vishwakarma, M. (2018). Hydrogen embrittlement in different materials: A review. *International Journal of Hydrogen Energy*, *43*(46), 21603–21616. doi:10.1016/j.ijhydene.2018.09.201

Eberle, U., Felderhoff, M., & Schueth, F. (2009). Chemical and physical solutions for hydrogen storage. *Angewandte Chemie International Edition*, *48*(36), 6608–6630. doi:10.1002/anie.200806293 PMID:19598190

Eiler, J. M. (2020). *The Once and Future Hydrogen Economy*. Academic Press.

European Hydrogen and Fuel Cell Technology Platform. (2005). *Strategic Research Agenda*. Available online: https://www.fch.europa.eu/sites/default/files/documents/hfp-sra004_v9-2004_sra-report-final_22jul2005.pdf

Felseghi, R. A., Carcadea, E., Raboaca, M. S., Trufin, C. N., & Filote, C. (2019). a). Hydrogen fuel cell technology for the sustainable future of stationary applications. *Energies*, *12*(23), 4593. doi:10.3390/en12234593

Felseghi, R. A., & Papp, A. (2014). Considerations regarding hybrid systems of power generator from renewable energy sources. *Studia Universitatis Babes-Bolyai, Ambientum, 59.*

Felseghi, R. A., Şoimoşan, T. M., Filote, C., & Răboaca, M. S. (2019, b). Considerations Regarding the Green Retrofitting of Residential Buildings From Human Wellbeing Perspectives. In Retrofitting for Optimal Energy Performance (pp. 143-175). IGI Global.

Filote, C., Felseghi, R. A., Raboaca, M. S., & Aşchilean, I. (2020). Environmental impact assessment of green energy systems for power supply of electric vehicle charging station. *International Journal of Energy Research*, *44*(13), 10471–10494. doi:10.1002/er.5678

Grimm, A., de Jong, W. A., & Kramer, G. J. (2020). Renewable hydrogen production: A techno-economic comparison of photoelectrochemical cells and photovoltaic-electrolysis. *International Journal of Hydrogen Energy*, *45*(43), 22545–22555. doi:10.1016/j.ijhydene.2020.06.092

Holladay, J. D., Hu, J., King, D. L., & Wang, Y. (2009). An overview of hydrogen production technologies. *Catalysis Today*, *139*(4), 244–260. doi:10.1016/j.cattod.2008.08.039

Hosseini, S. E., & Wahid, M. A. (2020). Hydrogen from solar energy, a clean energy carrier from a sustainable source of energy. *International Journal of Energy Research*, *44*(6), 4110–4131. doi:10.1002/er.4930

HyUnder Project. (n.d.). Available online: www.hyunder.eu

Iordache, I., & Stefănescu, I. (2010). Tehnici Avansate De Tratare a Apei ¸Si Separare a Hidrogenului [Advanced Techniques for Water Treatment and Hydrogen Separation]. Editura AGIR.

Ishaq, H., & Dincer, I. (2019). A comparative evaluation of three CuCl cycles for hydrogen production. *International Journal of Hydrogen Energy*, *44*(16), 7958–7968. doi:10.1016/j.ijhydene.2019.01.249

Ishaq, H., & Dincer, I. (2020). A comprehensive study on using new hydrogen-natural gas and ammonia-natural gas blends for better performance. *Journal of Natural Gas Science and Engineering*, *81*, 103362. doi:10.1016/j.jngse.2020.103362

Iulianelli, A., & Basile, A. (Eds.). (2014). *Advances in hydrogen production, storage and distribution.* Elsevier.

Kakoulaki, G., Kougias, I., Taylor, N., Dolci, F., Moya, J., & Jäger-Waldau, A. (2021). Green hydrogen in Europe–A regional assessment: Substituting existing production with electrolysis powered by renewables. *Energy Conversion and Management*, *228*, 113649. doi:10.1016/j.enconman.2020.113649

Kannah, R. Y., Kavitha, S., Karthikeyan, O. P., Kumar, G., Dai-Viet, N. V., & Banu, J. R. (2020). Techno-economic assessment of various hydrogen production methods-A review. *Bioresource Technology*, 124175. PMID:33022437

Kiani, R., Mirzaei, F., Ghanbari, F., Feizi, R., & Mehdipour, F. (2020). Real textile wastewater treatment by a sulfate radicals-Advanced Oxidation Process: Peroxydisulfate decomposition using copper oxide (CuO) supported onto activated carbon. *Journal of Water Process Engineering*, *38*, 101623. doi:10.1016/j.jwpe.2020.101623

Koyuncu, I., Yilmaz, C., Alcin, M., & Tuna, M. (2020). Design and implementation of hydrogen economy using artificial neural network on field programmable gate array. *International Journal of Hydrogen Energy*, *45*(41), 20709–20720. doi:10.1016/j.ijhydene.2020.05.181

Li, Y., Sun, Y., Qin, Y., Zhang, W., Wang, L., Luo, M., Yang, H., & Guo, S. (2020). Recent advances on water-splitting electrocatalysis mediated by noble-metal-based nanostructured materials. *Advanced Energy Materials*, *10*(11), 1903120. doi:10.1002/aenm.201903120

Linssen, J., & Hake, J. F. (2016). Hydrogen research, development, demonstration, and market deployment activities. *Hydrogen Science and Engineering: Materials, Processes, Systems and Technology*, 57-84.

Liu, W., Sun, L., Li, Z., Fujii, M., Geng, Y., Dong, L., & Fujita, T. (2020). Trends and future challenges in hydrogen production and storage research. *Environmental Science and Pollution Research International*, *27*(25), 31092–31104. doi:10.100711356-020-09470-0 PMID:32529621

Lyu, J., Kudiiarov, V., & Lider, A. (2020). An overview of the recent progress in modifications of carbon nanotubes for hydrogen adsorption. *Nanomaterials (Basel, Switzerland)*, *10*(2), 255. doi:10.3390/nano10020255 PMID:32024092

Maggio, G., Nicita, A., & Squadrito, G. (2019). How the hydrogen production from RES could change energy and fuel markets: A review of recent literature. *International Journal of Hydrogen Energy*, *44*(23), 11371–11384. doi:10.1016/j.ijhydene.2019.03.121

Makepeace, J. W., He, T., Weidenthaler, C., Jensen, T. R., Chang, F., Vegge, T., ... & David, W. I. (2019). Reversible ammonia-based and liquid organic hydrogen carriers for high-density hydrogen storage: Recent progress. *International Journal of Hydrogen Energy, 44*(15), 7746-7767.

Mandal, S., & Jana, A. K. (2020). Simulating reactive distillation of HIx (HI–H2O–I2) system in Sulphur-Iodine cycle for hydrogen production. *Nuclear Engineering and Technology*, *52*(2), 279–286. doi:10.1016/j.net.2019.07.033

McCarty, R. D., Hord, J., & Roder, H. M. (1981). *Selected properties of hydrogen (engineering design data)* (Vol. 168). US Department of Commerce, National Bureau of Standards. doi:10.6028/NBS.MONO.168

Mehrpooya, M., & Habibi, R. (2020). A review on hydrogen production thermochemical water-splitting cycles. *Journal of Cleaner Production*, *275*, 123836. doi:10.1016/j.jclepro.2020.123836

Mekhilef, S., Saidur, R., & Safari, A. (2012). Comparative study of different fuel cell technologies. *Renewable & Sustainable Energy Reviews*, *16*(1), 981–989. doi:10.1016/j.rser.2011.09.020

Midilli, A., Ay, M., Dincer, I., & Rosen, M. A. (2005b). On hydrogen and hydrogen energy strategies: I: current status and needs. *Renewable & Sustainable Energy Reviews*, *9*(3), 255–271. doi:10.1016/j.rser.2004.05.003

Midilli, A. D. N. A. N., Ay, M., Dincer, I., & Rosen, M. A. (2005a). On hydrogen and hydrogen energy strategies II: Future projections affecting global stability and unrest. *Renewable & Sustainable Energy Reviews*, *9*(3), 273–287. doi:10.1016/j.rser.2004.05.002

Molnarne, M., & Schroeder, V. (2019). Hazardous properties of hydrogen and hydrogen containing fuel gases. *Process Safety and Environmental Protection*, *130*, 1–5. doi:10.1016/j.psep.2019.07.012

Nuttall, W. J., & Bakenne, A. T. (2020). Towards a Hydrogen Economy. In *Fossil Fuel Hydrogen* (pp. 43–52). Springer. doi:10.1007/978-3-030-30908-4_4

Oruc, O., & Dincer, I. (2021). Assessing the potential of thermochemical water splitting cycles: A bridge towards for clean and sustainable hydrogen generation. *Fuel*, *286*, 119325. doi:10.1016/j.fuel.2020.119325

Palys, M. J., & Daoutidis, P. (2020). Using hydrogen and ammonia for renewable energy storage: A geographically comprehensive techno-economic study. *Computers & Chemical Engineering*, *136*, 106785. doi:10.1016/j.compchemeng.2020.106785

Pareek, A., Dom, R., Gupta, J., Chandran, J., Adepu, V., & Borse, P. H. (2020). Insights into renewable hydrogen energy: Recent advances and prospects. *Materials Science for Energy Technologies*, *3*, 319–327. doi:10.1016/j.mset.2019.12.002

Punekar, N. S. (2018). Hallmarks of an Enzyme Catalyst. In *ENZYMES: Catalysis, Kinetics and Mechanisms* (pp. 43–51). Springer. doi:10.1007/978-981-13-0785-0_5

Ren, J., Musyoka, N. M., Langmi, H. W., Mathe, M., & Liao, S. (2017). Current research trends and perspectives on materials-based hydrogen storage solutions: A critical review. *International Journal of Hydrogen Energy*, *42*(1), 289–311. doi:10.1016/j.ijhydene.2016.11.195

Ren, X., Dong, L., Xu, D., & Hu, B. (2020). Challenges towards hydrogen economy in China. *International Journal of Hydrogen Energy*, *45*(59), 34326–34345. doi:10.1016/j.ijhydene.2020.01.163

Safari, F., & Dincer, I. (2020). A review and comparative evaluation of thermochemical water splitting cycles for hydrogen production. *Energy Conversion and Management*, *205*, 112182. doi:10.1016/j.enconman.2019.112182

Sazali, N. (2020). Emerging technologies by hydrogen: A review. *International Journal of Hydrogen Energy*, *45*(38), 18753–18771. doi:10.1016/j.ijhydene.2020.05.021

Tabak, J. (2009). *Natural gas and hydrogen*. Infobase Publishing.

Valente, A., Iribarren, D., & Dufour, J. (2017). Life cycle assessment of hydrogen energy systems: A review of methodological choices. *The International Journal of Life Cycle Assessment*, *22*(3), 346–363. doi:10.100711367-016-1156-z

Van de Krol, R., & Grätzel, M. (2012). *Photoelectrochemical hydrogen production* (Vol. 90). Springer. doi:10.1007/978-1-4614-1380-6

Veziroğlu, T. N. (2000). Quarter century of hydrogen movement 1974–2000. *International Journal of Hydrogen Energy*, *25*(12), 1143–1150. doi:10.1016/S0360-3199(00)00038-0

Veziroğlu, T. N. (2004). IJHE thrust toward meeting the needs of implementation. *International Journal of Hydrogen Energy*, *1*(29), 15. doi:10.1016/j.ijhydene.2003.09.003

Wang, H., Kong, H., Pu, Z., Li, Y., & Hu, X. (2020). Feasibility of high efficient solar hydrogen generation system integrating photovoltaic cell/photon-enhanced thermionic emission and high-temperature electrolysis cell. *Energy Conversion and Management*, *210*, 112699. doi:10.1016/j.enconman.2020.112699

Wang, Z., Roberts, R. R., Naterer, G. F., & Gabriel, K. S. (2012). Comparison of thermochemical, electrolytic, photoelectrolytic and photochemical solar-to-hydrogen production technologies. *International Journal of Hydrogen Energy*, *37*(21), 16287–16301. doi:10.1016/j.ijhydene.2012.03.057

Weeda, M., & Segers, R. (2020). *The Dutch hydrogen balance, and the current and future representation of hydrogen in the energy statistics*. Academic Press.

Wieczorek, A. K., Krystian, M., & Zehetbauer, M. (2006). SPD processed alloys as efficient vacancy-hydrogen systems. *Diffusion and Defect Data, Solid State Data. Part B, Solid State Phenomena*, *114*, 177–182. doi:10.4028/www.scientific.net/SSP.114.177

Wulf, C., Reuß, M., Grube, T., Zapp, P., Robinius, M., Hake, J. F., & Stolten, D. (2018). Life Cycle Assessment of hydrogen transport and distribution options. *Journal of Cleaner Production*, *199*, 431–443. doi:10.1016/j.jclepro.2018.07.180

Yadav, D., & Banerjee, R. (2018). Economic assessment of hydrogen production from solar driven high-temperature steam electrolysis process. *Journal of Cleaner Production*, *183*, 1131–1155. doi:10.1016/j.jclepro.2018.01.074

Ying, Z., Wang, Y., Zheng, X., Geng, Z., Dou, B., & Cui, G. (2020). Experimental study and development of an improved sulfur–iodine cycle integrated with HI electrolysis for hydrogen production. *International Journal of Hydrogen Energy*, *45*(24), 13176–13188. doi:10.1016/j.ijhydene.2020.03.037

Yuzer, B., Selcuk, H., Chehade, G., Demir, M. E., & Dincer, I. (2020). Evaluation of hydrogen production via electrolysis with ion exchange membranes. *Energy*, *190*, 116420. doi:10.1016/j.energy.2019.116420

Zhiznin, S. Z., Vassilev, S., & Gusev, A. L. (2019). Economics of secondary renewable energy sources with hydrogen generation. *International Journal of Hydrogen Energy*, *44*(23), 11385–11393. doi:10.1016/j.ijhydene.2019.03.072

Zhou, L. (2005). Progress and problems in hydrogen storage methods. *Renewable & Sustainable Energy Reviews*, *9*(4), 395–408. doi:10.1016/j.rser.2004.05.005

ADDITIONAL READING

Ball, M., Basile, A., & Veziroglu, T. N. (Eds.). (2015). *Compendium of hydrogen energy: hydrogen use, safety and the hydrogen economy*. Woodhead Publishing.

Barbir, F. (2012). *PEM fuel cells: theory and practice*. Academic press.

Barbir, F., Basile, A., & Veziroglu, T. N. (Eds.). (2015). *Compendium of Hydrogen Energy: Hydrogen Energy Conversion*. Woodhead Publishing.

Graham, T. (2020). *Tomorrow's Energy, Renewables Alone? Complementary Resources for Tomorrow: Proceedings of Energy and Resources for Tomorrow 2019*. University of Windsor.

Gupta, R., Basile, A., & Veziroglu, T. N. (Eds.). (2016). *Compendium of Hydrogen Energy: Hydrogen Storage, Distribution and Infrastructure*. Woodhead Publishing.

Ito, K., Li, H., & Hao, Y. M. (2016). *Hydrogen Energy Engineering: A Japanese Perspective.*

Miranda, P. E. (Ed.). (2018). *Science and Engineering of Hydrogen-Based Energy Technologies: Hydrogen Production and Practical Applications in Energy Generation*. Academic Press.

Scipioni, A., Manzardo, A., & Ren, J. (Eds.). (2017). *Hydrogen economy: supply chain, life cycle analysis and energy transition for sustainability*. Academic Press.

Seelam, P. K., Rathnayake, B., Pitkäaho, S., Turpeinen, E., & Keiski, R. L. (2020). Overview on recent developments on hydrogen energy: Production, catalysis, and sustainability. In Current Trends and Future Developments on (Bio-) Membranes (pp. 3-32). Elsevier. doi:10.1016/B978-0-12-817110-3.00001-1

Sherif, S. A., Goswami, D. Y., Stefanakos, E. L., & Steinfeld, A. (Eds.). (2014). *Handbook of hydrogen energy*. CRC Press. doi:10.1201/b17226

Zohuri, B. (2019). Cryogenics and liquid hydrogen storage. In *Hydrogen Energy* (pp. 121–139). Springer. doi:10.1007/978-3-319-93461-7_4

KEY TERMS AND DEFINITIONS

Circular Economy: In the linear economy, raw natural resources are taken, transformed into products and get disposed of. On the opposite, a circular economy model aims to close the gap between the production and the natural ecosystems' cycles – on which humans ultimately depend upon. This means, on one hand, eliminating waste – composting biodegradable waste or, if it's a transformed and non-biodegradable waste, reusing, remanufacturing and finally recycling it. On the other hand, it also means cutting off the use of chemical substances (a way to help regenerate natural systems) and betting on renewable energy.

Energy System: A system primarily designed to supply energy services to end-users.

Fuel Cell: A device that continuously changes the chemical energy of a fuel (such as hydrogen) and an oxidant directly into electrical energy.

Hydrogen: A nonmetallic gaseous chemical element with atomic number 1 that is the simplest and lightest of the elements.

Hydrogen Economy: Is an envisioned future in which hydrogen is used as an energy vector for power systems, fuel for heat production, fuel for hydrogen vehicles, medium for energy storage, and for long-distance transport of energy. In order to phase out fossil fuels and limit global warming, hydrogen can be created from water using intermittent renewal sources such as wind and solar, and its combustion only releases water vapor to the atmosphere.

Renewable Energy: Often referred to as clean energy, comes from natural sources or processes that are constantly replenished.

Stationary Application: Means any application of a Fuel Cell System in which such Fuel Cell System is designed to be attached to a fixed site, being land, a building or other immobile structure for extended use at that site, and includes stationary electrical power systems which are transportable for temporary site power supply.

Sustainable Development: Economic development that is conducted without depletion of natural resources.

Chapter 2
Hydrogen Production From Waste and Renewable Resources

Amit Kumar Chaurasia
https://orcid.org/0000-0003-1786-3745
Indian Institute of Technology, Roorkee, India

Prasenjit Mondal
Indian Institute of Technology, Roorkee, India

ABSTRACT

Increasing population and rapid urbanization lead to degradation of the natural environment while waste generation and energy crisis are major challenges in the most developing country. Hydrogen is considered one of the most promising energy carriers and capable to replace fossil fuels and meet the world's energy demand and concomitantly reduce toxic emissions. Currently, the world produces around 50 million tonnes/year from the process (i.e., electrolysis of water, steam reforming of hydrocarbons, and auto-thermal processes), but these processes are not sustainable and economical due to energy requirements and waste/pollutants generation. These challenges required growing interest in renewable energy resources such as hydrogen as an energy carrier. Hydrogen production from renewable sources attracted recent research attention because of its potential for sustainability and diversity. Hydrogen can be produced by various thermal, chemical, and biological technologies that include steam reforming, electrolysis, biomass conversion, solar conversion, and biological conversion.

1. INTRODUCTION

Modernization, rapid industrialization and speedy urbanization with fast-growing population lead to degradation of the natural environment, waste generation, greenhouse gases emission, energy crisis, which are the major challenges in the developing countries (Doman et al., 2013). Global primary energy demand is increasing at a rate of 2.1% per year with projected to total increase of 56% by 2040. Fossil fuels are the primary fuels shares more than 85% of global energy demand. Conventional fuels are limited and non-renewable in nature, their utilization release the greenhouse gases emission mainly CO_2.

DOI: 10.4018/978-1-7998-4945-2.ch002

Further, global CO_2 emission from fossil fuel combustion and industrial process is estimated around 35.7 billion tones in 2014 (*EAA Annual Report 2015/2016*, n.d.). Today there is broad scientific consensus that human activities have altered our climate and climate change have become global challenges for our society and humankind. Mitigating these repercussions will demand sizeable reductions in carbon emissions by minimizing fossil fuels based energy and that can be achieve only if conscientious action is take now to change how our society produces hydrogen or renewable energy by processes its wastes (Hallenbeck & Benemann, 2002; Hosseini & Wahid, 2016a). These challenges required growing interest in renewables energy resources such as hydrogen as energy carrier. Indeed, one of the most promising approaches to these critical problems is the conversion of waste materials into renewable energy such as hydrogen gas (Chaurasia et al., 2019).

1.1 Need of Hydrogen From Waste and Renewable Energy Sources

Hydrogen are used to generating electricity, run hydrogen fuel cell vehicles for transport, and a variety of commercial/industrial uses, which has several sustainable benefits on our health, climate as well as our economy.

1. Fossil fuels are non-renewable resource
2. Fossil fuel has adverse effects on environment and climate
3. Energy from waste materials is economically sound and sustainable
4. Energy from waste is generally sustainable for the environment
5. Little to No Global Warming Emissions
6. Enhance Public Health with Environmental

1.2 Hydrogen Basic Properties

Hydrogen is an energy carrier not as energy source that can be used to store and deliver energy products from other sources. These are the properties of hydrogen are tabulated in Table 1, that contribute to make hydrogen as a combustible fuel. Hydrogen possess the high energy density as 122 KJ/g, which is around three times of gasoline fuel, two times more efficient in combustion engines, with only water as by products and there is no greenhouse gases or pollutants generation.

1.3 Hydrogen From Waste and Renewable Sources

Advancement in living standard and technologies increases the waste generation. For examples, India produces more than 62 million tons of municipal solid waste annually as reported in 2019 by MoEF India. Wastes are such materials that are not prime products and discarded after primary use has no further use in terms of production, transformation or consumption. Waste can be characterised by proximate analysis (it gives information about moisture, volatiles, ash and carbon content) and Ultimate analysis (it gives details about C, H, N and O, energy content). Waste to energy conversion technologies consist of several routes such as physical, chemical and biological route. These technologies have a variety of techniques to produce hydrogen energy from waste such as aanaerobic digestion (reduce complex compound in simpler chemical compound in the absence of oxygen), incineration, gasification and pyrolysis can be applied on municipal solid waste (Eker & Sarp, 2017). In pyrolysis heating of waste in the absence of oxygen,

Table 1. Some basic properties of hydrogen

Properties	Hydrogen
Atomic Number	1
Atomic Weight	1.0079
Melting Point	-259°C
Boiling Point	- 253°C
Vapour Density	0.08376 kg/m^3 (at 20 °C, 1 atm)
Specific Gravity	0.0696
Flashpoint	–253 °C
Valence	1
Higher calorific value	1,41,790 (kJ/kg)
Diffusivity and Flame speed	High and High
Octane number	130
Air diffusion coefficient	0.61 cm^2/s
Air/fuel stoichiometry	34.30 (kg/kg)
Stoichiometry mixture in air	29.53%
Specific heat	14.89 kj/kgK
Energy in explosion	58.82 kj/gTNT

which break the any compound into a mixture of liquid, gaseous fuel, residue etc. at high temperature. Gasification is the breakdown of biomass in gas fuel at high temperature with partial oxidation. Waste to energy conversation technologies offer sustainable energy with waste reduction, and reduce carbon dioxide emissions. Biological route of waste to energy generation offer several advantages as metals recovery, renewable incentives, and higher energy revenues.

1.4 Advantages of Hydrogen From Waste and Renewable Sources

1. Biological agents such as microbes are capable to degrade almost all types of waste. It's already occurring in nature. Its mean nothing is waste in this universe.
2. It is reduced waste and meets energy criteria that are the global concern.
3. It is reduced toxic and hazardous pollutants.
4. Minimization of toxic compounds.
5. Tolerant to change in waste composition initially may inactive but not halted

1.5 Hydrogen Energy and Hydrogen Fuel Cells in Stationary Application

Hydrogen offer regenerative and sustainable mobility choices by hydrogen fuelled vehicles. Hydrogen is a primary energy resources that could be converted in different form of energy with higher efficiency. It has high gravimetric energy density, stored in various ways (liquid hydrogen or as solid hydride), easy to transport in remote areas and hydrogen-based fuel cells have more than 60 percentage efficiency.

Table 2. Fuels comparison with hydrogen.

Fuel Type	Energy/Mass Ratio (J/kg)	Energy/Volume Ratio (J/m3)	Carbon Emission (kgC/kg Fuel)
Hydrogen (Liquid)	141.90	10.10	0.00
Hydrogen (Gas)	141.90	0.013	0.00
Gasoline	47.50	34.85	0.86
Biodiesel	37.00	33.00	0.50
Natural gas	50.00	0.04	0.46
Coal	30.00	-	0.50

However, hydrogen energy has some barriers which need to be addressed by research and technology development. Hydrogen burns with air; thus, it's had operational safety difficulties.

Hydrogen storage in the liquid mode is not economical and required very low temperature. The most of the current technologies of hydrogen production and conversion are not economical. Hydrogen fuel cells (HFCs) energy conversion cost are relatively high. The resources of hydrogen storage, transport, infrastructure and distribution are not well explored and required extensive research as well as high capital cost.

Stationery applications area of the hydrogen energy is a vast energy consumer that looking for alternatives energy sources to maximize efficiency with pollutant reduction. Hydrogen fuel cells (HFCs) generate energy through electrochemical conversions without combustion, thus, they generate lesser pollutants then conventional technologies (ABANADES, 2019). HFCs are highly efficient (produced more power per unit of fuel), fuel flexible, low maintenance, quiet operation, durability, scalability, less space requirement with zero GHGs emission. Hydrogen Fuel cells (HFC) can be utilized for stationary power units for primary power (hospitals, homes, businesses, telecommunications (data centre), fire station, grocery and retail store), combined heat and power (CHP) uses. HFCs can also generate additional heat, and its utilization increases the efficiency from 60 to 90% as compared to electric grid (Doman et al., 2013). Heat can be utilized by buildings, facilities and even cooling in absorption chiller. HFCs have the market applications as speciality vehicles, emergency backup power and primary power for critical loads. Specialty vehicles such as forklifts used in various airport. More than 6500 HFCs vehicles are sold from Hyundai Nexo, Toyata Mirai and Honda clarity in California USA. Many other countries already have produced the HFCs vehicles or in process to launch in 2020. Several HFCs buses (USA), heavy duty trucks (California), trains (Germany) are operation and some others country expecting to launch in coming years. Research and study promoted the use of hydrogen energy in almost all sectors as energy productions, storage and transportation, heat generation, electricity production, cooling purpose in domestic as well as industrial sector (Abe et al., 2019; Abrell et al., 2019). This also shown that hydrogen energy production technology is practical, economically viable and sustainable. It's probable that hydrogen energy will change the future of energy sectors.

1.6 World Energy Prospects of Hydrogen Energy and Hydrogen Fuel Cells in Stationary Application

Urbanization and economic growth are directly associated with energy consumption. World energy requirements are increasing with respect to time and estimated that 600 EJ in the year of 2050. Some recent study predicted that that renewable resources shares of 36% by 2025 and expected to 69% by 2050 of total global energy demand, in which hydrogen energy contributes 11% in 2025 and 34% in 2050 (Kadier et al., 2014; Xiao et al., 2013). When renewable energy supported by community that reduce the utilization or fossil fuels and coal can be drop by around 40% and 37%, as by 2030. Since data reported in 2019, the world produces more than 60 million tonnes/year of hydrogen by industries for commercial purposes (Cédric Philibert, 2017). About 85-88% of H_2 is produce through steam reforming of natural gas at refineries and some industries by water electrolysis. Hydrogen are mainly produce and consume by refineries as hydrogen approximately 96% is derive from fossil fuels, 47% from natural gas, 31% from hydrocarbons, 17% from coal and about 5% from electrolysis (Lamy & Millet, 2020). Various real HFCs pilot projects are being validated, analysed and reports about technology development with crucial challenges across the world. This analysis explored HFCs design, architecture and power range (4.9 kW up to 2.7 MW), while equipment has the powers rang of 0.6 kW to 390 kW. Mohan et al., 2017, produced 5 kg hydrogen per day since last 6 months from the organic waste of the IICT India campus Cafeteria and process further extended to produce hydrogen from municipal waste across many city of India ("Turning Organic Waste into Hydrogen," 2019). HFCs commercial markets show the increasing trends of technology transfer around 395,000 units in 2014 and 575,000 units in 2018 in the stationary fields of HFCs equipment. Among this Asia region has the highest number HFCs practical applications and Japan has the largest contributions in it while Europe show slower growth in last decade. Data reported in last 5 years shows PEM Fuel cells was dominant among other HFCs in the segment of stationery and transport (Felseghi et al., 2019). The role of HFCs in global energy system as the heat production for residential, commercial and industrial applications and transportation (Light vehicles, heavy-duty vehicles and public transport vehicles such as hydrogen powered trains). HFCs evades the land requirement and air index impacts of biofuels, and time consume in recharging of electric vehicles. While, electric vehicles are mature technology due to low cost and advancement in technology. Hydrogen powertrains are also in trail phase in Japan and expected to be launch in 2020 (*Global EV Outlook 2020 – Analysis - IEA*, n.d.). The governments of South Korea, France, and Switzerland announced in 2018 to building a hydrogen economy (Staffell et al., 2019). Japan has invested $349 M for making hydrogen set-up until the 2020 Summer Olympics in Tokyo. In April 2019, UK announced the first commercial plant to treat 25 tonnes of plastic in to hydrogen per day. Environmental regulations promote the hydrogen energy utilization requirement at refineries and industrial purpose across the world. Among these some successful renewables hydrogen energy transition into hydrogen economy as Swiss MARKAL model, TIMES, Energy PLAN of Switzerland (Abrell et al., 2019), THESIS model of UK and the Markal model of Italy, European HySociety Project, TAIGEM-EH of Taiwan, Danish energy system of Denmark and the dynamic framework of Austria (Hosseini & Wahid, 2016a).

2. TECHNOLOGIES FOR HYDROGEN PRODUCTION

Hydrogen are produced from a variety of waste and renewable sources by different biological processes that include thermal, chemical and biological process. However, to realise the hydrogen energy benefits, it can be produced by renewable resources not from fossil resources (Das et al., 2008).

Table 3. Comparisons of hydrogen production technologies.

Process	Advantages	Disadvantages	Efficiency	Cost ($/kg)
Steam Reforming	Commercialized stage	CO_x Production Unsteady yield	75-86	2.28
Oxidation	Commercialized stage	H_2 Production with value added products (coke and oil) production	61-74	1.48
Auto thermal Reforming	Commercialized stage	H_2 Production with CO_2, use of fossil fuels	60-70	1.49
Biological Photolysis	Utilize CO_2, to make O_2, Under development stage	Lower yields of hydrogen, light energy requirements	11-12	2.15
Dark Fermentation (DF)	Waste reduction with H_2 production, light andO_2 not required, Technology development required	Low efficiency and low yield Required high volume of digester	65-78	2.57
Photo Fermentation	Degrade almost any wastewater, light required, under technology development.	Low hydrogen yield, Low efficiency, required light, high volume of reactor needed, sensitive to O_2	0.1	2.83
Gasification	Waste reduction, low cost substrate, low pollutants generation, Commercialized stage	Impurities, unstable rate of H_2 production, geographical suitability	35-45	1.77-2.05
Pyrolysis	Utilized low cost substrate or waste materials, CO_2-neutral. Established technology	Impurities, unstable rate of H_2 production, geographical suitability	40-50	1.59-170
Thermolysis	Sustainable technology, O_2 as by-product, high cost, no pollutants generation	High capital and operation cost, corrosion issue.	25-50	7.98-8.40
Photolysis	Sustainable technology under development, O_2 as by-product, high cost, no pollutants generation	Light required, catalyst under development, low yield	0.06	8-10
Electrolysis	Commercialized and greener technology, O_2 as by-product, costly process	Stationery technology, high cost	62-78	10.30

All the hydrogen production technologies comparisons, advantages/disadvantages, efficiency with cost are listed in the Table 3.

Figure 1. The major hydrogen production technologies

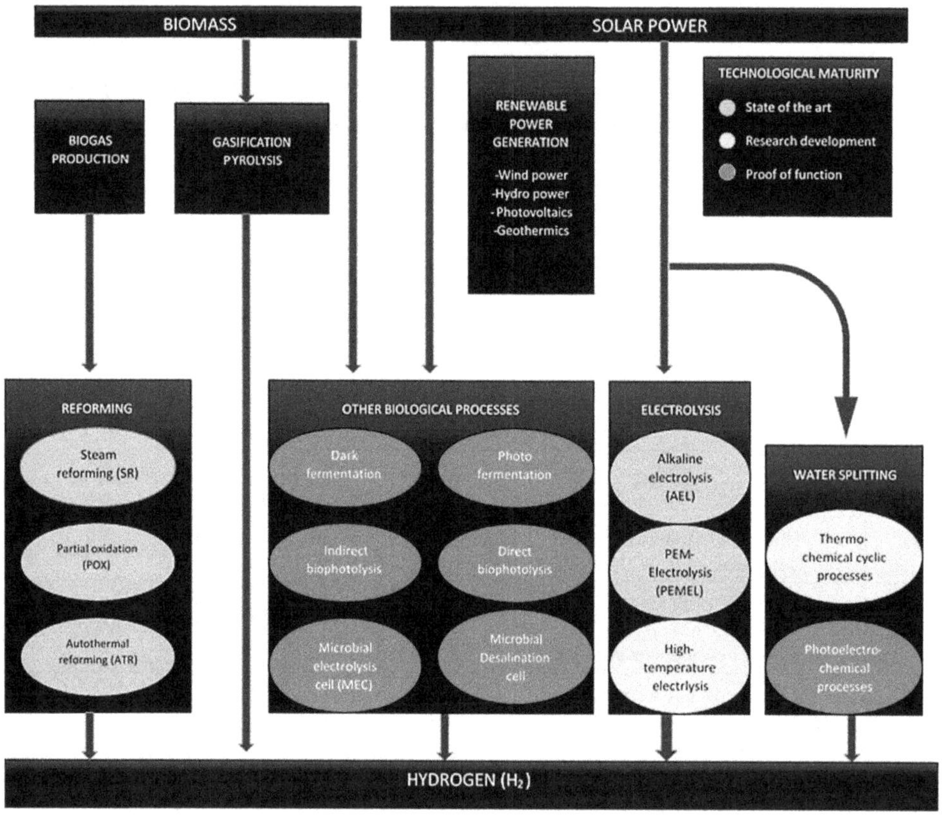

3. HYDROGEN FROM WATER

3.1 Electrolysis

Hydrogen can be produce by water electrolysis, in which water split by passing electricity through two electrodes anode and cathode to produced hydrogen gas (Chi & Yu, 2018a). Two electrodes (inert metal i.e. Pt. or Ir) placed in water are connected by a DC power source, where hydrogen is produced at cathode are half of the oxygen produced at anode in proportion to the electric current (Sapountzi et al., 2017). The typical electrolysis reactions are:

$$H_2O + 2F \rightarrow H_2 + 1/2O_2 \text{ (where F = 96485 °C)} \quad (1)$$

Reverse of eq. (1) is use in combustion phenomena or hydrogen fuel cells. However, there are many challenged need to be addressed such as in multiple cell circuit competing side reaction, lesser faradaic efficiency, over-potential losses of the electrode, lack of electro-catalyst (Lamy & Millet, 2020). A Commercial electrolyser has efficiencies of 55–75% with 70–53 kWh/kg hydrogen at 1 atmospheric pressure and 298 K (Chi & Yu, 2018a; Shiva Kumar & Himabindu, 2019a). Hydrogen production from water is well known and commercialize technologies consist of three methods as electrolysis, thermolysis, and photo-electrolysis. A variety of feedstocks and operating parameters are utilized in it although the principal remain the same in all such process (Kalamaras & Efstathiou, 2013). The widely used electrolyser is alkaline based electrolyser, operated in 60-80 °C with 20-30% electrolyte concentration of KOH/NaOH aqueous solution at less than 400 mA/cm^2 applied current (Lamy & Millet, 2020; Sapountzi et al., 2017). A commercial alkaline electrolyser has 30% KOH at around 80-90 °C with 55-75% energy recovery by utilising 4.49 kWh/m^3 of electricity. AWE has disadvantages of cannot start quickly, balance pressure between electrodes, slow loading response, difficult to adopt in variable nature and use of steady power input. In PEM electrolyser has electrode (Pt black, Nickel, rhodium etc.) or act as electrode catalysts, which are separated by a selective membrane (Polystyrene, Nafion 317) and it also act as gas separator) and operating current density is 10 A/cm2. PEM electrolyse has the advantages of fast response, compact design, high efficiency, high output pressure, simple plant balance, no alkaline fog and generate less pollution than alkaline electrolyser. The efficiency of the PEM electrolyser can be increased by addition of electro-catalyst and mediator compound. However, the cost of precious metals as electro-catalyst makes the process costlier than alkaline process. Additionally, innovative methods of synthesis of electro-catalyst that provide the better control over nanoscale morphology to improve the electro-catalyst performance and energy recovery (Chaurasia & Mondal, 2021; Rani et al., 2017). Some US study predicted the economics of upscaling PEM electrolyser that the cost will be reduced to one fourth when the hydrogen generation rate reached at 1000 kg/d. Solid oxide fuel cells are operating in reverse of alkaline electrolyser and has efficiency of 69-90% (90% efficiency when heat consumption was included) and can approached around 100%. However, SOEC face the challenges of corrosion, seals, thermal cycling and in other hand, PEM has higher efficiency than alkaline and no corrosion and seal issue of SOEC. In SOEC solid oxide is chemically and thermally stable and shows high anions transfer with ZrO3 doped with 8-mol% Y2O3 electrolyte (De Giz et al., 1995; Sapountzi et al., 2017; Staffell et al., 2019). The high-temperature electrolysis has improved the efficiency and attracted further research interest. However, high operating temperature is still the major challenges in material degradation and operation time. SOEC end products required further treatment to get high purity hydrogen. Moreover, these technologies is still face challenges of cast factors but use of nanotechnology (Ni, Fe, cobalt phosphate and some other nanomaterial) has great potential (Chaurasia & Mondal, 2021). Nanostructured α-Ni(OH)2 spheres showed the only 0.33 V over-potential at 10 mA/cm2 (small Tafel slope ~42 mV/dec) with improved durability and its comparable to RuO_2 catalysts (Chi & Yu, 2018a, 2018b). Further research is required about the durability of the electrode, electrolyte materials and membrane. Different electrolysis technology comparisons are given in Figure. 2. Proton Onsite and Siemens recently launches MW scale PEM electrolyser for industrial applications. Sunfire has developed the SOEC prototype for commercial use in Karlsruhe, Germany and several electrolysis projects were implemented in US and worldwide (Chi & Yu, 2018b; *Sunfire | Hydrogen*, n.d.). Moreover, electrolysis was old concepts and not commercially successful due to feasible technology and economic barriers such as capital cost.

3.2 Thermochemical Water Splitting

Thermolysis or thermochemical process are a promising way to produce hydrogen energy by utilizing heat from concentrated solar or nuclear from water with the efficiency of 50%. Thermochemical water splitting required high temperatures (above 2000 °C) from concentrated solar power/waste heat sources to produce hydrogen and oxygen directly from water. Water is decomposed at 2500°C but sustainable materials are not available (Nalakath et al., 2018; Safari & Dincer, 2020). Still it required the high temperature materials and process conditions. To reduce these ambiguities and eliminate the requirement

Figure 2. Typical electrolyser comparisons
(Reprint with copyright permission) (Chi & Yu, 2018a, 2018b).

	Low Temperature Electrolysis			High Temperature Electrolysis		
	Alkaline (OH^-) electrolysis		Proton Exchange (H^+) electrolysis		Oxygen ion(O^{2-}) electrolysis	
	Liquid	Polymer Electrolyte Membrane		Solid Oxide Electrolysis (SOE)		
	Conventional	Solid alkaline	H^+ - PEM	H^+ - SOE	O^{2-} - SOE	Co-electrolysis
Operation principles	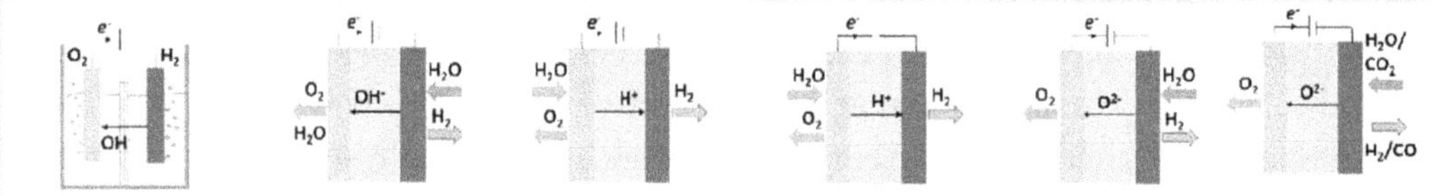					
Charge carrier	OH^-	OH^-	H^+	H^+	O^{2-}	O^{2-}
Temperature	20-80°C	20-200°C	20-200°C	500-1000°C	500-1000°C	750-900°C
Electrolyte	liquid	solid (polymeric)	solid (polymeric)	solid (ceramic)	solid (ceramic)	solid (ceramic)
Anodic Reaction (OER)	$4OH^- \rightarrow 2H_2O + O_2 + 4e^-$	$4OH^- \rightarrow 2H_2O + O_2 + 4e^-$	$2H_2O \rightarrow 4H^+ + O_2 + 4e^-$	$2H_2O \rightarrow 4H^+ + 4e^- + O_2$	$O^{2-} \rightarrow {}^1/_2O_2 + 2e^-$	$O^{2-} \rightarrow {}^1/_2O_2 + 2e^-$
Anodes	Ni > Co > Fe (oxides) Perovskites: $Ba_{0.5}Sr_{0.5}Co_{0.8}Fe_{0.2}O_{3-\delta}$, $LaCoO_3$	Ni-based	IrO_2, RuO_2, $Ir_xRu_{1-x}O_2$ Supports: TiO_2, ITO, TiC	Perovskites with protonic-electronic conductivity	$La_xSr_{1-x}MnO_3$ + Y-Stabilized ZrO_2 (LSM-YSZ)	$La_xSr_{1-x}MnO_3$ + Y-Stabilized ZrO_2 (LSM-YSZ)
Cathodic Reaction (HER)	$2H_2O + 4e \rightarrow 4OH^- + 2H_2$	$2H_2O + 4e \rightarrow 4OH^- + 2H_2$	$4H^+ + 4e^- \rightarrow 2H_2$	$4H^+ + 4e^- \rightarrow 2H_2$	$H_2O + 2e \rightarrow H_2 + O^{2-}$	$H_2O + 2e^- \rightarrow H_2 + O^{2-}$ $CO_2 + 2e^- \rightarrow CO + O^{2-}$
Cathodes	Ni alloys	Ni, Ni-Fe, $NiFe_2O_4$	Pt/C MoS_2	Ni-cermets	Ni-YSZ Subst. $LaCrO_3$	Ni-YSZ perovskites
Efficiency	59-70%		65-82%	up to 100%	up to 100%	-
Applicability	commercial	laboratory scale	near-term commercialization	laboratory scale	demonstration	laboratory scale
Advantages	low capital cost, relatively stable, mature technology	combination of alkaline and H^+-PEM electrolysis	compact design, fast response/start-up, high-purity H_2	enhanced kinetics, thermodynamics: lower energy demands, low capital cost		+ direct production of syngas
Disadvantages	corrosive electrolyte, gas permeation, slow dynamics	low OH^- conductivity in polymeric membranes	high cost polymeric membranes; acidic: noble metals	mechanically unstable electrodes (cracking), safety issues: improper sealing		
Challenges	Improve durability/reliability; and Oxygen Evolution	Improve electrolyte	Reduce noble-metal utilization	microstructural changes in the electrodes: delamination, blocking of TPBs, passivation		C deposition, microstructural change electrodes

of higher temperature gas separation, recent studied shows some other thermochemical cycles (hybrid sulphur and sulphur–iodine process, metal–metal or multivalent metal oxides, and hybrid Cu-Cl) (Lee et al., 2010; Nalakath et al., 2018; Safari & Dincer, 2020; Wang et al., 2019). Many studies from US, Japan and European countries proposed a number of different cycles but only few of them have ability for scale-up such as DOE USA solar thermochemical hydrogen production research program (STCH), Argonne national laboratory (2006), German aerospace research centre (DLR) and Paul Scherrer institute in Switzerland. Atomic Energy of Canada Limited started scaling up Cu-Cl thermochemical cycle in the year of 2007. Moreover, the comparative research revel that the vanadium-chlorine gives the high efficiency of 78% whereas in terms of global warming potential (GWP), while Sulfur-Iodine and hybrid sulphur cycles in the range of 0.48 - 0.51 kg CO_2·eq/kg H2, respectively (Acar & Dincer, 2019; Safari & Dincer, 2020). On the other hand, hybrid Cu-Cl cycle shows a good combination with other heat sources. Although thermochemical water splitting cycles (TWSCs) has a lot technology advancement but still in the needs there are many challenges such as hydrogen production complexity, cost and purity as well as it is storage. The major key challenges for TWSCs is the performances and durability of constituent substances for thermochemical cycling need to be enhanced, efficient and improved reactor configuration, well-suited for higher temperatures required to be explored. And the cost of the concentrating mirror systems for solar thermochemical systems needs to be reduced. International energy agencies (IEA

Paris, France) recommended some of the key strategies for the hydrogen technologies scale-up are as the role of hydrogen in the long-term energy scenarios, Adequate solutions to overcome the investment risks, extensive research and commercialization studies in terms of cost, technical and commercial difficulties identification and elimination, development of codes and standards, international collaboration for demographic barriers.

3.3 Photoelectrolysis

Photoelectrolysis is considered as one of the low-cost and efficient process to produced renewable hydrogen. Hydrogen is produced by water splitting in the presence of sun light and catalyst (doped semiconductors material as photovoltaics). When solar photon strikes the anode, generate an electron-hole pair and the hole decompose water at anode surface (Hosseini & Wahid, 2016b; Kalamaras & Efstathiou, 2013) to form hydrogen ions then hydrogen ions travel through the solution and diffuse at the cathode surface with electron to form hydrogen. The oxygen and hydrogen gas separated with the help of suitable membrane. Photoelectrolysis technologies are yet in experimental stages, while it demonstrates potential in energy recovery at lower costs. Photoelectrodes material such as Fe_2O_3, TiO_2 with some other semiconductor material and membrane materials enhanced the hydrogen production efficiency and shows good potential for future hydrogen economy (Awate et al., 2017). Till date the highest-performance were achieved on crystalline III–V semiconductors with exceptional optoelectronic qualities. Many literatures used and well described about transition metal oxide due to corrosion resistance and inexpensive. But these are not commercialized due to lack of the optoelectronic quality, which are required for higher photocurrents and hydrogen production reactions. Thin-film copper chacopyrites shows promising performance in some recent literature for photoelectrochemical hydrogen production (Mascaretti et al., 2017; Pinaud et al., 2013). But the cost of semiconductors material, membrane materials and their development are key challenges as well as efficiency determining parameter. The technical challenges of semiconductors materials are stability (aqueous solution or oxide are photochemically stable but their band gaps are high for efficient light absorption around ~3 eV), efficiency (band energy gap should lower than 2.2 eV), energetics, catalysts, semiconductor hybrid configuration, system designs featuring passive controls with low-cost (Pinaud et al., 2013). These challenges can be overcome by materials development and rational high throughput screening, standardized testing, hybrid design, reliable database of materials and their properties, accelerated testing protocols etc. Extensive research work and funding is needed to investigate and improve the efficiency of photoelectrolysis system by further research on electron transport catalyst, surface enhancement, surface catalyst for these systems.

4. BIOMASS CONVERSION

Hydrogen gas production from the conversion of biomass are considered as the sustainable technology. Hydrogen production from renewable feedstock such as biomass has several advantages on energy, environment and economy. The diverse range of the biomass such as energy crops, seasonal grass, vegetables and agricultural waste materials, forestry waste residue, industrial and community waste being used for the hydrogen production (Hendriks & Zeeman, 2009; Sittijunda & Reungsang, 2018). Several research and study claim that hydrogen from biomass is the renewable pathway which has maximum economical, technical and sustainable viability. Hydrogen production from biomass is considered as complex and multistep process, mainly reported for thermochemical and biological process. Thermochemical biomass conversion process is mainly consisting of combustion, pyrolysis, liquefication and gasification process.

Figure 3. The biomass-based hydrogen production technologies

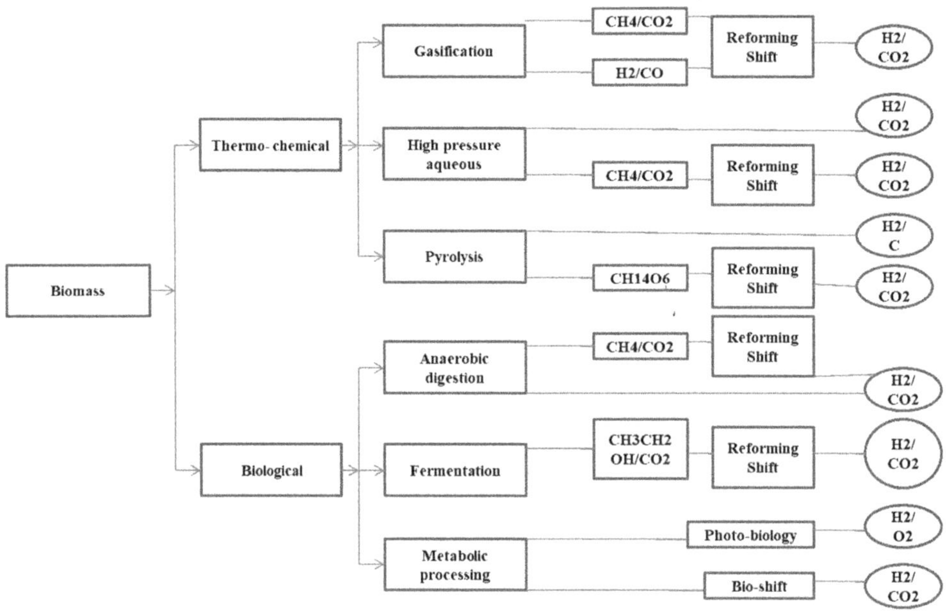

4.1 Gasification

Gasification is commercially utilised process based upon the partial oxidation of the biomass to hydrogen gas or biogas. Mixture should be vaporized and efficiency of gasification is relatively low in both the process of with catalyst or without catalyst. Gasification can be done by catalytic or indirect methods, in which biomass is processes at high temperature (above 1000 K) and pressure for their conversion in hydrogen mixture. The constituent part undergoes partial oxidation for the production of the gas and charcoal afterthought hydrogen, CO_X and CH_4 can be produced from the degradation of charcoal. After that gas mixture is subjected in membrane separation or gas separators. Biomass feedstock convert by steam gasification and super critical water gasification, which slightly differ from steam reforming and pyrolysis. The reactions may be written as:

$$\text{Biomass} + \text{heat} + \text{steam} \rightarrow H_2 + CO + CO_2 + CH_4 + \text{hydrocarbons} + \text{char} \quad (2)$$

Gasification of biomass required oxygen as compared to pyrolysis. The primary goals of gasification are to maximize the gaseous products but pyrolysis gives both bio-oil and charcoal. The key challenges of the gasification methods are the moisture contaminations and lower thermal efficiency. The moisture content should be less than 35% is considered suitable for gasification (Hosseini & Wahid, 2016b, 2016a; Monte et al., 2009; Ni et al., 2006). The gasification reactors are mainly consisting of fixed bed, fluidized bed and indirect gasifier. The gasification may produce some undesirable products such as tar aerosols and it's may undergone to the polymerization to a complex compound (unfavourable for steam reforming), production of ash content leads to sintering, slagging, fouling as well as agglomeration in the reactor.

The key parameters for the gasification process are such as biomass/substrate quality, temperature ranges/rates, particle size, catalyst types and finally steam-to-biomass ratio alter the hydrogen production and yield. Several study and research reported that biomass gasification can utilised a board category of feedstock, and gasification is sustainable and economically feasible process but need huge capital investment.

4.2 Pyrolysis and Copyrolysis

In this method, biomass materials heated at 500–800 °C at the pressure of 0.1–0.5 MPa in the absence of air/oxygen to get the bio-oils, gaseous components and charcoal. The reactions are as:

$$CnHm + \text{heat} \rightarrow nC + 0.5\, m\, H2 \quad (3)$$

The pyrolysis is further divide into low and high temperature process. Among these high temperature process offer high-energy products and fast pyrolysis products are solid, liquid and gas. Some literature reported that Pyrolysis and Copyrolysis are well-developed processes at commercial scale also. Another thermochemical process known as fast pyrolysis, in which bio-oil is thermally decomposed, followed by catalytic reforming of liquid or vapour mixture into hydrogen (Kalamaras & Efstathiou, 2013; 2010; Ni et al., 2006). The key factors of the biomass pyrolysis process are such as biomass type, catalyst, temperature and retention/residence time (Abdalla et al., 2018). H2 production by the pyrolysis process includes catalytic pyrolysis (gives 50 mole percentage of hydrogen), catalytic steam reforming of pyrolysis bio-oil by N2 and O2 optimization (yield decreases with reaction progress and obtained 80% hydrogen after reforming at 850 ∘ C). The important factors of the in-line steam reforming of volatiles from feedstocks by the fast pyrolysis (Naik et al., 2010) can be as the reforming parameters are the temperature, retention time and steam/feedstock ratios. Various literature shows that the different types of catalysts such as Ni based catalysts, calcium carbonates and Y-type zeolite, different metal oxides such as Al2O3, SiO2, ZrO2, TiO2 and Cr2O3, has the effect on the degradation of hydrocarbon in tar (Monte et al., 2009). Besides this Ru and Rh are the most effective catalysts as compared to Ni (noble metal) or other catalyst. Moreover, these noble metals are more expensive, so it cannot be employed due to cost factors. Pyrolysis process has the advantages of lower GHGs generation due to absence of water/air, also have good fuel upgradation and economy. It's also had flexibility of liquid fuel transport and study reported that it's already employed in distributed bio-refineries and some other industrial scale. Pyrolysis and Copyrolysis are well-developed processes at commercial scale also.

5. BIOLOGICALS HYDROGEN PRODUCTION TECHNOLOGIES

Bio-hydrogen production from biological processes favours the green technology for satisfying future hydrogen demands. Hydrogen produced from waste or organic material through the action of living organisms is call as bio-hydrogen and it's recognized as clean power source of the future. Biological methods have advantages as it does not liberate COX and other harmful gases or any waste generation into the environment. The undesirable products in the biohydrogen production process are determined by operational parameters, which including substrate type, microorganism types and their metabolic pathway, substrate loading rate, pH, temperature with geographical conditions (Chaurasia et al., 2019). Microbial community used in the biohydrogen process or in bioreactors is the efficiency determining parameters. The major biological process is classified in Fig.3., in which multistage or sequential dark and photo-fermentation and microbial electrolysis cells (MECs) process are a rather new approach for bio-hydrogen production (Chaurasia et al., 2019; Mondal et al., 2017; Shankar et al., 2017). All the bio-hydrogen production process is sole dependent upon the presence of hydrogen producing enzymes. Microbes either intracellular or extracellular produce these enzymes and their catalytic efficiency varies diversely but literature show that these enzymes are not limiting factor.

Figure 4. The biological hydrogen production technologies

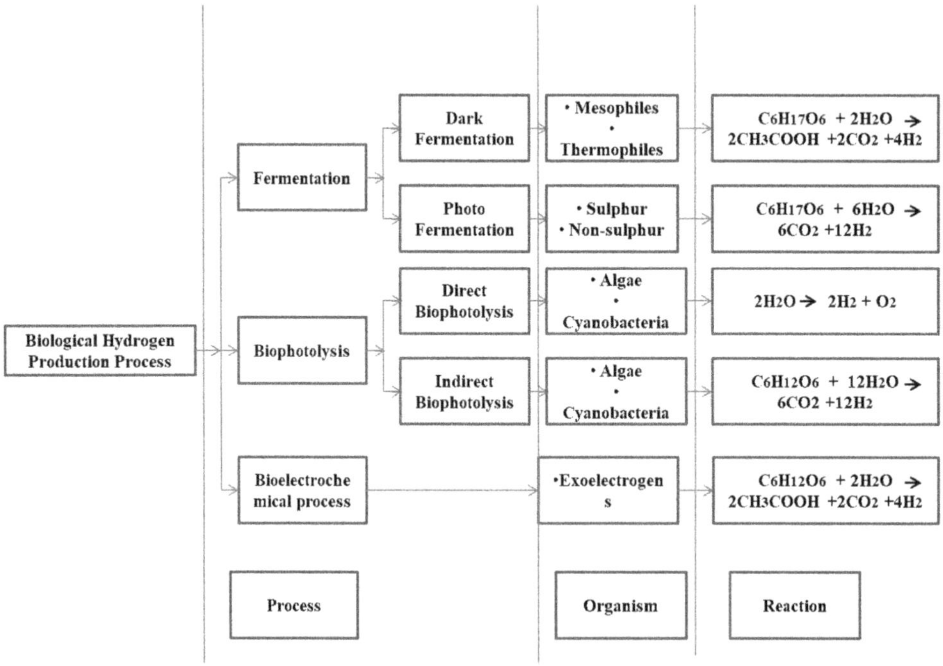

The challenges in the biological process are the raw material cost compared to hydrogen yield, which needs extensive work to make such process economical.

The major biological hydrogen production process can be summarized as:

1. Photolysis direct and indirect photolysis
2. Dark fermentation
3. Photo fermentation
4. Sequential two-stage or multi-stage integrated biological system
5. Microbial electrolysis cells (MECs)

5.1 Hydrogen Production by Bio-Photolysis

Bio-photolysis is a water-splitting process carry out by microorganisms such as Microalgae, Cyanobacteria (blue-green algae) and some higher plants to generate hydrogen in the presence of light as energy sources. Biological hydrogen production by photosynthetic microorganisms by using a simple solar reactor system such as a transparent closed box. These processes have low energy requirements and conversion efficiency is also low but it can be compensated by very low process cost. Bio-photolysis proceed in two distinctive way such as direct or indirect methods. The direct bio-photolysis use microalgae species such *Chlamydomonas reinhardtii* (by PSI, PSII and hydrogenase), while in indirect bio-photolysis cyanobacteria produce hydrogen by carbohydrate accumulation with dark fermentation.

Some study reported that green alga, *Scenedesmus,* produced molecular hydrogen due to production of hydrogenase enzyme and nitrogen fixing cyanobacteria *Anabaena cylindrica,* produced hydrogen due to production of nitrogenase enzyme in an argon atmosphere (Abdalla et al., 2018). Literature reported that higher hydrogen production in the unicellular green algae Chlamydomonas reinhardtii via direct bio-photolysis. Also, activity of hydrogenase was observed in Scenedesmus obliquus and few studies reported that slightly higher hydrogen yield efficiency in case of micro algae microorganism. In direct bio-photolysis can be enhanced by genetic engineering, optimizing light intensity, and/or combining some other process. Cyanobacteria produced hydrogen with the help of enzymes (nitrogenase and hydrogenase) during their metabolic functions. Highest rate of hydrogen production was reported in Anabaena species strains.

The research underway to design the experiments to improve the feasibility of biological hydrogen production by the use of photosynthetic microorganisms. To enhance the hydrogen production by bio-photolysis manipulate microorganisms in a manners that maximum light energy is use for hydrogen production, with the least energy transferred to cell maintenance and O_2 production. Other way is altering the microorganisms by genetic engineering that are less sensitive to oxygen or study reported that sulphate dosing could suppress production of oxygen.

5.2 Dark Fermentation

A complex fermentative process that converts organic substance rich waste into bio-hydrogen, in absence of light. Some specific bacterial strain that can cultivate in dark and produce hydrogen under anoxic or anaerobic conditions by the oxidation of an organic matters to releases electrons and H^+, then further producing hydrogen gas. Dark fermentation produced a mixed biogas (CH_4, CO) with higher quantity of hydrogen, whereas photolysis produced pure hydrogen. But dark fermentation is energy efficient process because it does not require light energy as well as produced higher energy, due to the degradation of mono sugar and carbohydrates. A variety of substrate and microorganism can be utilized to produce bio-hydrogen (Argun & Kargi, 2011; Eker & Sarp, 2017; P. C. Hallenbeck & Benemann, 2002). These organisms used their ferredoxin oxidoreductase and hydrogenase enzymes in acidic pH range (4–7) at temperature (30–39 °C, i.e. mesophilic microorganisms) (50–64 °C, i.e. thermophilic microorganisms). Some aerobic bacteria (*Alcaligenes* and *Bacillus*) also reported for hydrogen production with very low hydrogen yield. Clostridia produce H_2 in the exponential growth phase with two major metabolic pathways, acidogenic and solventogenic pathways. In acidogenic process rapid cell growth with high hydrogen, while in solventogenic low hydrogen production. Suitable physiological conditions and microorganism are key parameter to bio-hydrogen gas formation. Tolerant and mutant strain of *Clostridium* are capable to reduce organic acid formation with high yield of hydrogen gas production due alteration in their metabolic pathway. Some typical dark fermentation reaction is:

$$C_6H_{12}O_6 + 6H_2O \rightarrow 12H_2 + 6CO_2 \ \Delta G = -25.83 \text{ kJ/mol} \quad (4)$$

$$C_6H_{12}O_6 + 2H_2O \rightarrow CH_3CH_2COOH + 2H_2 + 2CO_2 \quad (5)$$

$$C_6H_{12}O_6 + 2H_2O \rightarrow CH_3COOH + 4H_2 + 2CO_2 \ \Delta G^o = -\ 206 \text{ kJ} \quad (6)$$

The range of hydrogen yield reported is 1.2-2.7 mole hydrogen per mole hexose in many bacterial strains but yield can be improved by redirecting metabolic pathway (Argun & Kargi, 2011). The metabolic pathway governs by pH, hydrogen partial pressure, temperature, enzyme, and nutrient concentrations. Yield can also be improved by using the mixed bacterial cultures and/or with suitable inducer, careful introduction of substrate concentration, heat shock pretreatment of substrate and acid/base pretreatment. The scale-up of the dark fermentation is highly dependent on types of microbes, substrate, operational parameters (pH, temperature, gas partial pressure, hydraulic retention time (HRT), nutrient availability or presence of inhibitors) that must be optimized in each specific case.

5.3 Photo Fermentation

Hydrogen production by photoheterotrophic fermentation process where light is use for the conversion of organic waste substance to bio-hydrogen. Anoxygenic microorganism absorb sunlight and perform electron transport by organic waste decomposition. Photo-fermentation is a series of bio-chemical conversion similar to anaerobic digestion but sugars or substrate initially added in small quantity to initiate the fermentation. Photo-fermentation has advantaged that oxygen does not inhibits the process. *Cyanobacteria,* algae are capable of photosynthesis were light energy drive the metabolic pathway to produce hydrogen energy. Some purple non-sulphur (PNS) bacteria e.g. genus *Rhodobacter* such as *Rhodobacter spheroids* etc. species and green sulphur bacteria has potential for the production of hydrogen under nitrogen-deficient conditions in a process catalysed by nitrogenase. Photo fermentation occurs at the temperature of 30-36 ^oC at pH 6.8-7.5 with 400-1000 nm light intensity (Assawamongkholsiri et al., 2018). Hydrogen production sole depends on the light intensity, microbial strain and carbon source.

$$C_3COOH + 2H_2O + \text{Light energy} \rightarrow CH_3COOH + 4H_2 + 2CO_2 \ \Delta G^o = +\ 104 \text{ kJ} \quad (7)$$

Moreover, the hydrogen yield in photo-fermentation is relatively lower than dark fermentation but hydrogen production from the similar carbon source are different accordingly bacterial strain. The light intensity strongly stimulates the expression and activity of nitrogenase as well as the photophosphorylation capacity of bacterial strain. Some authors also reported that the reduction of pigment of the microorganism by genetic manipulations to increase the light penetration as well as hydrogen production. The major disadvantages are low activity of nitrogenase enzyme, high energy requirements, and hydrogen re-oxidation. It can be overcome by the proper ratio of carbon to nitrogen constituents. High light intensity can increase the rate of hydrogen production but have the adverse effects on microorganism and it can be enhanced by genetic modification of the organism.

5.4 Sequential Dark and Photo-Fermentations or Multistage Fermentation

Multistage fermentations system can be utilized to maximize the hydrogen or energy recovery from a waste or renewable material. These methods consist of multi or two to three subsequent stage process such as dark fermentation followed by photo fermentation or vice versa or photo fermentation-dark fermentation followed by microbial electrolysis cells etc. configuration or in other arrangement. Sequential fermentation can be use in various arrangement of bio-chemical reactions. First stage fermentation effluent is used as substrate for subsequent fermentation process or configuration has the unique advantage and maximize the energy recovery from the waste.

Combined sequential dark and photo-fermentation process (over all reactions):

$$C_6H_{12}O_6 + 6H_2O \rightarrow 12H_2 + 6CO_2 \ \Delta G^o = + 3.2 \text{ kJ} \qquad (8)$$

Utilization of multistage system may offer some challenges, and it must be address by further research work and development. Few studies reported that multistage system in co-culture methods (suspension or immobilized culture) are sequential methods. More research and investigation are need to evaluate the viability or commercialization of such system to maximize energy recovery. The hydrogen yield can be maximized as most by combining dark fermentation with other process such as photo fermentation because the complete oxidation of substrate is not possible in a single process (complete oxidation of glucose into H_2 and CO_2 cannot occur due to thermodynamically unfavourable reactions). Therefore, for the maximum yield and continuous production of H_2 could be obtained by combining dark- and photo-fermentation process in a sequential manner.

6. MICROBIAL ELECTROLYSIS CELLS (MECS)

MECs is the sustainable technology that produce bio-hydrogen from the microbial decomposition of organic matter (wastewater, biomass) and other renewable resources. MECs is the extended version of microbial fuel cell (MFC) and a bio-electrochemical system (BESs) have great potential for energy generation with simultaneous sustainable treatment of any types of waste, attracted scientific attentions towards the technological development in last two decades. MECs emerging and resources recovery technology enabling to convert any biodegradable waste in to energy, fuel or valuable products but efficiency of MECs depends on the different operating conditions as well as different substrate. Electrochemically active microbes oxidize the organics present in waste materials and generate CO_2, electron and proton. The exoelectrogens microorganism transfer the electron to the anode and protons into the electrlytes, these protons transferred selectively through proton exchange membrane (PEM) into the cathode chamber (Chaurasia et al., 2019; Chaurasia & Mondal, 2021; Mondal et al., 2017; Shankar et al., 2017). The electron then transfers through circuit to the cathode electrode and join with protons in electrolytes. MECs required a cathode potential of >0.14 V for hydrogen generation under neutral physiological conditions. In MECs, widely used substrate is acetate, which converted through different electrode reactions as presented in Eq. 10-14.

$$\text{Overall: } CH_3COO^- + 4H_2O\ 2HCO^{3-} + H^+ + 4H_2\ (\Delta G\hat{r}=104.6kj/mole) \quad (11)$$

$$E_{eq} = -\Delta Gr/nF = (-104.6 \times 10^3)/8 \times 96485 = -0.14V \quad (12)$$

$$E^o_{acetate/CO2} = -0:289V,\ E_{H+/H2O} = +0:818V \text{ AND } E_{H+/H2} = -0:412V \text{ at NHE (pH= 7)} \quad (13)$$

MEC has positive ΔGr for acetate conversion under standard physiological conditions (Eq. 10-15) in neutral pH. Therefore, conversion of acetate or waste to hydrogen gas requires an external voltage of >0.14V to overcome the thermodynamic barrier. In general, the MEC performance is influence by the various factors such as overpotential of anode and cathode, initial pH, temperature, electrolyte, substrates, electrode surface area, and types of microorganism. MECs utilizes the microorganisms as biocatalysts and production of hydrogen at cathode required an additional applied voltage of 0.14 V (theoretical), which is lesser as compared to the electrolysis of water (1.24 V in theory). MECs emerges as an attractive source to produce hydrogen from any type of waste materials such as wastewaters, biomass etc. and has the wide range of application in industry due to treatment of wastewater, reuse of water and low energy requirement. In practical operation of MECs for hydrogen production, a higher applied voltage is required due to over-potentials of anode and cathode electrodes and the resistance offered by membrane in double chamber MECs. The increase of overpotential losses of electrode materials will reduced the current density as well as hydrogen yield and overall energy recovery of the MECs system. Therefore, minimization of the potential losses is a crucial challenge in the practical application of MECs. Cation exchange membrane (CEM) and anion exchange membrane (AEM) are usually employed in MECs system to enhanced hydrogen production. AEM is mostly employed in the MECs for hydrogen production due to low internal resistance. Membrane less MECs has also been offer the reduction in internal resistance. Although the MECs technology have to meet several developments towards scale-up and wide spread applications, which are including lower hydrogen production rate, resistance losses, complex design architecture, and expensive electrodes and membrane materials. MECs performance can be enhanced by the development of new effective cathodes materials, which have high catalytic performances in neutral geographical conditions, and have lover over-potential. MECs can also be utilized for the production of the valuable products such as chemical etc.

6.1 Electrodes Materials: Cathode

In early literature of laboratory or pilot researches, Pt (Platinum) is widely select as cathode in the MECs due to its high hydrogen production and electrode low overpotential losses. However, Pt electrode has high cost and poisoning of Pt by sulphides. These important drawbacks of Pt can limit the wider industrial application of MECs. In last decade, a large research attraction for the development of novel cathodes material, which must have high electrocatalytic activity and low-cost materials to replace platinum. Pt as cathode material were use in different fuel cell and electrochemical processes, molybdenum disulphide (MoS_2) and nickel phosphide (Ni_2P) are also be used in literature but only for highly acidic or alkaline conditions. It is also notable that research on MECs and MFCs focused on alkaline and PEM electrolysers, and major studied have been reports in alkaline and acidic medium. While MECs operation is, carry out at near to the neutral pH, low H^+ and OH^- ions. Experimental data obtained from these studied at higher or lower pH value cannot considered as base for design of suitable electro-catalyst. Several studies reported that first row transition metals could be employed as cathodes due to chemically stable, low potential losses, less waste generation. Stainless steel (SS) and Ni alloys are considered as promising materials for cathodes materials. Some laboratory scale research used SS as cathode material in MECs, because stainless steel is a known for low price and non-noble metal material available various shape and size. However, the rate of hydrogen yield and energy recovery of SS cathode material is the topic of further research. Some studied has also been used nickel as cathode material but result obtained on nickel cathode in MECs has higher current density and lower hydrogen evolution overpotential as compared to the SS. Therefore, SS has been employed as cathodes in various configuration such as plates, mesh, brushes and, felt. It can be concluded that Nickel as cathodes have a good reaction suitability as compared to a Pt-based cathode in MECs. Hydrogen production rate of 2 $m^3(H_2)/m^3$ per day was achieved at 0.6V applied voltage with 86% cathodic recovery, which was 33% more to the NiW cathodes and only 13% in the case of Pt. catalysts (Awate et al., 2017). Still nickel-based cathode materials need further optimization and upgradation. The future cathode materials in MECs must have high intrinsic electrocatalytic activity for the hydrogen production rate are the critical parameter. Besides this, electrodeposited materials due to various deposit configuration gives more active sides at the surface and results enhanced the electrocatalytic properties. The small amount of phosphorus (P) present at cathodes has changed the physical structure of the cathodes or electrodeposition of Ni-P and enhanced the hydrogen production rate. Electrodeposition of catalyst to make cathodes has very great effect on hydrogen production rate and because of simplicity and low cost (Chaurasia et al., 2019).

6.2 Anode

Anode has equal contribution as cathodes in the MECs for the enhancement of hydrogen production. Anode and anode materials also play vital role for the enhancing the energy efficiency of the MECs system but also affects the electrons generation, give high reduction potential for the proton. Various materials are reported in the literature such as graphene, graphene (sheet, foam), graphite oxides, graphene deposited stainless steels (SS), etc. Most literature shows that the carbon-based electrodes such as carbon cloth, fibres, plates or rods, and their granules have high efficacy and provide uniform structures were employed as anode in the MECs. Other anode materials such as activated carbon, surface modified SS, carbon fabricated nanotubes, metal anodes (such as Cu, Ag, Ni, Co, Ti), Pt have high electrocatalytic activity. Moreover, in metal anodes, metal corrosion is key challenges, so the choice of anode potential is essential. Anode felt as anode in MECs offer some resistance and it can be mainly due to the limitations in ion transfer as well as the anode and hindrance. Few reports show bioanode can also enhanced the MECs performance and were found to be very effective. A good candidate of anode must have high conductivity, large surface area, chemically stable in electrolytes, well suited biocompatibility with no adverse effects on microorganism, also possess good adaptability with resistant to biofouling.

6.3 Substrate

Type substrate greatly influence the MECs performances and energy recovery from the waste materials. Commonly, reported substrate are wastewater, glucose, acetates, agricultural and dairy waste residue, etc and these are known to be high potential substrate for the production of hydrogen in the MECs. Some reports show different substrates such as acetic acid, butyric acid, lactic acid, a mixture of two or more of it with glucose show the highest power density in the acetate. Microalgae are also used as substrate in MECs and show a good performance in the MECs. Moreover, as compared to other substrate microalgal show promising rate of hydrogen production and low-cost as well also. Some literature shows various complex feedstocks for hydrogen production in the MEC such as milk, why, glycerol, starch, and mixed substrates were promising. It seen that mixed substrate and milk produced higher current density and energy generation but starch and glycerol show lower performances in MEC. Food waste (FW) show better results in MECs as compared to anaerobic digestion (AD) for hydrogen production but lower degradation of organic matters (Kongjan et al., 2011; Li et al., 2019; Teixeira Franco et al., 2017). Such results are mainly due to complex nature of industrial waste and may need pretreatment but it lowers the cost also. Using buffer and pH control in the substrate also enhanced the MECs performances. Vegetables leaves also show higher hydrogen production than rice straw, it mainly due to the complex cellulosic compounds present in the rice straw. Also, the organic acids and reducing sugars show higher hydrogen production with high yields of hydrogen.

6.4 Microorganism

Microorganism are the most crucial factors of the MECs and microorganism are the determining factors of the MECs and MECs performances are sole dependent on the living constituent of the MECs reactors (Chaurasia et al., 2019; Chaurasia & Mondal, 2021; Mondal et al., 2017; Shankar et al., 2017). The most common microorganism used in MECs are Sphingobacteria, Ignavibacteria, and Betaproteobacteria, Deltaproteobacteria and Bacteroidia, Solitalea. A variety of microbes were used in the MECs are reported in the literature. Some literature reported that mediators (dyes, metals ions, co-enzymes) can also improve the MECs performances (Chaurasia et al., 2019; Liang et al., 2017; You, 2016). Mediators enhanced the transfer of the electrons; electron freely flow as well as electron generation because microorganism transfer the electrons extracellular and intracellular or some time together. Literature reports about electron transfer was high in nanowires or use of mediators (Borole & Lewis, 2017). Also, electron transport in the bacteria was not high while it is due to electron acceptor by anodophiles. Shewanella and Geobacter strains do not need any mediators and they are capable in electron transfer due to having pili (protein-rich filaments) known as nanowires (Borole & Lewis, 2017; Liang et al., 2017; Sapountzi et al., 2017; Shiva Kumar & Himabindu, 2019b). Genetically engineers' microbes specialized for MECs are also reported in the literature. Microbes are very sensitive to various geographical factors. It may also note that a combination of microorganism species such as two, three or more or may in together can be employed or geographical suitable microbes must be selected before the experimentation. Therefore, selection of microbes must be explored before the commercialized MECs technology.

6.5 Membrane

Membranes are used in MECs to provide to facilitated the diffusion to reduce the power consumption. The most common membranes are anion exchange membrane (AEM), proton exchange membrane (PEM), and ceramic membranes. Some membrane has negative impacts on microbial community and membrane has specific resistance, ionic conductivity (Chaurasia et al., 2019; Chi & Yu, 2018b; Sapountzi et al., 2017). Forward osmosis (FO) membrane and cation exchange membrane (CEM) were also reported in the literature as membrane in the MECs. Studied on gas diffusion has demonstrated that the FO and CEM have lower electron diffusion on their surface and can differ with dry/wet and electrolytes conditions,

Table 4. Efficiency of some renewable hydrogen production technologies

Sr. No.	Technology	Feed Stock	Efficiency	Current Scenario
1	Steam Reforming of bio-oil or hydrocarbon	Biomass	70%-85%	Commercial
2	Biomass feedstock gasification	Biomass	30%-55%	Commercial
3	Photolysis of water	Sunlight+ Water	12%	Long term
4	Dark fermentation of waste material	Biomass	61%-82%	Long term
5	Photo fermentation of waste material	Biomass +Sunlight	5-17%	Long term
6	MECs	Biomass+ Electricity	77-91%	Long term
7	Electrolyser and/or PEM	H2O + Electricity	45%-58%	Commercial
8	Solid oxide electrolysis cells	H2O + Electricity+ Heat	38%-58%	Long term
9	Photo electrochemical water splitting	H2O + Heat/ H2O + Sunlight	16-19%	Long term

whereas FO have high hydrogen production rate than CEM membranes (Chi & Yu, 2018a; Lamy & Millet, 2020). It's mainly due to higher cathodic hydrogen efficiency, freely flow of electrons, lower resistance.

CONCLUSION

The present chapters has reviewed the most of the hydrogen production process such as dark fermentation, AD, photo-fermentation, sequential process, gasification, electrolysis, photolysis and microbials electrolysis cells (MEC) with some others. However, most of the hydrogen production process not commercialized and it needs to be extensive research, international collaboration and commercial strategies must be explored to make such process economically viable.

REFERENCES

Abanades. (2019). Metal Oxides Applied to Thermochemical Water-Splitting for Hydrogen Production Using Concentrated Solar Energy. *ChemEngineering*, *3*(3), 63. doi:10.3390/chemengineering3030063

Abdalla, A. M., Hossain, S., Nisfindy, O. B., Azad, A. T., Dawood, M., & Azad, A. K. (2018). Hydrogen production, storage, transportation and key challenges with applications: A review. *Energy Conversion and Management*, *165*(April), 602–627. doi:10.1016/j.enconman.2018.03.088

Abe, J. O., Popoola, A. P. I., Ajenifuja, E., & Popoola, O. M. (2019). Hydrogen energy, economy and storage: Review and recommendation. International Journal of Hydrogen Energy, 44(29), 15072–15086. doi:10.1016/j.ijhydene.2019.04.068

Abrell, J., Eser, P., Garrison, J. B., Savelsberg, J., & Weigt, H. (2019). Integrating economic and engineering models for future electricity market evaluation: A Swiss case study. *Energy Strategy Reviews*, *25*, 86–106. doi:10.1016/j.esr.2019.04.003

Acar, C., & Dincer, I. (2019). Review and evaluation of hydrogen production options for better environment. *Journal of Cleaner Production*, *218*, 835–849. doi:10.1016/j.jclepro.2019.02.046

Argun, H., & Kargi, F. (2011). Bio-hydrogen production by different operational modes of dark and photo-fermentation: An overview. *International Journal of Hydrogen Energy*, *36*(13), 7443–7459. doi:10.1016/j.ijhydene.2011.03.116

Assawamongkholsiri, T., Reungsang, A., Plangkang, P., & Sittijunda, S. (2018). Repeated batch fermentation for photo-hydrogen and lipid production from wastewater of a sugar manufacturing plant. *International Journal of Hydrogen Energy*, *43*(7), 3605–3617. doi:10.1016/j.ijhydene.2017.12.119

Awate, B., Steidl, R. J., Hamlischer, T., Reguera, G., Yang, Y., Qin, M., Yang, X., He, Z., Marone, A., Ayala-Campos, O. R., & Trably, E., Carmona-Martïnez, A. A., Moscoviz, R., Latrille, E., Steyer, J. P., Alcaraz-Gonzalez, V., Bernet, N., Montpart, N., Baeza, M., … Cheng, S. (2017). Proton transfer in microbial electrolysis cells. *Journal of Power Sources*, *356*(4), 725–736. doi:10.1016/j.ijhydene.2016.09.166

Borole, A. P., & Lewis, A. J. (2017). Proton transfer in microbial electrolysis cells. *Sustainable Energy & Fuels*, *1*(4), 725–736. doi:10.1039/C7SE00034K

Cédric Philibert. (2017). *Producing industrial hydrogen from renewable energy*. International Energy Agency. https://www.iea.org/newsroom/news/2017/april/producing-industrial-hydrogen-from-renewable-energy.html

Chaurasia, A. K., Goyal, H., & Mondal, P. (2019). Hydrogen gas production with Ni, Ni–Co and Ni–Co–P electrodeposits as potential cathode catalyst by microbial electrolysis cells. *International Journal of Hydrogen Energy*. Advance online publication. doi:10.1016/j.ijhydene.2019.07.175

Chaurasia, A. K., & Mondal, P. (2021). *Simultaneous Removal of Organic Load and Hydrogen Gas Production Using Electrodeposits Cathodes in MEC*. Springer. doi:10.1007/978-981-15-6595-3_34

Chi, J., & Yu, H. (2018a). Water electrolysis based on renewable energy for hydrogen production. *Cuihua Xuebao. Chinese Journal of Catalysis*, *39*(3), 390–394. doi:10.1016/S1872-2067(17)62949-8

Chi, J., & Yu, H. (2018b). Water electrolysis based on renewable energy for hydrogen production. Cuihua Xuebao/Chinese Journal of Catalysis, 39(3), 390–394. doi:10.1016/S1872-2067(17)62949-8

Das, D., Khanna, N., & Nejat Veziroğlu, T. (2008). Recent developments in biological hydrogen production processes. Chemical Industry and Chemical Engineering Quarterly, 14(2), 57–67. doi:10.2298/CICEQ0802057D

De Giz, M. J., Tremiliosi-Filho, G., Gonzalez, E. R., Srinivasan, S., & Appleby, A. J. (1995). The hydrogen evolution reaction on amorphous nickel and cobalt alloys. *International Journal of Hydrogen Energy*, *20*(6), 423–427. doi:10.1016/0360-3199(94)00068-B

Doman, L. E., Arora, V., Metelitsa, A., Leahy, M., Barden, J. L., Ford, M., Jones, A., Brian Murphy, E. T., Lillis, K., Slater-Thompson, N., Behjat Hojjati, B., Peterson, D., Zaretskaya, V. V, Chase, N., Morris, M., Lindstrom, P., Geagla, A., Grissom, S., Maples, J., … Smith, C. L. (2013). *World energy demand and economic outlook Macroeconomic assumptions Liquid fuels Natural gas Transportation sector Energy-related carbon dioxide emissions*. www.eia.gov

EAA Annual Report 2015/2016. (n.d.). Retrieved January 6, 2020, from https://www.eaa.org.hk/en-us/Information-Centre/Publications/EAA-Annual-Report/EAA-Annual-Report-2015-2016

Eker, S., & Sarp, M. (2017). Hydrogen gas production from waste paper by dark fermentation: Effects of initial substrate and biomass concentrations. *International Journal of Hydrogen Energy*, *42*(4), 2562–2568. doi:10.1016/j.ijhydene.2016.04.020

Felseghi, R. A., Carcadea, E., Raboaca, M. S., Trufin, C. N., & Filote, C. (2019). Hydrogen fuel cell technology for the sustainable future of stationary applications. *Energies*, *12*(23), 4593. Advance online publication. doi:10.3390/en12234593

Global EV Outlook 2020 – Analysis - IEA. (n.d.). Retrieved October 21, 2020, from https://www.iea.org/reports/global-ev-outlook-2020

Hallenbeck, P. C., & Benemann, J. R. (2002). Biological hydrogen production; Fundamentals and limiting processes. *International Journal of Hydrogen Energy*, *27*(11–12), 1185–1193. doi:10.1016/S0360-3199(02)00131-3

Hendriks, A. T. W. M., & Zeeman, G. (2009). Pretreatments to enhance the digestibility of lignocellulosic biomass. *Bioresource Technology*, *100*(1), 10–18. doi:10.1016/j.biortech.2008.05.027 PMID:18599291

Hosseini, S. E., & Wahid, M. A. (2016a). Hydrogen production from renewable and sustainable energy resources: Promising green energy carrier for clean development. *Renewable & Sustainable Energy Reviews*, *57*, 850–866. Advance online publication. doi:10.1016/j.rser.2015.12.112

Hosseini, S. E., & Wahid, M. A. (2016b). Hydrogen production from renewable and sustainable energy resources: Promising green energy carrier for clean development. *Renewable & Sustainable Energy Reviews*, *57*, 850–866. Advance online publication. doi:10.1016/j.rser.2015.12.112

Hydrogen Production by Photoelectrolysis. (n.d.). Retrieved October 21, 2020, from https://bhuang02.tripod.com/photoelectrolysis.htm

Hydrogen Production from Biomass and Organic Waste — Fuel Cell & Hydrogen Energy Association. (n.d.). Retrieved January 18, 2020, from https://www.fchea.org/in-transition/2019/7/8/hydrogen-production-from-biomass-and-organic-waste

Kadier, A., Simayi, Y., Kalil, M. S., Abdeshahian, P., & Hamid, A. A. (2014). A review of the substrates used in microbial electrolysis cells (MECs) for producing sustainable and clean hydrogen gas. *Renewable Energy*, *71*, 466–472. doi:10.1016/j.renene.2014.05.052

Kalamaras, C. M., & Efstathiou, A. M. (2013). Hydrogen Production Technologies: Current State and Future Developments. *Conference Papers in Energy*, *2013*, 1–9. doi:10.1155/2013/690627

Kongjan, P., O-Thong, S., & Angelidaki, I. (2011). Performance and microbial community analysis of two-stage process with extreme thermophilic hydrogen and thermophilic methane production from hydrolysate in UASB reactors. *Bioresource Technology*, *102*(5), 4028–4035. doi:10.1016/j.biortech.2010.12.009 PMID:21216592

Lamy, C., & Millet, P. (2020). A critical review on the definitions used to calculate the energy efficiency coefficients of water electrolysis cells working under near ambient temperature conditions. In *Journal of Power Sources* (Vol. 447). Elsevier B.V. doi:10.1016/j.jpowsour.2019.227350

Lee, D., Ebie, Y., Xu, K., Li, Y., & Inamori, Y. (2010). Bioresource Technology Continuous H 2 and CH 4 production from high-solid food waste in the two-stage thermophilic fermentation process with the recirculation of digester sludge. *Bioresource Technology*, *101*(1), S42–S47. doi:10.1016/j.biortech.2009.03.037 PMID:19410447

Li, P., He, C., Yu, R., Shen, D., & Jiao, Y. (2019). Anaerobic Co-digestion of Urban Sewage Sludge with Agricultural Biomass. *Waste and Biomass Valorization*, 1–11. doi:10.100712649-019-00870-z

Liang, Z., Ahn, H. S., & Bard, A. J. (2017). A Study of the Mechanism of the Hydrogen Evolution Reaction on Nickel by Surface Interrogation Scanning Electrochemical Microscopy. *Journal of the American Chemical Society*, *139*(13), 4854–4858. doi:10.1021/jacs.7b00279 PMID:28276238

Mascaretti, L., Ferrulli, S., Mazzolini, P., Casari, C. S., Russo, V., Matarrese, R., Nova, I., Terraneo, G., Liu, N., Schmuki, P., & Li Bassi, A. (2017). Hydrogen-treated hierarchical titanium oxide nanostructures for photoelectrochemical water splitting. *Solar Energy Materials and Solar Cells*, *169*(May), 19–27. doi:10.1016/j.solmat.2017.04.045

Mondal, P., Kumari, P., Singh, J., Verma, S., Chaurasia, A. K., & Singh, R. P. (2017). Oil from Algae. In *Sustainable Utilization of Natural Resources* (pp. 213–253). CRC Press. doi:10.1201/9781315153292-8

Monte, M. C., Fuente, E., Blanco, A., & Negro, C. (2009). Waste management from pulp and paper production in the European Union. *Waste Management (New York, N.Y.)*, *29*(1), 293–308. doi:10.1016/j.wasman.2008.02.002 PMID:18406123

Naik, S. N., Goud, V. V., Rout, P. K., & Dalai, A. K. (2010). Production of first and second generation biofuels: A comprehensive review. Renewable and Sustainable Energy Reviews, 14(2), 578–597. doi:10.1016/j.rser.2009.10.003

Nalakath, H., Keskin, T., & Yazgin, O. (2018). ScienceDirect Biohydrogen production from autoclaved fruit and vegetable wastes by dry fermentation under thermophilic condition. *International Journal of Hydrogen Energy*. Advance online publication. doi:10.1016/j.ijhydene.2018.12.068

Ni, M., Leung, D. Y. C., Leung, M. K. H., & Sumathy, K. (2006). An overview of hydrogen production from biomass. *Fuel Processing Technology*, *87*(5), 461–472. doi:10.1016/j.fuproc.2005.11.003

Pinaud, B. A., Benck, J. D., Seitz, L. C., Forman, A. J., Chen, Z., Deutsch, T. G., James, B. D., Baum, K. N., Baum, G. N., Ardo, S., Wang, H., Miller, E., & Jaramillo, T. F. (2013). Technical and economic feasibility of centralized facilities for solar hydrogen production via photocatalysis and photoelectrochemistry. *Energy & Environmental Science*, *6*(7), 1983–2002. doi:10.1039/c3ee40831k

Rani, M., Shanker, U., & Chaurasia, A. K. (2017). Catalytic potential of laccase immobilized on transition metal oxides nanomaterials: Degradation of alizarin red S dye. *Journal of Environmental Chemical Engineering*, *5*(3), 2730–2739. Advance online publication. doi:10.1016/j.jece.2017.05.026

Safari, F., & Dincer, I. (2020). A review and comparative evaluation of thermochemical water splitting cycles for hydrogen production. In Energy Conversion and Management (Vol. 205). doi:10.1016/j.enconman.2019.112182

Sapountzi, F. M., Gracia, J. M., & Weststrate, C. J. (2017). Electrocatalysts for the generation of hydrogen, oxygen and synthesis gas. In Progress in Energy and Combustion Science (Vol. 58, pp. 1–35). Elsevier Ltd. doi:10.1016/j.pecs.2016.09.001

Shankar, R., Pathak, N., Chaurasia, A. K., Mondal, P., & Chand, S. (2017). Energy Production through Microbial Fuel Cells. In *Sustainable Utilization of Natural Resources* (pp. 353–380). CRC Press. doi:10.1201/9781315153292-12

Shiva Kumar, S., & Himabindu, V. (2019a). Hydrogen production by PEM water electrolysis – A review. *Materials Science for Energy Technologies*, *2*(3), 442–454. doi:10.1016/j.mset.2019.03.002

Shiva Kumar, S., & Himabindu, V. (2019b). Hydrogen production by PEM water electrolysis – A review. *Materials Science for Energy Technologies*, *2*(3), 442–454. doi:10.1016/j.mset.2019.03.002

Sittijunda, S., & Reungsang, A. (2012). Media optimization for biohydrogen production from waste glycerol by anaerobic thermophilic mixed cultures. *International Journal of Hydrogen Energy*, *37*(20), 15473–15482. doi:10.1016/j.ijhydene.2012.02.185

Sittijunda, S., & Reungsang, A. (2018). Methane Production from the Co-digestion of Algal Biomass with Crude Glycerol by Anaerobic Mixed Cultures. *Waste and Biomass Valorization*, *11*(5), 1873–1881. doi:10.100712649-018-0542-0

Staffell, I., Scamman, D., Velazquez Abad, A., Balcombe, P., Dodds, P. E., Ekins, P., Shah, N., & Ward, K. R. (2019). The role of hydrogen and fuel cells in the global energy system. *Energy & Environmental Science*, *12*(2), 463–491. doi:10.1039/C8EE01157E

Sunfire | Hydrogen. (n.d.). Retrieved October 21, 2020, from https://hydrogeneurope.eu/member/sunfire

Swaminathan, M. (2018). How can India's waste problem see a systemic change? *Economic and Political Weekly*. https://www.epw.in/engage/article/institutional-framework-implementing-solid-waste-management-india-macro-analysis?0=ip_login_no_cache%3D7d58c9a286a575de0525e5c5f79bc33a

Teixeira Franco, R., Buffière, P., & Bayard, R. (2017). Cattle manure for biogas production. Does ensiling and wheat straw addition enhance preservation of biomass and methane potential? *Biofuels*, 1–12. doi:10.1080/17597269.2017.1387751

Turning organic waste into hydrogen. (2019). *C&EN Global Enterprise, 97*(14), 26–29. doi:10.1021/cen-09714-feature3

Wang, T., Tang, Z., Guo, Y., Zhang, X., Yang, Q., Xu, B., & Wang, H. (2019). Comparison of Heavy Metal Speciation of Sludge During Mesophilic and Thermophilic Anaerobic Digestion. *Waste and Biomass Valorization*, 1–10. doi:10.100712649-018-00575-9

Xiao, L., Deng, Z., Fung, K. Y., & Ng, K. M. (2013). *Biohydrogen generation from anaerobic digestion of food waste.* doi:10.1016/j.ijhydene.2013.08.072

You, J. (2016). Waste and Wastewater Clean-up using Microbial. *Fuel Cells (Weinheim), 208*. http://eprints.uwe.ac.uk/26953

Chapter 3
Exploration of Ni–P–Based Catalytic Electrodes for Hydrogen Evolution Reaction

Ameen Sha M.
https://orcid.org/0000-0002-2466-1749
Muslim Association College of Engineering, India

Meenu P. C.
BITS Pilani, Hyderabad, India

ABSTRACT

High efficacy and industrial applicability on electrocatalytic hydrogen production is achieved by proper furnishing of components in the reaction cell. The idea on the basic mechanism of hydrogen evolution reaction (HER), efficient modification of available systems, and recent trends in the development strategies of suitable materials are very important to be explored to design novel systems for large-scale production. This review chapter discusses the scientific details on electrocatalytic HER and plausible materials used for catalyzing the reaction. And it outlines the trends in design and development of transition metal-based catalytic coating systems with a special focus on Ni-P alloy coating and scientific aspects of the methods and the materials used for the HER. On the whole, the discussion on HER and its catalytic systems provides an insight of their potential to be explored for enhanced energy production in hydrogen fuel cell technology for stationary and industrial applications.

INTRODUCTION

Hydrogen Fuel Cell Technology, for stationary application can be the best potential for Hydrogen energy with maximum energy efficiency but minimum or no environmental issues. Hydrogen, the fuel used in this technology should be continuously propelled to the cell for continuous energy production. This chapter aims to provide understanding of scientific aspects of hydrogen production techniques and the enhancement techniques especially by suitable catalyst material design and implementation. The thorough

DOI: 10.4018/978-1-7998-4945-2.ch003

knowledge of the aspects mentioned here will realize the enhanced hydrogen production for Hydrogen Fuel Cell Technology for stationary application. Electrocatalytic hydrogen evolution is considered as an effective and sustainable technique to convert electric energy to chemical energy. Hydrogen Evolution Reaction (HER) is regarded as the main step in electrochemical water splitting and the foundation for discovering the mechanism of many supplementary multi–electron transfer electrochemical phenomena. The experimental set up for electrochemical water splitting comprises of a Direct Current (DC) power source electrically connected to two electrodes, i.e. an anode and a cathode (with electrocatalysts) immersed in water. Hydrogen and oxygen are evolved from the cathode and anode respectively. Generally, for a water–splitting reaction to begin, excess energy as overpotential is required to overcome several activation barriers. If not, the reaction either won't happen or may occur at a relatively lower rate, which is due to the poor self-ionization power of water. However, efficacy of the reaction can be increased by: adding suitable electrolytes as a salt, an acid or a base; and with the usage of appropriate electrocatalysts. Targeting on efficient water splitting on a large scale basis, so far, many electrocatalysts have been developed accompanied by current substantial advances in discovering more on their electrochemical nature. From a fundamental point of view, an electrocatalyst, for better performance, must satisfy the following conditions;

- For the desired reaction i.e., for HER, the intrinsic overpotential must be low.
- Surface area of the electrocatalyst must be active enough facilitating good approachability to the reactants and suitably for the fast exclusion of products (Sapountzi et al., 2017)
- Possess high electrical conductivity so as to provide favorable pathways for electron transfer reactions.
- Possess chemical stability i.e., it should be compatible with the used electrolyte.
- Possess electrochemical stability i.e., it should not get corroded at high overpotential.
- Possess good mechanical stability (applicable at high temperature electrolysis).

All these conditions guarantee high overall system efficiency. Moreover, long term stability and cost effectiveness are also to be focussed to compete with the conventional non sustainable methods.

Electrocatalytic HER is the most widely explored electrochemical reaction in a feasible manner and has wide applications extending from corrosion and electrodeposition of metals to energy storage by means of hydrogen production. Among the various sustainable hydrogen production methods developed, electrocatalytic HER is considered to be the most efficient strategy owing to its supreme capacity and carbon free emission. It has a great benefit of producing extremely pure hydrogen i.e., with purity greater than 99.9%, perfect for some upgraded processes, mainly for the manufacture of electronic components (Pletcher & Li, 2011). Moreover, the theoretical and experimental studies of HER electrocatalysts with special attention on their electronic structure, surface electrochemistry and design need to be evaluated critically.

The idea about basic mechanism of HER, materials development for the same and recent trends in the techniques are very important to be explored, modification of useful systems efficiently and to design novel systems for large scale production. Transition based catalytic systems have shown its significant impact on the research and development in this field. In this context, this review chapter draws attention on scientific aspects of methods and materials used for HER, focusing on its mechanism and Ni based catalytic systems as electrodes which are highly significant for further exploration and industrial applications.

BACKGROUND

HER in the current scenario aims on the sustained production of hydrogen as an alternative clean fuel for several energy systems. But it suffers from inactive reaction kinetics owing to additional water dissociation step. Hence, the state of catalysts in acidic media will not be the same in alkaline media. Thereby, new technologies are adopted to enhance the efficiency of catalysts for HER. Strategies for overcoming the kinetic issues during alkaline HER include the tuning of active sites of electrocatalyst effectively by synthesizing materials with modified morphologies and electronic structures. Effective HER with cheaper, highly active, plentiful, environmentally benign and stable electrocatalysts is essential for the production of hydrogen in a cost effective manner. The major challenges for electrocatalytic HER are the complete utilization of surface active sites and efficient charge transfer kinetics. Optimizing the structure/morphology of various nano-catalysts and selection of suitable conducting supports are the alternatives to be followed.

Carbon as an electrocatalyst for HER has been explored since the beginning of the US space program. It has the advantage to be used as a supporting material promoting efficient transfer of electrons at its surface. Murthy et al. recently reported the involvement of carbon as the support material in enhancing the electrocatalytic HER by comparing the electrochemical parameters with that of an unsupported system (Murthy et al., 2018). Pt based catalysts have been predominant till now for HER at industrial scale which increased the cost of hydrogen production (Tilak et al., 1986; Vesborg et al., 2015; Xing et al., 2017). Hence, recently, worldwide research has been commenced to find alternative for Pt based catalysts (Eftekhari, 2017). A series of catalysts have been explored from microscale to nanoscale and homo to heterostructured catalysts of earth abundant metals: which include heteroatom doped carbon (Duan et al., 2017), transition metal based (Fe, Co, Ni, Mo, W etc.) (Abbas et al., 2017), transition metal based oxides (Chen et al., 2019), alloys (Rosalbino et al., 2008; Rosalbino et al., 2011), non–metal based (C, N, S, O, P etc.) (Mahmood et al., 2018), sulphides, phosphides and their alloys (Eftekhari, 2017; Elias & Hegde, 2016; Tang et al., 2016; Qu et al., 2017, Li et al., 2017) and hydroxide catalysts (Jia et al., 2017). Design of 3D catalytic surface emerged as a recent trend and promising outcome. It could offer active sites which are electrochemically active to achieve a pronounced increase in catalytic (Murthy et al., 2018; Ma et al., 2018). Yet another future perspective was engineering of multi–component electrocatalysts at molecular level facilitated large scale hydrogen production.

Critical research to enhance HER performance by electrolyte engineering (under acidic, neutral and basic conditions) is also going on aggressively (Shinagawa & Takanabe, 2017). In general, electrocatalytic HER in different electrolyte mediums follows different reaction pathways, i.e., in acid medium, availability of large number of protons enhances HER whereas in alkaline medium, it is moderately harder to attain at low over potentials (Liu et al., 2017) as it requires more energy for generating more protons by breaking water molecules (Subbaraman et al., 2011; Danilovic et al., 2012; Safizadeh et al., 2015). Numerous theoretical and experimental studies have been tried to enhance the mechanism of HER in alkaline electrolytes (Hu et al., 2003; Li et al., 2016). A fair amount of efforts have been proposed to establish new alkaline medium based catalysts. However, there are still many unanswered questions concerned with the activity of these electrocatalysts. Additional research surveys on electrolyte engineering are essential to further substantiate enhanced hydrogen production by water splitting (Shinagawa & Takanabe, 2017).Innovations and developments in the large scale production of hydrogen are crucially needed in the current scenario.

MAIN FOCUS OF THE CHAPTER

Basics and Mechanism of HER

Electrocatalytic hydrogen evolution from water splitting reaction in different media can be explored with a sharp insight into the basic mechanism, the thermodynamic aspect and chemical process of the reaction. All the aspect of electrocatalytic HER is significantly related to the characteristics of the electrode and the electrode/electrolyte interface.

Basics of Water Electrolysis and Electrical Double Layer

The reaction during water electrolysis can be represented as two half relations at two electrodes namely anode (negatively polarized) and cathode (positively polarized). If suitable potential is applied to the system by an external source, the ionic species in the electrolyte undergo reactions in different manner. In the case of water splitting, hydrogen ions are reduced to H_2 gas at cathode and H_2O is oxidized to O_2 at anode. Different Electrical Double Layer (EDL) models such as Helmholtz, Gouy Chapman and Stern can characterize the ionic migration between electrode/electrolyte interfaces.

Characteristic feature occurring in an EDL is the storage of charge at a small distance from the electrode surface. In 1879, Helmholtz proposed the simple EDL as an arrangement of opposite charges as similar to a parallel plate capacitor. A representation of the EDL formation at negatively charged electrode is presented along with the electrode set up and the potential trend of the EDL from the charged electrode surface is shown in Figure 1. The excess charge gets accumulated on the electrode surface during polarization. This excess charge is retained at a narrow distance from the surface and neutralized by the ionic species in some ionic zones of the electrolyte. The electrolyte section of the interface comprises of a compact layer containing inner and outer Helmholtz layers and a diffuse layer as shown in the Figure 1. The diffuse layer extends from the Outer Helmholtz Plane (OHP) to the separator. Simply, an EDL comprise of three portions (Park & Seo, 2011), (1) charged ions adsorbed on the surface, (2) oppositely charged ions (oppositely charged from that of the surface charge) layer (Stern layer) and (3) Layer of high concentration of counter ions (diffuse layer).

The electrical potential trend in different layers is also presented in Figure 1. The electrode surface or Stern layer has the highest potential. As the distance from electrode to electrolyte increases, the surface potential is decreased and attains zero potential at the boundary of the EDL. In the case of colloidal particles, surrounding liquid layer is attached to the surface. The boundary of the layer is called slipping plane and the electric potential corresponding to the plane is zeta potential. This is helpful for studying the interaction between the colloidal particles.

Basic Thermodynamics of HER

The feasibility of an electrochemical reaction can be expressed in terms of thermodynamic functions. The electrical work needed or carried out during a chemical reaction can be expressed as the product of cell potential and cell current. The cell potential is the potential required to carry out a reaction in specific conditions. The cell current means the total charge of electron transferred during the reaction. It can be calculated by multiplying the number of moles of electron transferred with the charge of 1 mole of electron (1F = 96487 C).

Figure 1. Presentation of electrical double layer components and potential generated during HER

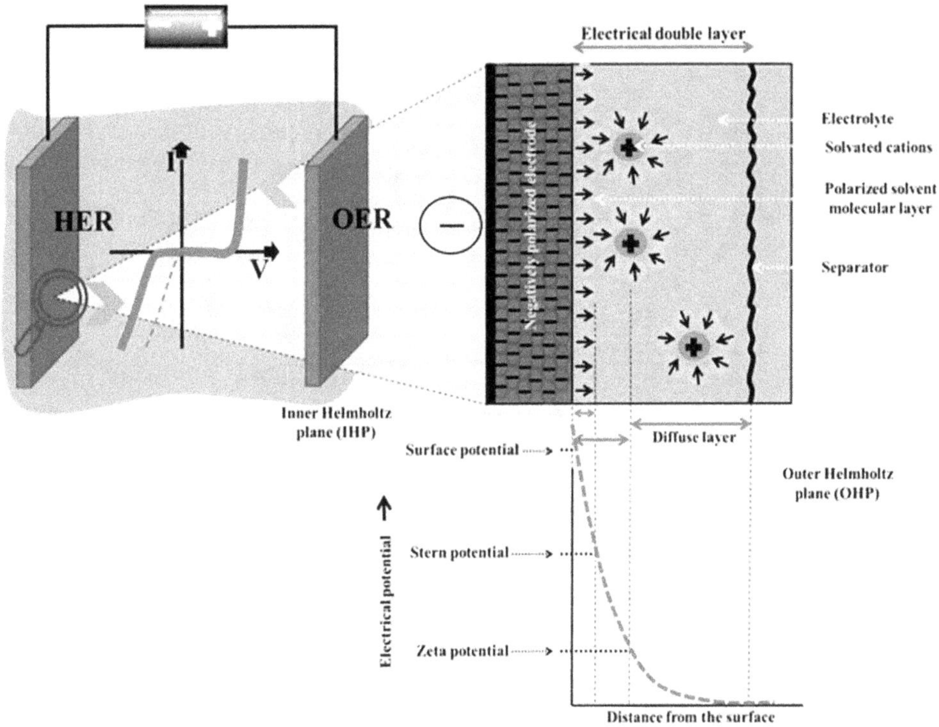

Maximum amount of work needed for an electrochemical reaction is considered as the free energy of the reaction. Free energy change is generally written as shown in Eq. 1.

$$\Delta G = -\,nFE_{cell} \tag{1}$$

Standard enthalpy (ΔH^0) and entropy (ΔS^0 of the HER during water splitting is about 286 and 0.163 kJ mol^{-1} respectively. Accordingly, the free energy of the reaction is equal to 237 kJ mol^{-1}. At these thermodynamic conditions, the reversible potential ($V_{rev} = \Delta G^0/nF$) and thermodynamic potential ($V_{th} = \Delta H^0/nF$) of the HER by water splitting at 25 °C are 1.23 and 1.48 eV respectively. These thermodynamic and electrochemical aspects indicate that the HER by water splitting is a high energy consuming one. The potential needed for the reaction varies according to the features of electrodes and electrolytes used. The skillful engineering of the requirements can minimize the extra potential.

Mechanism of Electrocatalytic HER

The kinetics of HER most probably involves the evolution of hydrogen from either H_3O^+ or water molecules itself. Hydrogen can be produced from acidic or basic electrolyte mediums with the aid of a suitable electrode. In most alkaline electrolysis, two mechanisms are proposed. The first one is of electrochemical discharge with adsorption of hydrogen followed by desorption via chemically or electrochemically. Hydrogen adsorption in alkaline and acidic medium is represented in Eqs. 2 and 3 respectively.

Figure 2. Hydrogen evolution reaction mechanism at the electrode surface in alkaline media

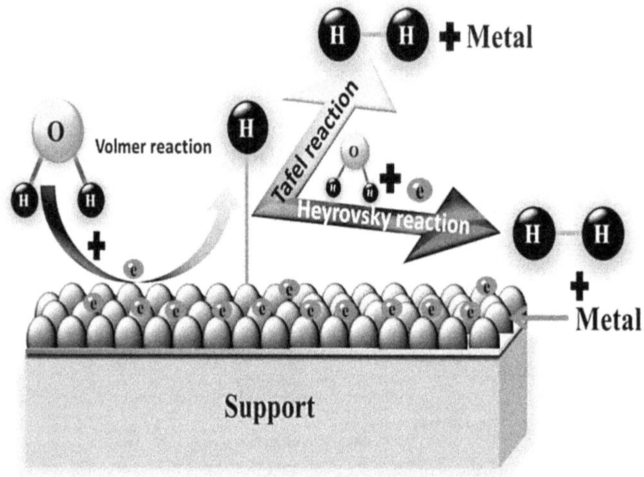

$H_2O + e^- + M \rightarrow M\text{–}H + OH^-$ (Volmer reaction in alkaline media) (2)

$H_2O + e^- + M \rightarrow M\text{–}H + OH^-$ (Volmer reaction in acidic media) (3)

'M' represents free metal surface and M–H represents hydrogen adsorbed on metal.

Desorption of hydrogen during the reaction can either follow Tafel or Heyrovsky reaction according to the reactions given in Eqs. 4 to 6.

$2M\text{–}H \rightarrow H_2 + 2M$ (Tafel reaction in alkaline media) (4)

$M\text{–}H + H_2O + e^- \rightarrow H_2 + 2M$ (Heyrovsky reaction in alkaline media) (5)

$M\text{–}H + H^+ + e^- \rightarrow H_2 + M$ (Heyrovsky reaction in media) (6)

In Volmer mechanism, a charge transfer process corresponding to a discharge process occurs. The Tafel reaction is associated with hydrogen atom–hydrogen atom recombination. The mechanism in a metal based electrode is symbolically represented in Figure 2.

The HER mechanism in oxide based electrodes have some modifications as hydration of the electrode surface in electrolyte significantly varies as compared to that in metal based electrodes. Alkaline HER mechanism is proposed in the oxide based electrolytes by many researchers. The most accepted mechanism is given in Eqs. 7 to 9.

$H_2O + e^- + M\text{–}OH \rightarrow M\text{–}H\text{–}OH + OH^-$ (7)

$$M\text{–}H\text{–}OH + e^- \rightarrow M\text{–}H + OH^- \quad (8)$$

$$M\text{–}H + H_2O \rightarrow M\text{–}OH + H_2 \quad (9)$$

The reaction mechanism occurring at the electrode can be identified from Tafel analysis based on the Tafel slope values. Critical electrochemical parameters obtained from the Tafel fitting of the Tafel plot is shown in Figure 3.

Figure 3. Representation of conventional Tafel plot and its fit with label of some useful parameters (Tafel slope and exchange current density)

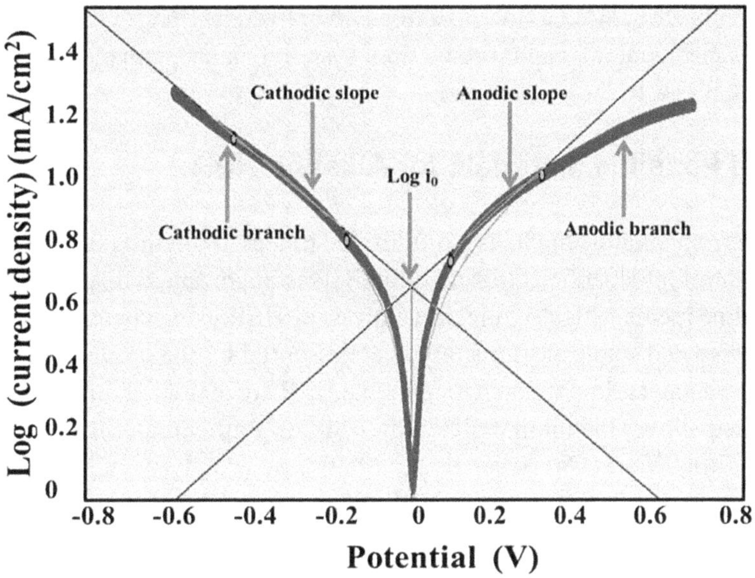

Materials and Methods Development for HER

Platinum group metals are the most active and stable catalysts for HER. However, for industrial applications, they are uneconomical owing to their limited availability and high cost. In this context, the modern electrocatalysis, on a global scale, aims on replacing these platinum group metals with low cost and active catalysts and research is going on worldwide aggressively. A series of catalysts have been explored from micro to nanoscale and homo to heterostructure basis. Major challenges for electrocatalytic HER are the complete utilization of surface active sites and efficient charge transfer kinetics. Hence, new strategies and technologies are adopted for overcoming the kinetic issues and to enhance the efficacy of catalysts for HER.

Development of Possible Methods for Hydrogen Production

A wide variety of electrodes and electrolytes are used in water electrolyzes. The hydrogen generation in electrolyzes follows small changes in different systems with the usage of different ionic electrolytes. Polymer electrolyte membrane, alkaline and solid oxide electrolyzers are some important categories of the electrolyte type (Millet, 2013). Innovative research on water electrolyzers is also conducted by NASA to endure man in space for prolonged time periods. To generate breathable oxygen on a spacecraft, an Oxygen Generation Subsystem (OGS) is essential. In addition, currently this is accomplished by water electrolysis, and the hydrogen byproduct can be used inside the Air Revitalization System (ARS) of the spacecraft to recover oxygen from decreased carbon dioxide. The scientists are trying different approaches for advancing this process of water electrolysis (Tong et al., 2020). Collectively, this advancement includes modifications on instrumentation and electronic parts, system design and analyses, usage of efficient current collectors, production of stable and active electrodes etc. The decisive aims of the possible method advancement are capital reduction, system balance, enhancing efficiency and energy and connecting compressor to the electrolyzer.

Development of Possible Materials for Alkaline HER

Hydrogen evolution from alkaline media is a promising strategy for hydrogen production with commercial feasibility. The development of electroactive catalysts is important to furnish a cathode for hydrogen evolution from alkaline media. Effective electrocatalysis of HER using cheaper, highly active, plentiful, environmentally benign and stable electrocatalysts are essential for the production of hydrogen in a cost effective manner. The catalysts investigated for alkaline HER are mainly of three categories such as noble metals and their based alloys (Platinum, Palladium, Iridium, Ruthenium, Silver, etc.); transition metals (Ni, Fe, Ti, Co, Mo, Mn, W and Cu) (Chen et al., 2019) based catalysts including their alloys (Fang et al., 2016; Csernica et al., 2017; Kuang et al., 2017; Shen et al., 2018; Gao et al., 2018), oxides (Zhang et al., 2018), hydroxides (Hu et al., 2017), sulfides (Anantharaj et al., 2016) selenides (Zhang et al., 2018 tellurides (Kosmala et al. 2018), nitrides (Peng et al., 2019), phosphides (Shao et al., 2017; Liu et al., 2018), carbides (Kou et al., 2019), borides (Liu et al., 2019); non–metal based catalysts and alloys of S, B, N, C, P (Mahmood et al., 2018). Research activities on materials developments are generally based on these categories.

Noble Metal Based Catalysts

Pt group based catalytic systems exhibit high exposed surface area and are widely used for alkaline HER. These catalytic systems also possess high stability during HER. Some of the important classes of noble metal based catalysts include Pt/C (Cheng et al., 2015). Pt/C catalyst exhibits high efficiency towards HER than commonly used Pt metals. Some transition metal and noble metal combinations, for example, Pt–RE (RE = Sm, Ho, Ce) are dispersed more due to their critical loading. This reduces the production cost of catalysis (Vesborg et al., 2015). Pt/Fe on Ni foam (Huang et al., 2017) and Pt–Ni–Co (Oh et al., 2016) alloys are some other promising catalysts investigated for enhanced hydrogen evolution activity. Glassy carbon substrate supported noble metal based catalyst such as PdC (Durst et al. 2014), $Ru/C_3N_4/C$ (Zheng et al. 2016) and RuO_2/Co_3O_4 (Subbaraman et al., 2011) show remarkable HER performance in alkaline media with comparatively lower overpotential at selected current densities. The cost of these noble metal based catalysts limit their widespread application in alkaline HER.

Transition Metal Based Catalysts

Transition metal based catalysts are widely in research for alkaline HER due to their great abundance, enhanced activity, stability and good accessibility. The research outcomes regarding the catalytic efficiency of transition metal based electrodes show their promising implementation in large scale hydrogen production. Transition metals and their alloys and compounds of some non–metals such as O, S, Se, N, P, C, and B can exhibit enhanced performance during catalytic HER. Transition metals such as Ni, Co, W, Fe etc. and their compounds exhibit promising electrolytic performance. Ni based catalyst composed with rGO (Wang et al. 2017), carbon quantum dots and CNTs (McArthur et al. 2014) show excellent performance in HER catalysis with excellent activity and stability. The activity of these catalysts is due to their characteristics such as large active surface area, excellent conductivity, efficient surface functionalization etc. (Yang et al., 2015). Doping of Ni with P and/or S is reported to have enhanced HER activity with low overpotential of about 24 mV at a current density of 10 mA cm^{-2} (Jin et al., 2019). Mn doped Ni also exhibited a low overpotential of 24 mV at 10 mA cm^{-2} for HER (Shao et al., 2018). Yet another electrocatalyst is W Single Atom Catalyst (SAC) with substrate that unveiled good performance and N doping to the metal has further enhanced its activity towards alkaline HER (Chen et al., 2018).

Transition metal alloys also exhibited remarkable electronic characteristics and high performance towards HER. One example is Ni_3Fe generated on N doped carbon fibers (Li et al., 2018). Some lower noble metal alloys of transition metals such as Ru–Co exhibited significant HER performance after encapsulating it into nitrogen doped graphene layers (Su et al. 2018). Ni–Co–Fe alloy developed for HER catalysis (Darband et al., 2019) displayed remarkable activity and excellent stability. Transition metal oxides such as CoO_2 (Eftekhari, 2017; Yang et al., 2018), VO_2 (Moatti et al., 2019), MnO_2 (Begum et al., 2019) and NiO (Zhang et al., 2018; Zhou et al., 2018) had decent catalytic activity and their activity was further enhanced by some modifications such as by hybridizing them with transition metal oxides (Shang et al., 2017; Gong et al., 2018; Zhou et al., 2018), transition metal phosphides (Zhang et al., 2018) and transition metal itself (Chen et al., 2018; Xiang et al., 2019). The enhanced activity of these catalysts was due to the formation of defects in them especially in the form of oxygen vacancies. Transition metal hydroxides such as $Ni(OH)_2$ (Yin et al., 2015; Yang et al., 2017; Yu et al., 2018; Wang et al., 2018) and $Co(OH)_2$ (Xing et al., 2017) coupled with Pt metal have an excellent activity due to their specific crystal structure.

Non–metal Based Catalysts

Non–metal catalysts also came into scenario along with metal based catalytic systems for enhancement of activity and stability. Purely non–metal based catalysts also exhibited favorable results for HER catalysis. Different carbon based materials such as C_3N_4 (Shalom et al., 2014), graphitic nano carbons (Lai et al., 2012), carbon and graphene micro tubes (Zhang et al., 2016) had been reported for remarkable catalytic activity in alkaline HER. The activity of C_3N_4 catalyst has been improved with some metal based material and modifications are prompted with Fluorine doped Tin Oxide (FTO), TiO_2 and ZnO_2 (Shalom et al., 2014). Chen et al. (2018) have reported that graphitic carbon based non–metal catalysts doped with non–metals exhibited a very good HER activity. The research outcomes regarding the non–metal based catalysts reveal that the materials are really promising for electrocatalytic HER due to their excellent electronic characteristics.

Developments and Advances in Electrocatalytic Electrodes and Materials

Electrocatalytic HER by water electrolysis is a widely explored method for hydrogen production. Water electrolysis can be conducted both in acidic and alkaline electrolytes. Generally, platinum based electrodes are used for the hydrogen evolution half reaction during water splitting with the aid of electricity. The activity and stability of platinum for electrocatalytic HER is significantly higher enough to sustain large current even with low platinum loadings to the electrode (Vesborg et al., 2015).

Platinum based coatings have been explored extensively for electrocatalytic HER. Many factors influence platinum based coatings that lead to its extensive implementation in this field. However, for industrial applications, they are uneconomical owing to their limited availability and high cost. In this context, the modern electrocatalysis, on a global scale, aims at replacing these platinum group metals with low cost and active catalysts/catalyst coatings. Developments and researches on non-noble electrocatalyst are going on aggressively with the main target to achieve the same characteristics of the noble metal based catalysts by engineering them with the transition metal based materials (Chen et al., 2019).

Criteria of Selection/Development of Catalytic Electrodes–Special Focus on Characteristics

An ideal catalyst for water splitting should have high efficacy, durability and cost effectiveness (Vesborg et al., 2015). This efficacy is measured as a function of overpotential and exchange current density. A catalyst with low overpotential and exchange current density values is considered as competent for the electrocatalytic application. The parameters that indicate the efficacy of the catalyst are directly related to their surface and the electrochemical characteristics. Durability indicates the stability of the catalyst during the reaction time. The long term immunity of the catalyst against its deactivation for catalytic applications can be monitored as a function of current or potential response. Chronoamperometry and linear sweep voltammetry are some of the electrochemical techniques that are used to study the stability of the catalysts. However, the durability of the catalyst at real system application is still under investigation. The main problems faced in this field are cost of the catalyst, chemical and technological expenses. Non–abundant materials such as Pt, Ru, and Re are expensive but still are explored in this field. The replacement of such expensive materials can lead to a cost effective hydrogen production.

The criteria of an ideal electrocatalyst can be achieved by giving special focus on electrode characteristics. Cost effective modifications of material incorporated electrodes have to be implemented for large scale production of hydrogen. The characteristics are schematically represented in Figure 4. As shown in figure, the surface characteristics such as surface morphology, topography, other physicochemical characteristics (roughness and porosity) and critical elemental composition are to be modified by means of different chemical techniques. These modifications can enhance the number of active sites and hence the electron transport mechanism in catalytic electrodes.

The intrinsic catalytic performance of an electrode mainly need modifications in some of its chemical and physical properties. Consequently, conductivity of the system would be enhanced. An electrochemical scientist is mostly concerned with the surface reactions at an electrode. For that, different techniques are to be used so that, the atoms, molecules and other nanoparticles could be clubbed with the electrode

Figure 4. Different surface tuning strategies and enhanced surface characteristics

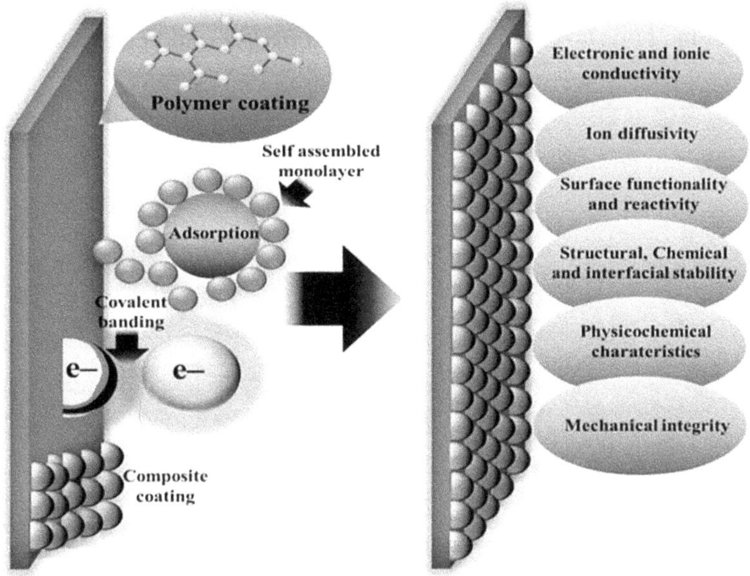

surface. The enhanced properties of the electrode provide high selectivity towards different reactions, corrosion and fouling resistance. Furthermore, any interference, during the surface reactions, affecting the overall catalytic performance could be avoided. Energy conversion devices are to be deployed to monitor the surface modifications and this principle can be used to study all probable electrochemical processes, corrosion mechanism and prevention methods in molecular electronics and electronic devices.

Different approaches used for the electrode modification includes chemisorption of functional groups to the substrate (electrode), covalent bonding of chemical species, polymer film coatings by other chemisorption for polymeric coating solution and modification by means of composite materials (Sun et al., 2018).

Ni–P Electrodes–Research and Developments for HER

Ni based coatings are widely explored with different substrates for many applications including electrocatalytic water splitting reactions (Shibli & Dilimon, 2008; Tang et al., 2010; Feng et al., 2014; Shibli et al., 2016; Wang et al., 2018). The advantage of the nickel based electrode is that the film can be easily developed and functional materials can be effectively incorporated into its matrix. Such electrodes with Mo and Fe based alloys have been explored for some applications such as wearable storage devices, electrocatalysis and sensors (Sahasrabudhe et al., 2018).

An important advantage of the Ni–P coating is that, it could be obtained over different substrates by using the techniques of electro and electroless plating. The former would result a coating with significant uniformity and compact morphology and the latter would give the one having remarkable physicochemical and electrochemical properties (Zhang et al., 2017).

Widely explored method of electroless Ni plating is adopted by many researchers for the development of Ni–P coating for electrocatalytic HER. The reactions at the time of deposition process can be optimized by different methods and hence the coating properties can be enhanced.

Figure 5. Schematic representation of electroless Nickel–Phosphorus (Ni–P) coating bath

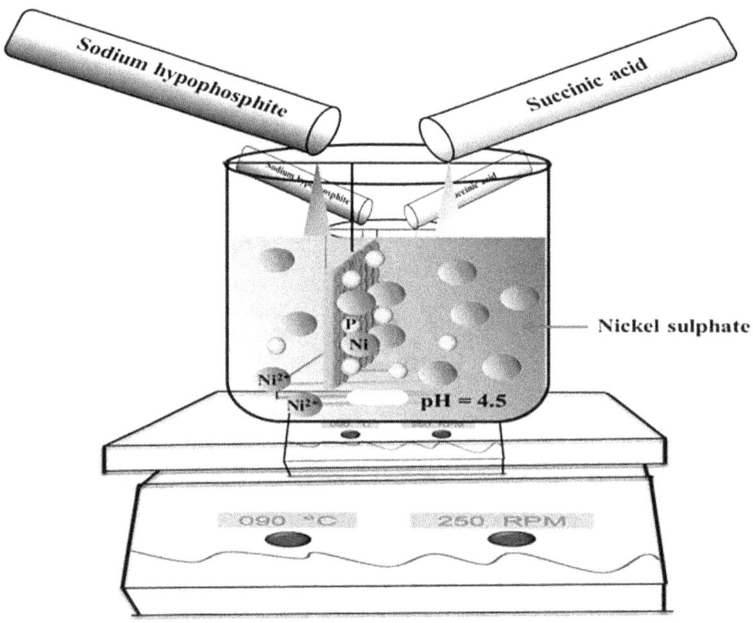

The electroless deposition of Ni–P can be performed on mild steel substrates (treated) and the Ni–P coated ones as electrodes have the ability to catalyze HER. Different functional materials incorporated Ni–P matrix also can enhance the HER efficiency. An electroless coating of Ni–P alloy is schematically represented in Figure 5. The reaction mechanism is given in Eqs. 10 to 13.

$$Mn^{+} + H_2PO_2^{-} + H_2O \rightarrow H_2PO_3^{-} + H_2 \quad (10)$$

$$H_2PO_2^{-} + H_2O \rightarrow CAS \rightarrow H_2PO_3^{-} + H_2 \quad (11)$$

$$H_2PO_3^{-} + H_{ad} \rightarrow H_2O + OH^{-} + P \quad (12)$$

$$3H_2PO_2^{-} \rightarrow H_2PO_3^{-} + H_2O + 2OH^{-} + 2P \quad (13)$$

The bare Ni–P electrodes can be modified through different modification techniques, including tuning the coating conditions and parameters. Post–treatment of coating, especially thermal treatments (Rabizadeh et al., 2017) and reinforcement of electrodes with different additives can also be performed.

Some of the recent literature reports concerning Ni–P coatings are cited below;

- Rana & Farhat (2019) reported the modification of Ni–P electrodes with graphene incorporation for enhancing its tribological characteristics
- Some significance surfactants added to coating bath to develop Ni–P electroless coating with incorporated TiC composite for high wetting characteristics have been explored by Czagany & Baumli (2019).
- An alloy of tungsten and Ni–P have been reported to finely tune the magnetic properties and microstructure of the Ni–P coating (Zhan et al., 2019).
- Ni–P based material proved to be promising for biological applications owing to its morphological effect (Zheng et al., 2019).

The Ni–P coated materials can also be modified through different methods to enhance their surface characteristics for different applications. For examples:

- Zhang et al. (2019) reported the synthesis of core shell Ni–P material via an iron thermal strategy and the synthesized core shell material, at its critical composition and thickness, exhibited an enhanced electrochemical performance.
- Diamond powder incorporated Ni–P coating with optimized P content has been developed by Kim et al. (2019) and it exhibited good surface stability against surface etching conditions.
- Effect of phosphorus content in Ni–P coating has been explored for microstructure development (Lin et al., 2019).
- The Ni–P based electrodes have been post treated by subjecting them to laser annealing with an aim to enhance their physicochemical and crystalline characteristics (Liu et al., 2017).

Issues, Controversies, Problems

Hydrogen is an ideal alternative to fossil fuels but during its implementation many practical issues related to cost and pollution come in to the scene. Nowadays the hydrogen production methods employed in industrial scale mainly exploit fossil fuel, hence results carbon emission. Appropriate carbon capture technology should be clubbed with such hydrogen production techniques to avoid carbonaceous pollution and their impacts. Electrolytic water splitting to produce hydrogen is an ideal method. The techniques mainly need water, a renewable solvent as the main reagent. The energy used to split water is electricity which mainly contributes to the cost of the technique. The use of electricity for this purpose cannot be completely avoided but can reduce its use. Well-engineered electrodes are required to reduce the demand of electricity during electrolysis.

Design and development of highly active and durable electrode system will leads to the cost effective industrial hydrogen production via electrolysis of water and other media. The developments in this field are mainly focused on noble metal based or that incorporated electrode system. This result in high cost of production hence unaffordable to the industry. Moreover metal based systems get corroded during hydrogen production from alkaline or acidic media. Available electrode development techniques are time and money consuming steps. Electroless deposition of functional materials is simple but effective electrode development technique can be adopted for similar system development.

SOLUTIONS AND RECOMMENDATIONS

Electrocatalytic hydrogen production from water splitting reaction is an ideal method for industrial hydrogen production. The method is green because it has the potential to produce extra pure hydrogen without any carbon emission. The use of electricity can be reduced by avoiding the possibility of over potential development during the water splitting reaction on the electrode surface. The major contribution of the over potential is due to the non-availability of electroactive sites on the electrode surface. Eventhough the Pt based catalytic systems can be implemented, their industrial utility is limited due to high cost. This could be overcome by the usage of phosphorus alloy of Ni coated on metal substrates to produce active surfaces. Electroless deposition process is an effective coating preparation strategy which produces compact and uniform coating with the catalyst material on the surface. The produced coatings can be effectively used for large scale electrocatalytic hydrogen production. Still it needs improvements in terms of catalytic efficiency and durability of the material. The techniques and the catalytic electrode systems can be properly tune to generate smart electrodes for electrocatalytic hydrogen production.

FUTURE RESEARCH DIRECTIONS

Electrode design and development for superior hydrogen production is vital to hydrogen energy implementation. As a novel catalytic electrode, Ni based electrodes are widely researched. Their character tuning is the focus point of majority of researchers. The content tuning, pre- and post- treatment methods and highly conductive material incorporation are some among the attractive areas for future research. The engineering of specific surface morphologies by proper channelization of the preparation strategy, say electroless deposition and other physical and chemical electrode treatment can also be implemented to produce highly active and durable coatings. The different basic knowledge of electroless deposition, various reinforcement techniques, reinforcing materials, their development and applications should be studied in detail and their detailed discussion will be important to the students and researchers for the further exploration of the method and to design related novel materials.

CONCLUSION

The present chapter draws an attention on different chemical and physical aspects of electrocatalytic hydrogen evolution reaction, methods and requirements for the development of electrode material for the reaction. Recent trends in the development of electrocatalytic coatings via electroless deposition of catalysts materials and the trends in design and development Ni-P alloy coating are discussed in detail. The available details of the present catalyst system indicate that it needs further exploration and research for their industrial implementations.

ACKNOWLEDGMENT

This research received no specific grant from any funding agency in the public, commercial, or not-for-profit sectors.

REFERENCES

Abbas, S. A., Iqbal, M. I., Kim, S. H., & Jung, K. D. (2017). Catalytic activity of urchin–like Ni nanoparticles prepared by solvothermal method for hydrogen evolution reaction in alkaline solution. *Electrochimica Acta*, *227*, 382–390. doi:10.1016/j.electacta.2017.01.039

Anantharaj, S., Ede, S. R., Sakthikumar, K., Karthick, K., Mishra, S., & Kundu, S. (2016). Recent trends and perspectives in electrochemical water splitting with an emphasis on sulfide, selenide, and phosphide catalysts of Fe, Co, and Ni: A review. *ACS Catalysis*, *6*(12), 8069–8097. doi:10.1021/acscatal.6b02479

Begum, H., Ahmed, M. S., & Jeon, S. (2019). Δ–MnO_2 nanoflowers on sulfonated graphene sheets for stable oxygen reduction and hydrogen evolution reaction. *Electrochimica Acta*, *296*, 235–242. doi:10.1016/j.electacta.2018.11.073

Chen, W., Pei, J., He, C. T., Wan, J., Ren, H., Wang, Y., & Zheng, X. (2018). Single Tungsten Atoms Supported on MOF–Derived N–Doped Carbon for Robust Electrochemical Hydrogen Evolution. *Advanced Materials*, *30*(30), 1800396. doi:10.1002/adma.201800396 PMID:29888491

Chen, W. T., Chan, A., Waterhouse, S. D., Llorca, J., Idriss, H., & Waterhouse, G. I. N. (2018). Performance comparison of Ni/TiO_2 and Au/TiO_2 photocatalysts for H_2 production in different alcohol–water mixtures. *Journal of Catalysis*, *367*, 27–42. doi:10.1016/j.jcat.2018.08.015

Chen, Z., Duan, X., Wei, W., Wang, S., & Ni, B. J. (2019). Recent advances in transition metal–based electrocatalysts for alkaline hydrogen evolution. *Journal of Materials Chemistry. A, Materials for Energy and Sustainability*, *7*(25), 14971–15005. doi:10.1039/C9TA03220G

Cheng, N., Stambula, S., Wang, D., Banis, M. N., Liu, J., Riese, A., & Botton, G. A. (2016). Platinum single–atom and cluster catalysis of the hydrogen evolution reaction. *Nature Communications*, *7*(1), 13638. doi:10.1038/ncomms13638 PMID:27901129

Csernica, P. M., McKone, J. R., Mulzer, C. R., Dichtel, W. R., Abruña, H. D., & DiSalvo, F. J. (2017). Electrochemical hydrogen evolution at ordered Mo_7Ni_7. *ACS Catalysis*, *7*(5), 3375–3383. doi:10.1021/acscatal.7b00344

Czagany, M., & Baumli, P. (2019). Effect of surfactants on the behavior of the Ni–P bath and on the formation of electroless Ni–P–TiC composite coatings. *Surface and Coatings Technology*, *361*, 42–49. doi:10.1016/j.surfcoat.2019.01.046

Danilovic, N., Subbaraman, R., Strmcnik, D., Chang, K. C., Paulikas, A. P., Stamenkovic, V. R., & Markovic, N. M. (2012). Enhancing the alkaline hydrogen evolution reaction activity through the bifunctionality of $Ni(OH)_2$/metal catalysts. *Angewandte Chemie International Edition*, *51*(50), 12495–12498. doi:10.1002/anie.201204842 PMID:23129151

Darband, G. B., Aliofkhazraei, M., Hyun, S., Rouhaghdam, A. S., & Shanmugam, S. (2019). Electrodeposition of Ni–Co–Fe mixed sulfide ultrathin nanosheets on Ni nanocones: A low–cost, durable and high performance catalyst for electrochemical water splitting. *Nanoscale*, *11*(35), 16621–16634. doi:10.1039/C9NR04529E PMID:31460535

Duan, J., Chen, S., & Zhao, C. (2017). Ultrathin metal–organic framework array for efficient electrocatalytic water splitting. *Nature Communications*, *8*(1), 15341. doi:10.1038/ncomms15341 PMID:28580963

Durst, J., Siebel, A., Simon, C., Hasche, F., Herranz, J., & Gasteiger, H. A. (2014). New insights into the electrochemical hydrogen oxidation and evolution reaction mechanism. *Energy & Environmental Science*, *7*(7), 2255–2260. doi:10.1039/C4EE00440J

Eftekhari, A. (2017). Electrocatalysts for hydrogen evolution reaction. *International Journal of Hydrogen Energy*, *42*(16), 11053–11077. doi:10.1016/j.ijhydene.2017.02.125

Elias, L., & Hegde, A. C. (2016). Synthesis and characterization of Ni–P–Ag composite coating as efficient electrocatalyst for alkaline hydrogen evolution reaction. *Electrochimica Acta*, *219*, 377–385. doi:10.1016/j.electacta.2016.10.024

Fang, M., Gao, W., Dong, G., Xia, Z., Yip, S., Qin, Y., & Ho, J. C. (2016). Hierarchical NiMo–based 3D electrocatalysts for highly–efficient hydrogen evolution in alkaline conditions. *Nano Energy*, *27*, 247–254. doi:10.1016/j.nanoen.2016.07.005

Feng, Z. A., Gabaly, F. E., Ye, X., Shen, Z. X., & Chueh, W. C. (2014). Fast vacancy–mediated oxygen ion incorporation across the ceria–gas electrochemical interface. *Nature Communications*, *4374*(1), 1–9. doi:10.1038/ncomms5374 PMID:25007038

Gao, W., Gou, W., Zhou, X., Ho, J. C., Ma, Y., & Qu, Y. (2018). Amine–modulated/engineered interfaces of NiMo electrocatalysts for improved hydrogen evolution reaction in alkaline solutions. *ACS Applied Materials & Interfaces*, *10*(2), 1728–1733. doi:10.1021/acsami.7b16125 PMID:29282973

Gong, W., Fugetsu, B., Wang, Z., Sakata, I., Su, L., Zhang, X., Ogata, H., Li, M., Wang, C., Li, J., Medina, J., Terrones, M., & Endo, M. (2018). Carbon nanotubes and manganese oxide hybrid nanostructures as high performance fiber supercapacitors. *Communications Chemistry*, *1*(1), 161–168. doi:10.103842004-018-0017-z

Hu, H., Qiao, M., Pei, Y., Fan, K., Li, H., Zong, B., & Zhang, X. (2003). Kinetics of hydrogen evolution in alkali leaching of rapidly quenched Ni–Al alloy. *Applied Catalysis A, General*, *252*(1), 173–183. doi:10.1016/S0926-860X(03)00416-2

Hu, J., Zhang, C., Jiang, L., Lin, H., An, Y., Zhou, D., & Yang, S. (2017). Nanohybridization of MoS_2 with layered double hydroxides efficiently synergizes the hydrogen evolution in alkaline media. *Joule*, *1*(2), 383–393. doi:10.1016/j.joule.2017.07.011

Huang, L., Hou, Y., Yu, Z., Peng, Z., Wang, L., Huang, J., & Li, Z. (2017). Pt/Fe–NF electrode with high double–layer capacitance for efficient hydrogen evolution reaction in alkaline media. *International Journal of Hydrogen Energy*, *42*(15), 9458–9466. doi:10.1016/j.ijhydene.2017.02.055

Jia, Y., Zhang, L., Gao, G., Chen, H., Wang, B., Zhou, J., Soo, M. T., Hong, M., Yan, X., Qian, G., Zou, J., Du, A., & Yao, X. (2017). A heterostructure coupling of exfoliated Ni–Fe hydroxide nanosheet and defective graphene as a bifunctional electrocatalyst for overall water splitting. *Advanced Materials*, *29*(17), 1700017. doi:10.1002/adma.201700017 PMID:28256771

Jin, H., Liu, X., Chen, S., Vasileff, A., Li, L., Jiao, Y., & Qiao, S. Z. (2019). Heteroatom–Doped Transition Metal Electrocatalysts for Hydrogen Evolution Reaction. *ACS Energy Letters*, *4*(4), 805–810. doi:10.1021/acsenergylett.9b00348

Kim, I., Kim, K., & Song, O. (2019). Surface etching of diamond powders with electroless Ni–P coating according to phosphorus content. *Diamond and Related Materials*, *97*, 107465. doi:10.1016/j.diamond.2019.107465

Kosmala, T., Coy, D. H., Komsa, H. P., Ma, Y., Krasheninnikov, A. V., Batzill, M., & Agnoli, S. (2018). Metallic twin boundaries boost the hydrogen evolution reaction on the basal plane of molybdenum selenotellurides. *Advanced Energy Materials*, *8*(20), 1–8. doi:10.1002/aenm.201800031

Kou, Z., Zhang, L., Ma, Y., Liu, X., Zang, W., Zhang, J., Huang, S., Du, Y., Cheetham, A. K., & Wang, J. (2019). 2D carbide nanomeshes and their assembling into 3D microflowers for efficient water splitting. *Applied Catalysis B: Environmental*, *243*, 678–685. doi:10.1016/j.apcatb.2018.11.008

Kuang, M., Wang, Q., Han, P., & Zheng, G. (2017). Cu, Co-Embedded N-Enriched Mesoporous Carbon for Efficient Oxygen Reduction and Hydrogen Evolution Reactions. *Advanced Energy Materials*, *7*(17), 1–8. doi:10.1002/aenm.201700193

Lai, J., Guo, H., Wang, Z., Li, X., Zhang, X., Wu, F., & Yue, P. (2012). Preparation and characterization of flake graphite/silicon/carbon spherical composite as anode materials for lithium–ion batteries. *Journal of Alloys and Compounds*, *530*, 30–35. doi:10.1016/j.jallcom.2012.03.096

Li, T., Luo, G., Liu, K., Li, X., Sun, D., Xu, L., Li, Y., & Tang, Y. (2018). Encapsulation of Ni_3Fe Nanoparticles in N–Doped Carbon Nanotube–Grafted Carbon Nanofibers as High–Efficiency Hydrogen Evolution Electrocatalysts. *Advanced Functional Materials*, *28*(51), 1805828. doi:10.1002/adfm.201805828

Li, X., Liu, P. F., Zhang, L., Zu, M. Y., Yang, Y. X., & Yang, H. G. (2016). Enhancing alkaline hydrogen evolution reaction activity through Ni–Mn_3O_4 nanocomposites. *Chemical Communications*, *52*(69), 10566–10569. doi:10.1039/C6CC04141H PMID:27500290

Li, Z., Xia, M., & Shao, G. (2017). The Effects of CeO_2 Nanorods and CeO_2 Nanoflakes on Ni–S Alloys in Hydrogen Evolution Reactions in Alkaline Solutions. *Catalysts*, *7*(7), 197. doi:10.3390/catal7070197

Lin, J. D., & Chou, C. T. (2019). The influence of phosphorus content on the microstructure and specific capacitance of etched electroless Ni–P coatings. *Surface and Coatings Technology*, *368*, 126–137. doi:10.1016/j.surfcoat.2019.04.009

Liu, Y., Liang, X., Gu, L., Zhang, Y., Li, G. D., Zou, X., & Chen, J. S. (2018). Corrosion engineering towards efficient oxygen evolution electrodes with stable catalytic activity for over 6000 hours. *Nature Communications*, *2609*(1), 1–10. doi:10.103841467-018-05019-5 PMID:29973591

Liu, Y., Zhang, J., Li, X., Zeng, Z., Cheng, X., Wang, Y., & Pa, M. (2019). Phosphorous Modified Metal Boride as High Efficiency HER Electrocatalyst. *International Journal of Electrochemical Science*, *14*, 6123–6132. doi:10.20964/2019.07.68

Liu, Z., Sajjad, S. D., Gao, Y., Kaczur, J. J., & Masel, R. I. (2017). An alkaline water electrolyzer with Sustainion membranes: 1A/cm^2 at 1.9V with Base Metal Catalysts. *ECS Transactions*, *77*(9), 71–73. doi:10.1149/07709.0071ecst

Ma, Z., Zhao, Q., Li, J., Tang, B., Zhang, Z., & Wang, X. (2018). Three–dimensional well–mixed/highly–densed NiS–CoS nanorod arrays: An efficient and stable bifunctional electrocatalyst for hydrogen and oxygen evolution reactions. *Electrochimica Acta*, *260*, 82–91. doi:10.1016/j.electacta.2017.11.055

Mahmood, N., Yao, Y., Zhang, J. W., Pan, L., Zhang, X., & Zou, J. J. (2018). Electrocatalysts for hydrogen evolution in alkaline electrolytes: Mechanisms, challenges, and prospective solutions. *Advancement of Science*, *5*(2), 1700464–1700487. doi:10.1002/advs.201700464 PMID:29610722

McArthur, M. A., Jorge, L., Coulombe, S., & Omanovic, S. (2014). Synthesis and characterization of 3D Ni nanoparticle/carbon nanotube cathodes for hydrogen evolution in alkaline electrolyte. *Journal of Power Sources*, *266*, 365–373. doi:10.1016/j.jpowsour.2014.05.036

Millet, P. (2013). 9 - Chlor-alkali technology: Fundamentals, processes and materials for diaphragms and membranes. *Woodhead Publishing Series in Energy*, *2*, 384–415. doi:10.1533/9780857097347.2.384

Moatti, A., Sachan, R., Cooper, V. R., & Narayan, J. (2019). Electrical Transition in Isostructural VO_2 Thin–Film Heterostructures. *Scientific Reports*, *9*(1), 1–10. doi:10.103841598-019-39529-z PMID:30816206

Murthy, A. P., Madhavan, J., & Murugan, K. (2018). Recent advances in hydrogen evolution reaction catalysts on carbon/carbon–based supports in acid media. *Journal of Power Sources*, *398*, 9–26. doi:10.1016/j.jpowsour.2018.07.040

Oh, A., Sa, Y. J., Hwang, H., Baik, H., Kim, J., Kim, B., Joo, S. H., & Lee, K. (2016). Rational design of Pt–Ni–Co ternary alloy nanoframe crystals as highly efficient catalysts toward the alkaline hydrogen evolution reaction. *Nanoscale*, *8*(36), 16379–16386. doi:10.1039/C6NR04572C PMID:27714051

Park, S. J., & Seo, M. K. (2011). *Interface science and composites* (1st ed.). Academic Press.

Peng, X., Pi, C., Zhang, X., Li, S., Huo, K., & Chu, P. K. (2019). Recent progress of transition metal nitrides for efficient electrocatalytic water splitting. *Sustainable Energy & Fuels*, *3*(2), 366–381. doi:10.1039/C8SE00525G

Pletcher, D., & Li, X. (2011). Prospects for alkaline zero gap water electrolysers for hydrogen production. *International Journal of Hydrogen Energy*, *36*(23), 15089–15104. doi:10.1016/j.ijhydene.2011.08.080

Qu, K., Zheng, Y., Jiao, Y., Zhang, X., Dai, S., & Qiao, S. Z. (2017). Polydopamine-Inspired, Dual Heteroatom-Doped Carbon Nanotubes for Highly Efficient Overall Water Splitting. *Advanced Energy Materials*, *7*(9), 1–8. doi:10.1002/aenm.201602068

Rabizadeh, T., Allahkaram, R. S., & Zarebidaki, A. (2010). An investigation on effects of heat treatment on corrosion properties of Ni–P electroless nano–coatings. *Materials & Design*, *7*(7), 3174–3179. doi:10.1016/j.matdes.2010.02.027

Rana, A. R. K., & Farhat, Z. (2019). Preparation and tribological characterization of graphene incorporated electroless Ni–P composite coatings. *Surface and Coatings Technology*, *369*, 334–346. doi:10.1016/j.surfcoat.2019.04.043

Rosalbino, F., Delsante, S., Borzone, G., & Angelini, E. (2008). Electrocatalytic behaviour of Co–Ni–R (R=Rare earth metal) crystalline alloys as electrode materials for hydrogen evolution reaction in alkaline medium. *International Journal of Hydrogen Energy*, *33*(22), 6696–6703. doi:10.1016/j.ijhydene.2008.07.125

Rosalbino, F., Maccio, D., Saccone, A., Angelini, E., & Delfino, S. (2011). Fe–Mo–R (R= rare earth metal) crystalline alloys as a cathode material for hydrogen evolution reaction in alkaline solution. *International Journal of Hydrogen Energy*, *36*(3), 1965–1973. doi:10.1016/j.ijhydene.2010.11.054

Safizadeh, F., Ghali, E., & Houlachi, G. (2015). Electrocatalysis developments for hydrogen evolution reaction in alkaline solutions–a review. *International Journal of Hydrogen Energy*, *40*(1), 256–274. doi:10.1016/j.ijhydene.2014.10.109

Sahasrabudhe, A., Dixit, H., Majee, R., & Bhattacharyya, S. (2018). Value added transformation of ubiquitous substrates into highly efficient and flexible electrodes for water splitting. *Nature Communications*, *9*(1), 1–14. doi:10.103841467-018-04358-7 PMID:29789569

Sapountzi, F. M., Gracia, J. M., Fredriksson, H. O., & Niemantsverdriet, J. H. (2017). Electrocatalysts for the generation of hydrogen, oxygen and synthesis gas. *Progress in Energy and Combustion Science*, *58*, 135. doi:10.1016/j.pecs.2016.09.001

Shalom, M., Gimenez, S., Schipper, F., Herraiz-Cardona, I., Bisquert, J., & Antonietti, M. (2014). Controlled carbon nitride growth on surfaces for hydrogen evolution electrodes. *Angewandte Chemie International Edition*, *53*(14), 3654–3658. doi:10.1002/anie.201309415 PMID:24574144

Shang, X., Chi, J. Q., Lu, S. S., Dong, B., Liu, Z. Z., Yan, K. L., & Liu, C. G. (2017). Hierarchically three–level $Ni_3(VO_4)_2$@$NiCo_2O_4$ nanostructure based on nickel foam towards highly efficient alkaline hydrogen evolution. *Electrochimica Acta*, *256*, 100–109. doi:10.1016/j.electacta.2017.10.017

Shao, Q., Wang, Y., Yang, S., Lu, K., Zhang, Y., Tang, C., & Li, Y. (2018). Stabilizing and Activating Metastable Nickel Nanocrystals for Highly Efficient Hydrogen Evolution Electrocatalysis. *ACS Nano*, *12*(11), 11625–11631. doi:10.1021/acsnano.8b06896 PMID:30481970

Shao, Y., Shi, X., & Pan, H. (2017). Electronic, magnetic, and catalytic properties of thermodynamically stable two–dimensional transition–metal phosphides. *Chemistry of Materials*, *29*(20), 8892–8900. doi:10.1021/acs.chemmater.7b03832

Shen, Y., Zhou, Y., Wang, D., Wu, X., Li, J., & Xi, J. (2018). Nickel–copper alloy encapsulated in graphitic carbon shells as electrocatalysts for hydrogen evolution reaction. *Advanced Energy Materials*, *8*(2), 1701759–1701765. doi:10.1002/aenm.201701759

Shibli, S. M. A., Anupama, V. R., Arun, P. S., Jineesh, P., & Suji, L. (2016). Synthesis and development of nano WO_3 catalyst incorporated Ni–P coating for electrocatalytic hydrogen evolution reaction. *International Journal of Hydrogen Energy*, *41*(24), 10090–10102. doi:10.1016/j.ijhydene.2016.04.156

Shibli, S. M. A., & Dilimon, V. S. (2008). Development of TiO_2 supported nano–RuO_2–incorporated catalytic nickel coating for hydrogen evolution reaction. *International Journal of Hydrogen Energy*, *33*(4), 1104–1111. doi:10.1016/j.ijhydene.2007.12.038

Shinagawa, T., & Takanabe, K. (2017). Towards versatile and sustainable hydrogen production through electrocatalytic water splitting: Electrolyte engineering. *ChemSusChem*, *10*(7), 1318–1336. doi:10.1002/cssc.201601583 PMID:27984671

Su, J., Yang, Y., Xia, G., Chen, J., Jiang, P., & Chen, Q. (2017). Ruthenium–cobalt nanoalloys encapsulated in nitrogen–doped graphene as active electrocatalysts for producing hydrogen in alkaline media. *Nature Communications*, *8*(1), 14969–14979. doi:10.1038/ncomms14969 PMID:28440269

Subbaraman, R., Tripkovic, D., Strmcnik, D., Chang, K. C., Uchimura, M., Paulikas, A. P., Stamenkovic, V., & Markovic, N. M. (2011). Enhancing hydrogen evolution activity in water splitting by tailoring Li^+–$Ni(OH)_2$–Pt interfaces. *Science*, *334*(6060), 1256–1260. doi:10.1126cience.1211934 PMID:22144621

Sun, Z., Zhang, J., Xie, J., Zheng, X., Wang, M., Li, X., & Tang, B. (2018). High–performance alkaline hydrogen evolution electrocatalyzed by a Ni_3N–CeO_2 nanohybrid. *Inorganic Chemistry Frontiers*, *5*(12), 3042–3045. doi:10.1039/C8QI00905H

Tang, C., Gan, L., Zhang, R., Lu, W., Jiang, X., Asiri, A. M., & Chen, L. (2016). Ternary $Fe_xCo_{1-x}P$ Nanowire Array as a Robust Hydrogen Evolution Reaction Electrocatalyst with Pt–like Activity: Experimental and Theoretical Insight. *Nano Letters*, *16*(10), 6617–6621. doi:10.1021/acs.nanolett.6b03332 PMID:27676264

Tang, S., Wang, C., & Sung, X. L. (2010). A study of corrosion resistance and formation characteristics of electroless Ni–P alloy coating on Al composite with a sample surface pretreatment. *Surface and Coatings Technology*, *205*(1), 43–49. doi:10.1016/j.surfcoat.2010.05.047

Tilak, B. V., Ramamurthy, A. C., & Conway, B. E. (1986) High performance electrode materials for the hydrogen evolution reaction from alkaline media. In *Proceedings of the Indian Academy of Sciences–chemical sciences* (Vol. 97, pp. 359–393). Springer.

Tong, W., Forster, M., Dionigi, F., Soren, D., Roghayeh, S. E., Peter, S., Alexander, J. C., & Pau, F. (2020). Electrolysis of low-grade and saline surface water. *Nature Energy*, *5*(5), 367–377. doi:10.103841560-020-0550-8

Vesborg, P. C., Seger, B., & Chorkendorff, I. B. (2015). Recent development in hydrogen evolution reaction catalysts and their practical implementation. *The Journal of Physical Chemistry Letters*, *6*(6), 951–957. doi:10.1021/acs.jpclett.5b00306 PMID:26262851

Wang, L., Li, Y., Xia, M., Li, Z., Chen, Z., Ma, Z., & Shao, G. (2017). Ni nanoparticles supported on graphene layers: An excellent 3D electrode for hydrogen evolution reaction in alkaline solution. *Journal of Power Sources*, *347*, 220–228. doi:10.1016/j.jpowsour.2017.02.017

Wang, Y., Zhuo, H., Zhang, X., Dai, X., Yu, K., Luan, C., & Gao, F. (2018). Synergistic effect between undercoordinated platinum atoms and defective nickel hydroxide on enhanced hydrogen evolution reaction in alkaline solution. *Nano Energy*, *48*, 590–599. doi:10.1016/j.nanoen.2018.03.080

Xiang, R., Duan, Y., Peng, L., Wang, Y., Tong, C., Zhang, L., & Wei, Z. (2019). Three–dimensional Core@Shell Co@CoMoO$_4$ nanowire arrays as efficient alkaline hydrogen evolution electro–catalysts. *Applied Catalysis B: Environmental*, *246*, 41–49. doi:10.1016/j.apcatb.2019.01.035

Xing, H., Yuan, Y., Huiyuan, Z., Yong, S., Bingxi, L., & Heping, T. (2017). Solar thermochemical hydrogen production using metallic oxides. *Energy Sources. Part A, Recovery, Utilization, and Environmental Effects*, *39*(3), 257–263. doi:10.1080/15567036.2013.878769

Xing, Z., Han, C., Wang, D., Li, Q., & Yang, X. (2017). Ultrafine Pt nanoparticle–decorated Co(OH)$_2$ nanosheet arrays with enhanced catalytic activity toward hydrogen evolution. *ACS Catalysis*, *7*(10), 7131–7135. doi:10.1021/acscatal.7b01994

Yang, L., Zhou, H., Qin, X., Guo, X., Cui, G., Asiri, A. M., & Sun, X. (2018). Cathodic electrochemical activation of Co$_3$O$_4$ nanoarrays: A smart strategy to significantly boost the hydrogen evolution activity. *Chemical Communications*, *54*(17), 2150–2153. doi:10.1039/C7CC09416G PMID:29424845

Yang, W., Zhang, J., Ma, Q., Zhao, Y., Liu, Y., & He, H. (2017). Heterogeneous Re action of SO$_2$ on Manganese Oxides: The Effect of Crystal Structure and Relative Humidity. *Scientific Reports*, *4550*, 1–14.

Yang, Y., Liu, J., Guo, S., Liu, Y., & Kang, Z. (2015). A nickel nanoparticle/carbon quantum dot hybrid as an efficient electrocatalyst for hydrogen evolution under alkaline conditions. *Journal of Materials Chemistry. A, Materials for Energy and Sustainability*, *36*(36), 18598–18604. doi:10.1039/C5TA04867B

Yin, H., Zhao, S., Zhao, K., Muqsit, A., Tang, H., Chang, L., & Tang, Z. (2015). Ultrathin platinum nanowires grown on single–layered nickel hydroxide with high hydrogen evolution activity. *Nature Communications*, *6*(1), 6430. doi:10.1038/ncomms7430 PMID:25728293

Yu, F., Zhou, H., Huang, Y., Sun, J., Qin, F., Bao, J., & Ren, Z. (2018). High–performance bifunctional porous non–noble metal phosphide catalyst for overall water splitting. *Nature Communications*, *9*(1), 2551. doi:10.103841467-018-04746-z PMID:29959325

Zhan, X., Daniil, M., Liu, X., Willard, M., Akolkar, R., & Ernst, F. (2019). Effect of tungsten alloying on magnetic properties of amorphous Ni–P. *Journal of Alloys and Compounds*, *786*, 742–749. doi:10.1016/j.jallcom.2019.01.109

Zhang, B., Wang, H. H., Su, H., Lv, L. B., Zhao, T. J., Ge, J. M., & Chen, J. S. (2016). Nitrogen–doped graphene microtubes with opened inner voids: Highly efficient metal–free electrocatalysts for alkaline hydrogen evolution reaction. *Nano Research*, *9*(9), 2606–2615. doi:10.100712274-016-1147-1

Zhang, B., Zhang, J., Shi, J., Tan, D., Liu, L., Zhang, F., & Han, B. (2019). Manganese acting as a high–performance heterogeneous electrocatalyst in carbon dioxide reduction. *Nature Communications*, *2980*(1), 1–8. doi:10.103841467-019-10854-1 PMID:31278257

Zhang, R., Ren, X., Hao, S., Ge, R., Liu, Z., Asiri, A. M., & Sun, X. (2018). Selective phosphidation: An effective strategy toward CoP/CeO$_2$ interface engineering for superior alkaline hydrogen evolution electrocatalysis. *Journal of Materials Chemistry. A, Materials for Energy and Sustainability*, *5*(5), 1985–1990. doi:10.1039/C7TA10237B

Zhang, T., Wu, M. Y., Yan, D. Y., Mao, J., Liu, H., Hu, W. B., & Qiao, S. Z. (2018). Engineering oxygen vacancy on NiO nanorod arrays for alkaline hydrogen evolution. *Nano Energy*, *43*, 103–109. doi:10.1016/j.nanoen.2017.11.015

Zhang, X., Gu, W., & Wang, E. (2017). Wire–on–flake heterostructured ternary $Co_{0.5}Ni_{0.5}$ P/CC: An efficient hydrogen evolution electrocatalyst. *Journal of Materials Chemistry. A, Materials for Energy and Sustainability*, *5*(3), 982–987. doi:10.1039/C6TA08094D

Zheng, L., Ma, S., Wang, Z., Zhang, Y. Q., Xu, X., & Chen, X. (2019). Ni–P nanostructures on flexible paper for morphology effect of nonenzymatic electrocatalysis for urea. *Electrochimica Acta*, *320*, 134586. doi:10.1016/j.electacta.2019.134586

Zheng, Y., Jiao, Y., Zhu, Y., Li, L. H., Han, Y., Chen, Y., & Qiao, S. Z. (2016). High electrocatalytic hydrogen evolution activity of an anomalous ruthenium catalyst. *Journal of the American Chemical Society*, *138*(49), 16174–16181. doi:10.1021/jacs.6b11291 PMID:27960327

Zhou, Q., Li, T. T., Qian, J., Hu, Y., Guo, F., & Zheng, Y. Q. (2018). Self–supported hierarchical CuO_x@ Co_3O_4 heterostructures as efficient bifunctional electrocatalysts for water splitting. *Journal of Materials Chemistry. A, Materials for Energy and Sustainability*, *29*(29), 14431–14439. doi:10.1039/C8TA03120G

Zhou, W., Lu, X. F., Chen, J. J., Zhou, T., Liao, P. Q., Wu, M., & Li, G. R. (2018). Hierarchical Porous Prism Arrays Composed of Hybrid Ni–NiO–Carbon as Highly Efficient Electrocatalysts for Overall Water Splitting. *ACS Applied Materials & Interfaces*, *45*(45), 38906–38914. doi:10.1021/acsami.8b13542 PMID:30360101

ADDITIONAL READING

Ameen Sha, M., Liju, E., Riyas, A. H., Bhagya, T. C., Meera, M. S., & Shibli, S. M. A. (2020). Surface engineering of CeO_2–TiO_2 composite electrode for enhanced electron transport characteristics and alkaline hydrogen evolution reaction. *International Journal of Hydrogen Energy*, *45*(27), 13789–13804. doi:10.1016/j.ijhydene.2020.03.048

Chinchu, K. S., Riyas, A. H., Ameen Sha, M., Geethanjali, C. V., Viswanathan, S. S., & Shibli, S. M. A. (2020). ZrO_2–CeO_2 assimilated electroless Ni-P anti-corrosion coatings. *Surfaces and Interfaces*, *21*, 100704. doi:10.1016/j.surfin.2020.100704

Shibli, S. M. A., & Ameen Sha, M. (2018). Development and characterization of electro active CeO_2–RuO_2 mixed oxide and its role in alkaline hydrogen evolution reaction. *Journal of Alloys and Compounds*, *749*, 250–261. doi:10.1016/j.jallcom.2018.03.274

Shibli, S. M. A., Ameen Sha, M., Anisha, B. L., Ponnamma, D., & Sadasivuni, K. K. (2018). Effect of phosphorus on controlling and enhancing electrocatalytic performance of Ni–P–TiO_2–MnO_2 coating. *Journal of Electroanalytical Chemistry (Lausanne, Switzerland)*, *826*, 104–116. doi:10.1016/j.jelechem.2018.08.021

Shibli, S. M. A., Chinchu, K. S., & Ameen Sha, M. (2019). Development of Nano–tetragonal Zirconia–Incorporated Ni–P Coatings for High Corrosion Resistance. *Acta Metallurgica Sinica. English Letters*, *32*(4), 481–494. doi:10.100740195-018-0823-4

Shibli, S. M. A., Riyas, A. H., Ameen. Sha, M., & Rajee, M. (2017). Tuning of phosphorus content and electrocatalytic character of CeO_2–RuO_2 composite incorporated Ni–P coating for hydrogen evolution reaction. *Journal of Alloys and Compounds*, *696*, 595–603. doi:10.1016/j.jallcom.2016.11.238

Sumi, V. S., Ameen Sha, M., Arunima, S. R., & Shibli, S. M. A. (2019). Development of a novel method of NiCoP alloy coating for electrocatalytic hydrogen evolution reaction in alkaline media. *Electrochimica Acta*, *303*, 67–77. doi:10.1016/j.electacta.2019.02.063

Sumi, V. S., Arunima, S. R., Deepa, M. J., Ameen Sha, M., Riyas, A. H., Meera, M. S., Viswanathan, S. S., & Shibli, S. M. A. (2020). PANI-Fe_2O_3 composite for enhancement of active life of alkyd resin coating for corrosion protection of steel. *Materials Chemistry and Physics*, *247*, 122881. doi:10.1016/j.matchemphys.2020.122881

Sumi, V. S., Meera, M. S., Ameen Sha, M., & Shibli, S. M. A. (2020). Effect of rGO on Fe_2O_3-TiO_2 composite incorporated NiP coating for boosting hydrogen evolution reaction in alkaline solution. *International Journal of Hydrogen Energy*, *45*(4), 2460–2477. doi:10.1016/j.ijhydene.2019.11.167

KEY TERMS AND DEFINITIONS

Electrocatalyst: A catalyst which can be used for electrochemical reactions to increase the rate of the reaction.

Electrode/Electrolyte Interface: The contact region of the electrode surface and electrolyte which extend to certain nanometer distance.

Electroless Deposition: Deposition of the atoms or species without the assistance of electricity.

Electrolyzer: An electrolyte or medium used for electrolysis.

Functional Material: A material with specific characteristics which have impact on selected reaction/s.

HER (Hydrogen Evolution Reaction): The term HER means electrolytic water splitting to produce hydrogen.

Ni-P Alloy: Nickel and phosphorus containing alloy.

Chapter 4
Overview Regarding Synthetic Gas Production by Biomass Gasification

Alin Corbu
National Research and Development Institute for Cryogenic and Isotopic Technologies – INC-DTCI-ICSI Râmnicu Vâlcea, Romania

Maria Simona Raboaca
https://orcid.org/0000-0002-7277-4377
National Research and Development Institute for Cryogenic and Isotopic Technologies, Râmnicu Vâlcea, Romania

Sudeep Tanwar
Institute of Technology, Nirma University, India

ABSTRACT

The renewable energy for the production of syngas (synthetic gas) by gasification of biomass has specific features and objectives. The present chapter aims to describe the specific features of renewable energy for the production of syngas by gasification of biomass. The specific objective refers to the study of the equipment for the production of synthetic gaseous fuels (syngas) by gasification of biomass. The "biomass" term can be defined generically as the biodegradable part of agricultural products, residues and waste, gathering both animal and plant substances, wood industry, and also the biodegradable component of urban and industrial waste. Biomass sources are the most found renewable resources on Earth, also including all the organic components resulted from the organism's metabolic process, being the first resource of energy man has used, since the discovery of fire.

DOI: 10.4018/978-1-7998-4945-2.ch004

INTRODUCTION

The renewable energy for the production of syngas (synthetic gas) by gasification of biomass has specific features and objective.

The present chapter aims to describe the specific features of renewable energy for the production of syngas (synthetic gas) by gasification of biomass. The specific objective refers to the study of the equipment used in the production of syngas (synthetic gaseous fuels) by gasification of biomass.

The "**Biomass**" term can be defined generically as the biodegradable part of agricultural products, residues and waste, gathering both animal and plant substances, wood industry and also the biodegradable component of urban and industrial waste. (Definition contained in Decision no. 1844 of 2005 on Promoting the use of biofuels and other renewable fuels). Biomass sources are the most found renewable resources on Earth, also including all the organic components resulted from the organsims metabolic proces, being the first resource of energy man has ever used, since the discovery of fire. *(Nunes, L. J. R., Causer, T. P., & Ciolkosz, D., 2020; Kargbo, H., Harris, J. S., & Phan, A. N., 2021; Pinales-Márquez, C. D., et.al., 2021)*

Biomass gasification can be defined generically as a thermochemical conversion of organic matter into a gas product (syngas), by using the process of partial oxidation. For this conversion to occur, special conditions are required such as high temperature and low air circulation, in order to prevent the complete biomass burning, but enough for the conversion. *(Chambon, C. L., et.al., 2020)*

By definition, **SYNGAS** is the final product of the process of gasification of organic solid matter as a result of its thermal decomposition. (*AlNouss, A., McKay, G., & Al-Ansari, T., 2020).*

BIOMASS ENERGY

General Aspects

Biomass accumulated in chemical form in matter of animal or plant origin represents some of the most spread resources on the planet. The main use is in the food industry, but is also supplies energy, building materials, paper, chemicals and medicines. Biomass resources have been converted into energy ever since the discovery of fire. The biomass notion implies multiple categories of products and by-products, such as waste coming from wood and agricultural industries and also, municipal waste *(Wang, Z., Bui, Q., & Zhang, B., 2020, a).*

According to European Union legislation, *"biomass represents the biodegradable fraction of waste products and agricultural residues (including vegetable and animal substances), forestry and related industries, as well as the biodegradable fraction of municipal and industrial waste"* .

Biomass is considered one of the basic renewable resources of the future that can be used on a small and large scale. For 3/4 of the world's population from developing countries, biomass is the most used energy source. The green plants, which are the basis of life on the planet, convert water and carbon dioxide found in the air into oxygen and organic matter using solar energy. Water from earth and carbon dioxide from the air are combined using photosynthesis and the result is represented by carbohydrates that form the constituent elements of biomass. *(Wang, Z., et.al, 2020,b)* The energy from the sun is stored through photosynthesis in the biomass` chemical bonds. Burning the biomass leads to the combination of oxygen from air with the plant carbon in order to produce water and carbon dioxide. *Gao, J., & Zhang,*

L., 2021). Given that the carbon dioxide is absorbed again by the plants, we can say that the process is cyclical (see Figure 1).

Figure 1. Carbon cycle in nature

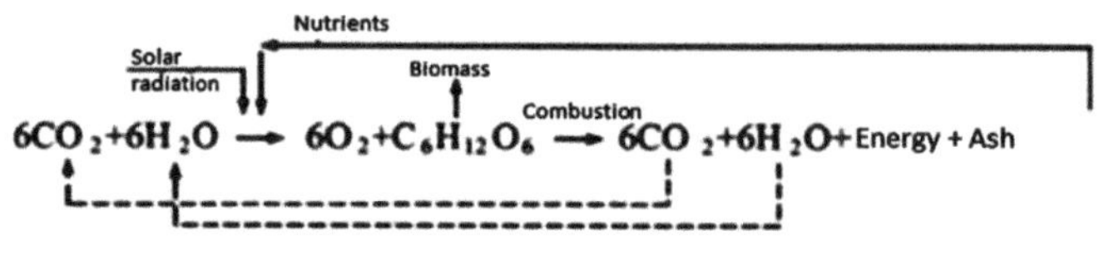

Fossilized biomass has been exploited as coal by man in the past hundred years. This fossilized biomass is the result of a chemical transformation occuring in a very long period of time, which converts the carbohydrate polymer fraction into a ligning fraction resembling chemical composition. Therefore, the supplementary chemical bonds in the carbons make it a fuel and a source of energy with higher concentration. Considering their time of formation, all fossil fuels – natural gas, coal and fuel oil are very old biomass. It is a process taking millions of years to complete, in which the planet burries its plants and turns them into valuable fuel. Even though fossil fuels have in their composition fresh biomass also, considering the long formation time, they are not defined as renewable sources of energy. *(Choi, S. H., & Manousiouthakis, V. I., 2020; Patuzzi, F., et.al., 2021; Drake, H., et.al., 2021)*

Different species of biomass have a varying chemical composition, but it can be stated that they contain in the dry state approximately 15-30% carbohydrates (glucides or sugars) and ligning ($C_{40}H_{44}O_6$). The carbohydrate component consists of several molecules of glucide connected together in polymers or long chains. The two main categories of carbohydrates are: cellulose ($C_6H_{10}O_5$, 40-45%) and hemicellulose (20-35%). The ligning component consists of molecules other than glucides. Cellulose in the form of long polymers is exploited by nature for building fibers which offers plant solidity. *(De, S., et.al., 2020; Bao, D. X., Yu, Y. Y., & Zhao, Q. M., 2020; Mudryk, K., et.al., 2021)*

Biomass has many advantages as an energy source. Biomass used as an energy source has a great economic and social impact. It can be used both for the production of electricity and heat, as well as for producing a wide range of products: liquid, solid and gaseous fuels, but also other products. Raw biomass can be found in abundance everywhere in different forms. In the last years new technologies have been developed in order to efficiently convert biomass into fuel. Even so, there are some biomass resources which cannot be used for energy purposes. Instead, they can be used for paper and chemicals, food or timber industries. This is why other priority utilizations must be considered when using biomass as an energy source.

Types of Biomass: Classification

Biomass is represented by: agricultural and forestry products and their vegetal residues, vegetable wate coming from the food industry, from cellulose and paper production from cellulose, cork waste, wood waste from constructions and demolitions etc. *(Vassilev, S. V., et.al, 2013; Lazzari, E., et.al., 2018; Tursi, A.,2019; Toro Delgado, J. A., 2020)*. (see Figure 2).

Figure 2. Types of solid organic fuels – biomass

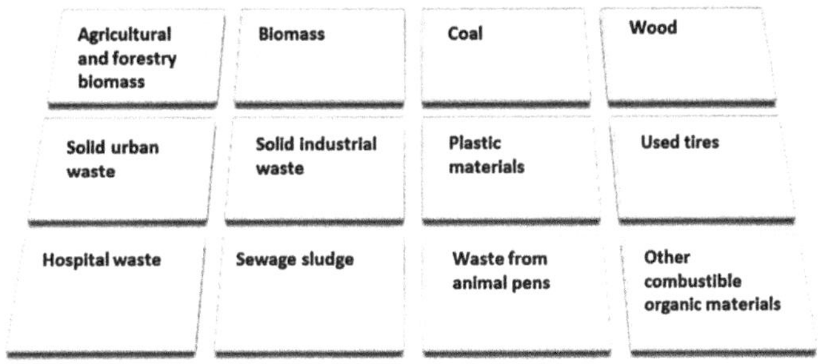

There is a wide variety of biomass resources (the main source being wood), as follows:

- energy crops:
  - trees with high growth: eucalyptus, willow, poplar;
  - agricultural plants: wheat, barley, rye, maize, rape, sugar cane, sugar beet, etc .;
  - perennial plants: miscanthus;
  - herbaceous plants with fast growth: Panicum virgatum, Switchgrass
- residues:
  - wood from the grooming of trees and constructions;
  - straw and cereal stems;
  - other residues from the processing of some food products (coffee, tea, olives, walnuts, sugar cane).
- waste and by-products from:
  - paper;
  - wood: talas, sawdust;
  - used animal and vegetable fats;
  - the organic component of urban waste.
- methane coming from manure and plant treatment wastewater.

Depending on origin, biomass can be classified as: primary, secondary, residual and fossil. *(Vassilev, S. V., et.al., 2010)*

Primary biomass is produced by the photosynthesis activity of plants, representing the ensemble of plant raw materials, with more or less rapid growth, used directly, or following a conversion process, in human nutrition, forage, different industries or for production. energy.

Secondary biomass is produced by heterotrophic beings, those using primary biomass, namely herbivorous and omnivorous animals. Of great importance are the residual products from the industrial or livestock farming activities.

Residual biomass is produced in human activities: straw, sawdust, waste from slaughterhouses, urban waste, etc.

Fossil biomass is represented by natural gas, oil and coal.

From a residue point of view (waste), biomass can be classified as follows: primary residues; secondary residues; tertiary residues.

Primary residues are produced from plants or from forest products. Such biomass is found "in the field" and must be gathered for later use.

Secondary residues are obtained through biomass processing for the production of food or other wood products, being available in paper mills, food industry, etc.

Tertiary residues become available after a biomass product has been used. They represent different wastes, including household waste, wood waste, waste water treatment, etc.

Energy biomass is the organic matter used to obtain biofuels (biofuels), electricity and heat.

Calorific Power of Biomass

Fuel characteristics of biomass. Research in the field has shown that biomass has a number of fuel characteristics, which determine its competing qualities in the energy market. Biomass as an energy resource is exploited and evidenced by the combustion plants adopted when burning different combustible materials.

From the use for the production of thermal energy point of view biomass is characterized by two types of analysis: thermogravimetric analysis in which the mass ratios of: volatile, fixed carbon and ash are highlighted; as well as the detailed chemical analysis in which the mass ratios of: carbon, oxygen, hydrogen, nitrogen, sulfur, chlorine and minerals are highlighted. The chemical composition of the biomass can vary to a large extent.

Multiple attempts have been made in order to create a correspondance between caloric power and composition. Lignin has a higher calorific value than cellulose due to its high oxidation degree. Other compounds, such as low oxidation hydrocarbons, increase the calorific power of biomass. The biomass calorific power is highly linked to the content of lignin. Therefore, the higher calorific power of a dry and devoid sample can be computed using the relation:

$$Qs = 88.9 \cdot (LC) + 16{,}821.8,\ kJ/kg \tag{1}$$

where: (LC) is the content of lignin related to the ash-free and dry state, %.

The higher calorific value of biofuels is computed according to the content of fixed cafbon, Cf (%) using the relation below:

$$Qs = 196\ Cf + 14.119,\ kJ/kg \tag{2}$$

In the current studies, fuel heat estimation formulas for different vegetable oils or lignin and cellulose materials have been developed as a function of their chemical composition. The Dulong modified formula can be used for solid biofuels, considering the content of carbon, C (%), oxygen, O(%), hydrogen, H (%) and nitrogen, N (%):

$$Qs = 33{,}500\ C + 142{,}300\ H - 15{,}400\ O - 14{,}500\ N,\ kJ/kg \tag{3}$$

The following relation is also described in current literature:

$$Qsanh = 349.1 \cdot Canh + 1{,}178.3 \cdot Hanh + 100.5 \cdot Sanh$$

$- 15.1 \cdot Nanh - 103.4 \cdot Oanh - 21.1 \cdot Aanh, kJ / kg$ (4)

where: Canh, Hanh, Sanh, Nanh, Oanh, Aanh are the percent contents of carbon, hydrogen, sulfur, nitrogen, oxygen and ash, respectively, related to the anhydrous state.

In Figure 3 the superior calorific values of the most used solid fuels are illustrated.

Figure 3. Calorific power of biomass

Biomass	Upper calorific power for anhydrous state, KJ/kg
Alfalfa stalks (clover)	18 400
Almond shells	19 400
Cotton stalks	15 800
Ground nut shells	15 700-20 000
Olive seeds	21 400
Sunflower seed shells	16 120
Sunflower stalks	21 800
Walnut shells	21 100
Manure	14 800
Mangal	31 800
Vegetable waste	12 600
Wheat straw	17 200-18 900
Rice straw	15 200
Rice seed shells	15 500-19 800
Wood	15 500
Corn stalks	15 700-16 200
Corn cobs	17 400
Tobacco stalks	16 400 (7% Humidity)
Grapevine strings	16 500 (7% Humidity)
Apple branches	15 200 (7% Humidity)

Analysis and Estimation of Biomass Production

Global Biomass Potential

The RES scenario (see Figure 4) shows that biomass brings the largest contribution to the total energy consumption - 80%, and by 2040 the biomass consumption will increase by about 60% compared to 2010. *(Popescu, A. I., & Necula, C.,2019*; Brás, I., Silva, *E., & De Lemos, L. T., 2020).*

But the use of biomass will decrease by 2040 by almost 50% due to the significant development of photovoltaic and wind energy. The European Union has officially adopted the Renewable Energy Directive 20-20-20, on the basis of which the Member States must ensure a growth of the contribution of

Figure 4. RES Scenario

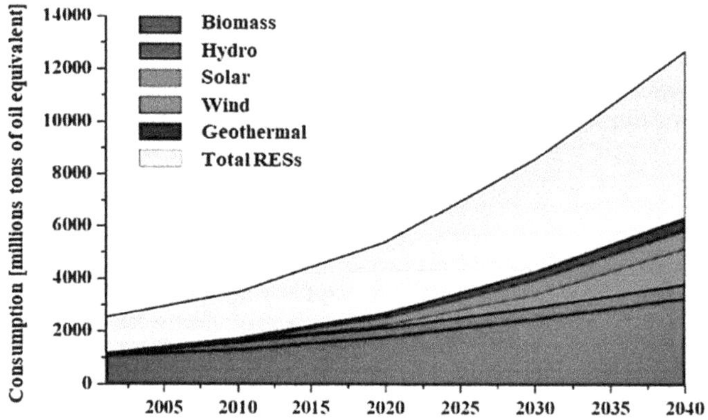

renewable energy sources in the supply of energy by 20% by 2020. For biomass the target is to contribute 25% in energy supply by 2020. Biomass has many benefits compared to classical fossil resources, namely: unlimited reserves, over 65 billion tons per year, neutral CO_2 emissions through plant growth (through photosynthesis), low NOx and SO_2 emissions due to the low nitrogen and sulfur content of plants, the possibility of producing not only electric and thermal energy, but also biofuels and synthetic chemical compounds through thermochemical conversion technologies - combustion, pyrolysis, liquefaction, gasification or biochemistry - anaerobic and aerobic fermentation, hydrolysis, transesterification. *(Bar-On, Y. M., Phillips, R., & Milo, R., 2018).*

In most member countries, targets for the replacement of conventional fuels with biofuels obtained from biomass have been set. By the end of 2010, 5.75% of the energy used in the transport sector had to be supplied by biofuels, in countries such as Austria, Germany, France, Sweden, Spain, Italy, England. Even if in some countries the proposed target could not be achieved, by 2020, the European Council has set a new target of 10%.

Biofuels obtained from biomass conversion fall into two broad categories: primary biofuels, used in unprocessed form to generate electricity and heat, respectively secondary biofuels resulting from conversion processes. Different scenarios predict that by 2030 biofuel production will increase 28 times compared to 2000 production, mainly due to biofuels of generation I and II.

Figure 5. Energy potential of biomass for Romania

Biomass Type	UM	Technical potential	Economically adjustable potential
Vegetable biomass (Forestry + Agricultural Crops)	TJ/year	471 000	289 500
	Thousands toe/year	11 249	6 915
Biogas	TJ/year	24 600	14 800
	Thousands toe/year	587	353
Urban waste – Bio fraction	TJ/year	22 800	13 700
	Thousands toe/year	544	327
Total	TJ/year	518 400	318 000
	Thousands toe/year	12 382	7 595

Biomass in Romania

«The study on the evaluation of the current energy potential of renewable energy sources in Romania» *(Enescu, A., et.al., 2019; Popescu, A. I., & Necula, C.,2019)*, indicates certain potentials of energy (Figure 5), respectically economically amenable for biomass:

Biomass potential for lots, regions and total is given in the following Figure 6.

Figure 6. Biomass potential of Romania

No.	Region	Forestry Biomass thousand tons/year TJ	Wood waste thousand tons/year TJ	Agric2ultural Biomass thousand tons/year TJ	Biogas million m^3/year TJ	Urban waste thousand tons/year TJ	TOTAL TJ
I	Delta Dunarii	-	-	-	-	-	-
		-	-	-	-	-	
II	Dobrogea	54	19	844	71	182	29987
		451	269	13422	1477	910	
III	Moldova	166	58	2332	118	474	81357
		1728	802	37071	2462	2370	
IV	Carpati	1873	583	1101	59	328	65415
		19552	8049	17506	1231	1640	
V	Platoul Transilvaniei	835	252	815	141	548	43757
		8721	3482	12956	2954	2740	
VI	Campia de Vest	347	116	1557	212	365	60906
		3622	1603	24761	4432	1825	
VII	Subcarpatii	1248	388	2569	177	1314	110198
		13034	5366	40849	3693	6570	
VIII	Campia de Sud	204	62	3419	400	1350	126639
		2133	861	54370	8371	6750	
TOTAL		4727	1478	12637	1178	4561	518439
		49241	20432	200935	24620	22805	

Current Context of Biomass Valorification

The current and most important priority is to efficiently valorificate biomass local resources by implementing feasible projects. Currently, wood biomass is traditionally used in Romania as firewood (due to the lack of natural gas supply system), which is utilised for food preparation and heating. The primary source is represented by low efficiency (approx. 20%) combustion plants with a high degree of pollution. Support through national / regional programs is needed in ordert to make the transition to modern boilers having an adequate efficiency (> 75%). *(Leng, L., et.al., 2020; Song, B., et.al., 2021; Hameed, Z., et.al., 2021)*

Another important aspect for extending the use of biomass in individual energy production applications is the provision of biomass sources (briquettes or pellets) at reasonable costs. In Romania, due to the fact that producers have ensured the export of pellets at very high costs, they practice practice prices far beyond the purchasing capability of domestic clients. The reason behind this discrepancy on the local market is the contrast between the cost of natural gas, which is the primary energy substitute of wood biomas, in Romania and in other European countries such as Austria, Germany or Italy. *(Pablo, G., et.al., 2020; Panaitescu, C., et.al., 2020; Negreanu-Pirjol, B. S., et.al., 2020; BACĂU, C., et.al., 2021; Tripa, S., & Indrie, L., 2021).*

Solutions to this situation can be in two directions: either the natural gas price alignment is expected - but with the disadvantage of affecting the mean purchasing power from Romania, or it will be used for integrated, superior and efficient biomass production / recovery systems locally.

TECHNOLOGIES FOR VALORIFICATION OF BIOMASS

General Aspects

Most of the biomass that is available for bioenergy projects comes in the form of unprocessed solid plant material, with an overall moisture content of about 50%. Human activity is associated with numerous resources of biomass: in particular, waste and residues from agriculture, the industrial sector, municipal waste, forestry and other industrial domains. All these biomass resources undergo processing using various technologies: direct combustion (electricity and / or heat generation, cogeneration plants), fermentation (sugars for alcohol, bioethanol), anaerobic digestion (cogeneration, for methane rich gas), oil extraction (bioethanol), pyrolysis (for manganese, gas and oils)) and gasification (for carbon monoxide CO and hydrogen rich *syngas*) (Figure 7, Figure 8, Figure 9).

Burning biomass is the easiest way to produce heat. This technology is called direct combustion. There are also other methods for converting biomass to energy, such as: combined combustion, gasification and modular systems.

Direct combustion or combustion: biomass undergoes a burning process inside a boiler in order to generate steam under pressure. The steam is then used to actuate a turbine connected to a power generator to produce electricity. This has a yield $\leq 30\%$.

Biomass gasification is currently considered as an alternative to conventional fuels. In the gasification process wood, coal and other materials belonging to the biomass category are used as raw material. Essentially, the gasification process consists of converting solid biomass into carbon oxide mainly through a thermochemical process. *(Schill, G. P., et.al., 2020; Pan, X., et.al., 2020; Casquero-Vera, J. A., et.al., 2021; Pokhrel, R. P., et.al., 2021)*

Figure 7. Biomass conversion processes

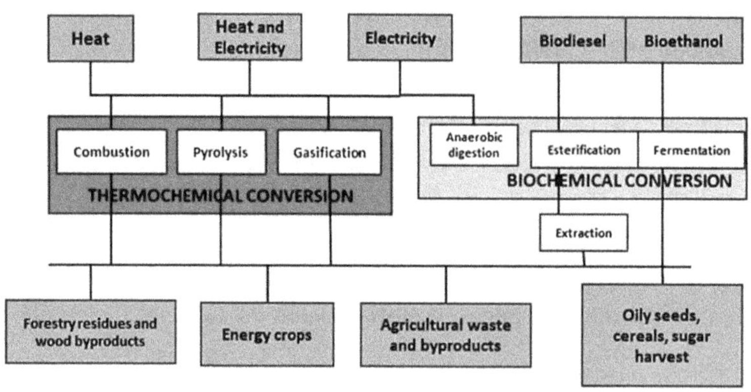

Figure 8. Diagram of the technologies for converting biomass into energy

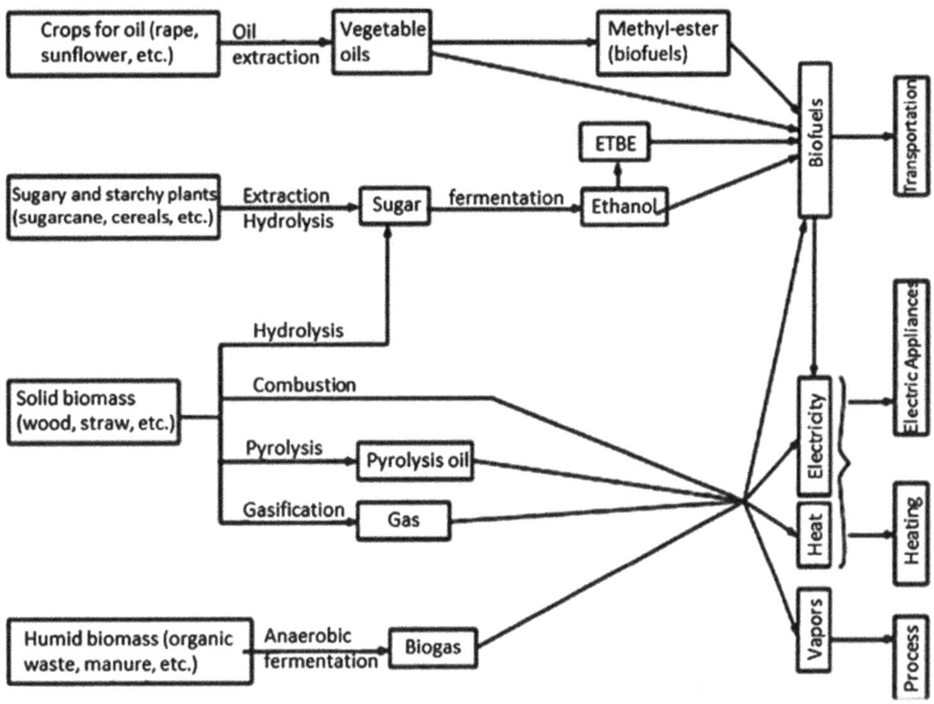

Figure 9. Methods for obtaining biofuels

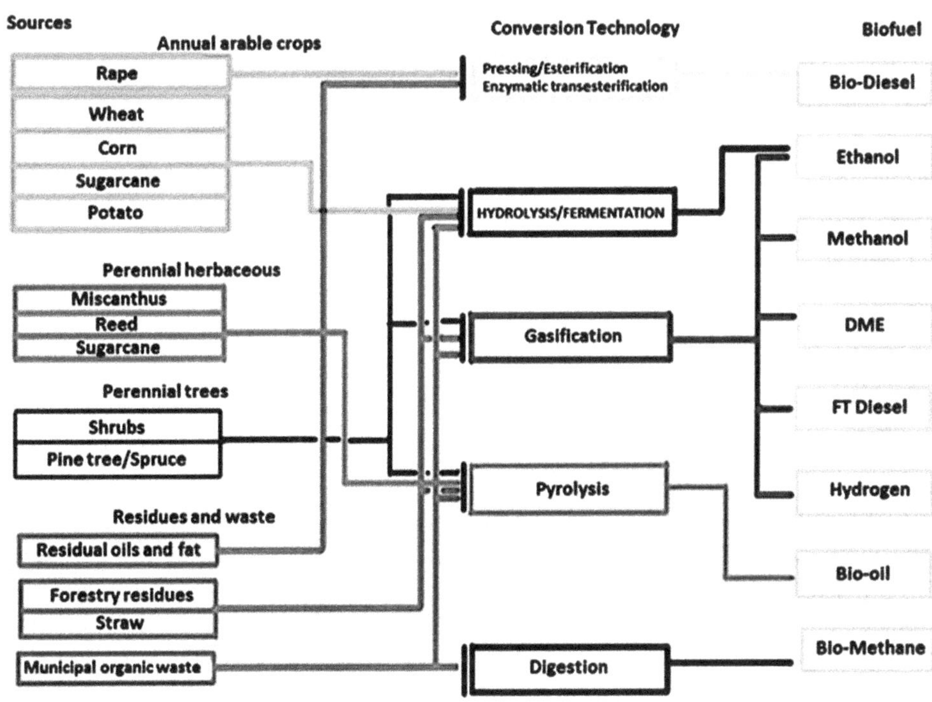

Solid biomass gasification takes place in enclosures at pressures slightly lower or equal to atmospheric pressure. The gas can be cleaned of impurities and used as a less polluting fuel for the operation of gas and steam turbines, internal combustion gas engines and / or for direct combustion in boilers replacing natural gas. *(Situmorang, Y. A., et.al., 2020)*

The superiority of this method is given by the fact that the decentralized energy conversion systems can operate efficiently even at small sizes. It has a yield $\leq 75\%$.

Combined combustion: part of the coal used by a coal-fired power station is replaced by biomass. Combined combustion is cheaper as no new or additional technology is required to incorporate biomass into coal. This has a yield $\leq 85\%$. *(Bajwa, D. S., et.al., 2018)*

For the production of thermal energy from biomass, the following procedures are currently used:

- burning in the layer in hot water boilers;
- gasification and combustion in the same enclosure, in hot water boilers;
- gasification in gas equipment and combustion in a heat exchanger.

Each process has advantages and disadvantages and is used depending on the specific application. To heat the air in convective dryers, where biomass is used as fuel, heat exchangers of the type are used: hot-air water; combustion-air gases. *(Destek, M. A., 2017).*

Biomass Combustion Technologies

Combustion is the oldest and most widely used biomass energy production technology. The efficiency of conversion into electricity is 20-25%. The biomass resources can be burned directly or toghether with coal (also known as co-firing). Co-firing is used in current boilers in order to diminish the carbon dioxide emissions. When designing the combustion system, there are some aspects to be considered: the characteristics of the resource to be burned, the regulations regarding the environment, the price and the capability of the convenient equipment. During combustion, the biomass goes through several phases. First the drying takes place, up to temperatures of 100ºC, then as the heating continues, pyrolysis and / or gasification takes place, followed by the actual combustion and liquefaction.

The limiting humidity of the biomass has to be lower than 60% for the combustion to be mantained. Moisture is the most significant characteristic of the biomass based on which the entire design of the combustion installation and process depend on. *(Li, Z., et.al., 2020)*

Biomass is different from coal in terms of the content of organic and inorganic matter, calorific power and physical characteristics. Biomass generally contains less iron, aluminium and carbon and more silica, oxygen and potassium, it has reduced friability, high content of water, low heat value and density. Biomass burning involves changing the combustion process in any plant, due to the composition of the biomass, especially the content of volatile substances. The calorific power of the biomass is considerably smaller than the calorific power of coal due to its high moisture and oxygen percent. *(Li, Y., et.al., 2017)*

It is highly advisable that solid biomass that will be used in industrial and commercial plants or for domestic use to undergo a pretreatment process such as: to be washed and dryed, reduced in size and compacted in order to obtain a greater uniformity, making it easier handling and reducing humidity to an acceptable level.

The most widely used solid biofuel is wood, which, as a raw material, can be found as: stems, logs, forest needles and leaves, sawdust, bark, furrows and logs coming from wood plants or recovered on

Figure 10. Home combustion installation with fixed grill

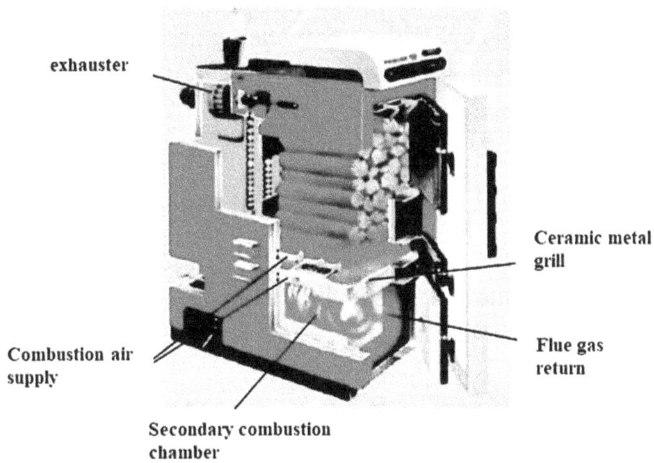

the construction sites. They can be used when directly possible as a fuel, or they can be processed into forms that are easier to transport, store and burn such as: pellets, briquettes and wood dust. *(Proskurina, S., et.al, 2019; Brand, M. A., et.al., 2021)*

Schematically, in Figure 10 and Figure 11, a domestic wood biomass combustion plant and step combustion reactions are presented.

Figure 11. Schematic of combustion in steps

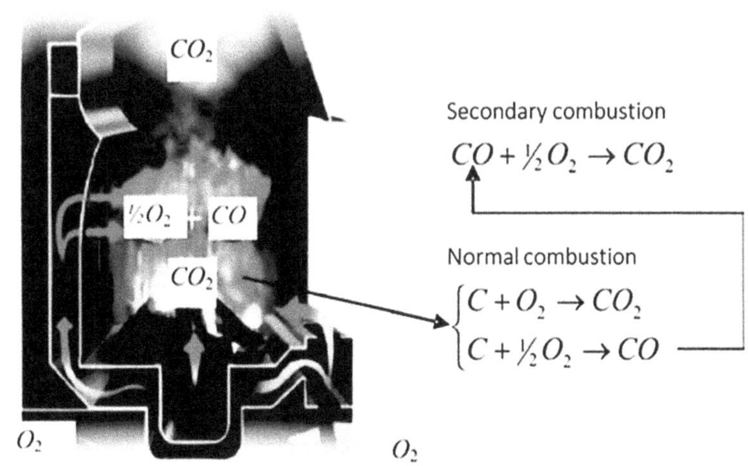

Firewood is forest fuel in the form of a tree trunk treated or not. The pellets are obtained by milling the sawdust, the bark, the chips, the stems and compacting the dust from the grinding process through a matrix. The friction heat is sufficient for ligning to be softened. In the tempering process, lignin turns to a solid structure and binds to the material. The form of the pellets can be spherical or cylindrical having a width lower than 25 mm.

The shape of the briquettes can be cylindrical or rectangular and it can be achieved by compacting chips, sawdust, bark or stumps in a cylinder compactor. Briquettes and pellets have an energy efficiency of approximately 17 GJ/ton with a humidity of 10% and density of 600-700kg/m^3. *(Sonkhaskar, Y. M., et.al., 2021; Adeleke, A. A., et.al., 2021)*

Gasification of Biomass

Usually solid fuel has in its composition elements like oxygen, carbon and hydrogen. In classical gasification the biomass is heated by combustion. There are four different processes in the reactor: drying, pyrolysis, oxidation and reduction.

In modern processes, usually the biomass drying is performed outside the reactor. There are, however, recent achievements, in which drying occurs in the reactor, at high speed, and the vapors produced are then used in the reduction process. Rapid pyrolysis technologies are known which result in liquid fuel, bio-oil and gasification technologies which result in gaseous fuels - synthetic or syngas.

Gasification is achieved by the breakdown of hydrocarbons (organic vapors) into synthetic gas (syngas) with careful control of the amount of oxygen present in the process. *(Karl, J., & Pröll, T., 2018; Shahabuddin, M., et.al., 2020; Boukis, N., & Stoll, I. K., 2021)*

The chemical reactions that take place are:

- Exothermic reactions:

Combustion: $\{\text{volatile biomass + coal / char}\} + O_2 \rightarrow CO_2$

Partial oxidation: $\{\text{volatile biomass + coal / char}\} + O_2 \rightarrow CO$

Methanization: $\{\text{volatile biomass + coal / char}\} + H_2 \rightarrow C_2H_2$

Water-gas switching: $CO + H_2O \rightarrow CO_2 + H_2$

CO methanization: $CO + 3H_2$

- Endothermic reactions:

Steam - Carbon: $\{\text{volatile biomass + coal / char}\} + H_2O \rightarrow CO + H_2$

Boudouard reaction: $\{\text{volatile biomass + coal / char}\} + CO_2 \rightarrow 2CO$.

The reactions occur at very high temperatures (500-1400ºC) and atmospheric pressure or at high pressures up to 33 bar.

The oxidant used may be pure oxygem, steam, air or a mixture. Air-based gasification typically produces high nitrogen content and therefore low heat output (4-6 MJ / m^3). Gasification based on the use of pure steam or oxygen (in the absence of air) produces a sunburn with a much higher calorific value (10-20 MJ / m^3).

Coal, natural gas or biomass (biomass conversion technology) can be used for synthesis. The first two technologies have been developed on an industrial scale since the beginning of the 20th century in Germany - Ruhrchemie, USA - Carthage Hydrocol, South Africa - SASOL, PetroSA, Malaysia - SHELL,

Qatar - Oryx GtL. The biomass conversion technology is developed only on a research-development scale in laboratory and pilot stations in Germany (Forschungszentrum Institute in Karlsruhe, companies Choren GmbH and Cutec GmbH in Freiberg and Clausthal-Zellerfeld), Austria (Technical University of Vienna, Institute of Vienna Research from Güssing), Finland (Neste Oil and Stora Enso companies from Neste Oil-Stora Enso from Vakaus) and USA (North Carolina Triangle Research Institute, Illinois Gas Technology Institute).

For gasification of biomass at temperatures above 800-1000 °C, the following were developed:

- *direct gasification reactors* - through a gasification agent that comes into contact and oxidizes biomass, with the formation of the energy required for the process. Direct gasification can be carried out in fixed layer reactors (countercurrent or equicurrent) or fluidized layer reactors. Countercurrent gasification reactors are the most popular reactors for thermochemical conversion of biomass. The principle of the reactor is to feed the biomass at the top, while the gasification agent (air, oxygen or steam) is introduced at the bottom, in counter current with the solid fuel. As pyrolysis, reduction and combustion reactions take place, the synthesis gas produced moves to the bottom of the reactor. This type of reactors have thermal efficiency but they tend to form slag. In the case of reactors in the equicenter, the solid phase and the oxidizing agent move in the same direction, the synthesis gas produced being discharged at the bottom of the reactor together with the ash. These types of reactors are effective for small biomass gasification plants, while for large installations the reactors used are in fluidized layer, in which the biomass is fluidized due to the oxidizing agent being introduced at high speed at the bottom of the reactor. The main advantage of this type of reactor is the possibility of using an active catalytic material that streamlines the gasification process.
- *indirect gasification reactors* - uses external energy source to provide the thermal energy required for biomass gasification. One such type of reactor is the double fluidized layer consisting of two reaction zones: the gas-fluidized, vaporized zone, in which the biomass is fed together with the active catalytic material (dolomite, olivine, calcite, silica, Ni catalyst or Fe on olivine support), dry and thermochemically converted into silage, respectively the combustion zone in which the coal remaining together with the catalytic material are transported to the combustion zone, where in the presence of air the combustion takes place. The released thermal energy is stored by the catalytic material which is separated from the combustion gas by a cyclone and fed back into the gasification zone.

For the purification of the synthesis gas resulting from the biomass gasification, both low temperature processes are used in the presence of solvents that absorb toxic compounds such as NH_3, HCN, H_2S, COS, HCl, and processes at high temperatures, in the presence of adsorbents such as ZnO, CuO or activated carbon.

- *reactors used in Fischer-Tropsch synthesis*. The choice of reactor for this type of synthesis has an influence on the operating parameters, catalytic performance, carbon monoxide conversion and hydrocarbon selectivity. Currently, fixed catalyst reactors, fluidized catalyst reactors, suspended catalyst reactors and microtubulation reactors are used. The most commonly used are fixed catalyst reactors, due to their high productivity at temperatures between 220-240ºC and pressures up to 45 bar, respectively the suspended catalyst reactors. The gas is fed to the bottom of the reactor through a gas dispenser for fluidization. The organic compounds resulting from the synthesis are evacuated by means of filters made of sintered material with a pore diameter below 5 µm. Due to the simple design, the design costs of the reactors are 20-40% lower than in the case of fixed-layer catalyst reactors.

Biomass Pyrolysis

It represents the biomass conversion technology through the process of thermal degradation of waste in the absence of air to produce pyrolysis fuel (bio-oil), charcoal (char) and syngas (for example, conversion of wood to charcoal). The operating process features are: operating temperature of 450-600ºC and gas retention time of 0.3-5.0 seconds. The biomass batch is processed before being introduced into the furnace: particle size 0.5-6 mm, humidity 3-10%, cellulose 48-50%, ash content 0.4-0.6%. Composition of the resulting products: organic liquid (bio-oil) 40-75% (including water), coal (char) 10-20%, synthetic gas 10-30%. *(Hu, X., & Gholizadeh, M., 2019; Wang, G., et.al., 2020)*

Schematized, a biomass pyrolysis plant is shown in Figure 12.

Figure 12. Schematic of biomass pyrolysis installation

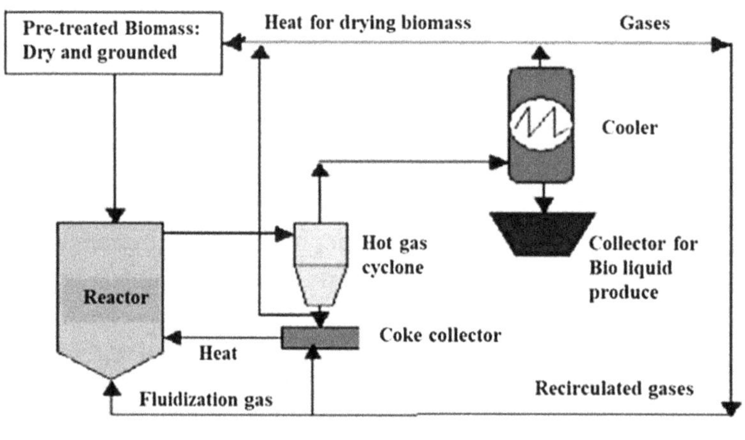

The most commonly used method is rapid pyrolysis. This consists of the rapid heating of the biomass at a well controlled temperature of about 500 ° C followed by the very rapid cooling (<2 sec) of the volatile substances formed in the reactor. It offers the unique advantage of producing a liquid that can be accumulated and transported. Although this method is under development, it already knows several configurations. The most used types of reactors are with stationary fluidized or globular flow, circulating fluidized layer, rotating cone, vacuum layer and Auger type.

The bio-oil produced through pyrolysis is miscible with water and consists of several oxygenated organic chemicals. Due to the complexity and nature of the bio-oil it has some less common properties (deposits, increased viscosity, decreased volatility and phase separation over time). Hence the need to optimize, perfect and develop pyrolysis plants for biomass in order to improve the quality of the bio-oil.

This can be done through physical methods: retention of carbon by filtration, emulsion with hydrocarbons, addition of solvent and chemical methods: reaction with alcohols and catalytic deoxygenation.

TYPES OF BIOFUELS

The main biofuels obtained by various technologies for processing and exploiting the biomass *(Jouzani, G. S., et.al., 2020; Chakraborty, D., et.al, 2020)* are presented in Table 1.

Table 1. Types of biofuel

Biofuel	What Is It?	Pros	Cons
Ethanol from agricultural biomass	Alcohol obtained by fermentation of cereals, technical plants and other plant sources	-fuel with high octane number, - reduced greenhouse gas emissions.	- difficulties in pipeline transport, - uses large quantities of agricultural biomass from food or feed.
Ethanol from lignocellulose	Alcohol obtained by converting lignocellulose biomass to fermentable carbohydrates followed by their fermentation to ethanol	- fuel with high octane number, - reduced greenhouse gas emissions, - doesn't use food or feed materials.	- difficulties in pipeline transport, - more expensive than ethanol from cereals.
Biogas	Gas mixture in which methane predominates, obtained by anaerobic fermentation of waste or other agricultural, household or industrial waste or by-products	- raw material is not valuable, - important role in waste management, - can be an energy source in rural communities, or poor areas of the globe.	- difficult to liquefy, - heterogeneous composition, depending on the raw material and technology.
Syngas	It is the final product of the process of gasification of organic solid matter as a result of its thermal decomposition.	- gasification based on the use of steam or pure oxygen (in the absence of air) produces a single gas with a high calorific value (10-20 MJ/m^3). - can be used in various fields.	- difficulty in pipeline transport, - requires CO_2 emission capture - not yet produced at high capacity.
Biodiesel	A fuel similar to diesel, obtained from vegetable oils	- reduces emissions, - engine lubricant.	- difficulty in pipeline transport, - not approved by all engine and car manufacturers.
Renewable Diesel	A fuel similar to diesel, obtained from vegetable fats and hydrocarbons	- meets the standards for very low sulfur diesel, - the addition of animal fats improves the ignition properties, - can be transported through pipes.	- higher emissions than biodiesel
Biobutanol	Alcohol fuel, similar to ethanol	- easier to transport, - less corrosive in pipes than ethanol	- not yet produced at high capacity.

Syngas – Characteristic, Properties, Uses

Synthesis gas (syngas) consists mainly of a mixture of hydrogen (H_2), carbon oxide (CO) and a small percentage of carbon dioxide (CO_2), water (H_2O), methane (CH_4), large hydrocarbons (C2+) and nitrogen (N_2) and other inert gases. *(Fiore, M., Magi, V., & Viggiano, A., 2020).*

Syngas is the final product of the process of gasification of organic solid matter as a result of its thermal decomposition. Following this process, solid, liquid and gaseous products are formed. The solid product has a carbon content greater than 76% and can be used directly for industrial purposes. Liquid products can be used as fuel in boilers, gas turbines or diesel engines. The gaseous product or the single product may have several applications depending on its properties. *(dos Santos, R. G., & Alencar, A. C., 2020; Zhao, X., et.al., 2020; Modi, M. A., & Patel, T. M., 2021)*

The quality of the single product depends on the type of gasification agent (air, steam or oxygen), the processing method and the operating process conditions. Usually, air is used as a gasification agent. The technology of oxygen/steam gasification and hydrogenation is also quite widespread. The use of steam as a catalyst positively influences the overall performance and efficiency of the solid biomass gasification process. The produced Syngas can have different values for the lower heat of combustion (Table 2).

Table 2. Gasification agent and value of the lower heat of combustion of syngas

	Gasification Agent	Lower Heat of Combustion, MJ/m³	
1	Air, steam/air	low	4-6
2	O_2 , steam	medium	12-18
3	H_2 , hydrogenation	high	40

Depending on the value of the lower heat of combustion, syngas can be used in several areas (Figure 13): direct production of electricity (low value); transport - combustible gases (low value); chemical industry - production of methane and methanol (medium and high value).

Hydrogen, a compound of synhas, in its pure state, can be used in fuel cells, which carry out its electrochemical conversion producing energy. Through the Fischer-Tropsch process, one mole of CO reacts with two moles of H_2 , thus forming primary alcohols (C_xH_{2x}). Methanol (CH_3OH) is obtained by catalytic reaction between CO and CO_2 with H_2 .

In most cases, the biomass gasification is carried out for the purpose of producing combustible gases (with a low or medium lower combustion heat value), which can be used subsequently:

- as a synthetic fuel for internal combustion engines due to their relatively good efficiency, about 40%;
- as an alternative fuel, secondary energy carrier for its electrochemical conversion into energy-efficient cogeneration equipment, plants or systems based on fuel cell technology, due to their exceptional efficiency, of about 80%.

Figure 13. Derivatives from syngas obtained through biomass

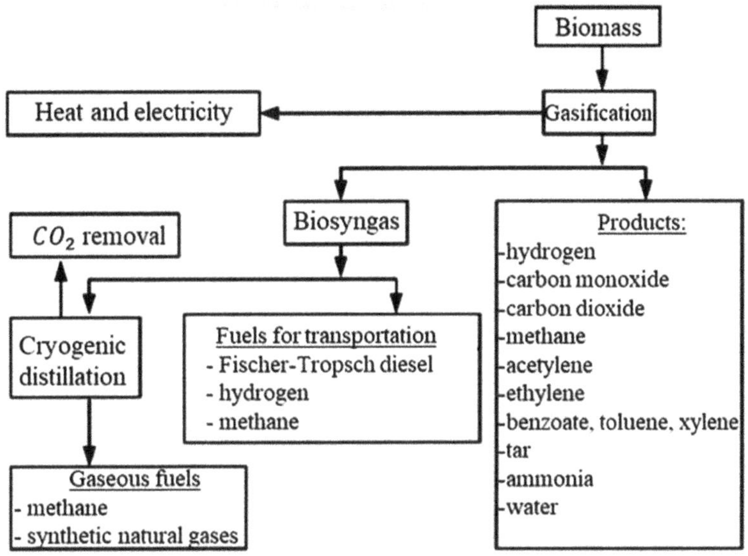

The production of electricity from syngas can also be achieved through gas turbine technology or the combined cycle technology with integrated gasification (IGCC), which allows biomass to be used with a rather high efficiency. *(Castillo Santiago, Y., et.al., 2021 Janbaz, M., & Behbahaninia, A., 2021; He, S., Li, S., & Gao, L., 2021)*

In Figure 14. a schematic of a conventional IGCC installation is presented. In industrial IGCC installations, the net electric efficiency is around 40%.

IGCC technology provides a real interest for the generation of electricity from biomass, but also from the point of view of environmental protection, due to the advantages it has over the conventional

Figure 14. Block diagram of a conventional IGCC Installation

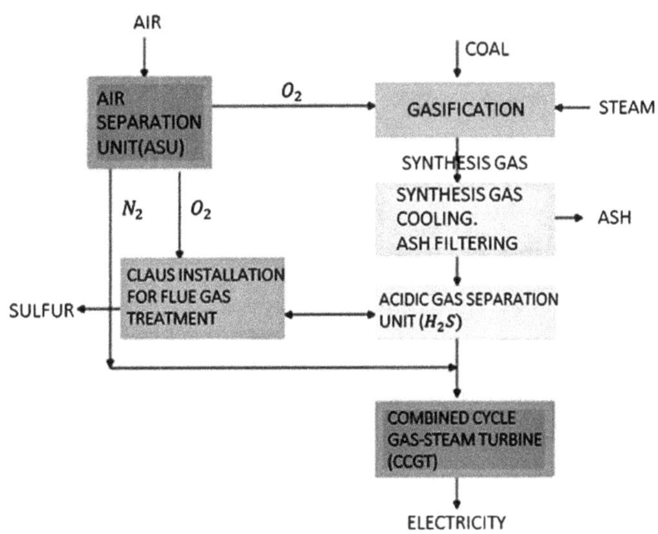

methods implemented in power installations, depending on the combustion of lignite or coal with steam generation which is further used by a turbine for the production of electricity.

One of the advantages of IGCC method is related to the remarcable fewer consequences on the environment. This is due to the presence of a synthesis gas desulfurization step.

Also, IGCC technology is flexible enough to allow the production of different vectors of energy, depending on current requirements, leading to superior economic and energy efficiencies.

An extremely essential aspect is the fact that IGCC method permits the carbon dioxide capture before combustion (pre-combustion capture) which has high efficiency and low costs compared to the ones in the case of combusted gas capture (post-combustion capture).

The implementation of IGCC method is growing exponentially, leading the majority of important turbines manufactures, like General Electric, Siemens, Alstom or Mitsubishi, to adapt their installations to use synthesis gas.

Figure 15 presents the block scheme of an IGCC installation with CO_2 storage and capture. In contrast to a conventional plant, the CO_2 capture plant also has a catalytic transformation step of water vapor with carbon monixide CO, CO_2 carbon dioxide and H_2 hydrogen. This conversion is intended to create a higer H_2 concentration in the synthesis gas as well as to convert the into capturable carbon dioxide CO_2 all elements with carbon in their composition. *(Ghiat, I., et.al., 2020; Chen, Z., et.al, 2020; Szima, S., et.al., 2021; Zhou, H., et.al., 2021)*

What is different in the above diagram is the fact that the unit which separates the acid gas deals also with the separation of H_2S and CO_2. The decarbonised synthesis gas at this stage is made predomin-

Figure 15. Block diagram of an electricity and H_2 cogeneration installation with CO_2 capture

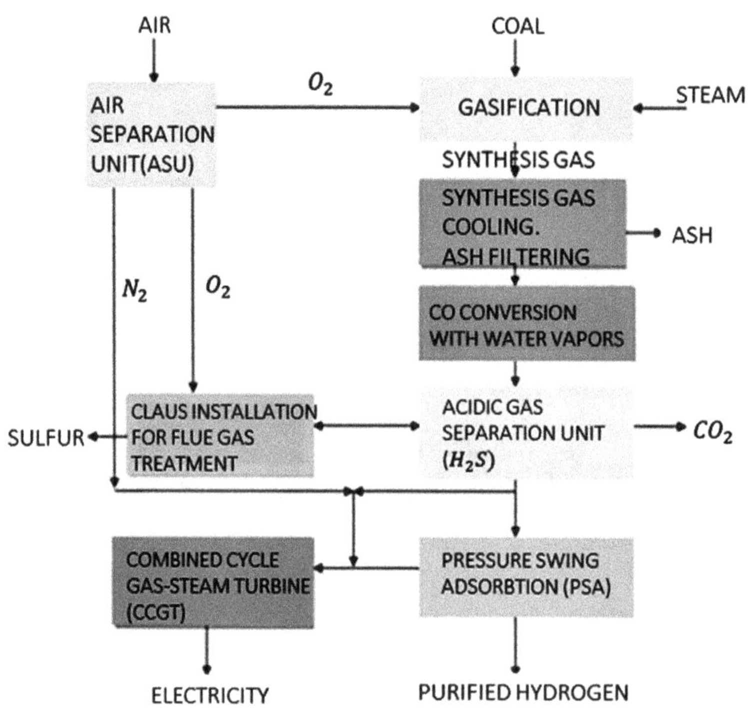

antely of hydrogen and is separated as follows: some of it goes to the adsorption - desorption unit with the modification of the pressure "Pressure Swing Adsorption - PSA" in order to obtain higher purity H_2 (> 99.9%), capable of being used both in petrochemical and chemical industries, but also for the stationary and automotive fuel cell technology, and the rest of it, combined with the waste gas resulting from the unit separating the acid gas, is further processed in the gas turbine with combined cycle gas-steam for the generation of electricity.

BIOMASS GASIFICATION TECHNOLOGY FOR THE PRODUCTION OF SYNGAS

Gasification systems transform solid biomass into a mixture of gases. The most important component of a solid biomass gasification installations is the gasifier, also known as the reactor. The plant is also equiped with a gas purification and conditioning system and a storage equipment for the resulting syngas. *(Sansaniwal, S. K., et.al., 2017; Safarian, S., et.al., 2020)*

The frst step in the biomass conversion is the process of fragmentation and drying, followed by the migration of the biomass towards the gasification reactor bunker (Figure 16).

Figure 16. Biomass gasification process

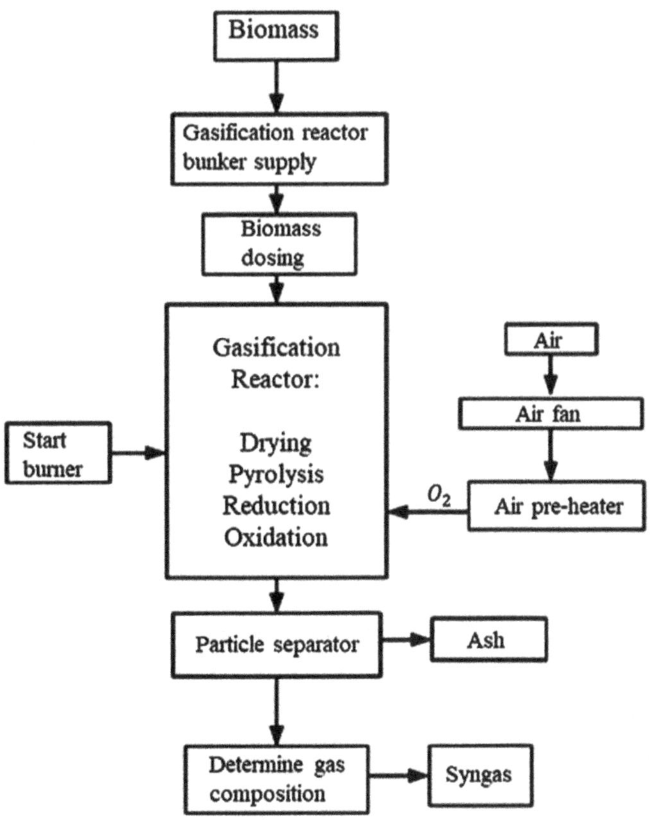

The reactor is fed via a metering unit. With the help of the starter burner, which operates on the basis of a fossil fuel, the process of oxidation of the biomass takes place. Air is used as a gasification agent. It is initially enriched with oxygen, after which it is heated and introduced into the system. In the reactor, the biomass is subjected to the following processes: drying, pyrolysis, reduction and oxidation, after which the synthesis gas and ash are obtained. Using a funnel, the obtained gas is cleaned of the solid particles and then passed through a swirling cyclone, where the fine particles are separated. The ash is removed from the process with the aid of a dispenser, and then stored in a special collector.

Schematized, the production process of the syngas utilizing biomass as raw material is shown in Figure 17. In order to obtain the finished product (the synthesis gas) from the raw material (biomass) a number of technological processes are required, from the processing of the biomass in the pre-feeding stage of the installation, the gasification of the biomass, to the purification of the gas in order to ensure the quality parameters for the supply of the refined synthesis gas to users.

A series of chemical reactions take place in the gasification process:

Figure 17. Steps of the production of syngas from biomass gasification

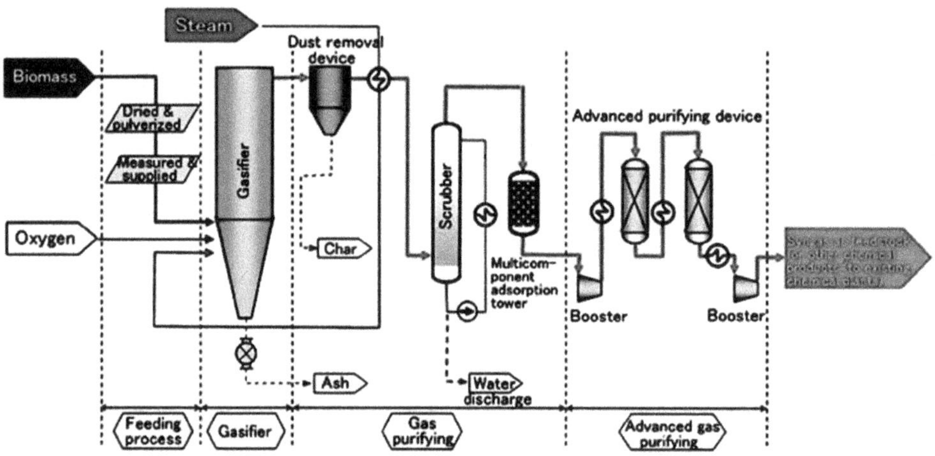

Biomass used in gasification contains besides hydrogen, oxygen and carbon, other components such as halogens, nitrogen or sulfur. These elements also transform in the reactions taking place, therefore nitrogen converts into NH_3 and HCN, and sulfur into H_2 and COS (carbonyl sulfide). If the sulfur compounds are not eliminated, they will be released into the air as sulfur oxides (SOx). In order to avoid polluting the athmosphere with SOx, gasification technologies are provided with a synthesis gas purification step, in which COS is converted to H_2S by using one of the chemical reactions presented.

In order to limit climate change, the aim is to reduce carbon dioxide emissions by capturing and storing it. This is possible in the generation of electricity by gasification of solid biomass, using an IGCC (Integrated Gasification Combined Cycle) plant with CO_2 capture.

Figure 18. Chemical reactions

- *Combustion reactions*

$C + \frac{1}{2}O_2 \rightarrow CO$ $\Delta H_1 = -111\ MJ/Kmol$

$CO + \frac{1}{2}O_2 \rightarrow CO_2$ $\Delta H_2 = -283\ MJ/Kmol$

$H_2 + \frac{1}{2}O_2 \rightarrow H_2O$ $\Delta H_3 = -242\ MJ/Kmol$

- *Boudouard reaction*

$C + CO_2 \leftrightarrow 2CO$ $\Delta H_4 = +172\ MJ/Kmol$

- *Carbon reaction with water vapors*

$C + H_2O \leftrightarrow CO + H_2$ $\Delta H_5 = +131\ MJ/Kmol$

- *Methanation reactions*

$C + 2H_2 \leftrightarrow CH_4$ $\Delta H_6 = -75\ MJ/Kmol$

$CO + 3H_2 \leftrightarrow CH_4 + H_2O$ $\Delta H_7 = -206.3\ MJ/Kmol$

- *CO reaction with water vapors*

$CO + H_2O \leftrightarrow CO_2 + H_2$ $\Delta H_8 = -41\ MJ/Kmol$

- *Pyrolysis reactions*

$4C_nH_m \rightarrow mCH_4 + (4n - m)C$ $\Delta H < 0$

$COS + H_2 \leftrightarrow H_2S + CO$ $\Delta H_9 = +7\ MJ/Kmol$

$COS + H_2O \leftrightarrow CO_2 + H_2S$ $\Delta H_{10} = -34\ MJ/Kmol$

Classic Gasification Reactors

Over 140 gasification installations exist worldwide, 90 of them being found in the USA and their number is expected to increase by 70% by 2020 *(Ferreira, S., et.al., 2019; Mayer, F., et.al., 2019; Thomson, R.,*

Figure 19. Countercurrent gasification reactor

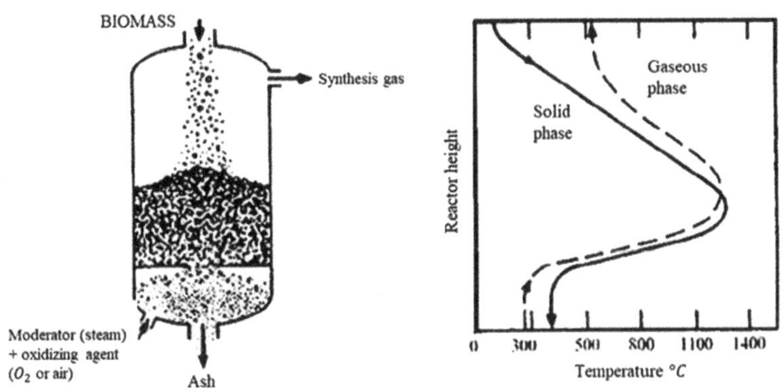

et.al, 2020). At the base of these installations are a various number of reactors which can be grouped in three broad classes:

- **Counter-current gasification reactors** (Moving-bed gasifiers) are the first implemented modern biomass reactors. The countercurrent reactor, which can be seen in Figure 19, shows at the top the solid fuel supply, and at the lower part, the gas phase (the oxidizing agent and the moderator) is introduced in countercurrent with the fuel. As it is consumed, the fuel moves to the bottom of the reactor. *(Mani Kalyani, A., & VM, J., 2020).*

Figure 20. Fluidized layer gasification reactor

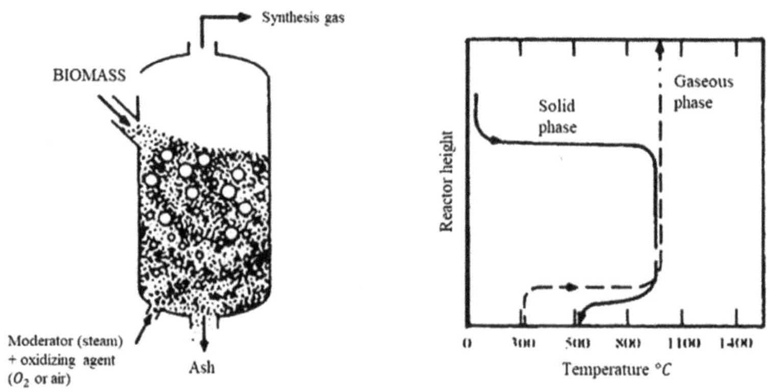

- **Fluidized layer gasification reactor** (Fluidized-bed gasifiers) ensures an optimal blending between the fuel and the oxidizing agent. The oxidizing agent which is represented by air or oxygen, is directed through a mass of solid fuel particles at certain speeds in order to fluidize the solid components. It is recommended for reactive starting materials such as lignite or biomass. Figure 20. Presents a fluidized layer reactor, as well as the corresponding temperature profile. *(Rasmussen, N. B., & Aryal, N., 2020).*

Figure 21. Entrained-floe gasifier

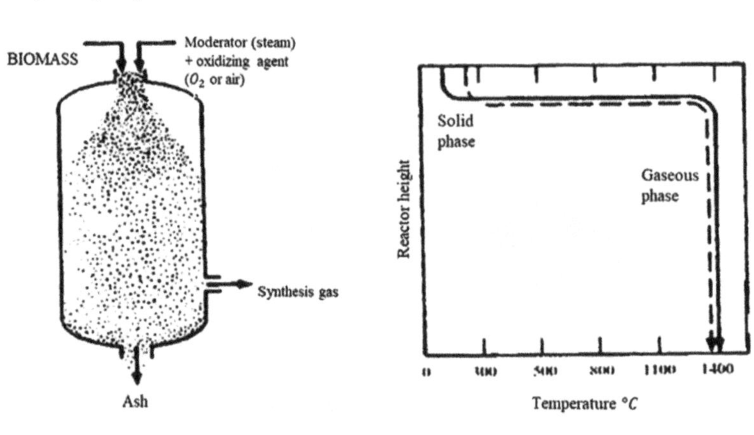

- **Entrained-flow gasifiers**, in these reactors, the solid and the gas states move on the same path. Equivalent gasification reactors can be utilised for raw components which are less reactive. Figure 21 presents the reactor together with the corresponding profile of temperatures. (*Mularski, J., Pawlak-Kruczek, H., & Modlinski, N., 2020).*

The technological equipment for the gasification of biomass and organic waste is based on the main constructive element one of the typologies of the reactors outlined above, having, depending on the case, horizontal or vertical axis, being static or rotating, but generally cylindrical.

Inside the reactors there are a number of constructive elements that allow the different stages of the process to be separated. Thus, a reactor can be composed of several areas in which the reactions of the gasification process are carried out separately and controlled (see the schematic representation in Figure 22). In each of these areas there are stirring equipment, which homogenizes the products in the processing stage.

Figure 22. Reactions of the gasification process

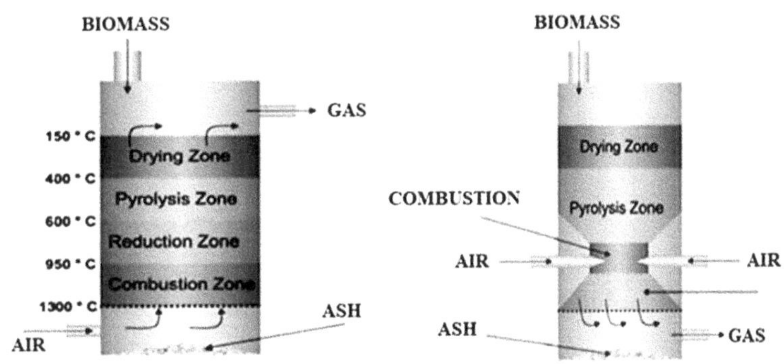

The outside of the reactor is covered with a mantle that allows the control and balancing of the endothermic and exothermic stages of the reactor. At the ends of the reactor there are cylindrical collars that allow the placement of fixed caps, through which all the components of the process are introduced, allowing their separation. These collars present rolling paths on traction rollers, which allow a slow rotation of the reactor to homogenize the process elements and advance to each of the reaction areas.

At the last end of the cylinder there are gaps that allow a better evacuation of the gases produced and a lower evacuation of the ash or unreacted solid products. The sewerage of these gases takes place through a surrounding bell, the top of which extracts the gases, which are redirected to the separation stage of the fine particles. This separation is achieved with the help of special multi-cycle separators for high temperatures and high performance. Subsequently the gases are brought to a heat exchanger, where a first cooling takes place. Then, the gases are cooled to about 100 ° C in order to filter and separate the last particles, smaller than 5 microns in special filters. Thereafter, a final cooling and condensation of the water vapor from the reaction is performed. Finally, the gas is brought to the ramp collector, being qualitatively appropriate for the final use in conversion equipment for generating electricity and heat.

Installation for the Production of Syngas Through Gasification

In a plant for the production of syngas through biomass gasification technology, in addition to the reactor construction elements, for the optimal functioning of the biomass gasification process, the following are required: (Mednikov, A. S., 2018; Oh, G., et.al, 2019; Tyurina, E. A., et.al.,2020; Pradecta, M. R., et.al., 2021)

- system for separating volatile particles,
- ash recovery system,
- system of recirculation of the tar generated in the process,
- cooling system of the generated synthesis gas,
- air intake,
- control system,
- temperature and flowmeter sensors,
- gas chromatograph, control of the generated gas,
- installation and wiring of the control elements,
- installation and wiring of motors.

The principle of operation of a plant for the production of the syngas is illustrated in Figure 23.

Figure 23. Syngas production

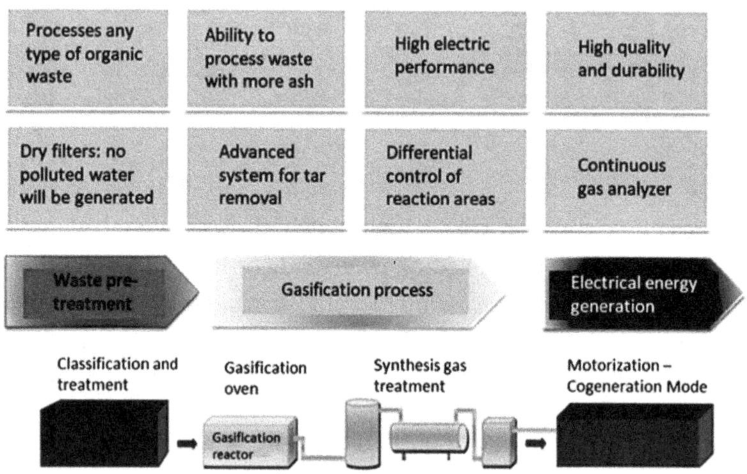

The current tendency is to provide high reliability and easy practical applicability of these types of plants, so that all these equipment of the gasification plant, implicitly the processes of biomass conversion and the generation of the synthesis gas can be pre-installed in modular containers, which has the following advantages: the reduction of the space required for location, easy installation, transport and handling of modular installations with low installation cost.

The main elements that form part of such an installation are:

Biomass Supply System

The pretreated residue or biomass is discharged into a first large capacity bunker, from where it is transported through a screw system to a weighing station that has a continuous information role regarding the admissible quantity of raw material. Then the pretreated biomass material is transported to the second bunker, which is coupled to a dosing and feeding axle of the reactor which performs, as a first step, the dehydration of the raw material. The power supply is automatically controlled by a level sensor to ensure continuous dosing. By construction, the installation also ensures the sealing of the reactor at this stage to avoid unwanted parasitic air intakes. Both dosing and feed axles are driven by power gearboxes equipped with a variable speed system.

Reactor

The reactors are made of special refractory metal alloys so that they are resistant to high operating temperatures and also resistant to corrosion. Inside each reactor a physical or chemical process takes place determined by the content of the processed raw materials. Thus, the processes that take place are as follows:

- internal reactions for homogenization of the raw material and transport,
- drying and pyrolysis reactions, which are provided inside the reactor specially designed and manufactured in order to support the reactor covers to ensure the tightness necessary to carry out the processes in an appropriate way,
- partial combustion and char gasification, an area where the reactor is extremely well insulated with refractory materials resistant to heat shock and high temperatures.

Inside the reactor a combustion chamber with dual function is provided for: the preparation and preheating of the entire installation (start-up) after maintenance stops, at the same time the pyrolysis gases are transformed into permanent combustible gases. Designing this equipment requires sizing the following elements:

- the room itself, made of steel sheet lined inside with refractory material,
- burner fueled by natural gas and controlled, forced air,
- fan that provides the necessary air for the burner chamber,
- quantity of air needed for the combustion of natural gas,
- air and natural gas intake ramps approved, ethically regulated components for their operation under control and safety conditions.

In the installation after the ignition of the burner the operating parameters are automated by the proper control of the gas/air inlet and regulated according to the desired temperature. When the thermal profile of the process has been realized, the system begins to feed the processed raw material – processed biomass. A safety temperature probe ensures that the energy released by the combustion of gas does not cause a dangerous rise in temperature at the downstream elements and equipment. The automation and control system automatically ensures the efficiency and safety for the entire process inside the equipment. Devices specially designed to mitigate and absorb travel in various directions are included to control the proper running movements of the reactors. The reactors are thermally insulated with high temperature resistant fiberglass to minimize surface heat loss.

Ash Evacuation-Recovery System

In the last section of the gasification reactor the ash disposal area is laid. This output of the reactor is realized directly by means of a honeycomb-type valve that passes over an extraction trough provided with a system of chains and lifting plates to drive the ash to the final discharge point. During the evacuation process, the gasification process is sealed by the gutter system. To prevent clogging and securing movement, the gutter is powered by an oversized power motor to ensure continuous ash removal.

Multi-Cyclone System for Cleaning Synthetic Gas

The process gases from the reactions carried out inside the reactor are directed towards multicyclones in order to be cleaned by the solid particles. The multicyclone, of circular section, comprises several cyclones mounted in parallel on the same base plate. The flow is equally distributed between all cyclones. Each is provided downstream with a special propeller for imprinting the circular motion of the gas. The solid particles, due to the centrifugal force effect, are forced to move to the walls and then fall gravitationally. The clean gas flows rise through the central pipe of the cyclones being captured through an outlet. At the bottom of the cyclone body, there is a reservoir in which the flying ash is collected.

Because the temperature of the inlet gas is high, the whole set of multicyclones is made of refractory steel. The multicyclone is anchored at 4 points for stability, and the system is thermally insulated with ceramic fiber at high temperatures to minimize surface heat loss.

Heat Exchangers

- Steam superheater: it is a superheater for steam before being injected into the combustion process in the reactor (gasification). It performs the heating of the fluid gases from the combustion chamber at a temperature above 1,000°C.
- Heat exchangers for air: they are responsible for the preheating and superheating of the combustion air required for combustion reactions through the heat exchange recovered from the synthesis gas generation process. It allows air to reach temperatures close to the reaction temperatures.
- Water heat exchanger: This circular section exchanger is attached to the air through a tube that receives the partially cooled gas.

Scrubber for Purification of Synthetic Gas

The gas, already cooled by the exchangers, is processed to obtain the quality parameters through a scrubber with a high efficiency, which fulfills the following functions:

- spraying water on the inlet channels of the scrubber,
- the gas flow enters a water pipe provided with a special screen that imprints forces so that the gases are immersed in the liquid. The same deflector produces forces on the back wall of the water, and the gases leave the liquid,

- filtering through drop separators,
- evacuation of purified gases.

Command and Control System

The installation is designed for continuous operation automatically. The entire system for the production of syngas by gasification of biomass is assisted by a central computer where a SCADA monitoring and control system is installed.

Pneumatic System – Air Circuit

The pneumatic circuit is connected to the general compressed air network. At one point, the air is distributed to all the elements that consume it. So:

- Proportional valve that opens the combustion gas passage, which is driven by a pneumatic engine.
- Pneumatic valves installed at the primary and secondary air inlets to the burner chamber and the combustion chamber.

Other Auxiliary Elements

In addition to the main basic components of the biomass conversion system that make up the group itself, the following auxiliary elements are required for its operation:

- Generator or auxiliary alternator motor,
- Air filters,
- Metal filters for synthesis gas,
- Coalescent filter for synthesis gas,
- Chiller (cooling and final condensation),
- Fire extinguishers,
- Water pipes,
- Cooling water, air cooler,
- Heat exchanger for exhaust gases,
- Exhaust chimney,
- Electric control panels.

CONCLUSION

In order to improve the security of energy supply and at the same time meet the environmental demands, it is emphasized the need for renewable energy resources to become an increasingly important part of the energy production structure.

The following chapter aims to describe the solutions for the recovery of biomass and of the synthesis gas production systems through the biomass gasification.

- worldwide there is an important potential of biomass that can be used energetically and that can contribute to ensuring a significant amount of non-polluting thermal energy and electricity.
- the structure and the physico-chemical characteristics of the biomass are important arguments for its re-evaluation as a combustible material with low pollutant potential and as a renewable source of energy.
- there is a need for an ecological approach to the process of combustion of fuels in general, due to their pollutant potential, gasification representing a practical, clean solution.
- gasification and production of syngas represents the ideal solution, from an ecological point of view, of energy recovery of biomass, by increasing its calorific capacity, limiting the polluting emissions and the appearance of ash deposition on the heat exchangers inside the boilers.
- the gasification plants have a quite high efficiency of biomass energy recovery, reaching values of at least 80% by using them for two separate purposes: the production of the synthesis gas and the burning of it for the production of technological steam.

REFERENCES

Adeleke, A. A., Odusote, J. K., Ikubanni, P. P., Lasode, O. A., Malathi, M., & Paswan, D. (2021). Essential basics on biomass torrefaction, densification and utilization. *International Journal of Energy Research, 45*(2), 1375–1395. doi:10.1002/er.5884

AlNouss, A., McKay, G., & Al-Ansari, T. (2020). Production of syngas via gasification using optimum blends of biomass. *Journal of Cleaner Production, 242*, 118499. doi:10.1016/j.jclepro.2019.118499

Bacău, C., Mateoc-Sirb, N., Ciolac, R., Mateoc, T., Teodorescu, R. I., & Tabără, V. (2021). *Study on the production and use of biomass energy in the Timis County*. Academic Press.

Bajwa, D. S., Peterson, T., Sharma, N., Shojaeiarani, J., & Bajwa, S. G. (2018). A review of densified solid biomass for energy production. *Renewable & Sustainable Energy Reviews, 96*, 296–305. doi:10.1016/j.rser.2018.07.040

Bao, D. X., Yu, Y. Y., & Zhao, Q. M. (2020). Evaluation of the chemical composition and rheological properties of bio-asphalt from different biomass sources. *Road Materials and Pavement Design, 21*(7), 1829–1843. doi:10.1080/14680629.2019.1568287

Bar-On, Y. M., Phillips, R., & Milo, R. (2018). The biomass distribution on Earth. *Proceedings of the National Academy of Sciences of the United States of America, 115*(25), 6506–6511. doi:10.1073/pnas.1711842115 PMID:29784790

Boukis, N., & Stoll, I. K. (2021). Gasification of Biomass in Supercritical Water, Challenges for the Process Design—Lessons Learned from the Operation Experience of the First Dedicated Pilot Plant. *Processes (Basel, Switzerland), 9*(3), 455. doi:10.3390/pr9030455

Brand, M. A., Rodrigues, T. M., da Silva, J. P., & de Oliveira, J. (2021). Recovery of agricultural and wood wastes: The effect of biomass blends on the quality of pellets. *Fuel, 284*, 118881. doi:10.1016/j.fuel.2020.118881

Brás, I., Silva, E., & De Lemos, L. T. (2020). Feasibility of using municipal solid wastes rejected fractions as fuel in a biomass power plant. *Environment Protection Engineering*, *46*(2). Advance online publication. doi:10.37190/epe200204

Casquero-Vera, J. A., Lyamani, H., Titos, G., Minguillón, M. C., Dada, L., Alastuey, A., Querol, X., Petäjä, T., Olmo, F. J., & Alados-Arboledas, L. (2021). Quantifying traffic, biomass burning and secondary source contributions to atmospheric particle number concentrations at urban and suburban sites. *The Science of the Total Environment*, *768*, 145282. doi:10.1016/j.scitotenv.2021.145282

Castillo Santiago, Y., Martínez González, A., Venturini, O. J., & Yepes Maya, D. M. (2021). Assessment of the energy recovery potential of oil sludge through gasification aiming electricity generation. *Energy*, 215.

Chakraborty, D., Efthi, J. H., Khanom, M., & Mahbubul, I. M. (2020). Prospective and Challenging Issues of Biofuels. *EDU Journal of Computer and Electrical Engineering*, *1*(1), 4–10. doi:10.46603/ejcee.v1i1.5

Chambon, C. L., Karia, T., Sandwell, P., & Hallett, J. P. (2020). Techno-economic assessment of biomass gasification-based mini-grids for productive energy applications: The case of rural India. *Renewable Energy*, *154*, 432–444. doi:10.1016/j.renene.2020.03.002

Chen, Z., Fang, J., Xu, C., Xia, Z., Yan, K., & Li, X. (2020). Carbon dioxide hydrate separation from Integrated Gasification Combined Cycle (IGCC) syngas by a novel hydrate heat-mass coupling method. *Energy*, *199*, 117420. doi:10.1016/j.energy.2020.117420

Choi, S. H., & Manousiouthakis, V. I. (2020). On the carbon cycle impact of combustion of harvested plant biomass vs. fossil carbon resources. *Computers & Chemical Engineering*, *140*, 106942. doi:10.1016/j.compchemeng.2020.106942

De, S., Mishra, S., Poonguzhali, E., Rajesh, M., & Tamilarasan, K. (2020). Fractionation and characterization of lignin from waste rice straw: Biomass surface chemical composition analysis. *International Journal of Biological Macromolecules*, *145*, 795–803. doi:10.1016/j.ijbiomac.2019.10.068 PMID:31739019

Destek, M. A. (2017). Biomass energy consumption and economic growth: Evidence from top 10 biomass consumer countries. *Energy Sources. Part B, Economics, Planning, and Policy*, *12*(10), 853–858. doi:10.1080/15567249.2017.1314393

dos Santos, R. G., & Alencar, A. C. (2020). Biomass-derived syngas production via gasification process and its catalytic conversion into fuels by Fischer Tropsch synthesis: A review. *International Journal of Hydrogen Energy*, *45*(36), 18114–18132. doi:10.1016/j.ijhydene.2019.07.133

Drake, H., Ivarsson, M., Heim, C., Snoeyenbos-West, O., Bengtson, S., Belivanova, V., & Whitehouse, M. (2021). Fossilized anaerobic and possibly methanogenesis-fueling fungi identified deep within the Siljan impact structure, Sweden. *Communications Earth & Environment*, *2*(1), 1–10. doi:10.103843247-021-00107-9

Enescu, A., Andrei, H., Ion, V., Diaconu, E., & Angelescu, N. (2019, October). Analysis and Modeling of Biomass Plant Energy Efficieny. In *2019 6th International Symposium on Electrical and Electronics Engineering (ISEEE)* (pp. 1-6). IEEE.

Ferreira, S., Monteiro, E., Brito, P., & Vilarinho, C. (2019). A holistic review on biomass gasification modified equilibrium models. *Energies*, *12*(1), 160. doi:10.3390/en12010160

Fiore, M., Magi, V., & Viggiano, A. (2020). Internal combustion engines powered by syngas: A review. *Applied Energy*, *276*, 115415. doi:10.1016/j.apenergy.2020.115415

Gao, J., & Zhang, L. (2021). Does biomass energy consumption mitigate CO2 emissions? The role of economic growth and urbanization: Evidence from developing Asia. *Journal of the Asia Pacific Economy*, *26*(1), 96–115. doi:10.1080/13547860.2020.1717902

Ghiat, I., AlNouss, A., McKay, G., & Al-Ansari, T. (2020). Biomass-based integrated gasification combined cycle with post-combustion CO2 recovery by potassium carbonate: Techno-economic and environmental analysis. *Computers & Chemical Engineering*, *135*, 106758. doi:10.1016/j.compchemeng.2020.106758

Hameed, Z., Aslam, M., Khan, Z., Maqsood, K., Atabani, A. E., Ghauri, M., Khurram, M. S., Rehan, M., & Nizami, A. S. (2021). Gasification of municipal solid waste blends with biomass for energy production and resources recovery: Current status, hybrid technologies and innovative prospects. *Renewable & Sustainable Energy Reviews*, *136*, 110375. doi:10.1016/j.rser.2020.110375

He, S., Li, S., & Gao, L. (2021). Proposal and energy saving analysis of novel methanol–electricity polygeneration system based on staged coal gasification method. *Energy Conversion and Management*, *233*, 113931. doi:10.1016/j.enconman.2021.113931

Hu, X., & Gholizadeh, M. (2019). Biomass pyrolysis: A review of the process development and challenges from initial researches up to the commercialisation stage. *Journal of Energy Chemistry*, *39*, 109–143. doi:10.1016/j.jechem.2019.01.024

Janbaz, M., & Behbahaninia, A. (2021). Thermodynamic analysis of Integrated Gasification Combined Cycle power plant: Energy, Exergy. *Jixie Gongcheng Xuebao*, *50*(4), 45–49.

Jouzani, G. S., Aghbashlo, M., & Tabatabaei, M. (2020). Biofuels: Types, Promises, Challenges, and Role of Fungi. In *Fungi in Fuel Biotechnology* (pp. 1–14). Springer. doi:10.1007/978-3-030-44488-4_1

Kargbo, H., Harris, J. S., & Phan, A. N. (2021). "Drop-in" fuel production from biomass: Critical review on techno-economic feasibility and sustainability. *Renewable & Sustainable Energy Reviews*, *135*, 110168. doi:10.1016/j.rser.2020.110168

Karl, J., & Pröll, T. (2018). Steam gasification of biomass in dual fluidized bed gasifiers: A review. *Renewable & Sustainable Energy Reviews*, *98*, 64–78. doi:10.1016/j.rser.2018.09.010

Lazzari, E., Schena, T., Marcelo, M. C. A., Primaz, C. T., Silva, A. N., Ferrão, M. F., Bjerk, T., & Caramão, E. B. (2018). Classification of biomass through their pyrolytic bio-oil composition using FTIR and PCA analysis. *Industrial Crops and Products*, *111*, 856–864. doi:10.1016/j.indcrop.2017.11.005

Leng, L., Zhang, W., Leng, S., Chen, J., Yang, L., Li, H., Jiang, S., & Huang, H. (2020). Bioenergy recovery from wastewater produced by hydrothermal processing biomass: Progress, challenges, and opportunities. *The Science of the Total Environment*, *748*, 142383. doi:10.1016/j.scitotenv.2020.142383 PMID:33113702

Li, Y., Zhou, L. W., & Wang, R. Z. (2017). Urban biomass and methods of estimating municipal biomass resources. *Renewable & Sustainable Energy Reviews*, *80*, 1017–1030. doi:10.1016/j.rser.2017.05.214

Li, Z., Li, Z., Tong, X., Zhang, J., Dong, L., Zheng, Y., Ma, W., Zhao, L., Wang, L., Wen, L., Dang, Z., Tuvshintogtokh, I., Liang, C., & Li, F. Y. (2020). Climatic humidity mediates the strength of the species richness–biomass relationship on the Mongolian Plateau steppe. *The Science of the Total Environment*, *718*, 137252. doi:10.1016/j.scitotenv.2020.137252 PMID:32325613

Mani Kalyani, A., & VM, J. (. (2020). Mechanisms governing the "gasification to char oxidation transition" in counter-current flame propagation in packed beds: Insights from single-particle experiments. *Particulate Science and Technology*, 1–5. doi:10.1080/02726351.2020.1767245

Mayer, F., Bhandari, R., & Gäth, S. (2019). Critical review on life cycle assessment of conventional and innovative waste-to-energy technologies. *The Science of the Total Environment*, *672*, 708–721. doi:10.1016/j.scitotenv.2019.03.449 PMID:30974361

Mednikov, A. S. (2018). A review of technologies for multistage wood biomass gasification. *Thermal Engineering*, *65*(8), 531–546. doi:10.1134/S0040601518080037

Modi, M. A., & Patel, T. M. (2021). Syngas Assessment from Plastic Waste Using Artificial Neural Network—A Review. *Machine Learning for Predictive Analysis*, 203-208.

Mudryk, K., Jewiarz, M., Wróbel, M., Niemiec, M., & Dyjakon, A. (2021). Evaluation of Urban Tree Leaf Biomass-Potential, Physico-Mechanical and Chemical Parameters of Raw Material and Solid Biofuel. *Energies*, *14*(4), 818. doi:10.3390/en14040818

Mularski, J., Pawlak-Kruczek, H., & Modlinski, N. (2020). A review of recent studies of the CFD modelling of coal gasification in entrained flow gasifiers, covering devolatilization, gas-phase reactions, surface reactions, models and kinetics. *Fuel*, *271*, 117620. doi:10.1016/j.fuel.2020.117620

Negreanu-Pirjol, B. S., Negreanu-Pirjol, T., & Paraschiv, G. M. (2020). Residual Marine Algae Biomass-An Important Raw Material for Obtaining a Soil Biostimulator-Regenerator. *European Journal of Medicine and Natural Sciences*, *4*(3), 74–85.

Nunes, L. J. R., Causer, T. P., & Ciolkosz, D. (2020). Biomass for energy: A review on supply chain management models. *Renewable & Sustainable Energy Reviews*, *120*, 109658. doi:10.1016/j.rser.2019.109658

Oh, G., Ra, H. W., Yoon, S. M., Mun, T. Y., Seo, M. W., Lee, J. G., & Yoon, S. J. (2019). Syngas production through gasification of coal water mixture and power generation on dual-fuel diesel engine. *Journal of the Energy Institute*, *92*(2), 265–274. doi:10.1016/j.joei.2018.01.009

Pablo, G., Gomes-Dias, J. S., Rocha, C. M., Romaní, A., Garrote, G., & Domingues, L. (2020). Recent trends on seaweed fractionation for liquid biofuels production. *Bioresource Technology*, *299*, 122613. doi:10.1016/j.biortech.2019.122613 PMID:31870706

Pan, X., Ichoku, C., Chin, M., Bian, H., Darmenov, A., Colarco, P., Ellison, L., Kucsera, T., da Silva, A., Wang, J., Oda, T., & Cui, G. (2020). Six global biomass burning emission datasets: Intercomparison and application in one global aerosol model. *Atmospheric Chemistry and Physics*, *20*(2), 969–994. doi:10.5194/acp-20-969-2020

Panaitescu, C., Stoicescu, M., Ponea, M. G., Nancu, D., & Nguyen, D. N. (2020). Biomass valuation in the context of sustainable agricultural development in Romania. *Ekonomika Poljoprivrede*, 3.

Patuzzi, F., Basso, D., Vakalis, S., Antolini, D., Piazzi, S., Benedetti, V., Cordioli, E., & Baratieri, M. (2021). State-of-the-art of small-scale biomass gasification systems: An extensive and unique monitoring review. *Energy*, *223*, 120039. doi:10.1016/j.energy.2021.120039

Pinales-Márquez, C. D., Rodríguez-Jasso, R. M., Araújo, R. G., Loredo-Treviño, A., Nabarlatz, D., Gullón, B., & Ruiz, H. A. (2021). Circular bioeconomy and integrated biorefinery in the production of xylooligosaccharides from lignocellulosic biomass: A review. *Industrial Crops and Products*, *162*, 113274. doi:10.1016/j.indcrop.2021.113274

Pokhrel, R. P., Gordon, J., Fiddler, M. N., & Bililign, S. (2021). Impact of combustion conditions on physical and morphological properties of biomass burning aerosol. *Aerosol Science and Technology*, *55*(1), 80–91. doi:10.1080/02786826.2020.1822512

Popescu, A. I., & Necula, C. (2019). Determination of the Calorific Power of Densified Solid Biofuels. *Hidraulica*, (3), 68–75.

Pradecta, M. R., Majid, A. I., & Widyaparaga, A. (2021, February). Syngas Production via Supercritical Gasification of Leachate: A Review. *Journal of Physics: Conference Series*, *1772*(1), 012056.

Proskurina, S., Junginger, M., Heinimö, J., Tekinel, B., & Vakkilainen, E. (2019). Global biomass trade for energy—Part 2: Production and trade streams of wood pellets, liquid biofuels, charcoal, industrial roundwood and emerging energy biomass. *Biofuels, Bioproducts & Biorefining*, *13*(2), 371–387. doi:10.1002/bbb.1858

Rasmussen, N. B., & Aryal, N. (2020). Syngas production using straw pellet gasification in fluidized bed allothermal reactor under different temperature conditions. *Fuel*, *263*, 116706. doi:10.1016/j.fuel.2019.116706

Safarian, S., Unnthorsson, R., & Richter, C. (2020). Techno-economic and environmental assessment of power supply chain by using waste biomass gasification in Iceland. *BioPhysical Economics and Sustainability*, *5*(2), 1–13. doi:10.100741247-020-00073-4

Sansaniwal, S. K., Pal, K., Rosen, M. A., & Tyagi, S. K. (2017). Recent advances in the development of biomass gasification technology: A comprehensive review. *Renewable & Sustainable Energy Reviews*, *72*, 363–384. doi:10.1016/j.rser.2017.01.038

Schill, G. P., Froyd, K. D., Bian, H., Kupc, A., Williamson, C., Brock, C. A., Ray, E., Hornbrook, R. S., Hills, A. J., Apel, E. C., Chin, M., Colarco, P. R., & Murphy, D. M. (2020). Widespread biomass burning smoke throughout the remote troposphere. *Nature Geoscience*, *13*(6), 422–427. doi:10.103841561-020-0586-1

Shahabuddin, M., Alam, M. T., Krishna, B. B., Bhaskar, T., & Perkins, G. (2020). A review of producing renewable aviation fuels from the gasification of biomass and residual wastes. *Bioresource Technology*, *312*, 123596. doi:10.1016/j.biortech.2020.123596 PMID:32507633

Situmorang, Y. A., Zhao, Z., Yoshida, A., Abudula, A., & Guan, G. (2020). Small-scale biomass gasification systems for power generation (< 200 kW class): A review. *Renewable & Sustainable Energy Reviews*, *117*, 109486. doi:10.1016/j.rser.2019.109486

Song, B., Lin, R., Lam, C. H., Wu, H., Tsui, T. H., & Yu, Y. (2021). Recent advances and challenges of inter-disciplinary biomass valorization by integrating hydrothermal and biological techniques. *Renewable & Sustainable Energy Reviews*, *135*, 110370. doi:10.1016/j.rser.2020.110370

Sonkhaskar, Y. M., Nikhade, G. R., Dharmik, S., Deshmukh, U., & Dhote, P. (2021). Development of Briquette Cum Pellet Making Machine. In *Advances in Mechanical Engineering* (pp. 391–397). Springer. doi:10.1007/978-981-15-3639-7_46

Szima, S., del Pozo, C. A., Cloete, S., Chiesa, P., Alvaro, Á. J., Cormos, A. M., & Amini, S. (2021). Finding synergy between renewables and coal: Flexible power and hydrogen production from advanced IGCC plants with integrated CO2 capture. *Energy Conversion and Management*, *231*, 113866. doi:10.1016/j.enconman.2021.113866

Thomson, R., Kwong, P., Ahmad, E., & Nigam, K. D. P. (2020). Clean syngas from small commercial biomass gasifiers; a review of gasifier development, recent advances and performance evaluation. *International Journal of Hydrogen Energy*, *45*(41), 21087–21111. doi:10.1016/j.ijhydene.2020.05.160

Toro Delgado, J. A. (2020). *Biomass classification by cluster analysis based on compositional data* (Bachelor's thesis, Uniandes).

Tripa, S., & Indrie, L. (2021). Households textile waste management in the context of a circular economy in Romania. *Environmental Engineering and Management Journal*, *20*(1), 81–87. doi:10.30638/eemj.2021.009

Tursi, A. (2019). A review on biomass: Importance, chemistry, classification, and conversion. *Biofuel Research Journal*, *6*(2), 962–979. doi:10.18331/BRJ2019.6.2.3

Tyurina, E. A., Elsukov, P. Y., & Mednikov, A. S. (2020). Synthetic liquid fuels: prospects for innovative technologies based on underground coal gasification. In *E3S Web of Conferences* (Vol. 209, p. 03026). EDP Sciences.

Vassilev, S. V., Baxter, D., Andersen, L. K., & Vassileva, C. G. (2010). An overview of the chemical composition of biomass. *Fuel*, *89*(5), 913–933. doi:10.1016/j.fuel.2009.10.022

Vassilev, S. V., Baxter, D., Andersen, L. K., & Vassileva, C. G. (2013). An overview of the composition and application of biomass ash. Part 1. Phase–mineral and chemical composition and classification. *Fuel*, *105*, 40–76. doi:10.1016/j.fuel.2012.09.041

Wang, G., Dai, Y., Yang, H., Xiong, Q., Wang, K., Zhou, J., Li, Y., & Wang, S. (2020). A review of recent advances in biomass pyrolysis. *Energy & Fuels*, *34*(12), 15557–15578. doi:10.1021/acs.energyfuels.0c03107

Wang, Z., Bui, Q., & Zhang, B. (2020a). The relationship between biomass energy consumption and human development: Empirical evidence from BRICS countries. *Energy*, *194*, 116906. doi:10.1016/j.energy.2020.116906

Wang, Z., Bui, Q., Zhang, B., & Pham, T. L. H. (2020b). Biomass energy production and its impacts on the ecological footprint: An investigation of the G7 countries. *The Science of the Total Environment*, *743*, 140741. doi:10.1016/j.scitotenv.2020.140741 PMID:32758837

Zhao, X., Babu Joseph, J. K., & Ozcan, S. (2020). Biogas Reforming to Syngas: A Review. *iScience*, *23*(5), 101082. doi:10.1016/j.isci.2020.101082 PMID:32380422

Zhou, H., Meng, W., Wang, D., Li, G., Li, H., Liu, Z., & Yang, S. (2021). A novel coal chemical looping gasification scheme for synthetic natural gas with low energy consumption for CO2 capture: Modelling, parameters optimization, and performance analysis. *Energy*, •••, 120249. doi:10.1016/j.energy.2021.120249

KEY TERMS AND DEFINITIONS

Biomass: Term can be defined generically as the biodegradable part of agricultural products, residues and waste, gathering both animal and plant substances, wood industry and also the biodegradable component of urban and industrial waste.

Biomass Energy: Biomass accumulated in chemical form in matter of animal or plant origin represents some of the most spread resources on the planet.

Biomass Gasification: A thermochemical conversion of organic matter into a gas product (syngas), by using the process of partial oxidation. for this conversion to occur, special conditions are required such as high temperature and low air circulation, in order to prevent the complete biomass burning, but enough for the conversion.

Biomass Pyrolysis: Represents the biomass conversion technology through the process of thermal degradation of waste in the absence of air to produce pyrolysis fuel (bio-oil), charcoal (char) and syngas (for example, conversion of wood to charcoal).

Syngas: Final product of the process of gasification of organic solid matter as a result of its thermal decomposition.

Chapter 5
Energy Storage Systems: From Classic to Hydrogen

Ana-Maria Nasture
https://orcid.org/0000-0003-0298-5774
National Research and Development Institute for Cryogenic and Isotopic Technologies, Râmnicu Vâlcea, Romania

Maria Simona Raboaca
https://orcid.org/0000-0002-7277-4377
National Research and Development Institute for Cryogenic and Isotopic Technologies, Râmnicu Vâlcea, Romania

Laurentiu Patularu
National Research and Development Institute for Cryogenic and Isotopic Technologies, Râmnicu Vâlcea, Romania

Ciprian Lupu
University Politehnica of Bucharest, Romania

ABSTRACT

Energy storage is a vital component in the chain of production-distribution-consumption of energy, even more so if the energy comes from a source that is intermittent and/or is not controllable as is the case with for example solar energy and wind energy. For many people, the term energy storage is the storage of electricity in batteries, as it is the most commonly found way of storing energy. In addition to classic batteries, there are other energy storage alternatives from a primary source for later use. The most valuable forms of energy storage are the ones that can both take over and release the energy on demand, in the form of electricity, such that, in the end, the electrical energy is transformed into thermal or mechanical energy. In stationary applications, energy can be stored in various forms such as batteries, ultracapacitors, or tanks of hydrogen, water, and different types of materials. This chapter will evaluate each form of energy storage.

DOI: 10.4018/978-1-7998-4945-2.ch005

INTRODUCTION

In recent years there has been a major worldwide concern for the change the source of energy from fossil fuels to renewable sources. The end user (be it a home or industrial user) has changed his profile in terms of energy consumption. The need for energy storage will certainly grow substantially. This is not just due to the natural growth of all the technologies in which storage is an important component, but also to important changes in the energy production and utilization landscape (Robert A. Huggins, 2016).

An important objective in the European Strategic Energy Technology Plan (SET Plan) is energy efficiency and decarbonization. This objective is attempting to be reached using research and innovation in production and storage of renewable energies (e.g. solar, wind, geothermal, hydro, hydrogen). Due to these SET Plan objectives, Energy Storage Systems are an important factor in meeting Europe's need for clean and efficient energy.

The trend being set is the usage of renewable sources everywhere, however these sources are not able to deliver constant 24/7 energy (e.g. Photovoltaic plant on a cloudy day or a non-windy day for wind turbines). Another present issue is that the increased number of devices electrically connected to the grid, including stationary applications, highly increase the chances of causing an imbalance. At the same time, the price of energy is varying in different power markets. In a scenario with the possibility of the main source having a power failure due to lack of constant generation, the need for a backup solution greatly increases. These four issues (Smoothing the flow of power; Increasing utilization of power-generation, transmission and distribution assets; Smoothing out the costs, and necessity of a backup solution) have the same valid answer: energy storage.

Historically, companies, grid operators, independent power providers, and utilities have invested in energy-storage devices to provide a specific benefit, either for themselves or for the grid. As storage costs fall, ownership will broaden and many new business models will emerge. (P. D'Aprile, J. Newman, D. Pinner, 2016)

New technologies and new materials for energy storage are presented every day. Their purpose is to increase efficiency, life-time, and to provide environmentally-friendly solutions.

According to the European Market Monitor on Energy Storage (EMMES), installed energy storage will grow up to 5.5 GWh. (EMMES, 2019) Figure 1, shows the growth of electrical, electrochemical and mechanical storage, from the year 2016 until now, with a prediction for year 2020 (with the exception of pumped hydro storage).

As the level of electrification increases, in parallel with the increasing number of decentralized energy sources, it raises the question of the technical flexibility of the systems of electricity supply, storage and distribution. stationary applications storage capacity can integrate variable sources of renewable energy.

Energy storage systems (for both stationary and mobile applications) must provide maximum benefit to the end-users of electricity and must be accepted by the electricity industry. Therefore, in order to achieve these two most important objectives, economic factors, technical factors, research and design factors for energy storage applications must always be taken into consideration. It is clear that each storage application will have its own specifications, but these three interrelated factors must lead to a global-scale spread of the proposed application and, ultimately, to its grid-integration use.

The usual method of storing energy for stationary applications is in pumped hydro, and its represented in 2018 around 94% of all storage systems. But EU is also focusing on other storage systems for stationari applications (batteries and hydrogen). The most important key issue factors regarding the batteries storage systems are: raw materials, battery cell manufacturing, reaserch and inovation, qualified human

Figure 1. Electrical energy storage capacity annually installed (MWh)
(The European Market Monitor on Energy Storage, 2019)

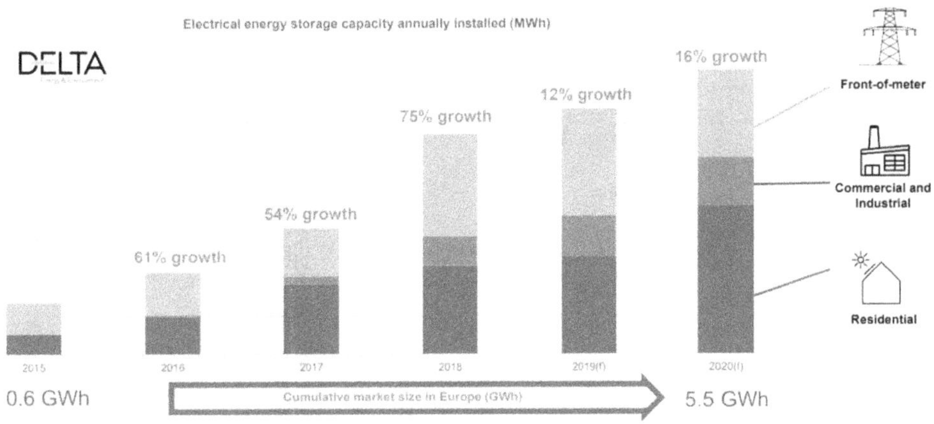

resources. In 2017 Commission created the European Battery Alliance to follow up these major key factors along with other secondary key factors.

The usual method of storing energy for stationary applications is in pumped hydro, and its represented in 2018 around 94% of all storage systems. But EU is also focusing on other storage systems for stationary applications (batteries and hydrogen). The most important key issue factors regarding the batteries storage systems are: the raw materials, battery cell manufacturing, research and innovation, and qualified human resources. In this case, in 2017 Commission created the European Battery Alliance to follow up these major key factors along with other secondary key factors.

We must point out that the EU Commission awarded grants totalling €1.34 billion from Horizon 2020 to 396 projects linked to energy storage on the grid and for low-carbon mobility with 25% of money for battery projects and 37% for hydrogen or fuel cell projects (see Figure 2) (European Court of Auditorium, 2019), in order to create the economical and development frames for researchers and companies.

In this paper explains different energy storage technologies for both stationary and mobile applications. For each type of energy storage, the operating principle, the strengths and the weaknesses will be presented, both from a constructive point of view and from an economic and environmental protection point of view. Also new storage alternatives will be presented, such systems being in a way a measure of humanity's innovation and inventiveness.

Types of Energy Storage Technologies

Energy storage systems (ESS) can be classified, according to the form energy, into mechanical, electrochemical, chemical, electrical, thermal as it is presented in Table 1.

A special category of EES is the Hybrid ESS, which combine different ESS from the five main categories, for example: Battery – Ultracapacitor, Battery-Fuel Cells, Battery – Compressed Air, Fuel Cells -Ultracapacitors, just to name a few.

Figure 2. Horizon 2020 energy storage projects.
(ECA, based on Commission's data, April 2019)

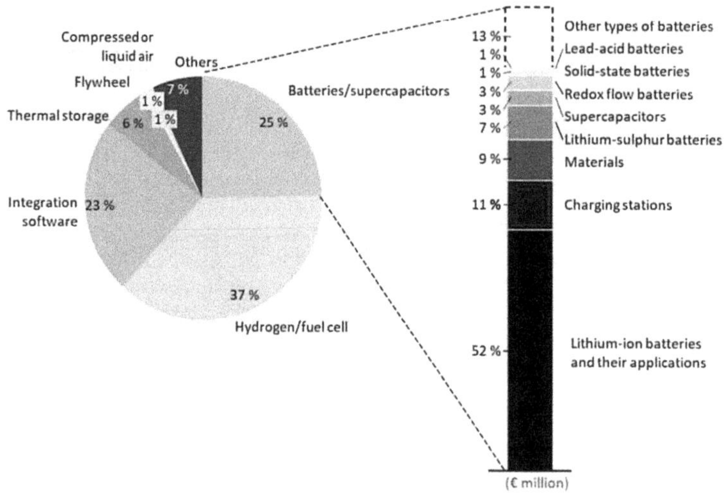

1. Mechanical Storage Systems

The **M**echanical **S**torage **S**ystems **(MSSs)** the most widespread systems through the world. Not only that are delivering high quantity of energy for a long time, but also, are the oldest systems in the world for store and convert the mechanical energy into another form of energy (these days into electricity). These storage systems are covering around 99% of global storage capacity. The most popular is **P**umped **H**ydro **S**torage **S**ystem (**PHSS**), flow by **C**ompressed **A**ir **E**nergy **S**torage (**CAES**), and, recently **F**lywheel **E**nergy **S**torage (**FES**) become more and more attractive solution (Fujihara T, HImano H, Oshima K, 1998).

Table 1. Energy storage system classification

<table>
<tr><th rowspan="2">Mechanical Storage</th><th colspan="2">Electrochemical Storage</th><th rowspan="2">Chemical Storage Fuel Cells</th><th rowspan="2">Electromagnetic Energy Storage</th><th rowspan="2">Thermal Storage</th></tr>
<tr><th>Secondary Batteries</th><th>Flow Batteries</th></tr>
<tr><td>Pumped Hydro</td><td rowspan="2">Ambient and High Temperature Lithium</td><td rowspan="2">Redox Flow</td><td>Alkaline Membrane Cells</td><td rowspan="2">Ultracapacitor</td><td>Latent Heat</td></tr>
<tr><td>Compressed Air</td><td>Proton Exchange Membrane</td><td>Sensible Heat</td></tr>
<tr><td rowspan="5">Flywheel</td><td>Sodium-Beta</td><td rowspan="5">Hybrid Flow</td><td rowspan="3">Solid Oxide Cells</td><td rowspan="5">Superconducting Magnetic Coil</td><td rowspan="5">Thermo-Chemical Sorption</td></tr>
<tr><td>Metal-Air</td></tr>
<tr><td>Lead Acid</td></tr>
<tr><td>Nickel based</td><td rowspan="2">Microbial Cells</td></tr>
<tr><td>Zinc-Halogen</td></tr>
</table>

1.1 Pumped Hydro Storage System (PHSS)

A pumped hydro storage is a hydroelectric system that use water that was pumped and stored into an upper reservoir from a reservoir the has a lower elevation. In this case water is the storage form of energy, because when energy is demanded a hydro turbine is activated, and using gravity, water is converted into energy. PHSS represents over 94 per cent of global electricity storage by capacity (IHA, 2018). There are two types of PHSS:

- Close-loop PHSS – when water is recirculating between lower elevation reservoir and the high elevation reservoir.
- Open-loop PHSS – when one of the reservoirs (upper or lower, usually the upper reservoir) is file up by a natural flow of water

From the point of view of advanced pumped storage technologies, there are two important categories:

- Adjustable speed pump-turbines – adjusting speed of pump-turbine in function of head variation and obtaining best efficiency of the hydro turbine.
- Ternary pumped storage units - on a single shaft of motor/generator are using separate pump and turbine (illustrated in Figure 3).

Evan that PHS plants efficiency is around 80%, have a very long lifetime, and have (in batteries terms) unlimited cycle of charging/discharging, the major disadvantages are: topographical dependence due to the fact that is necessary a level difference between the two water reservoir, and large land used for the water reservoir and therefore, the impact on the local environment, and the loss of natural habitats for flora, fauna and humans.

The energy coming from a PHSS can be transferred into the grid and later on to the end user (i.e domestical or industrial).

1.2 Compressed Air (CASS)

In CASS systems, electricity from a wind farm or from PVs is used to compress and store air in an underground cavern, under water or in tanks. In this way, using this technology, the intermittent energy, supplied from the renewable sources, is stored for a peak energy demand. When a demand of energy appears, the air is decompressing and activate a gas turbine that will generate electricity.

Like PHS plants, CASS require a certain geological topography, but unlike Pumped Hydro Storage System, CASS can be used in more suitable locations (abandoned mines, depleted natural gas reservoir, etc.) without affecting the landscape and natural habitats.

In a simple explanation the surrounding air is compressed using a compressor, and stored into a finite volume at high pressure. Air is released, preheated, and used as fuel for a high-pressure turbine, which will convert the kinetic energy into electrical energy, when is required (an energy peak demand).

When the air is compressed from atmospheric pressure to 70 bar, a thermal discharge take place. The air temperature can be in a range of 150 °C and 177 °C in the compression stage and must be lowered up to 50 °C in order that the air can be stored. To achieve this temperature, traditionally are used intermittent coolers (Diabatic CASS or D-CASS - Figure 4). On the turbine stage, when air is decompressed, the loss of the heat energy is compensated, reheating the compressed air before the turbine chamber

Figure 3. Typical conventional pumped hydro storage system
(IRENA (2017), Electricity Storage and Renewables: Costs and Markets to 2030, International Renewable Energy Agency, Abu Dhabi.)

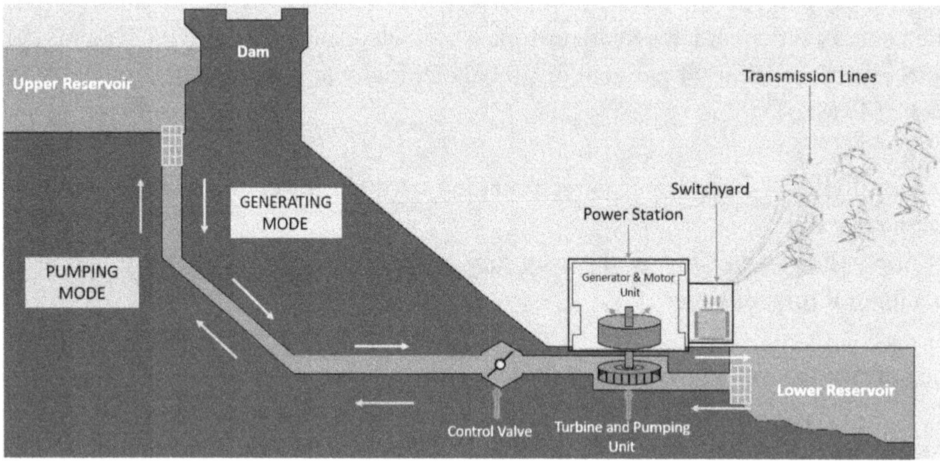

using natural gas fuel (which is polluting) or using a heat recuperation from the compression stage. The efficiency of a D-CASS is around 50%.

A modern approach is by thermally storing the heat generated in the compression stage, before to store air in the cave. Using adiabatic expiation, the compressed air is heated-up by extracting heat from the thermal storage, and the used in the turbine chamber. This type of energy storage is the Adiabatic CASS (A-CASS or Advanced Adiabatic CASS – AA-CASS), and its theoretically efficiency is 100%. In reality is predicted to be around 70-75%, due to the investable thermal losses.

Figure 4. A-CASS using electricity from a photovoltaic farm to store air into a cavern and generate electricity to the grid on peak energy demands
(A. Kere et al (2014), Dynamic behavior of a sensible-heat based thermal energy storage)

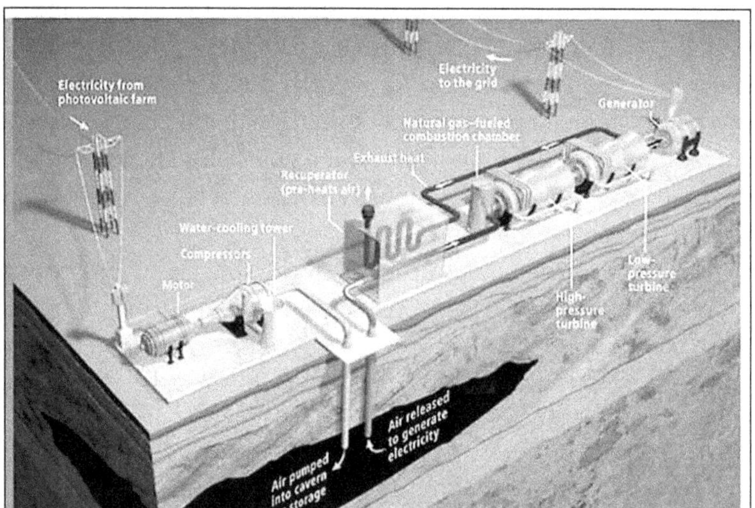

The Near-isothermal CASS system in which the air is compresses/expand near to a large thermal mass. In this way the generated heat from the compressing process is absorbed in a series of parallel fins, and in the expansion process, the heat is released rapidly in order to maintain the temperature at a stabile point. The efficiency of Near-Isothermal CASS is expected to be around 90-95%.

1.3 Flywheel

Flywheel energy storage systems (FESS) have an efficiency of 80%, and, in essence is like a battery, storing kinetic energy (using a motor) for a later use as an electrical energy (using the same motor as a generator). The flywheel is spinning using permanent magnets and reluctance (motor/generator usually on the same shaft). From electrical point of view, if the engine acts like a motor, the rotation speed of the flywheel is increasing, until it has the maximum velocity and, in this way, the kinetic energy is stored. The power electronics convert the necessary power from the grid to achieve the maximum rotation speed. In this step, the energy is maintained constant by suppling idle losses in machine. When a peak energy demand appears, the engine will act as a generator, and will transform kinetic energy into electrical energy, and the rotation speed is decreasing.

Figure 5. Flywheel Energy Storage System – main components
(European Energy Storage Technology Development Roadmap -Update 2017)

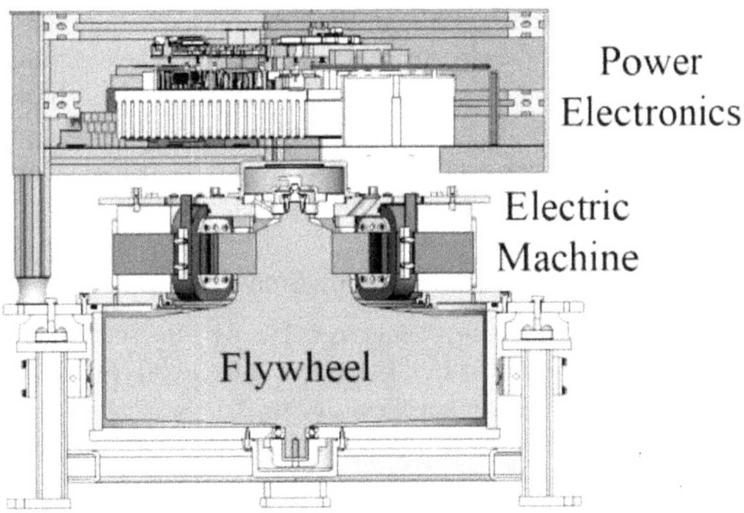

The stored energy in flywheel (E) is given by Equation 1, and is proportional to moment of inertia (J) and the square of angular velocity (ω) (De Oliveira J.G., 2011), (EASE EERA, 2017):

$$E = \frac{1}{2} J\omega^2 \tag{1}$$

Based on minimum and maximum angular velocity (ω_{min} and ω_{max}), the useful energy can be written like in Equation 2:

$$E = \frac{1}{2}\omega_{\max}^2 J\left(1 - \frac{\omega_{\min}^2}{\omega_{\max}^2}\right) \qquad (2)$$

From given equation results that the E (stored energy) can be increased either by increasing ω (the angular velocity) or by increasing J (i.e. the moment of inertia). In this way we can classify Flywheel into (Liu, H.; Jiang, J., 2007):

- Low speed FESS, up to 10,000 rpm, where the wheel is made from metal material, and is using a levitation system (most commonly magnetic type) which offsets its weight. Being so heavy, this system is use in stationary applications
- High speed FESS, up to 100,000 rpm, where wheel is made from composite materials, are lighter, and are using magnetic bearing all being put together into a vacuum chamber. The vacuum chamber it is used in order to eliminate frictional losses, but also to protect the wheel from external disturbances.

FESS are found mainly in stationary applications such as UPS, frequency and voltage regulation.

2. Electrochemical Energy Storage Systems

Electrochemical **E**nergy **S**torage **S**ystems (EESS) are using the method to store electricity under a chemical form. Both electrical and chemical energy have the same carrier, the electron, in this situation the losses are reduced due to the conversion from one form to another. The most important characteristics for storage systems are: the energy density, power density, charging-discharging cycles and the life-time. The energy density indicates how much energy can be store into the batteries. On the other hand, the power density indicates the maximum power output that batteries can generate.

2.1 Ambient and High Temperature Lithium Energy Storage Systems

Li-ion batteries are used on large scale. This popularity is given by the small size, weight, high energy and power density. At the beginning of '60s Bell Labs created the first types of Li-ion Batteries.

We must mention that the Nobel Prize for chemistry in 2019 was awarded to John Goodenough (University of Texas at Austin), M. Stanley Whittingham (Binghamton University) and Akira Yoshino (Meijo University), who "created a rechargeable world" (The Royal Swedish Academy of Sciences), for their work in researching and developing the lithium-ion battery.

Back in 1970s, in Exxon Mobil Corp. laboratory, Whittingham laid the ground framework for the lithium-ion battery. Ten years later, Goodenough used $LiCoO_2$ cathode material. This was a key point to create a lightweight battery with a higher energy density. Yoshino created the world's first rechargeable lithium-ion battery where the anode material was Lithium free, and the lithium ions migrate from the LiCoO2 cathode into the anode during charging. After this first step he used carbonaceous material for the anode, and in 1985 he obtained for the first modern lithium-ion battery (LIB).

Sony has launched, in the beginning of '90s (based on Yoshno's patent), the first commercial lithium-ion rechargeable batteries. Considering that it has been almost 30 years since they were released, lithium-ion batteries have not yet been surpassed by a new generation of battery technology.

A quick view at the statistical data reveal that the technology mix remains largely unchanged. Lithium-ion battery storage continued to be the most widely used, making up nearly 85% of all new capacity installed (Luis Munuera, Alberto Toril, 2018) (Figure 6).

Figure 6. Lithium-ion continues to dominate among all storage technologies excluding pumped hydro. *(Energy storage - International Energy Agency, 2018)*

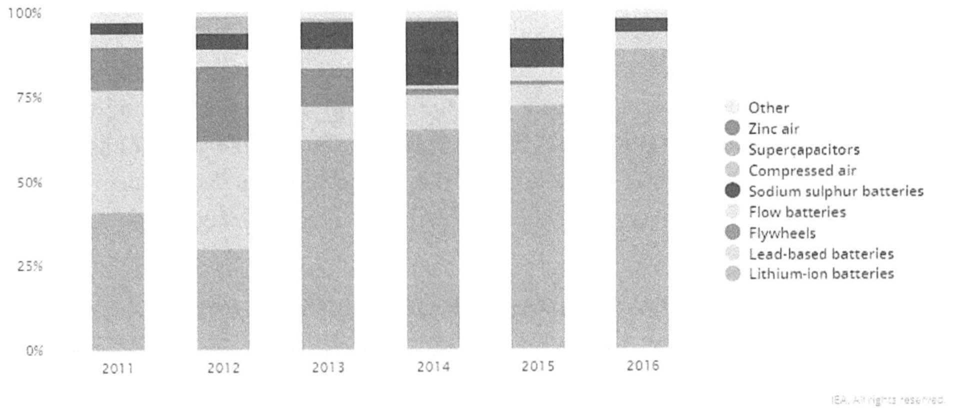

Lithium batteries can be classified by temperature, as shown in Table 2.

From the construction point of view there are three major components: negative and positive electrodes, and electrolyte. The anode material is made from carbon (widely spread graphite, in recent research) while cathode is made from metal-oxide (most popular Li-CoO_2 - lithium cobalt oxide, Li-$FePO_4$ - lithium iron phosphate or Li-Mn_2O_4 - lithium manganese oxide). The electrolyte is a lithium salt in an organic solvent (Silberberg, M. 2006). (typically, ethylene carbonate or diethyl carbonate containing complexes

Table 2. Lithium-Ion Batteries classification from operating temperature point of view

Ambient and High Temperature Lithium Energy Storage Systems		
Classification	**Type**	**Overall Reaction**
High Temperature	Li-Al/FeS (monosulfide) 1.3 VDC at 400 °C	2Li-Al + FeSΔLi_2S + Fe + 2Al
	Li-Al/FeS_2 (disulfide)	2Li-Al + FeS_2ΔLi_2FeS_2 + 2Al Or 6Li-Al + $3Li_2FeS_2$Δ$6Li_2S$ + 3Fe + 6Al
Ambeint Temperature	Li-Po	$Li_{1-x}MO_2 + Li_xC_n$Δ$LiMO_2 + C_n$
	Li-CoO_2	$LiC_6 + CoO_2$Δ$C_6 + LiCoO_2$
	Li-Mn_2O_4	$Li_xMn_2O_4 + Li_yC_6$ Δ$Li_{x+\delta}Mn_2O_4 + Li_{y-\delta}C_6$
	Li-$FePO_4$	$LiFePO_4$+ C_6 ΔLiC_6 +$FePO_4$
	Li-$NiMnCoO_2$	
	Li-$NiCoAlO_2$	
	Li_4-Ti_5O_{12}	$Li_4Ti_5O_{12} + 3Li^+ + 3e^-$ Δ$Li_7Ti_5O_{12}$

of lithium ions that are using salts such as lithium hexafluorophosphate $LiPF_6$, lithium hexafluoroarsenate monohydrate $LiAsF_6$, lithium perchlorate $LiClO_4$, lithium tetrafluoroborate $LiBF_4$, and lithium triflate $LiCF_3SO_3$) (Anil Kumar, Rakesh K. Gupta, 2018). During charging and discharging the electrochemical roles are switching between anode and cathode. In Figure 7 are presented two types of LIBs (a) is the Cylindrical battery LIB, and (b) is a Prismatic LIB structure.

As a working principle, the lithium ion (Li^+) rinses, during the discharge stage, from the anode to the cathode, and during the charging phase, the same ion travels back from the cathode to the anode. In lithium-ion batteries, the materials used for the anode and cathode are made from materials that can "absorb" or intercalate Li^+ (in the sense of retaining ions in the structure of the host material) without being significantly disturbed. Usually, the lithium-ion batteries (LIB), at the beginning of their life, are full discharged. The Li^+, are trapped into anode material and the LIB is not producing electricity.

To start the chemical reaction in said the LIB it must be charged. In the step of charging at the LIB's cathode occurs an oxidation reaction. That means that the cathode loses a part of electrons (electrons are negatively charged), and in order to maintain the equilibrium, Li^+ are relisted in the electrolyte solution.

Figure 7. Lithium-ion batteries construction - (a) Cylindrical LIBs, (b) Prismatic LIBs
(Gowtham S., Lithium ion battery – 2016)

(a) *(b)*

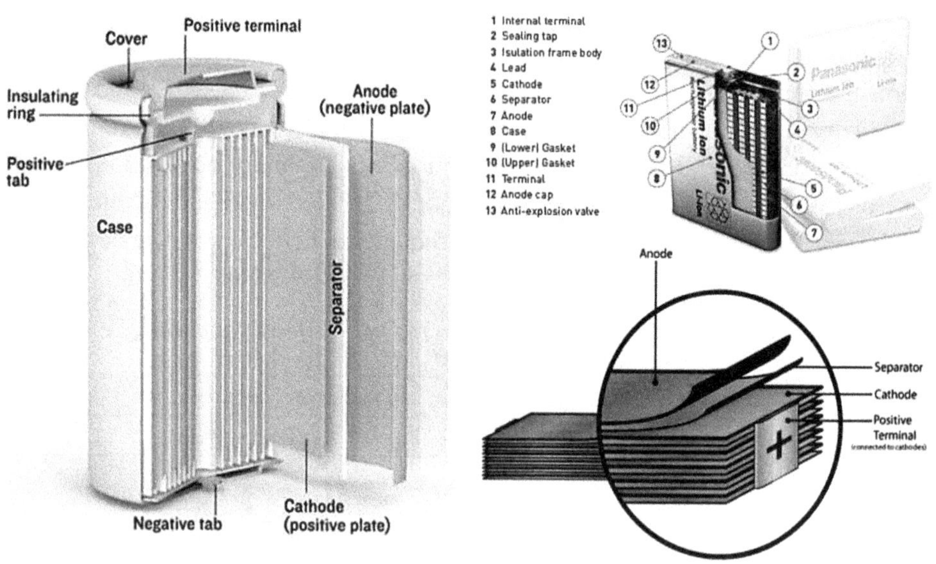

The Li+ goes to anode and are trapped into graphite with electrons presented in Figure 8. The processes revers in the discharge phase. From anode the Li+ are released and travel back, through electrolyte solution to the cathode where again are trapped. In Figure 9 is presented the discharging cycle of a LIB.

These cycles go one and one until the graphite anode eventually brakes apart, and the LIB performance become lower and lower until will stop working.

A comparation based on materials is proving the performance changings that occurs in different types of Lithium-Ion batteries (Figure 10, and Figure 11).

Figure 8. Rechargeable Lithium-ion batteries (LIB) – Charging phase

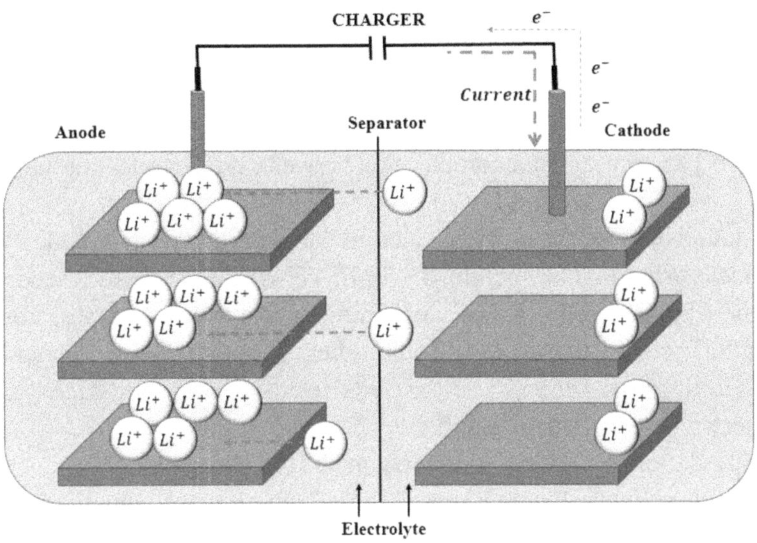

2.2 Sodium-beta Batteries

These batteries are using beta-alumina (β''-Al_2O_3) solid electrolyte which, at high temperatures, values both good Na^+ conductivity and also electrical isolation, and can be classify into sodium-sulfur (Na-S), that was designed in the 1960s (credited by Ford Motors Co.) and sodium-metal batteries, introduced in 1980s. (Akhil AA, Huff G, Currier AB, Kaun BC, Rastler DM, Chen SB, Cotter AL, Bradshaw DT, Gauntlett, 2013) (Electrical Energy Storage. White paper. The International Electrotechnical Commission, 2011) (Linden D, Reddy TB., 2001). These days NGK Insulators, Ltd. and Tokyo Electric Power Co. (Tepco) are the promoters for these next-generation Na-S (Sodium-Sulfur) batteries. As it was showed

Figure 9. Rechargeable Lithium-ion batteries (LIB) – Discharging phase

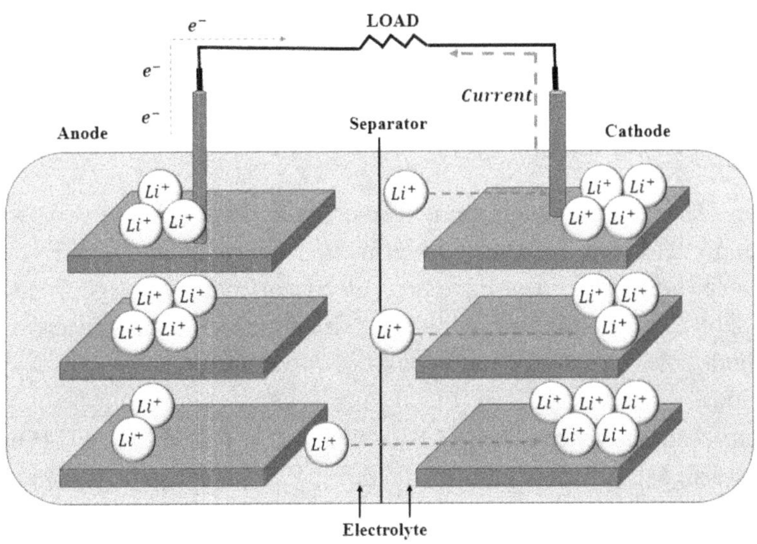

LIB are very popular, but the increasing demand and having low Li salts resources, it become a necessity to find an alternative solution. The sodium batteries are one of the candidates for "after LIB" era.

Sodium is wildly spread resource, it has a low cost, and the performances are comparative with the one of LIBs (H. Yu, Y. Ren, D. Xiao, S. Guo, Y. Zhu, Y. Qian, L. Gu, H. Zhou, 2015)(X. Hu, Z. Li, Y. Zhao, J. Sun, Q. Zhao, J. Wang, Z. Tao, J. Chen,, 2017) (H. Su, S. Jaffer, H. Yu, Energy Storage Materials, 2016). In Figure 12 is presented a comparation between different technologies types of batteries and Na-S batteries.

From the construction of view, the positive electrode of Na-S batteries is made by molten sulfur, and the negative electrode is using molten sodium. The solid ceramic electrolyte is separating the two active materials and is conducting the sodium ions, depending on the operating state, from anode to cathode or on the reverse path. There are two types of sodium-beta batterie (Figure 13 a and b):

- The Na-S battery where cathode is sulphur
- ZEBRA (Zeolite Battery Research Africa renamed Zero Emission
- Battery Research Activity) batteries where cathode is sodium-metal halide (most used nickel chloride NiCl2). Molten sodium tetrachloroaluminate (NaAlCl4A) it is used as secondary liquid electrolyte improving the performances of ZEBRA batteries.

Figure 14a illustrate the operating principle as it fallows: during discharge, Na^+ passes through the electrolyte and is combined with sulphur, forming sodium polysulfides.

The electrons path is through the external circuit and thus provides a potential difference. In the revers process of charging, Na+ from the polysulfide are released, and the ions migrate back through electrolyte and recombines to create Na. The most important aspect is that, these reactions need to be done in molten salts, so the temperature must be keep around 280 °C. To accomplish this challenge Na-S includes electric heater (for keeping reactants molten), and thermal insulation (in this way reducing the heat loss). The same principle is applied for ZEBRA (Figure 14b), in this case the negative electrode is liquid sodium, and the positive electrode is metal chloride, usually $NiCl_2$, and β''-Al_2O_3 is the electrolyte and a separator.

The overall electrochemical reaction in Na-S battery is $Na + xS \leftrightarrows Na_2Sx$, where x is 3–5, and for ZEBRA is $NiCl_2 + 2Na \leftrightarrows Ni + 2NaCl$ (in this case E=2.58V)

The thermal management problems combined with the thermal sealing problems (W. Coors, C. Boxley, M. Robins and A. Eccleston, 2012) are the now days researchers main key-focus interest.

2.3 Metal-air Batteries

In Metal-air batteries (MAB) the anode is made from a metal (Li, Ca, Mg, Fe, Al, and Zn) and cathode inexhaustible air supply. This type of battery in terms of safety present a high risk because Lithium in combination with air and humidity can produce fire. (Electrical Energy Storage. White paper, IEC, 2011).

In Figure 15 is shown the working principle. At the anode the metal is transforming into metallic ions, and at the cathode side oxygen from air is passing through a GDL (gas diffusion layer) transforming into hydroxide ions.

The overall electrochemical reaction in MABs is $4M + nO_2 + 2nH_2O \Delta 4M(OH)_n$ where M can be Li, Na, K (alkali metals), Mg (alkaline earth metals), Fe, Zn, Al, and n is the oxidation number of the metal ion.

Figure 10. Comparation between most spread types of Lithium-Ion Batteries (batteryuniversity.com)

Type	**Lithium Cobalt Oxide**	**Lithium Manganese Oxide**	**Lithium Nickel Manganese Oxide**
Short form	Li-cobalt	Li-manganese	NMC
Abbreviation	$LiCoO_2$ (LCO)	$LiMn_2O_4$ (LMO)	$LiNiMnCoO_2$ (NMC)
Cathode	$LiCoO_2$	$LiMn_2O_4$	$LiNiMnCoO_2$
Anode	Graphite	Graphite	Graphite
Nominal voltage	3.60V	3.70V (3.80V)	3.60V (3.70V)
Full charge	4.20V	4.20V	4.20V (or higher)
Full discharge	3.00V	3.00V	3.00V
Minimal voltage	2.50V	2.50V	2.50V
Specific Energy	150–200Wh/kg	100–150Wh/kg	150–220Wh/kg
Charge rate	0.7–1C (3h)	0.7–1C (3h)	0.7–1C (3h)
Discharge rate	1C (1h)	1C, 10C possible	1–2C
Cycle life (ideal)	500–1000	300–700	1000–2000
Thermal runaway	150°C (higher when empty)	250°C (higher when empty)	210°C (higher when empty)
Applications	Mobile phones, tablets, laptops, cameras	Power tools, medical devices, powertrains	E-bikes, medical devices, EVs, industrial
Comments	High energy, limited power. Market share has stabilized.	High power, less capacity; safer than Li-cobalt; often mixed with NMC to improve performance.	High capacity and high power. Market share is increasing. Also, NCM, CMN, MNC, MCN
Hexagonal spider graphic of performance	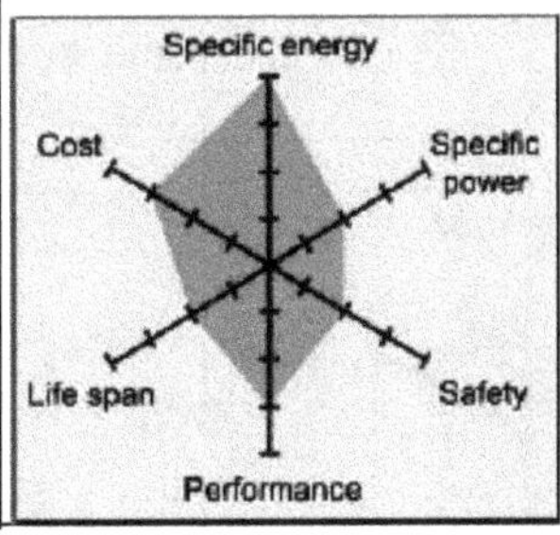	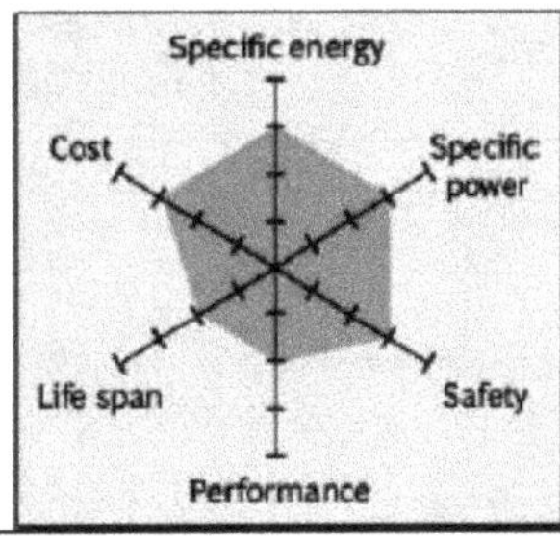	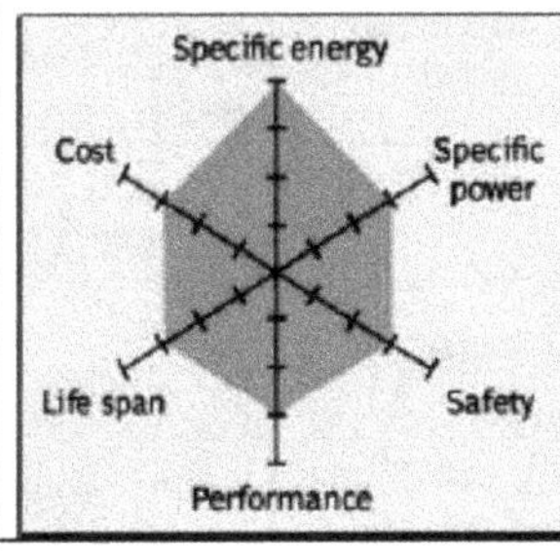

Figure 16 highlight the theoretical energy density for different types of metal-air batteries. In comparation with Li-ion batteries Theoretical energy density of MABs is from 3 to 30 times greater. The development of MABs begun at the end of 1870s (Maiche designed the first Zi-air battery), but latter on the development process the problems occurred with the metal anodes, the catalyst and the electro-

Figure 11. Comparation between most spread types of Lithium-Ion Batteries
(batteryuniversity.com)

Chemistry	Lithium Iron Phosphate	Lithium Nickel Cobalt Aluminum Oxide	Lithium Titanate Oxide
Short form	Li-phosphate	Li-aluminum	Li-titanate
Abbreviation	$LiFePO_4$ (LFP)	$LiNiCoAlO_2$ (NCA)	Li_2TiO_3 (common) (LTO)
Cathode	$LiFePO_4$	$LiNiCoAlO_2$	Li-manganese or NMC
Anode	Graphite	Graphite	Li_2TiO_3 (titanate)
Nominal voltage	3.20, 3.30V	3.60V	2.40V
Full charge	3.65V	4.20V	2.85V
Full discharge	2.50V	3.00V	1.80V
Minimal voltage	2.00V	2.50V	1.50V (est.)
Specific Energy	90–120Wh/kg	200-260Wh/kg	70–80Wh/kg
Charge rate	1C (3h)	1C	1C (5C max)
Discharge rate	1C (25C pule)	1C	10C possible
Cycle life (ideal)	1000–2000	500	3,000–7,000
Thermal runaway	270°C (safe at full charge)	150°C (higher when empty)	One of safest Li-ion batteries
Applications	Stationary with high currents and endurance	Medical, industrial, EV (Tesla)	UPS, EV, solar street lighting
Comments	Flat discharge voltage, high power low capacity, very safe; elevated self-discharge.	Highest capacity with moderate power. Similar to Li-cobalt.	Long life, fast charge, wide temperature range and safe. Low capacity, expensive.
Hexagonal spider graphic of performance	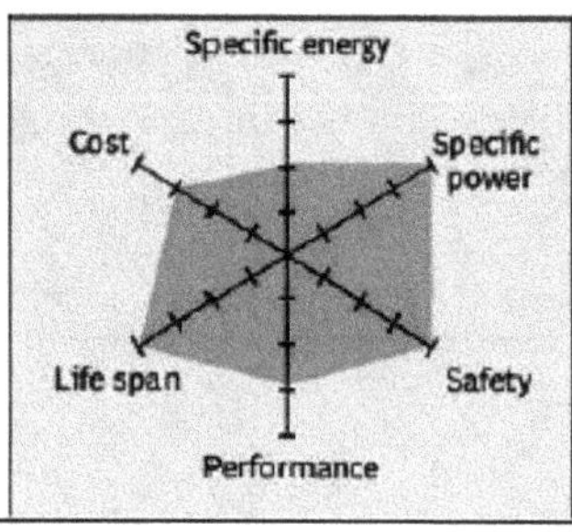	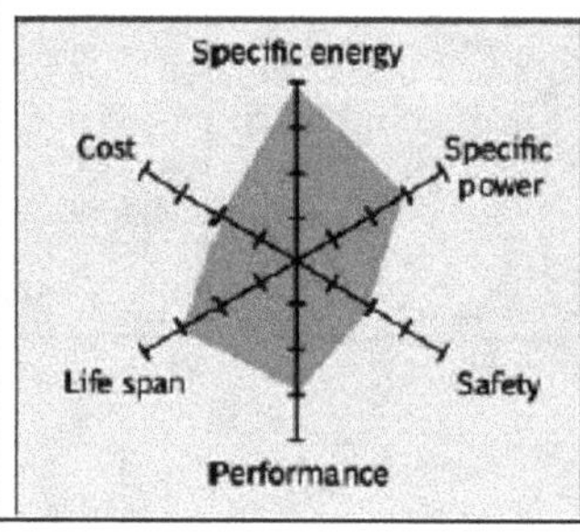	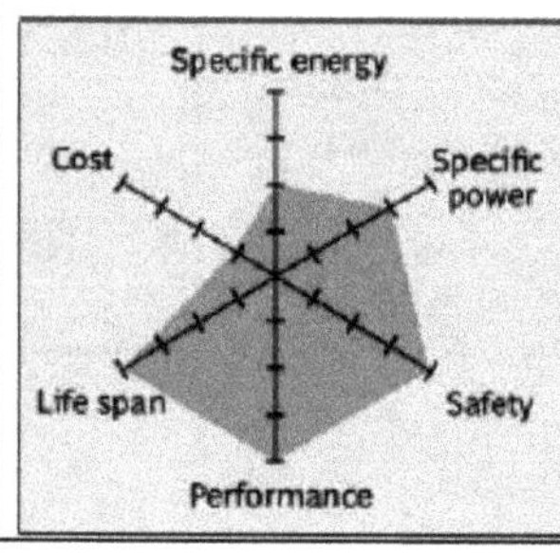

Figure 12. Comparation between Na-S battery technology and other types of batteries technologies

Battery technology	Development status	Output range	Capacity range	Response Time [milliseconds]	Operating Temperature Range [°C]	Energy density [Wh/kg]	Lifetime [years]	Applications
Sodium-sulfur	Commercialized	360 kWh to tens of MWh	50 kW to 100MW	1	287-360	100 to 120	15	UPS; Peak shaving; Load shifting; Spinning reserve
Zinc-bromide	Early commercialization	>50 kWh (no upper limit)	>50 kW (no upper limit)	4	20-50	75 to 85	10 to 20	UPS; Peak shaving; Load shifting; Integration of intermittent generators; Power plant black start
Vanadium-redox	Early commercialization	20 kWh to 10 MWh	5 kW to 10MW	10-20	< 40	50	15	
Nickel-cadmium	Fully commercialized	Largest to date 14.5 MWh	Largest to date 26 MW	1	Optimal at 25	35	15 to 20	Bridge applications; Integration of intermittent generators
Lead-acid	Commercialized	Up to tens of MWh	<1 kW to tens of MW	5-8	Optimal at 25	35	15 to 20	UPS; Peak shaving; Load shifting;

Figure 13. a) Na-S batterie structure; b) ZEBRA batteries structure
a) (NGK INSULATORS, LTD, https://www.ngk.co.jp/ 2019); b) (Karina B. Hueso, Michel Armand, and Teofilo Rojo, 2013)

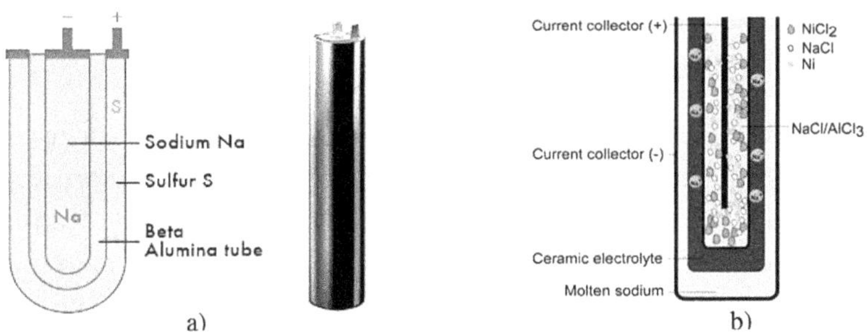

Figure 14. a) Na-S batterie operating cycles; b) ZEBRA batteries operating cycles
a) (NGK INSULATORS, LTD, https://www.ngk.co.jp/ 2019); b) (Karina B. Hueso, Michel Armand, and Teofilo Rojo, 2013)

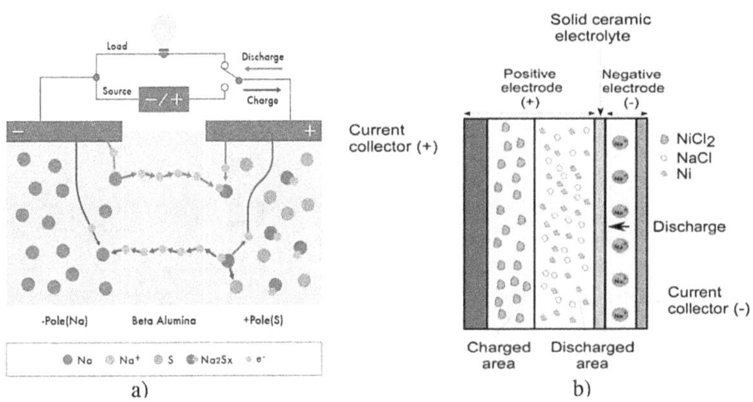

Figure 15. MAB with a) non-aqueous electrolyte and b) with aqueous electrolyte working principle

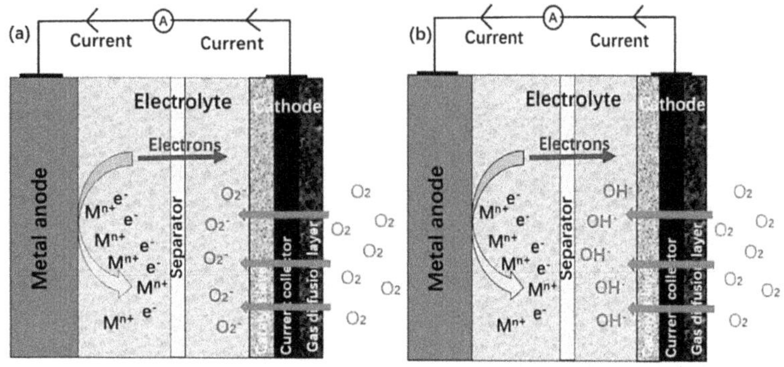

lytes becomes barriers in implementing these types of MABs on large-scale. The main objective is to replace LIB, because, is generally accepted that Lithium-ion batteries are approaching to the superior performance limit and, in the future, eventually improves over LIB technologies can add maximum 30% of energy density. Because of the technical issues above mentioned, and adding financial issue (the cost must be less than 100 $/kWh), even if serious steps forward are made in the improving MAB technologies, is not certain when the replacement will take place. (Y. Li, and J. Lu, 2017, 28. Li, F. J.; Zhang, T.; Zhou, H. S.; Yang, B, 2013).

2.4 Lead Acid Batteries

Gaston Planté invented in 1859 the lead acid battery and in 1912, when Cadillac used this type of battery for the starter motor, the lead acid batteries became very popular and wide spread. Lead-acid batteries structure is made of a lead (IV) oxide cathode, a sponge metallic lead anode lead-antimony alloy with lead (IV) oxide pressed into it (or lead-calcium alloy), and a sulphuric acid solution electrolyte usually sulphuric acid (H_2SO_4). According to the functioning the lead acid batteries can be classified as it fallows:

Figure 16. Theoretical Energy density for different MABs
(Y. Li, and J. Lu, 2017)

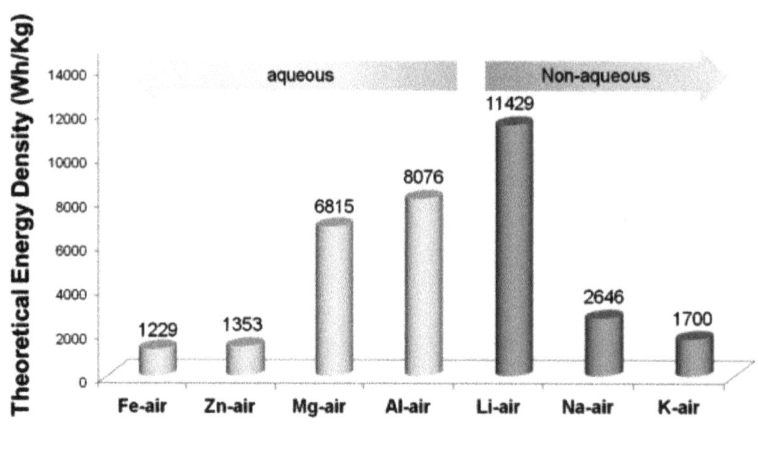

- Starting Batteries – Used to start and run engines. These types of batteries deliver a very large current in a very short time, discharging by about 2-5%.
- Deep Cycle Batteries – are used to store energy (solar, mini-hydro), electric vehicles and emergency backup. Over thousands of charging and discharging cycles, these batteries discharge around 80% of their capacity. In this category we can mention: *Wet Cell (Flooded)* with two variants: serviceable and maintenance-free, in which the electrolyte is liquid, and *Sealed lead acid batteries* with two variants **Gel Cell** (Gelled Electrolyte) also being known as valve-regulated lead acid (VRLA) that is using a mixture between silica gelling and electrolyte and has a form of semi-stiff paste, and **Absorbed Glass Mat** (AGM, Absorbed Electrolyte) that using a glass fibre mat immersed in electrolyte.
- Marine Batteries – is a battery between deep cycle and starting batteries although some are true deep cycle batteries.

From operational point of view the chemical energy play the role of storing the electrical energy in the charging phase, and in the discharging phase chemical energy is converted back into electricity (Figure 17). The chemical overall reactions during charge and discharge are as it follows:

$$PbO_2 + Pb + 2H_2SO_4 \underset{charge}{\overset{discharge}{\rightleftarrows}} 2PbSO_4 + 2H_2O$$

Negative plate overall reaction is:

$$Pb + SO_4^{2-} \underset{charge}{\overset{discharge}{\rightleftarrows}} PbSO_4 + 2e^-$$

Positive plate overall reaction is:

$$PbO_2 + SO_4^{2-} + 4H^+ + 2e^- \underset{charge}{\overset{discharge}{\rightleftarrows}} PbSO_4 + 2H_2O$$

Because is inexpensive (in terms of $/watt), has a good performance at cold temperature, good safety, and ease of recycling (around 99% recycling rate), lead acid battery continues to remains a viable choice. Battery experts considered that the limitations of this type of battery are given by the use of lead. The new tendency is to improve the active material, adding carbon to the negative electrode. This reduces sulfur, the conductivity is clearly improved and the recharging time is decreasing.

In Table 3 are put together the new technologies in Lead-acid batteries field.

Figure 17. Lead-acid battery – charging and discharging process

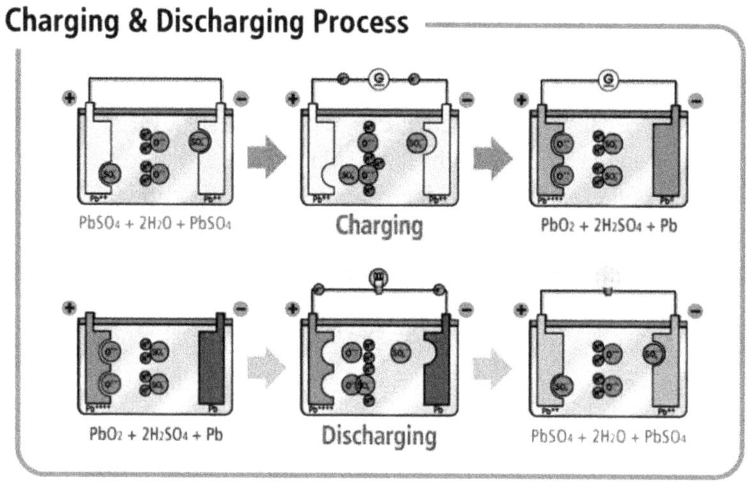

2.5 Nickel Based Batteries

Nickel-based batteries have a long history. In 1899 Waldmar Jungner invented the nickel-cadmium battery, but because the battery was too expensive, he tried to migrate from cadmium to iron. Encountering numerous technical issues, Jungner abandoned his research. In 1901, Thomas Edison continue the development on nickel-iron battery. Edison lost the bet on electric cars (gasoline cars are entering on the market), and also, he was very disappointed when car manufacturers decided to use lead-acid batteries instead of nickel-iron for SLI system (starter, light and ignition). The idea was not abandoned, and improves ware made over the years, until 1947 the first modern NiCd battery was made. It was very popular and the first choice for portable devices (power tools, video cameras, radios, medicals equipment).

Table 3. New types of LEAD-ACID Batteries

Battery	Performance	Actual Stage
Advanced Lead-carbon (ALC)	Rapid charging on regenerative breaking; Low-cost; Operates at sub-freezing; No need of active cooling;	Experimental. Steel need to correct in the discharge phase the rapid voltage drop
Firefly Energy	Operate for extended time in partial-states-of-charge; lighter the classical lead-acid batteries on the market; longer life-time; comparable to NiMH	On the market. Firefly Batteries Pvt. Ltd., India is producing this type of battery since 2014
Battery	Performance	Actual stage
Altraverda Bipolar	Based on lead; Ebonex® (titanium sub-oxide ceramic structure) for the grid and an AGM separator; comparable with NiCd at a specific energy of 50–60Wh/kg;	Under development
CSIRO Ultrabattery	Combines asymmetric ultracapacitor with lead-acid battery; improve the lifetime; cost reduction	Tested in Honda Insight HEV and tested in rural India by the Institute of Transformative Technologies
EEStor	Based on a modified barium titanite ceramic powder; The specific energy is clam to be up to 280Wh/kg; no deep-cycle wear-down, 3–6 minutes charge time;	Research is continuing.

The Nickel-based batteries structure (Figure 18b) is formed from the positive electrode (made by an active material that include nickel oxyhydroxide), the metal negative electrode (Fe, Cd, Zn, MH, or H_2), and an alkaline electrolyte (H. Chen, T. N. Cong, W. Yang, C. Tan, Y. Li, Y. Ding, 2009). These types of Nickel-based batteries are compared in Figure 19.

Figure 18. Nickel-cadmium battery a) charging and discharging process b) internal structure

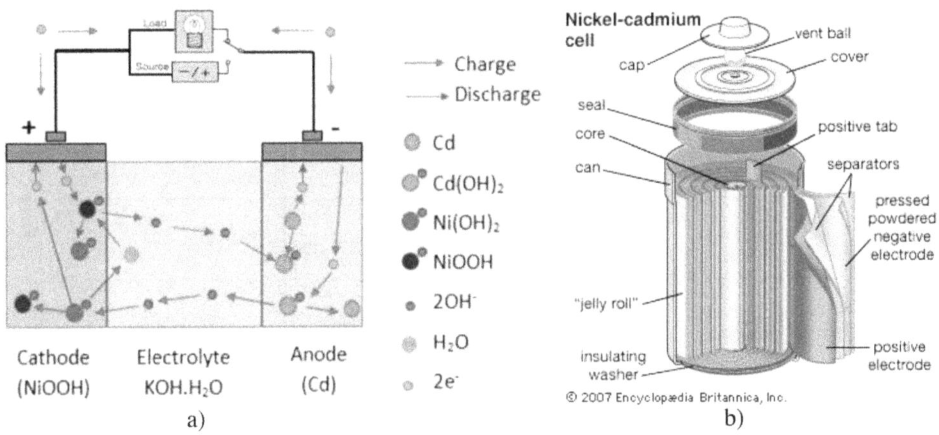

Operating principle (Figure 18a) when the NiCd battery is fully charged, $Ni(OH)_4$ is the positive plate and Cd is the negative plate. In the discharge mode KOH (potassium hydroxide) dissociate into K^+ (potassium) and OH^- (hydroxyl) ions. The K^+ travel to anode where $Ni(OH)_4$ is converted to $NI(OH)_2$. Similar OH- travels to the cathode where Cd is converted to $Cd(OH)_2$ (cadmium hydroxide) . In the charging mode, the process is reveres (e.g., K^+ go to the cathode and OH^- to the anode) and the When the battery is put on charging, the hydroxyl (OH-) ions move towards the anode, whereas the potassium ions (K+) move towards the cathode, anode and cathode has at full charge there previous chemical composition. Overall chemical reaction is:

$$Cd_{(s)}+2NiO(OH)_{(s)}+2H_2O_{(l)}\rightarrow Cd(OH)_{2(s)}+2Ni(OH)_{2(s)}$$

Ni–Cd batteries have so-called "memory effect", meaning that charging and discharging at the same state the Ni-Cd batteries remembers the points where recharge begun. During a new cycle of the battery drops voltage at that "memorised" point as if the battery had been discharged. In reality, the capacity of the battery is very little affected.

2.6 Zinc-Halogen Batteries

Charles Renard invented in 1884 Zinc/Chloride (Zn/Cl_2) battery and C. S. Bradley invented in 1885 the Zinc/Bromine battery. Due to technical limitations for that specific period of time (e.g. for Zn/Br (1) zinc tendency to form dendrites when deposition and (2) in aqueous zinc bromide electrolyte, bromine has a high solubility; for Zn/Cl_2 (1) complex structure, (2) piping issues, (3) heating, and (4) corrosive issues), these types of batteries were put into a stand-by state. Later on, with the evolution of materials

Figure 19. Comparation between Nickel-based batteries
(Batteryuniversity.com, 2016)

Chemistry	Nickel-cadmium (NiCd)	Nickel-metal-hydride (NiMH)	Nickel-iron (NiFe)	Nickel-zinc (NiZn)	Nickel-hydrogen (NiH)
Cathode	Nickel	Nickel	Oxide-hydroxide;	Similar to NiCd; uses alkaline electrolyte and nickel electrode	Nickel electrodes, hydrogen electrodes, in pressurized vessel
Anode	Cadmium	Hydrogen-absorbing	Irone with potassium hydroxide electrolyte		
Nominal voltage			1.20V	1.65V	1.25V
Specific Energy	45–80Wh/kg	60–120Wh/kg	50Wh/kg	100Wh/kg	40–75Wh/kg
Charge rate	Can be above 1C	0.5–1C	Not defined	Regular charge	Not defined
Discharge rate	Can be above 1C	1C	Moderate	Relative high power	Not defined
Cycle life (full DoD)	1,000	300–500	20 years in UPS	200–300	Very long cycle life (>70,000 partial)
Maintenance	Full discharge every 3 months (memory)	Full discharge every 6 months	Not defined	Not defined	Maintenance free; low self-discharge
Failure modes	Memory reduces capacity, reversible	Memory (less affected than NiCd)	Overcharge causes dry-out	Short cycle life due to dendrite growth	Minimal corrosion
Packaging	A, AA, C, also in fractional sizes	A, AA, AAA, C, prismatic	Not defined	AA and others	Custom made; each cell costs >$1,000
Environment	Broad temperature range. Toxic	Considered non-toxic	Poor performance when cold	Good temperature range	Operates at –28°C to 54°C
History	1899, sealed version made commercial in 1947	Research started in 1967, commercial in the 1980s; derived from nickel-hydrogen	In 1901,Thomas Edison patented and promoted NiFe in lieu of lead acid; failed to catch on for ICE, EV	In 1901, Thomas Edison was awarded the U.S. patent for the NiZn battery	Problems with instabilities in 1967 caused a shift from NiMH to NiH
Applications	Main battery in aircraft (flooded), wide temperature range	Hybrid cars, consumer, UPS	German V-1 flying bombs, V-2 rockets; railroad signaling, UPS, mining	Renewed interest to commercial market with Improvements	Exclusively satellites; too expensive for terrestrial use
Comments	Robust, forgiving, high maintenance. Only battery that can be ultrafast charged with little stress	More delicate than NiCd; has higher capacity; less maintenance	In 1990, Cd was substituted with Fe to save money. High self-discharge and	High power, good temperature range, low cost but high self-discharge and short	Uses a steel canister to store hydrogen at 8,270kPa (1,200psi)

and piping systems, research on new materials, and the 1973 oil crises, these types of batteries were put back on the table to the researchers. Exxon created special a department of Research and Engineering in order to revive these types of batteries having the declarative purpose to develop a battery. In Figure 20 is presented the schematic of Exxon three-cells Zn/Br_2 battery.

As it can be seen, the electrolyte solution (aqueous solution of zinc bromide) is circulated using pumps. In charging state at the negative electrode is deposit Zn and Br_2 is deposit on the positive electrode. In discharge state, zinc ions are formed at the negative electrode and bromide ions are formed the positive electrode. The separator does not allow diffusion of bromine to zinc reservoir.

The overall chemical reaction it can be written like:

Figure 20. Three-cell zinc / bromine
(Courtesy of Exxon Research and Engineering Co. and Sandia National Laboratories.)

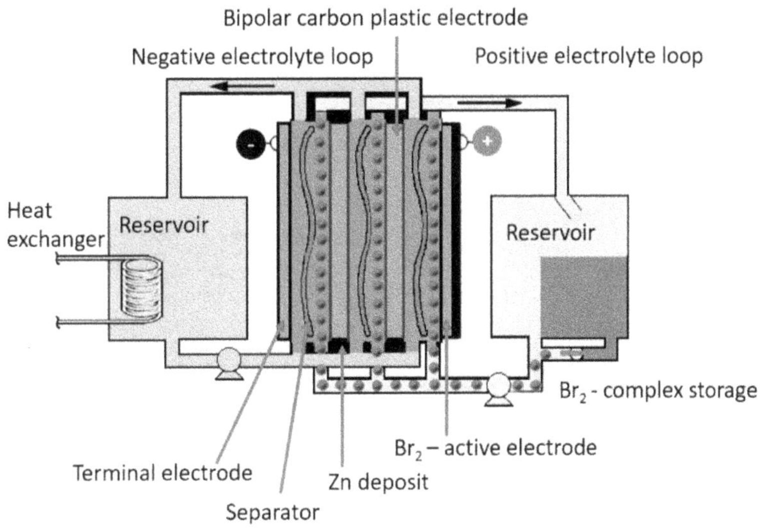

$$ZnBr_2(aq) \underset{charge}{\overset{discharge}{\rightleftarrows}} Zn + Br_2(aq)$$

The advantages of Zn/Br_2 battery are: good energy efficiency and specific energy, low cost and easily accessible materials, capable of fast charging, 100% discharge dose not affect the battery, flexibility in system design, adequate power density for most applications.

The disadvantages and future work: the auxiliary system required for the circulation systems and for the temperature control, really carefully in designing such a system - SAFETY FIRST, power capability must be improved.

However, both types of batteries were again put in a con shadow because of the reduce efficiency, high risk in case of crash, and the necessity, in the recharging phase, for highly trained personnel. (Linden D, Reddy TB, 2015), (P. T. Moseley, J. Garche, 2015) Now days, researchers are still looking for technical solutions to improve this concept and to overpass all the safety and maintenance problems. (Gobinath Pillai Rajarathnam, 2016).

2.7 Redox Flow Battery and Hybrid Flow Batteryes

The energy of the Redox Flow Batteries (RFB) and in the Hybrid Flow Batteries (HFB) is defined by the size of the storage tank. This aspect is very important, giving to these types of ESSs the flexibility and scalability properties. The basic principle is to store and release energy using a reversible electrochemical reaction. Like in Zinc-halogen batteries, two independent loops are used, one for each electrolyte, and also share the same advantages and disadvantages.

Rongke Power Co. Ltd. and UniEnergy Technologies (UET) created a joint-venture in order to deliver in Dalian, China the biggest RFB 200-MW/800-MWh vanadium energy storage project. Unlike

Figure 21. Dalian, China 200-MW/800-MWh vanadium energy storage project, Twitter #Rongke, 2018

Li-ion battery the RFB it has a lower power but this disadvantage is compensated by the lifetime, and in stationary application is very reliable and safe. Figure 22 shows the basic concept of RFB and HFB.

The VRB (Vanadium Redox Battery) was patented by the University of New South Wales in Australia in 1986, aldose it can be said that was this type of battery was researched (without successful) by Pissoort in the 1930s, NASA, and separately by Pellegri and Spaziante in the 1970s. In 1980's Maria Skyllas-Kazacos. The first successful demonstration of the all-vanadium redox flow battery which employed vanadium in a solution of sulfuric acid in each half was by at the University of New South Wales in the 1980s. The overall reaction for VRB is

At the positive electrode:

$$V^{4+} \underset{charge}{\overset{discharge}{\rightleftarrows}} V^{5+} + e^{-}$$

At the negative electrode:

Figure 22. Basic concept of a) RFB and b) HFB
(J. Winsberg, T. Hagemann, T. Janoschka, M. D. Hager U. S. Schubert, 2016)

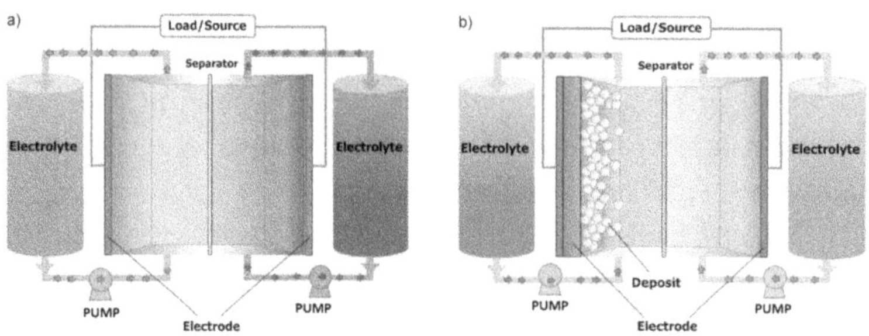

$$V^{2+} \underset{charge}{\overset{discharge}{\rightleftarrows}} V^{3+} + e^{-}$$

Of course, the performances of the RFB or HFB depends on the selected negative and positive electrolytes (all-vanadium, hydrogen/bromine, all-chromium, all-cobalt, zinc/cerium, sodium/iodine, all-copper, iron/air, zinc/polyiodide, iron/bromine, polythiophene, iron/vanadium, aqueous-based lithium/bromine, a metal-free organic–inorganic aqueous, and others), the stability in supporting anion or cation, and bipolar plates materials. The researchers are in a continued race for new materials and the improvements of the performances and safety for RFB and HFB.

3. Chemical Storage Fuel Cell

The **C**hemical **S**torage **S**ystems (**CSS**) use the excess electricity in a chemical reaction to create chemical compounds (molecules or atoms) that can subsequently be re-electrify to cover energy-demand. The major difference between Batteries and CSS-FC (Chemical Storage Systems Fuel Cell) is that the Battery is working in a close loop (except MAB), where oxidants are finishing resources, instead FC is working in an open loop, and the oxidants are supply externally.

One of the most attractive chemical compounds is **Hydrogen** that it become an energy carrier. The Hydrogen can not be found on earth in pure state, but using different process techniques it can be extracted from fossil fuels, natural gas, coal, biomass, water electrolyses, and from technological processes as a by-product. (IEA for Japan G20, 2019).

There are servel types of electrolysers, depending on the type of electrolyte:

- **Alkaline electrolysers** (Figure 23) that have low cost, with an efficiency between 63–70%, it uses an alkaline electrolyte (KOH Aqueous Solution) and operating pressure between 1-30 barg, operating temperature 60-80 °C, commercially available

Figure 23. A Nel Aalkaline Atmospheric Electrolyser, delivering up to 3880 Nm³/h at 200barg H_2
(nelhydrogen.com, 2019)

Figure 24. Scalable system with 4 PEMEl stacks
(AREVA H2Gen, 2019)

- **PEMEl** (Proton Exchange Membrane Electrolysers – Figure 24), high cost due to the usage of noble PEMEl materials, pure water as electrolyte, with an efficiency between 56–60%, operating pressure between 30-80 barg, and temperature between 50-80 °C, commercially available.

- **SOEC** (**S**olid **O**xide **E**lectrolysis **C**ells – Figure 25), can also operate in revers phase (converting hydrogen back to electricity), has an efficiency between 74–81%, it is using ceramics as the electrolyte and low-cost materials, operating pressure 1 brag, operating temperature 650–1000 °C, yet not commercially.

Figure 25. SOFC / SOEC test system
(Fiaxell SOFC Technologies, 2019)

- **MEC** (**M**icrobial **E**lectrolyser **C**ell – Figure 26), using wastewater or organic matter it has two benefits: generate electricity and treat the wastewater for reuse. MFC is still a technology in research due to the following limitation: low power density, ohmic and activation losses, bacterial metabolic loses, and high initial cost.

Figure 26. Single chamber MEC shown with gas collection tube (top), Ag/AgCl reference electrode (extending from the front), cathode connection (left clip) and brush anode connection (right clip), Call and Logan
(2008, Environ. Sci. Technol.)

Hydrogen can be used in different types of applications like combustible for FCEVs (Fuel Cell Electrical Vehicles), can be turned back into electricity using FC (Fuel Cells), combined with CO2 it can be obtained SNG (Synthetic Natural Gas), can be used as raw material in other technological processes (oil refining, semiconductor industry).

From the construction point of view a PEMEL cell has two chambers separated from each other by Membrane Electrode Assembly (MEA) formed by a PEM (Proton Exchange Membrane) usually Nafion®, on witch where heat pressed the electrodes (usually Platinum based catalyst on cathode chamber and Iridium based catalyst in the anode chamber). Because is permitting only to protons to pass, practically electrically insulates the electrodes from each other. A GDL (Gas Diffusion Layer, Ti or Carbon based) and Flow-Plates are sandwiching the MEA (Figure 27 – PEMEL structure).

The overall chemical reaction is:

$$H_2O \overset{electricity}{\rightarrow} H_2 + \frac{1}{2}O_2$$

10 reasons why PEM Electrolyser is very attractive: 1) high over-all energy efficiency; 2) high rate of hydrogen production; 3) high pressure operation ; 4) high purity of produced H_2 gas; 5) low gas crossover;

Figure 27. PEM Electrolyser components and thermodynamics of water electrolysis

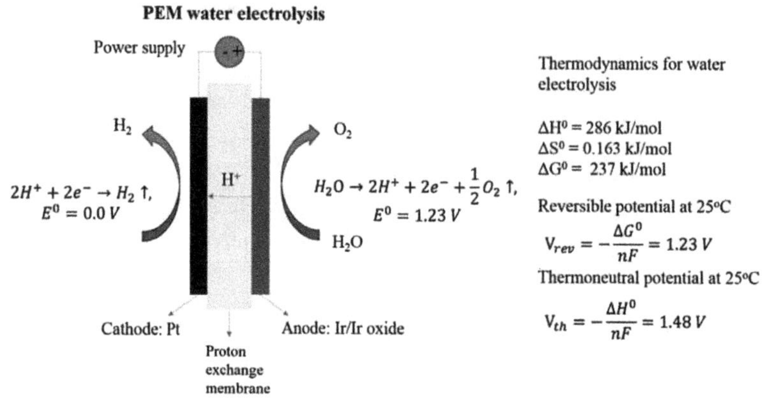

6) control over electrical power variation; 7) small footprint and compact design; 8) high safety level; 9) easy handling; 10) easy maintenance

10 things that must be improved: 1) cost; 2) performances; 3) over-all efficiency; 4) mechanical; chemical and thermal stability; 5) auxiliary components like dryers; filters, ultra-pure water product cost, power-electronics; 6) net power delivered; 7) resistance at corrosion; 8) durability; 9) maturity of the technology; 10) standardisation

Researchers all over the globe are trying to improve PEMEL technology, from the geometry and materials of bipolar flow-plate to the MEA materials. For example, recently (November 2019) it was reported an one-dimensional bunched platinum-nickel (Pt-Ni) alloy nanocages with a Pt-skin structure for the oxygen reduction reaction in fuel cells. The nanocages display high mass activity (3.52 amperes per milligram platinum) and specific activity (5.16 milliamperes per square centimetre platinum)—nearly 17 and 14 times higher respectively as compared with a commercial platinum on carbon (Pt/C) catalyst (X. Tian, X. Zhao, Y. Q. Su, L. Wang, H. Wang., D. Dang., B. Chi., H. Liu., E.J.M. Hensen, X. W. Lou.,. B. Y. Xia, 2019). The international team of researchers has published their work in Science in November 15, 2019, and there are already other researchers that wants use the new catalyst to develop a small foot-print size electrolyser of 10 MW in the future.

4. Electricity Storage Systems

In this type of Energy Storage, the electric energy is stored in a form of:

- electrostatic field using capacitors that have two-layers usual named double-layer capacitor (DLC) or Supercapacitors
- magnetic field using a superconducting coil SMES (Superconducting Magnetic Energy Storage)

Supercapacitors (SC – Figure 28) can be used like buffers in acceleration, when the power energy required is high, these systems can deliver the needed energy in a very short time (because of the low inner resistance). Unlike the batteries, SC can operate in a wide range of temperature and humidity, and have a equivalent of 10 years lifetime of continuous operation. Even that the specific power density is higher than in a LIB, the specific power is much lower comparing to LIB.

Figure 28. Construction of Supercapacitor

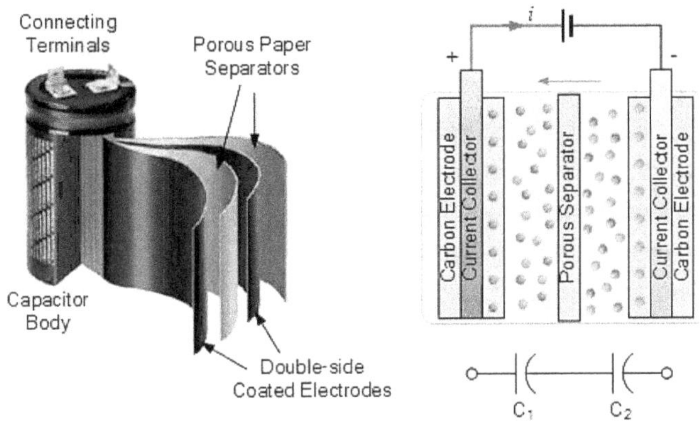

Figure 29 shows that the Ultracapitors are very similar in design terms like an electrochemical storage system. We have practical two electrodes made from carbon graphs separated by a porous paper membrane that split the Ultracapacitor in tow electrode chambers and allow the ions to pass from side to side. The existence of liquid or wet electrolyte in Ultracapitors would give us the false impression that we will have a chemical reaction, but in this case, there are no chemical reaction involved, but the energy is stored in an electrical field between the two electrodes. During loading, electrolyte ions are drawn to the electrodes and are virtually hung on the porous surface of the electrodes. During discharge, electrons are transferred to the external circuit, and ions are removed from the electrode and evenly distributed in the electrolyte. Again, we draw attention to the fact that this is not a chemical reaction and therefore the charging /discharging cycle is much faster and the Ultracapacitors performance are influenced by the surface of electrodes and electrolyte composition.

On the other hand, Superconducting Magnetic Energy Storage (SMES) systems store energy in the magnetic field. The SEMS system has four main components (Figure 28): the coil (usually made from a superconducting material like niobium-titanium); the cryogenic vessel in which the coil is immersed into coolant liquid helium or nitrogen (in this way is reached the superconducting state of the coil mate-

Figure 29. Superconducting Magnetic Energy Storage structure system - Global Superconducting Magnetic Energy Storage (SMES)
Systems Market 2019: Industry Analysis and Forecast 2024

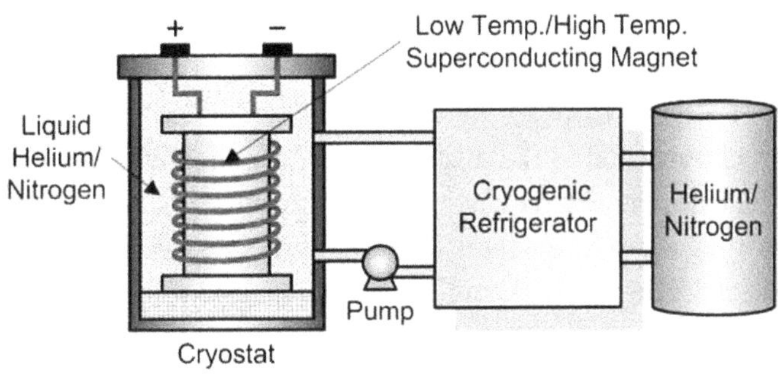

rial); refrigeration system that will create the optimum temperature near to 0°K for helium or nitrogen; the power condition system.

From the operational point of view (Figure 30), the energy is introduced into a SEMS using a DC power supply. When the desired power level has been reached, the supply can be disconnected. The energy is then stored in the magnetic field of the superconducting coil and can be released on demand, using an inverter, back to the network. Two aspects must be considered: the positive aspect – the energy can be stored indefinitely, without quantifiable losses and the negative aspect - in which we have 2-3% energy losses (on charge or discharge phase) resulted from the control system, and the power condition system.

Disadvantages of SMES are typically: the high investment costs (all the components are very expensive) and the need of permanent coolant cooling. The advantages are: high round-trip efficiency (around 95%), short charging time; short response time and high and quality power delivery.

SMES systems are used in UPSs, grid stabilization, in conjunction with batteries for hybrid EV applications, in conjunction with intermittent energy sources (Photovoltaic Panels or Windmill).

Figure 30. Working principle of SMES

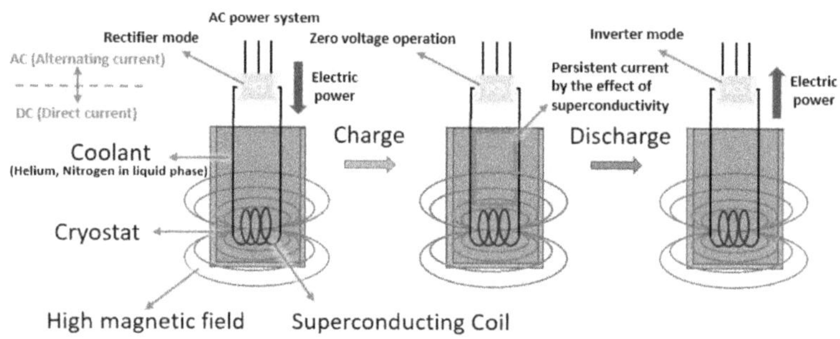

5. Thermal Energy Storage (TES)

Thermal energy storage (TES) is an advanced energy technology and is used in stationary applications, especially when it comes to water heating and climate control in buildings (heating and cooling). The power supply of a TES system (charging phase) is stored in different types of material (storage phase) and then released on request (discharge phase) to the end user. The materials used to store the energy.

Based on the materials used, the temperature of operation, and a few other parameters, there are three methods to store the energy in TES:

- Sensible heat storage (SHS). The storage medium is liquid or solid and is adequate for low-temperature applications.
- Latent heat storage (LHS). In this case it is used phase change materials (PCM). In charging state, the material become liquid, and in the discharging phase the material become solid.
- Thermochemical heat storage (TCS) where the energy can be store either using a) chemical reactions (where in charging phase the heat is used to dissociate two chemical compounds AB into A and B, and in the discharging phase A and B are recombined releasing the heat), or using sorption processes (in charging phase zeolite and water are separated; the heat of reaction is released in the

discharging phase when the reactants recombine). These types of storages are expensive and still in research.

FUTURE RESEARCH DIRECTIONS

The market for electricity is constantly increasing, so is necessary that the electrical grid to provide the necessary energy for consumers especially in times of high demand. Given the expected growth of this market, the use of energy storage systems in different forms at a low cost becomes absolutely necessary.

Some examples of start-ups which have managed to build new energy storage systems or have managed to improve the performance of existing storage systems using new technologies:

- Energy Vault They developed a long-term energy storage system (using the same principles as in PHS but replacing water with composite blocks made of low-cost materials. These systems (Figure 31) do not need a special topology to be located and costs are low.

Figure 31. Energy Vault's long-term energy storage system
Energy Vault's website 2019 https://energyvault.com/

- bGen™ from Brenmiller Energy, Israel is a patented high-temperature thermal energy storage unit stored in solid (crushed rock) material (Figure 32).

Figure 32. bGen™ energy storage system
Brenmiller Energy's website 2019 https://www.bren-energy.com/

- Ergosup, France has developed HyRiS Lab (Figure 33) for producing high pressure hydrogen on site for refueling hydrogen fuel vehicles.

Figure 33. HyRiS Lab from Ergrosup's website 2019
https://www.ergosup.com/

- Battrion, Switzerland is using new electrode materials Alligned graphite® Technology to reduces the loading time for the EV by about 50%, increases reliability and operational safety by reducing the heat generated during loading/unloading cycles.

- e-Zn Inc, Canada, created a Zn Reactor, which is an electrochemical cell with three distinctive structure: a charging section – in the charging state zinc metal is formed at the top of the structure; in the middle the zinc metal is stored; in the discharge phase the bottom of the structure it is used to dissolve zinc realising energy. Can be used for Electrical vehicle charging stations.

CONCLUSION

In this chapter we intended to present a brief overview of the energy storage systems methods and the working principles. No method is perfect, each having its advantages and disadvantages. In the stationary applications, and not only, combinations between various energy storage systems often lead to the desired results (Battery + UC, Battery + Flywheel, Battery + FC, Battery + FC + UC, Battery + PCM, PCM+FC+UC, etc.).

Researchers and developers around the world are seeking to use increasingly cheap and easily accessible materials to reduce costs. Various mechanical configurations, new chemical recipes for various compounds, new process installations, and new monitoring, communication and control algorithms, are also tested to ensure future energy needs. All these research aims at improving the following key issues: increase of storage capacity, improvement of charging/discharging times, increase of storage period, efficiency improvement, safety in exploitation, lowering the environmental impact, high recycling rate, and lowering overall costs.

REFERENCES

Akhil, A. A., Huff, G., Currier, A. B., Kaun, B. C., Rastler, D. M., Chen, S. B., Cotter, A. L., Bradshaw, D. T., & Gauntlett, W. D. (2013). *DOE/EPRI 2013 Electricity Storage Handbook in Collaboration with NRECA, SANDIA REPORT, SAND2013-5131*. Sandia National Laboratories. www.sandia.gov/ess

Batteries, R.-F. (2016). *From Metals to Organic Redox-Active Materials*. doi:10.1002/anie.201604925

Braff, W. A., Bazant, M. Z., & Buie, C. R. (2013). Membrane-less hydrogen bromine flow battery. *Nature Communications*, *4*, 1–6.

Carmo, M., Fritz, D. L., Mergel, J., & Stolten, D. (2013). A comprehensive review on PEM water electrolysis. *International Journal of Hydrogen Energy*, *38*, 4901–4934.

Chen, H. (2009). Thang Ngoc Cong, Wei Yang, Chunqing Tan, Yongliang Li, Yulong Ding Progress in electrical energy storage system: A critical review. *Progress in Natural Science*, *19*, 291–312.

Coors, W., Boxley, C., Robins, M., & Eccleston, A. (2012). *US Pat., 061823A2*. US Patent Office.

Cunha, Á., Martins, J., Rodrigues, N. & Brito, F. P. (2014). Vanadium redox flow batteries: a technology review. *Int. J. Energy Res*. doi:10.1002/er.3260

D'AprileP.NewmanJ.PinnerD. (2016). Retrieved from https://www.mckinsey.com/business-functions/sustainability/our-insights/the-new-economics-of-energy-storage

De Oliveira, J. G. (2011). *J.G. Power Control Systems in a Flywheel based All-Electric Driveline* (PhD thesis). Uppsala Universitet, Uppsala, Sweden.

EASE EERA. (2017). *Storage Technology Development Roadmap 2017*. Author.

Egan, D. R., Pnce de Leon, C., Wood, R. J. K., Jones, R. L., Stokes, K. R., & Walsh, F. C. (2013). Developments in Electrode Materials and Electrolytes for Aluminium-Air Batteries. *Journal of Power Sources*, *236*, 293–310.

Electrical Energy Storage. (2011). *White paper.* The International Electrotechnical Commission.

European Commission. (n.d.). *Industrial policy*. Retrived from: https://ec.europa.eu/growth/industry/policy/european-battery-alliance_en

European Court of Auditorium. (2019). *EU support for energy storage*. Briefing Paper.

European Market Monitor On Energy Storage (EMMES). (2019). *EMMES Edition 3.0*. Retrieved from https://www.delta-ee.com/EMMES

Fujihara, T. (1998). Development of pump turbine for seawater pumped-storage power plant. *Hitachi Rev*, *47*(5), 199–202.

Garthwaite. (2011). *How ultracapacitors work (and why they fall short)*. https://gigaom.com/

Hebner, R., Beno, J., & Walls, A. (2002). Flywheel batteries come around again. *IEEE Spectrum*, *39*, 46–51.

Hosseini, S. E., & Wahid, M. A. (2016). Hydrogen production from renewable and sustainable energy resources: Promising green energy carrier for clean development. *Renewable & Sustainable Energy Reviews*, *57*, 850–866.

Hu, X., Li, Z., Zhao, Y., Sun, J., Zhao, Q., Wang, J., Tao, Z., & Chen, J. (2017). Article. *Science Advances*, *3*, e1602396.

Huggins, R. A. (2016). *Energy Storage Fundamentals, Materials and Applications*. Springer.

IHA working paper (December 2018): "The world's water battery: Pumped hydropower storage and the clean energy transition"

IRENA. (2017). *Electricity Storage and Renewables: Costs and Markets to 2030*. International Renewable Energy Agency.

Jayachandran, A. (n.d.). Tata Nano could spawn electric, hybrid & air-powered variants – Report. *Indian Autos Blog*. Retrieve from: https://indianautosblog.com/tata-nano-could-spawn-electric-hybrid-air-variants-p256298

Karina, B. (2013). High temperature sodium batteries: Status, challenges and future trends. *Energy & Environmental Science*.

Kerea, A., Goetza, V., Pya, X., Olivesa, R., Sadikia, N., & Mercier, E. (2014). Dynamic behavior of a sensible-heat based thermal energy storage. *Energy Procedia*, *49*, 830–839.

Kilkis, B. (1989). *Energy Storage Systems* (1st ed.). Kluwer Academic Publishers.

Kumar, A. (2018). *Rakesh K. Gupta: Fundamentals of Polymer Engineering* (3rd ed.). CRC Press.

LaClair, T., Gao, Z., Abdelaziz, O., & Wang, M. (2016). *Thermal Storage System for Electric Vehicle Cabin Heating - Component and System Analysis*. SAE Technical Paper 2016-01-0244.

Lee, J. S., Tai Kim, S., Cao, R., Choi, N. S., Liu, M., Lee, K. T., & Cho, J. (2011). Metal-Air Batteries with High Energy Density: Li-Air versus Zn-Air. *Advanced Energy Materials*, *1*, 34–50.

Li, F. J., Zhang, T., & Zhou, H. S. (2013). Challenges of Non-Aqueous LiO2 Batteries: Electrolytes, Catalysts, and Anodes. *Energy & Environmental Science*, *6*, 1125–1141.

Li, Y., & Jun, L. M. A. B. (2017). Will They Be the Future Electrochemical Energy Storage Device of Choice? *ACS Energy Letters*, *2*, 1370–1377.

Linden, D., & Reddy, T. B. (2001). *Handbook of batteries* (3rd ed.). McGraw-Hill, Inc.

Liu, H., & Jiang, J. (2007). Flywheel energy storage—An upswing technology for energy sustainability. *Energy and Building*, *39*, 599–604.

Lohse-Busch, H., Duoba, M., Rask, E., & Stutenberg, K. (2013). *Ambient Temperature (20°F, 72°F and 95°F) Impact on Fuel and Energy Consumption for Several Conventional Vehicles, Hybrid and Plug-In Hybrid Electric Vehicles and Battery Electric Vehicle.* SAE Technical Paper 2013-01-1462.

Manla, E., Nasiri, A., Rentel, C. H., & Hughes, M. (2010). Modeling of zinc-bromide energy storage for vehicular applications. *IEEE Transactions on Industrial Electronics*, *57*(2), 624–632.

Marshall, A., Børresen, B., Hagen, H., Tsypkin, M., & Tunold, R. (2007). Hydrogen production by advanced proton exchange membrane (PEM) water electrolysers—Reduced energy consumption by improved electrocatalysis. *Energy*, *32*, 431–436.

Millet, P., Mbemba, M., Grigoriev, S. A., Fateev, V. N., Aukauloo, A., & Etievant, C. (2011). Electrochemical performances of PEM water electrolysis cells and perspectives. *International Journal of Hydrogen Energy*, *36*, 4134–4142.

Millet, P., Ngameni, R., Grigoriev, S. A., Mbemba, N., Brisset, F., & Ranjbari, A. (2010). PEM water electrolyzers: From electrocatalysis to stack development. *International Journal of Hydrogen Energy*, *35*, 5043–5052.

MunueraL.TorilA. (2018). Retrieved from: https://www.iea.org/tcep/energyintegration/energystorage/

Narayanan, S. R., Prakash, G. K. S., Manohar, A., Yang, B., Malkhandi, S., & Kindler, A. (2012). Materials Challenges and Technical Approaches for Realizing Inexpensive and Robust Iron-Air Batteries for Large-Scale Energy Storage. *Solid State Ionics*, *216*, 105–109.

Patrick, T. (2015). *Electrochemical Energy Storage for Renewable Sources and Grid Balancing*. Elsevier.

Rajarathnam. (2016). *The zinc/bromine flow battery: Fundamentals and novel materials for technology advancement.* Academic Press.

Sanz, L. (2015). Study and characterization of positive electrolytes for application in the aqueous all-copper redox flow battery. *Journal of Power Sources, 278*, 175–182.

Silberberg, M. (2006). *Chemistry: The Molecular Nature of Matter and Change* (4th ed.). McGraw-Hill Education.

Su, H., Jaffer, S., & Yu, H. (2016). Article. *Energy Storage Materials, 5*, 116–131.

Tian, X., Zhao, X., Su, Y. Q., Wang, L., & Wang, H. (2019). Engineering bunched Pt-Ni alloy nanocages for efficient oxygen reduction in practical fuel cells. *Science*.

Tie, S. F., & Tan, C. W. (2013). A review of energy sources and energy management system in electric vehicles. *Renewable & Sustainable Energy Reviews, 20*, 82–102.

Vincent & Scrosati. (1997). *Modern Batteries*. Elsevier Science.

Weber, A. Z. (2013). Hydrogen/Bromine Flow Batteries. *Proceedings of the 2013 Electrical Energy Storage Applications & Technologies (EESAT) Biennial International Conference*.

Xing, X., Zhang, D., & Li, Y. (2015). A non-aqueous all-cobalt redox flow battery using 1,10-phenanthrolinecobalt(II) hexafluorophosphate as active species. *Journal of Power Sources, 279*, 205–209.

Yu, H., Ren, Y., Xiao, D., Guo, S., Zhu, Y., Qian, Y., Gu, L., & Zhou, H. (2015). *Angewandte Chemie International Edition in English*. Academic Press.

Chapter 6
On-Board and Off-Board Technologies for Hydrogen Storage

Madhavi Konni
https://orcid.org/0000-0002-2022-5528
Department of Basic Sciences and Humanities, Dadi Institute of Engineering and Technology, India

Saratchandra Babu Mukkamala
Department of Chemistry, Institute of Science, GITAM University (Deemed), India

Manoj Kumar Karnena
https://orcid.org/0000-0001-9241-2990
Department of Environmental Science, Institute of Science, GITAM University (Deemed), India

ABSTRACT

Future energy systems will be determined by the increasing relevance of renewable energy resources due to global warming, energy crisis, and pollution. Hydrogen is considered one of the promising alternative fuels to replace oil, but its storage remains a significant challenge. The main hydrogen storage technologies can be broadly classified as physical, chemical, and hybrid methods. The physical methods rely on compression and liquefaction of hydrogen, and currently, compressed hydrogen storage is the most mature technology that is commercially available. The chemical methods utilize materials to store hydrogen, and hydrogen can be extracted by on-board regenerable or off-board regenerable chemical reactions depending on the type of material. Hybrid methods take advantage of both physical and chemical storage methods. The most prominent hybrid method is the cryo-adsorption hydrogen storage which utilizes physisorption-based porous materials. The chapter describes these technologies and discusses alternative/novel hydrogen storage material technologies.

DOI: 10.4018/978-1-7998-4945-2.ch006

I. INTRODUCTION

Due to emerging Economies, global energy demand is increasing continuously. Relic fuels supply nearly 80% of energy demand worldwide; further, if reliance on Relic fuels is continued, the principal source of this energy may have foreseeable Imminent (Economics, 2018). Global Warming, Energy crisis and Pollution are the major problems facing by the current modern society due to the excessive utilization of Fossil Fuels. Finite production of fossil fuels possess Bell-shaped Curve with reference to the theory of Hubbert's peak (Maggio & Cacciola, 2012; Houghton, 2005; Asif & Muneer 2007; Barbir et al.1990).

After Discovery of Certain reservoir for fossil fuel production it is leading to increase exponentially, then the rate of production decreases reaching peak point of depletion of fossil fuel and reduction in feasibility of extraction (Hubbert, 1956). So, it is understood that oil and other relic fuels, get exhausted in the near future. In the conveyance sector, oil is most important source of energy i.e., used excessively by mankind and according to the current estimates, the fuel that is likely to be depleted before end of twenty-first century is oil (Agrawal et al. 2007; Kerr, 2011).

A huge share in the world trade was contributed by oil economy, and any significant changes in it might take many years before it is completely comprehended; hence, it is domineering to act in present day towards finding cleaner replacements to oil. Even though the exhaustion of oil is vital concern which is required to be addressed at possible earliest, much more insistent issue requiring immediate attention is global warming. Increase in temperature of earth precisely due to anthropogenic emissions of CO_2 which is caused by use of excessive fossil fuel is Global warming (Asif & Muneer 2007). Sea level rise and extreme weather are the impacts of global warming which in turn can cause worldwide environmental change and colossal community. Being time-sensitive problem an effective measure for prevention of irreversible changes in atmosphere needs to be designed and executed immediately (Yi,1996). Oil spills and air pollution caused by combustion are also a grave problems requiring attention. Particulates like NO_x, CO_x (where X=1 or 2), SO_z (where Z= 2 or 3) and benzene are emitted from mobile platforms are harmful to health. Densely populated areas are responsible for problematic air pollution. On the other hand, Aquatic life and Surrounding communities have long devastating effects due to large scale oil spills. Lastly, excessive utilization of fossil fuels threatens energy security. Oil reserves are usually present in unsteady areas of the world (Mendelssohn et al. 2012). The countries importing oil are facing the problems in meeting oil demands as a consequence of disrupted supply from the time to time. Due to this disruption failure in supply of oil, the countries will have grave economic, political and social impacts. Several countries have their individual natural reserves which are diminutive energy security and another long-term energy security policy to drive away from oil in near future (Asif & Muneer 2007).

From the preceding conversation, it is clearly evident that the alternative fuels are necessary to protect mother earth which also helps in reducing risks related with extreme oil use. Due to owing their unique property of high energy density, fuels derived from oil have extensively been used in transportable applications and becoming a formidable task to replace this oil (Asif & Muneer 2007).

Hydrogen is understood to be among the strongest contenders, that replaces the oil in transportable applications. Though hydrogen might not be an energy source similar to oil, it acts as an energy carrier, connotation hydrogen availability on earth is not direct, which requires its production through other sources. At present, a noteworthy share of hydrogen energy is produced from methane through the process of "steam methane reforming (SMR)". Amongst other processes for the production of hydrogen is through water electrolysis; as this process uses production of electricity from renewable energy technologies since there is a possibility of hydrogen production from renewable energy sources.

After the production of hydrogen, transportation, distribution to end-user and storage for later use are some important stages that are needed to be addressed. The three stages of production, transportation and storage of hydrogen are called as hydrogen economy.

II. HYDROGEN ECONOMY

Hydrogen economy has been proposed as an alternative system to the oil economy in the early 1970s. The main motivation of hydrogen economy being non emittance of pollutants but only H_2O when it is used in an internal combustion engine or in fuel cells as divergent to employing hydrogen carbons. In order to combat environmental issues along with global warming hydrogen must be produced from renewable energy sources as debated in the introduction.

Figure 1. Hydrogen Economy
(Redrawn with permission from Crabtree et al. 2004)

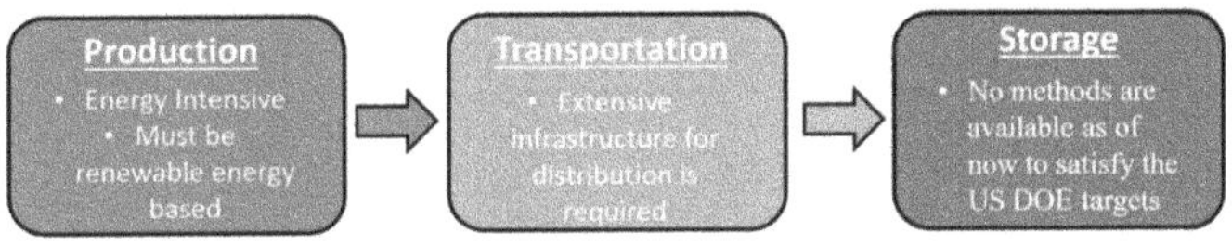

The economy of hydrogen comprises of 3 parts as shown in the above Fig.1 (Crabtree et al. 2004). Production of hydrogen through the process of SMR is cost-effective and most efficient (83%) option available today (Abbas & Daud, 2010). While, other methods of production like photocatalytic water splitting have an efficiency of 10-14% H_2 production and these are not yet competitive with SMR (Gerboni & Salvador,2009).

Issues with the production of hydrogen are:

- The process should be High energy-intensive
- The raw material should be renewable energy resource.

Extensive distribution systems are generally required for the transportation of hydrogen; it might be transported through tube trailers, pipelines and liquids (Felderhoff et al. 2007) by cryogenic tanks and as well as Cylinders in gaseous form.

Hydrogen has low energy density but has high energy capacity. Hydrogen energy density can be increased by liquefication and compression but these are cost-effective processes. For instance, to liquefy the hydrogen one-third of hydrogen's lower heating value is required. Therefore, on the basis of cost-effectiveness, end-use storage technology and safety, the medium of transportation is selected i.e., either gaseous or liquid (Lattin & Utgikar, 2007; Konni et al. 2020).

In some reports, it has been revealed that Hydrogen storage distribution cost is nearly 15 times much expensive than a liquefied hydrocarbon fuel. From the preceding conversation, it can be understood that elevated costs included in production and transportation are cause for deprived hydrogen economy. Nevertheless, the hydrogen storage step is the main obstacle in comprehending hydrogen economy (US

DOE, 2009). Till date, none of hydrogen storage technology is available which can fulfil the targets of the Department of Energy (DOE) 2020 (US DOE, 2009). Due to its noteworthy energy losses in each step Figure.2 (Bossel et al. 2009), hydrogen economy is criticized. However recent research in this field is providing hope for the Hydrogen economy to overcome the disadvantages that are facing in the current technology (Shinnar,2003; Høyer & Holden, 2007; Eberle, et al. 2009; Konni et al. 2020; Konni et al. 2019).

Hydrogen Storage Classification

See Figure 2.

Figure 2. Classification of Hydrogen Storage Technologies.

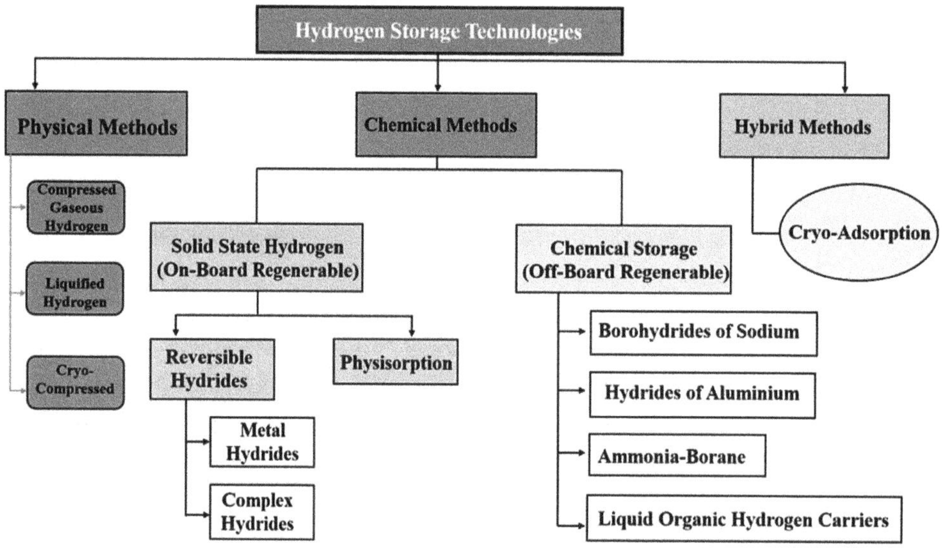

III. HYDROGEN STORAGE

Hydrogen Storage Technologies Classification

Technologies for hydrogen storage can be sketchily classified under three main classes:

1. Methods of physical nature
2. Materials-based hydrogen storage (also called chemical methods), and
3. Advanced or hybrid methods

Methods included under the class physical nature include liquefied hydrogen, compressed gaseous hydrogen and cryo-compressed hydrogen. Chemical methods include chemical storage and solid-state storage. Further, hybrid method of storage consists of cryo-adsorption which adopts both physical and chemical methods.

Hydrogen will not react with storage media in physical method, whereas in the chemical method, it will interact with storage media by the robust covalent/ionic bonds or by feeble Vander Waals forces. Hydrogen storage materials are included in chemical methods always to store the hydrogen. Since 2000s these methods drew attention, but due to they are in physical storage methods.

Further, chemical methods are classified into:

1. Solid-state storage and
2. Chemical storage

The key dissimilarity between the above two is that regeneration of solid-state materials can be done merely through charging with hydrogen, while chemical storage materials require to be regenerated in centralized facilities off-board (Eberle et al. 2009). Moreover, key elements for market of fuel cell vehicles are on-board regenerable materials (Chalk & Miller, 2006). A hybrid process with combination of physical and chemical methods, Cryo-adsorption possess advantage of physisorption, compression and liquefied material-based technologies for hydrogen storage.

IV. DOE HYDROGEN STORAGE TARGETS

DOE targets aimed at hydrogen storage systems intended for light-duty vehicles is summarized in Table 1 (Ganguly et al. 2019; Konni et al. 2017). Volumetric and gravimetric capacities are the present most stringent doe targets.

In addition, fuel purity, reversibility (i.e., operational cycle life) and fast kinetics (i.e., system full time) are problematic for complex hydrides. Besides storage materials weight, balance plant elements like storage tank, the control system, tubing, valves are also required to be accounted for. Two times of gravimetric storage capacity with that of the targets of system level is given by DOE as the main thumb rule towards achieve the good targets. Recently, kg H_2/kg are gravimetric hydrogen storage capacity units as given by DOE presented in table 1, previously gravimetric hydrogen capacity was expressed in wt.%. (Read et al. 2007; Konni et al. 2018).

If the system comprises hydrogen storage material, weight of the hydrogen storage material must be added to the weight of the system. Both the units of gravimetric hydrogen storage capacity (i.e., kg H_2/kg system and wt.%) are employed in the entire chapter.

V. RECENT STATUS OF HYDROGEN STORAGE

In table 2 the present status of storage technologies for Hydrogen is summarised. The capacity of different technologies spans is in the range of 1.9 to 6.8% (Gravimetric) and 13 - 39 g/L (Volumetric capacity). Chemical methods for hydrogen storage fall behind over the physical methods. Volumetric capacity targets are easier when compared to the gravimetric capacities and this can be observed in table 2. usually, lowest storage capacity is observed in solid storage materials. A weight percentage, of range 1.9 – 2.5% can only store by Complex hydrides and the physisorption materials like activated sorbents can store hydrogen up to 3wt%. Chemical hydrides can perform superior than solid-state store materials and store hydrogen up to 3.5 wt.%. 5.5 wt.% of hydrogen can store at 70Mpa in the compressed hydrogen storage

Table 1. US DOE hydrogen storage targets for on-board hydrogen storage systems for light-duty vehicles

Storage Parameter	Units	2020	Ultimate
System gravimetric capacity	Kg H_2/kg system	0.055	0.075
System volumetric capacity	Kg H_2/L system	0.040	0.070
Operating ambient temperature	^oC	-40/60 (sun)	-40/60 (sun)
Min/max delivery temperature	^oC	-40/85	-40/85
Operational cycle life	Cycles	1500	1500
Well to power plant efficiency	%	60	60
System fill time (5 kg H_2)	min	3.3	2.5
Fuel purity	% H_2	99.7% (dry basis)	
Loss of useable H_2	(g/h)/kg H_2 stored	0.05	0.05

(Ganguly et al. 2019).

tank. Cryo compressed and Liquefied hydrogen technologies are the best technologies to store hydrogen in terms of gravimetric storage capacity.

1. Physical Methods Of Hydrogen Storage

a. Compressed Gaseous Hydrogen Storage

The compressed gaseous form of hydrogen storage is one of the convenient storage methods available today. At 15-20 Mpa; steel cylinder tank is used for hydrogen storage and volumetric density of 10-12 kg/m^3 and gravimetric density of 1.5 wt.% can be reached by these methods (Tarasov & Lototskii, 2007). The DOE targets for 2020 i.e., 5.2 wt.% and 40 kg/m^3 (table 1) can be achieved but it requires High pressures. But increase in pressure will reduce the thickness of the Steel tank and this is not a solution for enhancing the storage capacity of hydrogen and it is also found that gravimetric hydrogen storage

Table 2. Current status of hydrogen storage technologies

Hydrogen Storage Method		Gravimetric Capacity (wt %)	Volumetric Capacity (g/L)
Physical storage	Compressed (350 bar)	2.8-3.8	16-18
	Compressed (750 bar)	2.6-4.4	19-25
	Liquid	4.8-6.8	31-39
Chemical storage	Complex hydride	1.9-2.5	16-28
	Carbon (porous)	2.9-3.1	13-15
	Chemical hydride	2.6-3.5	22-29
Hybrid storage	Cryo-compressed	5.0-5.8	28-38
Targets	2020	5.2	40
	Ultimate	7.0	70

(Read et al. 2007, Baroutaji et al. 2019)

capacity decrease with increasing pressure. Therefore, to achieve the targets of DOE, storage tank materials which are lighter are necessarily required. The carbon fibre reinforced composite tanks (figure 3) are developed to store the hydrogen at two different pressure levels of 35 and 70 Mpa. In order to reach the hopes of users on par with gasoline-powered vehicles hydrogen-fuelled vehicles should have a range of 300 miles and this requires storage of 4–7 kg of hydrogen, depending on the vehicle size (Tarasov & Lototskii, 2007; Hua et al. 2007).

Figure 3. Quantum Technologies type IV compressed gaseous hydrogen storage tank
(Reprinted from Ref. (von Helmolt & Eberle, 2007) with permission from the Journal of Power Sources. Copyright Elsevier)

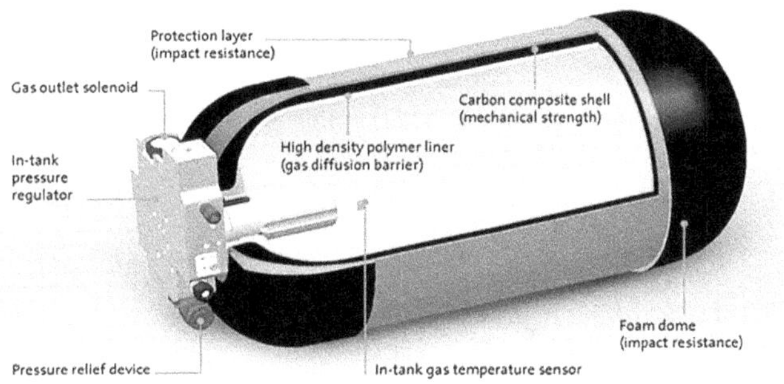

The main components of storage tanks are as follows:

- **Liner**: Hydrogen can permeate through many materials due to its smaller size. Hydrogen permeation from the tank can be prevented and eliminated by the liner. Polymeric materials are generally used as a liner when compared to the metallic liner materials due to its lightweight (Felderhoff et al.2007).
- **Shell**: These are the components of a storage tank loaded with bearing components (Hua et al. 2011).
- **Protection Layer**: this the outer most protective layer of the tank and also useful to withstand the impacts,
- Temperature sensor vehicle electrical interface, pressure sensors, pressure relief device, pressure regulator, vent line ports, connector, refuelling port, stone shield and fill line check valve are the other essential components required to balance the plant.
- **General considerations:** 1 litre of liquid hydrogen occupies 780 L of gas at normal temperature and pressure. Boiling temperature of hydrogen is at 1 atm of 20.2 K. Therefore, the boiling rate and compressibility are the two important considerations while designing and implementing the Cryo-method.

Storage tanks of capacity 35 and 70 Mpa does not satisfy the targets given by the DOE while performing a technical assessment. It is to be noted that 5.5 wt.% (35Mpa) system possess higher gravimetric capacity that 5.2 wt.% (75Mpa). However, 17.6 kg/m^3 of 70Mpa has high volumetric capacity than 26.3 kg/m^3 (35Mpa) System (Hua et al. 2011).

An amount of compression work is the major drawback of the compressed gaseous hydrogen storage. Compression work is 12 to 15% lower than the heating values of hydrogen which is essential to reach 35 and 70Mpa. With increasing pressure, the significance of compression work lowers as shown in figure 3.To keep the temperatures and pressures at safe levels and to fill the hydrogen as much as possible during the compression of hydrogen, the hydrogen tank needs to be cooled. The energy penalty will be added during the cooling process. Estimates state that compression is greater than that of liquefaction cost for hydrogen (Zhou, 2005).

Figure 4. Hydrogen compression work as a function of pressure
(Reprinted with permission from Ref. (Felderhoff et al. 2007). Copyright Royal Society of Chemistry)

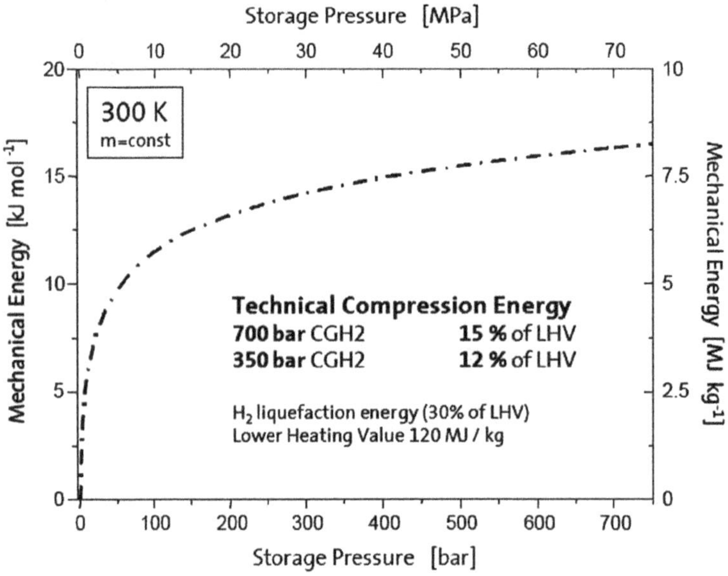

Storage technologies for gaseous hydrogen usually include storing at room temperature only (i.e., 300 K). Cryo-compressed technologies like keep hydrogen at high pressures in cryogenic temperatures in order to attain greater storage capacity. Gaseous storage is considered as most practical option upon its shortcomings by automotive industry which is the only technology that has been adopted commercially (i.e., Toyota Mirai) thus far (Hamilton et al. 2009).

b. Liquefied Hydrogen Storage

Liquified hydrogen storage is another physical storage method. As shown in Table 2, high volumetric and gravimetric densities can be achieved by this method. Comparing to the compressed hydrogen storage, liquified hydrogen storage pressures are lower (<1 Mpa) and this helps in eliminating the load-bearing (in terms of Cost) carbon fibre reinforced composite materials generally utilized in compressed hydrogen storage. Drawbacks of this storage system are boil-off losses and liquefaction energy penalty. Components of typical liquid hydrogen storage tank are presented in Fig. 4. The hydrogen storage tanks are insulated in between the double-walled vessels.

The temperature between liquefied hydrogen (20K) and ambient (300K) are generally different, the inner vessels are insulated with the multilayers consists of altering metal foil layers (vacuum super insulation) and padding materials and this helps in minimizing ambient heat gain.

Multilayer insulation design is showed in Figure 5

Figure 5. (a) Liquid hydrogen tank design by Linde and (b) the vacuum super insulation
(Reprinted from Ref. (von Helmolt Eberle, 2007) with permission from Journal of Power Sources. Copyright Elsevier)

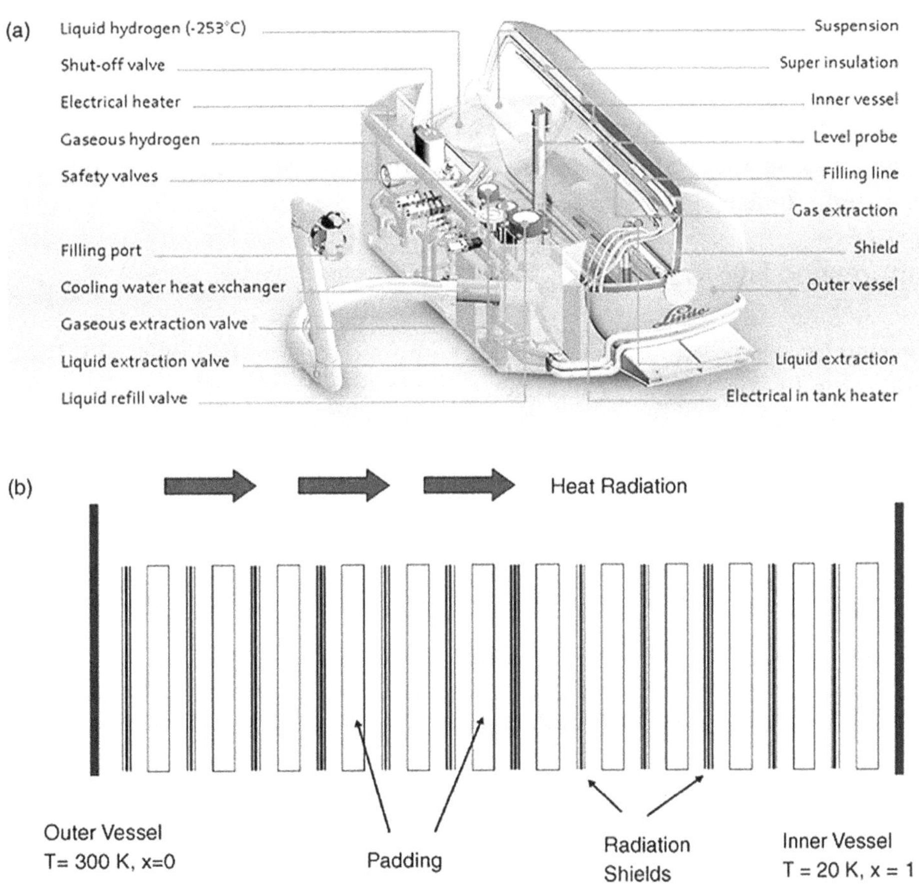

The efficiency of insulation (Manoharan et al. 2019) will be enhanced by the space between vessels. Insulation acts as the most important component in the liquid hydrogen tank. While designing a tank for all heat transfer mode are necessary to be examined. Radiation heat transfer is efficient in between ambient and liquid hydrogen as conduction occurs through pipes, ports and mountings.

The shape of the tank is additional important aspect in the thermal design of a liquid hydrogen tank. Cylindrical tanks will minimize the heat flow and this results in minimization of surface to volume ratio. In the current vehicle designs cylindrical tanks will occupy more space. Conformable tanks are preferable in case of cylindrical tanks in some cases, but these tanks show inferior thermal preferences when compared to cylindrical tanks. In addition, conformable tank manufacturing is more labour-intensive process and costly (von Helmolt & Eberle, 2007).

Elimination of input heat to the current storage tank though with vacuum super insulation is impossible; due to the heat input the hydrogen inside the tank evaporates continually resulting in increase of pressure inside the tank constantly. Excess of any hydrogen is required to be emitted to atmosphere, catalytically burned or captured (Felderhoff et al. 2007). If the pressure of the tank inside reaches around 1Mpa. Hydrogen not utilized or excess hydrogen is called boil-off losses. "Dormancy" is time period between parking the vehicle to the venting process. Longer dormancy periods are favoured, owing to longer dwell times and lesser loss through adiabatic or isothermal processes.

Applying innovative cooling methods will enhance the dormancy and the normal state-of-the-art tank designs possess dormancy period of 3 days. In comparison with the other storage, methods liquefied hydrogen storage shows the best volumetric and gravimetric storage densities (Manoharan et al. 2019). The liquefaction cost of hydrogen is 30% lower having value of hydrogen as shown in Figure. 4.

Overall cost of liquid hydrogen storage effortlessly surpass cost of gaseous hydrogen storage when considered in combination with boil-off losses. At present, liquid hydrogen storage is the lone attractive method for short-term storage (i.e., space applications) due to the above-mentioned limitations.

c. Cryo-Compressed Hydrogen Storage

High pressures and Cryogenic temperatures (< 77k) are used to store hydrogen in this method. This method overcomes the drawbacks of other hydrogen storage methods. The major flaw of cryo-compressed Hydrogen Storage is low volumetric and low gravimetric densities (at high pressures even). Energy densities can be enhanced by cooling the hydrogen to cryogenic temperatures at immense pressures, by avoiding the penalty associated with Liquefaction. Maximum 1 MPa pressure (low pressure) can be stored by liquid hydrogen tanks and this leads to the shortfall of dormancy period. Dormancy period of these tanks can be increased by enhancing the pressures. In comparison with low-pressure tanks, vaporised hydrogen can be stored in high-pressure tanks longer. The liquid hydrogen and cryo- compressed hydrogen helps in charging the cryo-compressed tank in a dual-phase region. Resilience of this nature makes storage system apt for diverse needs; for example, liquid hydrogen can be used when range of driving is more important than tariff (Sdanghi et al. 2019).

According to the assessments, this storage system can reach the DOE (Ahluwalia et al. 2010) targets i.e., volumetric, gravimetric densities and dormancy, but shortfalls in reaching the cost of manufacturing and well-to-Wheel efficiency. Amongst other technologies of hydrogen storage Cryo-Compression storage shows grater volumetric and gravimetric densities (see Table 1). In addition to that, it also eliminates and reduces the boil-off losses in comparison with liquified hydrogen Storage. Cryo-compression hydrogen storage have evolved to be a promising hydrogen technology in comparison with others.

2. Chemical Methods of Hydrogen Storage

Preceding discussion makes it evident that liquid hydrogen storage and compressed hydrogen storage (i.e., Physical Methods) have many drawbacks resulting in search for improved hydrogen storage methods. Till 2000s, researchers mostly focussed on metal hydrides and efforts on chemical storage methods were very limited. In the early 2000s, awareness of environmental problems was increased and research on hydrogen storage has gained more importance.

Material- based hydrogen storage is also called as chemical methods of hydrogen storage which are founded on interaction of hydrogen with hydrogen storage materials. The hydrogen interaction with hy-

Figure 6. Cryo-Compressed Hydrogen Storage.
Reprinted from Ref. (Barthelemy et al. 2017) with permission from the International Journal of Hydrogen Energy. Copyright Elsevier

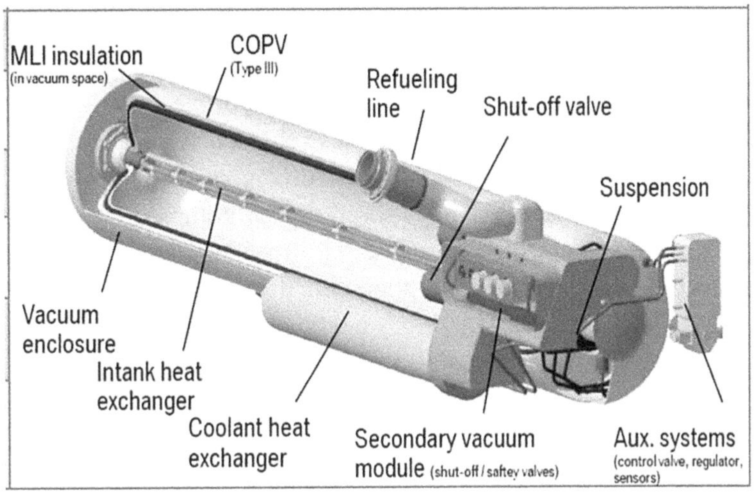

drogen storage materials in strength can be ionic and covalent bonds(strong), Kubas interaction(moderate) or van der Waals forces (weak) (Hoang & Antonelli, 2009). The operation condition like charging/discharging, pressure, temperatures etc., can be dictated by the strength of the interactions in the hydrogen storage method. In comparison with compressed or liquified hydrogen storage, hydrogen can be stored more efficiently in firm materials like $LiBH_4$, (complex hydride) can store 18.4 wt.% (gravimetric) and 121 kg/m^3 (volumetric), respectively (Züttel et al. 2003).

The plant components like a storage tank, safety features, cooling /heating equipment, does not take into consideration in materials-based storage capacity. Hydrogen storage capacity is half of the material storage capacity and it is a basic thumb rule (Read et al. 2007).

Chemical hydrogen storage methods can be categorized into

1. on-board regenerable and
2. off-board regenerable

The difference between these two storage materials has unfavourable thermodynamics or kinetics or both. Hydrogen on board vehicles can be recharged by the solid-state storage materials, whereas off-board can be regenerated. Off-board regeneration is more difficult in comparison with on-board regeneration. Off-board regeneration is costly and impractical. Charging of On-board regenerable materials can be done directly using hydrogen and doesn't require any further intermediate steps. The overall cost and fuel cycles will be increased in off-board regeneration as it requires processed spent fuels in the central facility.

a. Solid-State Hydrogen Storage (On-board Regenerable Materials)

The on-board regenerable materials are classified into two types based on the interactions and strength between hydrogen and storage materials

1. reversible hydrides (based on strong interactions) and
2. physisorption materials (based on weak interactions).

Reversible hydrides are further classified into metal hydrides and complex hydrides. Complex hydrides (Jain et al. 2010) and Porous materials (Thomas, 2007) consist of a significant portion of hydrogen storage materials and that can regenerate onboard hydrogen storage. The complex Hydrides have high gravimetric and Volumetric densities and these characteristics help in storing a high amount of hydrogen. Withal, they experience sluggish kinetics, high operating temperatures and reversibility. Contrarily, pours materials that are based on physisorption have complete reversibility and fast kinetics; nonetheless, at cryogenic temperatures they can store significant amount of hydrogen because the material matrix of porous nature i.e., with less adsorption enthalpy have weak interaction with hydrogen at room temperatures (Felderhoff et al. 2007).

Reversible Hydrides

Metal Hydrides

The old class metal hydrides with reversible hydrides are AB_5 ($LaNi_5$) and AB_2 ($ZrMn_2$) – type alloys that date back to 1960s (Zaluska et al. 2001). The metal Hybrids that are broadly studied are listed in Table 3. Movement of hydrogen atoms into the crystal matrix and its interaction with metals by dissociation from Metal Hybrids. AB_2 and AB_5 are termed as intermetallic hybrids, as their interstitial crystal lattice (Chandra et al. 2006) contain hydrogen atoms. Prime disadvantages with these hybrids are low gravimetric densities ($LaNi_5H_6$); require high temperatures for operation (MgH_2) and great price of rare earth metals (i.e., titanium and Lanthanum).

The metal hydrides consist of less gravimetric densities (1-2 wt.%) and not appropriate for a mobile application, but appropriate for the stationary application (Chandra et al. 2006); due to their favourable operating conditions and reversibility of hydrides. MgH_2 hydride among other Metal Hydrides gained significant attention because of its high gravimetric capacity (7.6 wt.%). Magnesium hydrides show both covalent and ionic characters; and also has immense operating temperature (Table 3) (Noritake et al. 2002). The studies on MgH_2 focussed on reducing desorption temperatures and increasing kinetics by destabilization by various metal oxides and transition metals (Oelerich et al. 2001; Liang et al. 1999).

Complex Hydrides

Bogdanović & Schwickardi, 1997 doped the Ti with $NaAlH_4$ and made a complex hydride and bought attention to the hydrogen storage community. They added 2 mol% of Ti to $NaAlH_4$ and showed that adding this small amount can regenerate the complex hydrides by applying high hydrogen pressure. High gravimetric capacity for storage of hydrogen can be attained only by doping the metals that are lighter than Mg (i.e., Li and Na). Thus, during the last decade researchers focussed on light metal complex hydrides. These hydrides show both covalent and ionic bonding characters. In $LiNH_2$, Covalent characters are shown by N-H bonding in NH_2 (anion), whereas Mixed characteristics i.e., covalent and ionic characters are shown by Li-N.

Theoretical gravimetric densities of complex hydrides: $LiBH_4$(18.4 wt.%) (Orimo et al. 2005), Li_3N (11.5 wt.%) (Chen et al. 2002), $LiAlH_4$ (10.6 wt.%) (Chen et al. 2001) and $NaAlH_4$ (7.5 wt.%) (Bogdanović & Schwickardi, 1997; Song & Guo, 2006) and their reversible capacities of gravimetric

Figure 7. Schematic representation of current approaches to store hydrogen with materials.
Reprinted from Ref. (Sun et al. 2018) with permission from Energy Storage Materials. Copyright Elsevier

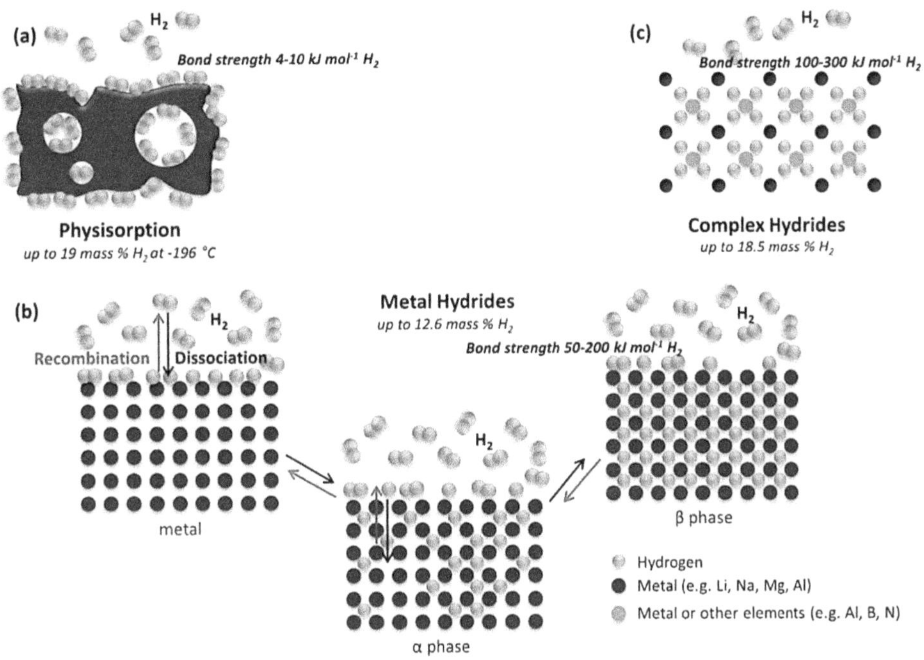

are about 4-5 wt.% under experimental operating temperatures and pressures (Sakintuna et al. 2007). The most studied complex hydrides of their reversible capacities are LiBH4–LiNH2–MgH_2 (4–8 wt.%, 200–280°C) (Sudik, 2008), $LiNH_2$–MgH_2/LiH (4–6 wt.%, 200–250°C) (Luo, 2004; Nakamori & Orimo, 2004), $NaAlH_4$ (3–5 wt.%, 120–150°C) (Resan et al. 2005) and $LiAlH_4$ (3–5 wt.%, 150–200°C) (Chen et al. 2001). Studies on these hydrides are mainly focussed on noxious toxic/poisonous gas emissions, sluggish kinetics, destabilization issues (Sakintuna et al. 2007; Borgschulte et al. 2011; Ichikawa et al. 2004; Demirocak et al. 2013).

Table 3. Extensively studied metal Hydrides

Metals	Hydrides	Capacity (wt.%)	Temperature for 1 Bar H_2 (°C)
Mg	MgH_2	7.66	279
Mg_2Ni	Mg_2NiH_4	3.59	255
FeTi	$FeTiH_2$	1.89	-8
$ZrMn_2$	$ZrMn_2H_2$	1.77	440
$LaNi_5$	$LaNi_5H_6$	1.37	12

(Principi et al. 2009)

Physisorption Based Materials

Porous Materials

Porous materials have large surface area and materials based on physisorption (Morris et al. 2019; Mohan et al. 2019), these materials include inorganic and organic ordered (Zeolites) and Amorphous (activated carbon) structures (Rowsell & Yaghi, 2004; Davis, 2002; Zhang & Cooper, 2005). These materials have various applications in ion exchange (Davis, 2002), gas storage (Morris et al. 2019), biomaterials (Karageorgiou & Kaplan, 2005) catalysis (Corma,1997), purification (Toyoda et al. 2003), and separation (Nakanishi & Tanaka, 2007; Horcajada et al. 2006). Based on the pore dimensions these materials are classified. Pores are separated into three categories on the basis of their measurements in accordance to International Union of Pure and Applied Chemistry; micropores (<2 nm), mesopores (2–50 nm), and macropores (>50 nm) (Ssing et al. 1985).

High surface area novel porous materials development like polymers of intrinsic porosity (PIMs) (McKeown & Budd, 2006), metal-organic frameworks (MOFs) (Rowsell & Yaghi, 2004), covalent organic frameworks (COFs) (Han et al. 2008), porous aromatic frameworks (PAFs) (Ben et al. 2009), templated carbons (Yang et al. 2007), conjugated microporous polymers (CMPs) (Jiang et al. 2008) and hyper crosslinked polymers (Germain et al. 2006), during the previous decade lead to extensive research attempts on storage materials (Morris et al., 2019). Owing to their reversibility, debauched sorption kinetics and capable gravimetric storage capacity these materials have been studied extensively.

Hydrogen storage in the above mentioned porous materials having large surface areas is comprehended by the interface of H_2 on porous material surface via physisorption (i.e., adsorption) and mainly on the basis of weak van der Waals forces (Rzepka et al. 1998). At cryogenic temperatures, a good amount of hydrogen uptake can be achieved by physisorption method. The storage capacity of hydrogen at room temperatures are usually less than 1 wt.% at modest temperatures (100 atm) (Thomas, 2007; Kajiura et al. 2003). The adsorption enthalpy pore size, pore volume and surface area the key parameters for hydrogen storage capacity (Bhatia & Myers, 2006; Frost et al. 2006). The adsorption enthalpy is low in porous materials (<6 KJ/mol) (Schmitz et al. 2008) and for uptake of hydrogen at room temperatures need to be improved (20–40 kJ/mol) (Jhi & Ihm, 2011).

Multiple alteration procedures like optimizing pore size (Bhatia & Myers, 2006) increasing surface area and nobbling with impure atoms (B, N, alkali/transition metals (Chung et al. 2008; Goddard and Han, 2009; Li & Yang, 2006; Zhou et al. 2008) will tune the enthalpy or improve the hydrogen uptake capacity.

Therefore, many researchers focussed on improving the adsorption enthalpy (i.e., creating open metal sites in MOFs) (Zhou et al. 2008), increasing surface areas (Farha et al. 2012), and fine-tuning pore size-dimensions of these materials (Katoch et al. 2019).

Chemical Hydrogen Storage (Off-board Regenerable Materials)

The term Off-board Regenerable materials in this section are used to describe, the materials that cannot be redeveloped simply (as the thermodynamics and Kinetics for these applications are unfavourable). Chemical Hydrides is another term that is frequently used in the literature to describe the materials with the same class.

Examples of chemical Hydrides used for chemical hydrogen storage: Sodium borohydrides ($NaBH_4$) and Aluminium hydrides (AlH_3). (but there are not places under the reversible hydride group due to their difficulty in reliability). Most studied hydrogen materials in the literature available are ammonia-borane (NH_3 _ BH_3 or simply AB) (Staubitz et al. 2010), $NaBH_4$ (Netskina et al. 2019), AlH_3 (Graetz et al. 2011) and Cyclohexane-benzene (liquid organic hydrogen carriers) (Crabtree, 2008; Teichmann et al. 2011). In detail about these materials are discussed below. Further, Methanol's (alcohol), Hydrocarbons (Eberle et al. 2009) and ammonia (NH_3) (Klerke et al. 2008) also reflected as chemical hydrogen storage materials. Main problem associated with such materials as energy carriers are, they will not provide solutions for sustainable development as these materials are based on relic fuels.

Borohydrides of Sodium

Hydrogen is released from sodium borohydride ($NaBH_4$) through hydrolysis reaction as reported by subsequent reaction path where half of the hydrogen is produced from water:

$$NaBH_4 + 2H_2O \rightarrow NaBO_2 + 4H_2$$

These borohydrides have 10.8 wt.% storage capacity of hydrogen by theoretically; Withal, in operating conditions, 2.9 – 7.5 wt.% storage capacities were reported (Netskina et al. 2019). The major issues linked with these hydrides with water system ($NaBH_4$–H_2O system) are the solubility of $NaBH_4$ and $NaBO_2$ is low in Water and this requires more quantity of water that can be stowed on-board and slow rate of reaction for hydrogen generation necessitates a catalyst to enhance the speed of reaction. However, over a longer duration, the catalyst durability may be induced and performance of the reaction may be degraded. These are the reason DOE restricted the above hydrides (as aqueous solutions) not to be used for hydrogen storage (Netskina et al. 2019).

Hydrides of Aluminium

Hydrides of aluminium (AlH_3- Aluminium hydride) has 148 kg/m^3 Volumetric and 10.1 wt.% gravimetric storage capacities of hydrogen. Hydrogen is liberated from AlH_3 with small amount of input heat (~7 kJ/mol) in accordance with endothermic reaction (Graetz, 2009).

$$AlH_3 \rightarrow Al + \frac{3}{2}H_2$$

Hydrogen delivery can be fastened by manipulating the size of crystallite, as crystallite with smaller size will decompose the AlH_3 faster and this will help in increasing the rate of reaction. Pressure (~700 MPa) is nearly required for regeneration of AlH_3 which is a major challenge at room temperatures. The Electrochemical hydrogenation and ethereal reactions (with $LiAlH_4$ and $AlCl_3$) are the other alternative ways for AlH_3 Regeneration.

Ammonia-Borane (AB)

A gravimetric storage capacity (theoretical) of 19.6 wt.% was achieved with this Ammonia-Borane (white crystalline solid) which is a non-explosive and non-flammable substance (Staubitz et al. 2010). Extraction of Hydrogen from AB can be achieved by many ways such as thermolysis (with or without the presence of a catalyst), hydrolysis and decomposition (Huang & Autrey, 2012).

Thermolysis and Hydrolysis were discussed briefly:

nNH_3BH_3 (s) → nNH_3BH_3

nNH_3BH_3 (l) → $[NHBH]_n$ (s) + nH_2 (g)

$[NHBH]_n$ (s) → $[NB]_n$ (s) + nH_2 (g)

The thermal decomposition of Ammonium-Borane requires onset temperature of 70^{0}C and maximum is 110^{0}C. (~1.1 mol H_2 can be yielded by 1 mol of Ammonium-Borane) and this decomposition will release a by-product polyaminoborane ($[NH_2BH_2]$ n) (Peng & Chen, 2008). Trace amounts of ammonia and borazine will be liberated by the decomposition of ammonium-borane and these are required for operating proton exchange membrane fuel cell and needs to be captured or eliminated (Demirocak et al. 2013).

Other concern of thermolysis of AB is induction duration before release of hydrogen; for example, AB starts to produce significant quantities of hydrogen just after 3 h at a temperature of 85^{0}C (Peng & Chen, 2008). The major concern with these substances (during thermolysis) is an induction period to release hydrogen a minimum 85^{0}C preheat for 3h is required. By mixing ionic liquids (Bluhm et al. 2006) with ammonium-borane can overcome the induction period or by doping the ammonium-borane with porous materials also decrease this problem (Gutowska et al. 2005). The overall storage capacity of hydrogen will be retarded by the above methods as they add dead weight to the system.

Ammonium-borane Hydrolysis process is mainly depending on pH. The aqueous Ammonium-borane is highly stable in basic and neutral solutions. However; by the interaction of aqueous ammonium-borane with an acid solution, the rate of reaction is fastened and will be hydrolysed rapidly.

Hydrogen should be available on-demand in the current world; so, reacting acidic solution with aqueous Ammonia-borane will liberate more amount of hydrogen. Prime problem of producing hydrogen by a hydrolytic pathway through ammonium-borane is production of highly stable compounds i.e., borate (B-O). Due to this stable bond formation regenerating of Ammonia-borane needs extensive energy and requires high capital (Stephens et al. 2007).

Liquid Organic Hydrogen Carriers (LOHC)

LOHC carriers of energy and the hydrogen can be unloaded and loaded with LOHC in cyclic way (Teichmann et al. 2011). A significant manufacturing benefit is there in carrying liquid as a hydrogen carrier rather than solid. Initially, the end-users are easily benefited and can be transported easily with the existing infrastructure and required quantities can be pumped easily to reaction chamber where it is rehabilitated into heat. Heating of whole hydrogens storage materials is not required due to this reason (Crabtree, 2008). Extensive studies on LOHC like N-ethyl carbazole and cyclohexane-benzene was done according to the literature available (Crabtree, 2008; Teichmann et al. 2011). Comparing LOHC with Chemical hydrogen storage materials, LOHC has a 6-8 wt.% (theoretical) hydrogen storage which is relatively low. Works are ongoing on this type of carriers and in detail especially on regenerating methods of LOHC (Crabtree, 2008; Teichmann et al. 2011).

Hybrid Methods of Hydrogen Storage

To overcome the shortcomings of different storages technologies, hybrid methods (use two or more different methods) are used and this will help in enhancing the storage. Cryo-adsorption is one of the main methods of the hybrid method. Physical-chemical process of hydrogen storage, Cryo-adsorption takes benefit of liquefied, compressed and physisorption materials based hydrogen storage technologies.

Cryo-Adsorption Hydrogen Storage

Hydrogen is stored at high pressures at cryogenic temperatures in physisorption based materials. The storage capacity of the system is mainly related to porous materials adsorption. To enhance the storage capacity pore size, pore-volume, surface area and heat are critical (Frost et al. 2006). Several researchers have investigated the performance of cryo-adsorption of different materials such as MOFs (Hirscher, 2011), zeolite (Li et al. 2009) and Activated carbon (Ahluwalia & Peng, 2009). Instead of empty tank for storage of hydrogen, porous materials will enhance the storage capacity at same temperatures and pressures. Hydrogen storage capacities of high-pressure porous materials presented in literature are the difference between saturation and the zero pressure (Figure 8).

It needs to recognize that applied operating environments are different from the real storage capacities of hydrogen in the system. As an example, pressure inside tank is certainly not zero as the fuel cell always operates at a pressure between 0.2-0.3 MPa (Wang et al. 2003). So, always there is a small amount of hydrogen remains in the tank. Temperature swing is the only way to mitigate this problem. Temperature

Figure 8. Schematic of on-board cryo-adsorption system.
Reprinted from Ref. (Ahluwalia & Peng,2009) with permission from the International Journal of Hydrogen storage. Copyright Elsevier

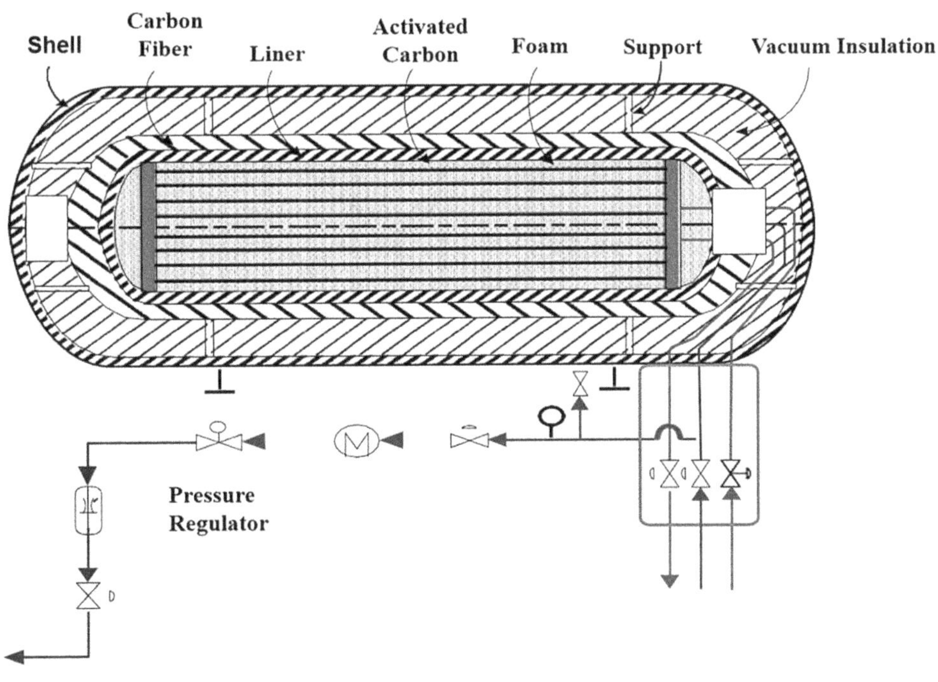

swing means enhancing temperature of tank by means of heat to desorb as abundant hydrogen as probable (Ahluwalia & Peng, 2009).

VI. CHALLENGES OF HYDROGEN STORAGE TECHNOLOGIES

Major challenges of hydrogen storage technologies are listed below:

1. **Compressed Hydrogen Storage**: High energy Penalty. For compression, a Lower heating value (up to 15%) is required. Storage tanks for tolerating high pressures are made by Carbon nanofibres which are very expensive.
2. **Liquefied hydrogen storage**: liquefaction requires quite great energy; lower heating values (up to 30%) are required. To minimise the heat gain from the surroundings the liquefaction tank must be super insulated. Frequent boils-off from the vent of the tank.
3. **Solid-state hydrogen storage**: as discussed earlier there are two chief technologies available for solid-state storage i.e., reversible hydrides and physisorption materials. Physisorption materials can store hydrogen only at cryogenic temperatures, so while filling the tank it needs to cool. Reversible hydrides require high temperatures and suffer from degradation over cycles and also emits toxic gases while discharging (NH_3).
4. **Chemical hydrogen storage**: off-board regeneration is required due to unfavourable kinetic and it is a costly process.

SUMMARY

From the beginning of the industrial era, modern society is uninterruptedly getting advantages from the exploiting of relic fuels undoubtedly. Though, relic fuels are the principal energy source (easy to store, High density. etc.) is a limited energy source. It is imperative to note that extensive utilization of relic fuels causes various environmental problem, one of the major concerns in the present scenario is global warming and climate change due to undue man-made CO_2 emissions. So, it is clearly evident that human community need to find an environmentally friendly alternative energy fuels alternative to relic fuel to move towards sustainable environment. Transportation sector majorly depends on oil that is anticipated to be extended before end of twenty-first century. CO_2 emissions emitted from automobiles are considered as a main source of CO_2 and there is no proven technology to capture the CO_2 that is economically viable. In near future, to mitigate climate change and to stop the reduction of oil it is needed to substitute with energy carrier which is environmentally sustainable. Amongst many options available for sustainable energy carrier, hydrogen is considered to be a promising candidate. This will not emit any noxious gases and free from carbon; when it is burned in internal combustion engines. Similar to the oil economy, hydrogen economy is a system that includes storage, production and transportation. At present, steam reformation methods are utilizing for the production of hydrogen. Renewable energy sources like wind or photovoltaic (PV) powered water electrolysis is required to produce hydrogen. Major challenges are being faced by the transportation of hydrogen are it requires an extensive networking system for distribution of energy which is so expensive.

Hydrogen storage is the last step in hydrogen economy and bottleneck in this economy. Many automotive industries are setting targets for the storage of hydrogen for vehicular application and collaborating with different countries in the world. The greatest rigorous targets are volumetric and gravimetric hydrogen storage capacities, resulting in comparison of different hydrogen storage technologies often times compared only on the basis of energy density irrespective of operating temperature, cost and efficiency. Storage technologies for hydrogen are comprised of solid-state, liquefied, compressed and chemical hydrogen storage technologies. The physical storage methods include liquified hydrogen and compressed storage systems, whereas chemical methods include solid-state and chemical hydrogens storage technologies. The key difference between physical and chemical processes is that hydrogen does not interact with the storage media in physical methods, while hydrogen interacts with the storage media in chemical methods. The conventional methods of hydrogen storage are liquified and compressed storage systems and these technologies have been used for decades. To enhance the gaseous storage, carbon fibre manufacturing methods with low cost were developed to reduce the costs of the tank. To minimise the boil-off losses vacuum super insulations are used and novel cooling technologies also enhanced the dormancy period. Compressed gaseous storage systems are commercialized and most successful technology among all other so far. Due to the shortcomings of liquified systems, majority of the research efforts concentrated on on-board regenerable (solid-state) and off-board regenerable (off-board regenerable). In comparison to solid chemical storage, more attention was gained by solid-state storage as they are reducing the costs concerned with spent fuels. Reversible hydrides and physisorption (porous) materials are the two subclasses of Solid-State storages materials.

A high temperature is required for the operation of reversible hydrides as they have reversibility issues. Porous materials store important quantity of hydrogen at cryogenic temperature, but the major drawback with these systems is price related with cooling. The challenge associated with solid-state materials is it is too difficult to bring the operating temperatures to the closer room temperatures. Hydrolysis process or thermolysis process is utilized to synthesis hydrogen in chemical hydrogen storage materials. By-products released during these processes are more stable to be renewed on-board merely charging by hydrogen required. The drawbacks associated with this material are complexity in regeneration and its cost. In conclusion, in this chapter advantages and disadvantages of hydrogen storage technologies and materials used for storage are presented in detail. Expedition for discovery of a benign fuel substitute to oil is still in process revealing hydrogen to be one of the best choices available in the present scenario. Further research is essential to progress hydrogen storage field frontward. Market diffusion of hydrogen storage technologies in near future is closely knotted to development of enriched hydrogen storage materials.

REFERENCES

Abbas, H. F., & Daud, W. W. (2010). Hydrogen production by methane decomposition: A review. *International Journal of Hydrogen Energy*, *35*(3), 1160–1190. doi:10.1016/j.ijhydene.2009.11.036

Agrawal, R., Singh, N. R., Ribeiro, F. H., & Delgass, W. N. (2007). Sustainable fuel for the transportation sector. *Proceedings of the National Academy of Sciences of the United States of America*, *104*(12), 4828–4833. doi:10.1073/pnas.0609921104 PMID:17360377

Ahluwalia, R. K., Hua, T. Q., Peng, J. K., Lasher, S., McKenney, K., Sinha, J., & Gardiner, M. (2010). Technical assessment of cryo-compressed hydrogen storage tank systems for automotive applications. *International Journal of Hydrogen Energy*, *35*(9), 4171–4184. doi:10.1016/j.ijhydene.2010.02.074

Ahluwalia, R. K., & Peng, J. K. (2009). Automotive hydrogen storage system using cryo-adsorption on activated carbon. *International Journal of Hydrogen Energy*, *34*(13), 5476–5487. doi:10.1016/j.ijhydene.2009.05.023

Asif, M., & Muneer, T. (2007). Energy supply, its demand and security issues for developed and emerging economies. *Renewable & Sustainable Energy Reviews*, *11*(7), 1388–1413. doi:10.1016/j.rser.2005.12.004

Barbir, F., Veziroğlu, T. N., & Plass, H. J. Jr. (1990). Environmental damage due to fossil fuels use. *International Journal of Hydrogen Energy*, *15*(10), 739–749. doi:10.1016/0360-3199(90)90005-J

Baroutaji, A., Wilberforce, T., Ramadan, M., & Olabi, A. G. (2019). Comprehensive investigation on hydrogen and fuel cell technology in the aviation and aerospace sectors. *Renewable & Sustainable Energy Reviews*, *106*, 31–40. doi:10.1016/j.rser.2019.02.022

Barthelemy, H., Weber, M., & Barbier, F. (2017). Hydrogen storage: Recent improvements and industrial perspectives. *International Journal of Hydrogen Energy*, *42*(11), 7254–7262. doi:10.1016/j.ijhydene.2016.03.178

Ben, T., Ren, H., Ma, S., Cao, D., Lan, J., Jing, X., ... Qiu, S. (2009). Targeted synthesis of a porous aromatic framework with high stability and exceptionally high surface area. *Angewandte Chemie International Edition*, *48*(50), 9457–9460. doi:10.1002/anie.200904637 PMID:19921728

Bhatia, S. K., & Myers, A. L. (2006). Optimum conditions for adsorptive storage. *Langmuir*, *22*(4), 1688–1700. doi:10.1021/la0523816 PMID:16460092

Bluhm, M. E., Bradley, M. G., Butterick, R., Kusari, U., & Sneddon, L. G. (2006). Amineborane-based chemical hydrogen storage: Enhanced ammonia borane dehydrogenation in ionic liquids. *Journal of the American Chemical Society*, *128*(24), 7748–7749. doi:10.1021/ja062085v PMID:16771483

Bogdanović, B., & Schwickardi, M. (1997). Ti-doped alkali metal aluminium hydrides as potential novel reversible hydrogen storage materials. *Journal of Alloys and Compounds*, *253*, 1–9. doi:10.1016/S0925-8388(96)03049-6

Borgschulte, A., Callini, E., Probst, B., Jain, A., Kato, S., Friedrichs, O., Remhof, A., Bielmann, M., Ramirez-Cuesta, A. J., & Züttel, A. (2011). Impurity gas analysis of the decomposition of complex hydrides. *The Journal of Physical Chemistry C*, *115*(34), 17220–17226. doi:10.1021/jp205566q

Bossel, U., Eliasson, B., & Taylor, G. (2003). The future of the hydrogen economy: Bright or bleak? *Cogeneration and Distributed Generation Journal*, *18*(3), 29–70. doi:10.1080/15453660309509023

Chalk, S. G., & Miller, J. F. (2006). Key challenges and recent progress in batteries, fuel cells, and hydrogen storage for clean energy systems. *Journal of Power Sources*, *159*(1), 73–80. doi:10.1016/j.jpowsour.2006.04.058

Chandra, D., Reilly, J. J., & Chellappa, R. (2006). Metal hydrides for vehicular applications: The state of the art. *JOM*, *58*(2), 26–32. doi:10.100711837-006-0005-0

Chen, J., Kuriyama, N., Xu, Q., Takeshita, H. T., & Sakai, T. (2001). Reversible hydrogen storage via titanium-catalyzed LiAlH4 and Li3AlH6. *The Journal of Physical Chemistry B*, *105*(45), 11214–11220. doi:10.1021/jp012127w

Chen, P., Xiong, Z., Luo, J., Lin, J., & Tan, K. L. (2002). Interaction of hydrogen with metal nitrides and imides. *Nature*, *420*(6913), 302–304. doi:10.1038/nature01210 PMID:12447436

Chung, T. M., Jeong, Y., Chen, Q., Kleinhammes, A., & Wu, Y. (2008). Synthesis of microporous boron-substituted carbon (B/C) materials using polymeric precursors for hydrogen physisorption. *Journal of the American Chemical Society*, *130*(21), 6668–6669. doi:10.1021/ja800071y PMID:18454522

Corma, A. (1997). From microporous to mesoporous molecular sieve materials and their use in catalysis. *Chemical Reviews*, *97*(6), 2373–2420. doi:10.1021/cr960406n PMID:11848903

Crabtree, G. W., Dresselhaus, M. S., & Buchanan, M. V. (2004). The hydrogen economy. *Physics Today*, *57*(12), 39–44. doi:10.1063/1.1878333

Crabtree, R. H. (2008). Hydrogen storage in liquid organic heterocycles. *Energy & Environmental Science*, *1*(1), 134–138. doi:10.1039/b805644g

Davis, M. E. (2002). Ordered porous materials for emerging applications. *Nature*, *417*(6891), 813–821. doi:10.1038/nature00785 PMID:12075343

Demirocak, D. E., Srinivasan, S. S., Ram, M. K., Kuhn, J. N., Muralidharan, R., Li, X., Goswami, D. Y., & Stefanakos, E. K. (2013). Reversible hydrogen storage in the Li–Mg–N–H system–The effects of Ru doped single walled carbon nanotubes on NH3 emission and kinetics. *International Journal of Hydrogen Energy*, *38*(24), 10039–10049. doi:10.1016/j.ijhydene.2013.05.176

Eberle, U., Felderhoff, M., & Schueth, F. (2009). Chemical and physical solutions for hydrogen storage. *Angewandte Chemie International Edition*, *48*(36), 6608–6630. doi:10.1002/anie.200806293 PMID:19598190

Economics, B. E. (2018). BP Energy Outlook. Academic Press.

Farha, O. K., Eryazici, I., Jeong, N. C., Hauser, B. G., Wilmer, C. E., Sarjeant, A. A., Snurr, R. Q., Nguyen, S. B. T., Yazaydın, A. Ö., & Hupp, J. T. (2012). Metal–organic framework materials with ultrahigh surface areas: Is the sky the limit? *Journal of the American Chemical Society*, *134*(36), 15016–15021. doi:10.1021/ja3055639 PMID:22906112

Felderhoff, M., Weidenthaler, C., von Helmolt, R., & Eberle, U. (2007). Hydrogen storage: The remaining scientific and technological challenges. *Physical Chemistry Chemical Physics*, *9*(21), 2643–2653. doi:10.1039/b701563c PMID:17627309

Frost, H., Düren, T., & Snurr, R. Q. (2006). Effects of surface area, free volume, and heat of adsorption on hydrogen uptake in metal– organic frameworks. *The Journal of Physical Chemistry B*, *110*(19), 9565–9570. doi:10.1021/jp060433+ PMID:16686503

Ganguly, G., Halder, D., Banerjee, A., Basu, S., & Paul, A. (2019). Exploring the Crucial Role of Solvation on the Viability of Sustainable Hydrogen Storage in BN-fullerene: A Combined DFT and Ab initio Molecular Dynamics investigation. *ACS Sustainable Chemistry & Engineering*, *7*(11), 9808–9821. doi:10.1021/acssuschemeng.9b00211

Gerboni, R., & Salvador, E. (2009). Hydrogen transportation systems: Elements of risk analysis. *Energy*, *34*(12), 2223–2229. doi:10.1016/j.energy.2008.12.018

Germain, J., Hradil, J., Fréchet, J. M., & Svec, F. (2006). High surface area nanoporous polymers for reversible hydrogen storage. *Chemistry of Materials*, *18*(18), 4430–4435. doi:10.1021/cm061186p

Goddard, W. A., & Han, S. S. (2009). *U.S. Patent Application No. 12/150,046*. US Patent Office.

Graetz, J. (2009). New approaches to hydrogen storage. *Chemical Society Reviews*, *38*(1), 73–82. doi:10.1039/B718842K PMID:19088966

Graetz, J., Reilly, J. J., Yartys, V. A., Maehlen, J. P., Bulychev, B. M., Antonov, V. E., Tarasov, B. P., & Gabis, I. E. (2011). Aluminum hydride as a hydrogen and energy storage material: Past, present and future. *Journal of Alloys and Compounds*, *509*, S517–S528. doi:10.1016/j.jallcom.2010.11.115

Gutowska, A., Li, L., Shin, Y., Wang, C. M., Li, X. S., Linehan, J. C., ... Gutowski, M. (2005). Nanoscaffold mediates hydrogen release and the reactivity of ammonia borane. *Angewandte Chemie International Edition*, *44*(23), 3578–3582. doi:10.1002/anie.200462602 PMID:15880751

Hamilton, C. W., Baker, R. T., Staubitz, A., & Manners, I. (2009). B–N compounds for chemical hydrogen storage. *Chemical Society Reviews*, *38*(1), 279–293. doi:10.1039/B800312M PMID:19088978

Han, S. S., Furukawa, H., Yaghi, O. M., & Goddard Iii, W. A. (2008). Covalent organic frameworks as exceptional hydrogen storage materials. *Journal of the American Chemical Society*, *130*(35), 11580–11581. doi:10.1021/ja803247y PMID:18683924

Hirscher, M. (2011). Hydrogen storage by cryoadsorption in ultrahigh-porosity metal–organic frameworks. *Angewandte Chemie International Edition*, *50*(3), 581–582. doi:10.1002/anie.201006913 PMID:21226135

Hoang, T. K., & Antonelli, D. M. (2009). Exploiting the Kubas interaction in the design of hydrogen storage materials. *Advanced Materials*, *21*(18), 1787–1800. doi:10.1002/adma.200802832

Horcajada, P., Serre, C., Vallet-Regí, M., Sebban, M., Taulelle, F., & Férey, G. (2006). Metal–organic frameworks as efficient materials for drug delivery. *Angewandte Chemie International Edition*, *45*(36), 5974–5978. doi:10.1002/anie.200601878 PMID:16897793

Houghton, J. (2005). Reports on Progress in Physics. *Global Warming, 68*, 1343-1403.

Høyer, K. G., & Holden, E. (2007). Alternative fuels and sustainable mobility: Is the future road paved by biofuels, electricity or hydrogen. *International Journal of Alternative Propulsion*, *1*(4), 352–369. doi:10.1504/IJAP.2007.013329

Hua, T. Q., Ahluwalia, R. K., Peng, J. K., Kromer, M., Lasher, S., McKenney, K., Law, K., & Sinha, J. (2011). Technical assessment of compressed hydrogen storage tank systems for automotive applications. *International Journal of Hydrogen Energy*, *36*(4), 3037–3049. doi:10.1016/j.ijhydene.2010.11.090

Huang, Z., & Autrey, T. (2012). Boron–nitrogen–hydrogen (BNH) compounds: Recent developments in hydrogen storage, applications in hydrogenation and catalysis, and new syntheses. *Energy & Environmental Science*, *5*(11), 9257–9268. doi:10.1039/c2ee23039a

Hubbert, M. K. (1956, January). Nuclear energy and the fossil fuel. In *Drilling and production practice*. American Petroleum Institute.

Ichikawa, T., Hanada, N., Isobe, S., Leng, H., & Fujii, H. (2004). Mechanism of novel reaction from LiNH2 and LiH to Li2NH and H2 as a promising hydrogen storage system. *The Journal of Physical Chemistry B*, *108*(23), 7887–7892. doi:10.1021/jp049968y

Jain, I. P., Jain, P., & Jain, A. (2010). Novel hydrogen storage materials: A review of lightweight complex hydrides. *Journal of Alloys and Compounds*, *503*(2), 303–339. doi:10.1016/j.jallcom.2010.04.250

Jhi, S. H., & Ihm, J. (2011). Developing high-capacity hydrogen storage materials via quantum simulations. *MRS Bulletin*, *36*(3), 198–204. doi:10.1557/mrs.2011.32

Jiang, J. X., Su, F., Trewin, A., Wood, C. D., Niu, H., Jones, J. T., Khimyak, Y. Z., & Cooper, A. I. (2008). Synthetic control of the pore dimension and surface area in conjugated microporous polymer and copolymer networks. *Journal of the American Chemical Society*, *130*(24), 7710–7720. doi:10.1021/ja8010176 PMID:18500800

Kajiura, H., Tsutsui, S., Kadono, K., Kakuta, M., Ata, M., & Murakami, Y. (2003). Hydrogen storage capacity of commercially available carbon materials at room temperature. *Applied Physics Letters*, *82*(7), 1105–1107. doi:10.1063/1.1555262

Karageorgiou, V., & Kaplan, D. (2005). Porosity of 3D biomaterial scaffolds and osteogenesis. *Biomaterials*, *26*(27), 5474–5491. doi:10.1016/j.biomaterials.2005.02.002 PMID:15860204

Katoch, A., Goyal, N., & Gautam, S. (2019). Applications and advances in coordination cages: Metal-Organic Frameworks. *Vacuum*, *167*, 287–300. doi:10.1016/j.vacuum.2019.03.038

Kerr, R. A. (2011). *Peak oil production may already be here*. Academic Press.

Klerke, A., Christensen, C. H., Nørskov, J. K., & Vegge, T. (2008). Ammonia for hydrogen storage: Challenges and opportunities. *Journal of Materials Chemistry*, *18*(20), 2304–2310. doi:10.1039/b720020j

Konni, M., Dadhich, A. S., & Mukkamala, S. B. (2017). Impact of surface modifications on hydrogen uptake by Fe@ f-MWCNTs and Cu@ f-MWCNTs at non-cryogenic temperatures. *International Journal of Hydrogen Energy*, *42*(2), 953–959. doi:10.1016/j.ijhydene.2016.09.085

Konni, M., Dadhich, A. S., & Mukkamala, S. B. (2019). Evaluation of surface changes at the interface between TiO2 nanoparticles and COOH-MWCNTs on hydrogen adsorption capability. *Nano-Structures & Nano-Objects*, *18*, 100304. doi:10.1016/j.nanoso.2019.100304

Konni, M., Karnena, M. K., & Mukkamala, S. B. (2020). Hydrogen Storage Capacity in Ni/Pd@ f-MWCNTS Decorated Graphene Oxide/Cu-BTC Composites at Room Temperatures: A Sustainable Cleaner Energy Production. *International Journal of Surface Engineering and Interdisciplinary Materials Science*, *8*(1), 1–12. doi:10.4018/IJSEIMS.2020010101

Konni, M., Kumar Karnena, M., & Babu Mukkamala, S. (2020). Enhanced Room-Temperature Synthesis of Li@ f-MWCNTs for Hydrogen Storage Application. *ChemistrySelect*, *5*(14), 4292–4296. doi:10.1002lct.202000489

Konni, M., & Mukkamala, S. B. (2018). Hydrogen uptake behavior of Cr 2 O 3 nanoparticle decorated f-MWCNTs at non-cryogenic temperatures. *Journal of the Indian Chemical Society*, *95*, 393–397.

Lattin, W. C., & Utgikar, V. P. (2007). Transition to hydrogen economy in the United States: A 2006 status report. *International Journal of Hydrogen Energy*, *32*(15), 3230–3237. doi:10.1016/j.ijhydene.2007.02.004

Li, J., Wu, E., Song, J., Xiao, F., & Geng, C. (2009). Cryoadsorption of hydrogen on divalent cation-exchanged X-zeolites. *International Journal of Hydrogen Energy, 34*(13), 5458-5465.

Li, Y., & Yang, R. T. (2006). Significantly enhanced hydrogen storage in metal– organic frameworks via spillover. *Journal of the American Chemical Society*, *128*(3), 726–727. doi:10.1021/ja056831s PMID:16417355

Liang, G., Huot, J., Boily, S., Van Neste, A., & Schulz, R. (1999). Catalytic effect of transition metals on hydrogen sorption in nanocrystalline ball milled MgH2–Tm (Tm= Ti, V, Mn, Fe and Ni) systems. *Journal of Alloys and Compounds*, *292*(1-2), 247–252. doi:10.1016/S0925-8388(99)00442-9

Luo, W. (2004). (LiNH2–MgH2): A viable hydrogen storage system. *Journal of Alloys and Compounds*, *381*(1-2), 284–287. doi:10.1016/j.jallcom.2004.03.119

Maggio, G., & Cacciola, G. (2012). When will oil, natural gas, and coal peak? *Fuel*, *98*, 111–123. doi:10.1016/j.fuel.2012.03.021

Manoharan, Y., Hosseini, S. E., Butler, B., Alzhahrani, H., Senior, B. T. F., Ashuri, T., & Krohn, J. (2019). Hydrogen Fuel Cell Vehicles; Current Status and Future Prospect. *Applied Sciences (Basel, Switzerland)*, *9*(11), 2296. doi:10.3390/app9112296

McKeown, N. B., & Budd, P. M. (2006). Polymers of intrinsic microporosity (PIMs): Organic materials for membrane separations, heterogeneous catalysis and hydrogen storage. *Chemical Society Reviews*, *35*(8), 675–683. doi:10.1039/b600349d PMID:16862268

Mendelssohn, I. A., Andersen, G. L., Baltz, D. M., Caffey, R. H., Carman, K. R., Fleeger, J. W., Joye, S. B., Lin, Q., Maltby, E., Overton, E. B., & Rozas, L. P. (2012). Oil impacts on coastal wetlands: Implications for the Mississippi River Delta ecosystem after the Deepwater Horizon oil spill. *Bioscience*, *62*(6), 562–574. doi:10.1525/bio.2012.62.6.7

Mohan, M., Sharma, V. K., Kumar, E. A., & Gayathri, V. (2019). Hydrogen storage in carbon materials—A review. *Energy Storage*, e35.

Morris, R. E., Wheatley, P. S., Warrender, S., & Duncan, M. (2019). *U.S. Patent No. 10,357,037*. Washington, DC: U.S. Patent and Trademark Office.

Nakamori, Y., & Orimo, S. I. (2004). Destabilization of Li-based complex hydrides. *Journal of Alloys and Compounds*, *370*(1-2), 271–275. doi:10.1016/j.jallcom.2003.08.089

Nakanishi, K., & Tanaka, N. (2007). Sol–gel with phase separation. Hierarchically porous materials optimized for high-performance liquid chromatography separations. *Accounts of Chemical Research*, *40*(9), 863–873. doi:10.1021/ar600034p PMID:17650924

Netskina, O. V., Tayban, E. S., Ozerova, A. M., Komova, O. V., & Simagina, V. I. (2019). Solid-State NaBH4/Co Composite as Hydrogen Storage Material: Effect of the Pressing Pressure on Hydrogen Generation Rate. *Energies*, *12*(7), 1184. doi:10.3390/en12071184

Noritake, T., Aoki, M., Towata, S., Seno, Y., Hirose, Y., Nishibori, E., Takata, M., & Sakata, M. (2002). Chemical bonding of hydrogen in MgH 2. *Applied Physics Letters*, *81*(11), 2008–2010. doi:10.1063/1.1506007

Oelerich, W., Klassen, T., & Bormann, R. (2001). Metal oxides as catalysts for improved hydrogen sorption in nanocrystalline Mg-based materials. *Journal of Alloys and Compounds*, *315*(1-2), 237–242. doi:10.1016/S0925-8388(00)01284-6

Orimo, S. I., Nakamori, Y., Kitahara, G., Miwa, K., Ohba, N., Towata, S. I., & Züttel, A. (2005). Dehydriding and rehydriding reactions of LiBH4. *Journal of Alloys and Compounds*, *404*, 427–430. doi:10.1016/j.jallcom.2004.10.091

Peng, B., & Chen, J. (2008). Ammonia borane as an efficient and lightweight hydrogen storage medium. *Energy & Environmental Science*, *1*(4), 479–483. doi:10.1039/b809243p

Principi, G., Agresti, F., Maddalena, A., & Russo, S. L. (2009). The problem of solid state hydrogen storage. *Energy*, *34*(12), 2087–2091. doi:10.1016/j.energy.2008.08.027

Read, C., Thomas, G., Ordaz, C., & Satyapal, S. (2007). US Department of Energy's system targets for on-board vehicular hydrogen storage. *Material Matters*, *2*(2), 3.

Resan, M., Hampton, M. D., Lomness, J. K., & Slattery, D. K. (2005). Effects of various catalysts on hydrogen release and uptake characteristics of LiAlH4. *International Journal of Hydrogen Energy*, *30*(13-14), 1413–1416. doi:10.1016/j.ijhydene.2004.12.009

Rowsell, J. L., & Yaghi, O. M. (2004). Metal–organic frameworks: A new class of porous materials. *Microporous and Mesoporous Materials*, *73*(1-2), 3–14. doi:10.1016/j.micromeso.2004.03.034

Rzepka, M., Lamp, P., & De la Casa-Lillo, M. A. (1998). Physisorption of hydrogen on microporous carbon and carbon nanotubes. *The Journal of Physical Chemistry B*, *102*(52), 10894–10898. doi:10.1021/jp9829602

Sakintuna, B., Lamari-Darkrim, F., & Hirscher, M. (2007). Metal hydride materials for solid hydrogen storage: A review. *International Journal of Hydrogen Energy*, *32*(9), 1121–1140. doi:10.1016/j.ijhydene.2006.11.022

Schmitz, B., Müller, U., Trukhan, N., Schubert, M., Férey, G., & Hirscher, M. (2008). Heat of Adsorption for Hydrogen in Microporous High-Surface-Area Materials. *ChemPhysChem*, *9*(15), 2181–2184. doi:10.1002/cphc.200800463 PMID:18821561

Sdanghi, G., Maranzana, G., Celzard, A., & Fierro, V. (2019). Review of the current technologies and performances of hydrogen compression for stationary and automotive applications. *Renewable & Sustainable Energy Reviews*, *102*, 150–170. doi:10.1016/j.rser.2018.11.028

Shinnar, R. (2003). The hydrogen economy, fuel cells, and electric cars. *Technology in Society*, *25*(4), 455–476. doi:10.1016/j.techsoc.2003.09.024

Song, Y., & Guo, Z. X. (2006). Electronic structure, stability and bonding of the Li-NH hydrogen storage system. *Physical Review. B*, *74*(19), 195120. doi:10.1103/PhysRevB.74.195120

Ssing, K., Everett, D., Haul, R., Moscou, L., Pierotti, R., Rouquerol, J., & Siemieniewsks, T. (1985). Reporting physisorption data for gas/solid system. *Pure and Applied Chemistry*, *57*(4), 603–619.

Staubitz, A., Robertson, A. P., & Manners, I. (2010). Ammonia-borane and related compounds as dihydrogen sources. *Chemical Reviews*, *110*(7), 4079–4124. doi:10.1021/cr100088b PMID:20672860

Stephens, F. H., Pons, V., & Baker, R. T. (2007). Ammonia–borane: The hydrogen source par excellence? *Dalton Transactions (Cambridge, England)*, (25), 2613–2626. doi:10.1039/B703053C PMID:17576485

Sudik, A., Yang, J., Halliday, D., & Wolverton, C. (2008). Hydrogen Storage Properties in (LiNH2) 2–LiBH4–(MgH2) X Mixtures (x= 0.0– 1.0). *The Journal of Physical Chemistry C*, *112*(11), 4384–4390. doi:10.1021/jp711393k

Sun, Y., Shen, C., Lai, Q., Liu, W., Wang, D. W., & Aguey-Zinsou, K. F. (2018). Tailoring magnesium-based materials for hydrogen storage through synthesis: Current state of the art. *Energy Storage Materials*, *10*, 168–198. doi:10.1016/j.ensm.2017.01.010

Tarasov, B. P., Lototskii, M. V., & Yartys', V. A. (2007). Problem of hydrogen storage and prospective uses of hydrides for hydrogen accumulation. *Russian Journal of General Chemistry*, *77*(4), 694–711. doi:10.1134/S1070363207040329

Teichmann, D., Arlt, W., Wasserscheid, P., & Freymann, R. (2011). A future energy supply based on Liquid Organic Hydrogen Carriers (LOHC). *Energy & Environmental Science*, *4*(8), 2767–2773. doi:10.1039/c1ee01454d

Thomas, K. M. (2007). Hydrogen adsorption and storage on porous materials. *Catalysis Today*, *120*(3-4), 389–398. doi:10.1016/j.cattod.2006.09.015

Toyoda, M., Nanbu, Y., Kito, T., Hiranob, M., & Inagaki, M. (2003). Preparation and performance of anatase-loaded porous carbons for water purification. *Desalination*, *159*(3), 273–282. doi:10.1016/S0011-9164(03)90079-8

US DOE. (2009). *Technical System Targets: Onboard Hydrogen Storage for Light-Duty Fuel Cell Vehicles*. Author.

von Helmolt, R., & Eberle, U. (2007). Fuel cell vehicles: Status 2007. *Journal of Power Sources*, *165*(2), 833–843. doi:10.1016/j.jpowsour.2006.12.073

Wang, L., Husar, A., Zhou, T., & Liu, H. (2003). A parametric study of PEM fuel cell performances. *International Journal of Hydrogen Energy*, *28*(11), 1263–1272. doi:10.1016/S0360-3199(02)00284-7

Yang, Z., Xia, Y., & Mokaya, R. (2007). Enhanced hydrogen storage capacity of high surface area zeolite-like carbon materials. *Journal of the American Chemical Society*, *129*(6), 1673–1679. doi:10.1021/ja067149g PMID:17243684

Yi, H. S. (1996). *Climate change 1995: Economic and social dimensions of climate change: Contribution of Working Group III to the second assessment report of the Intergovernmental Panel on Climate Change* (Vol. 2). Cambridge University Press.

Zaluska, A., Zaluski, L., & Ström-Olsen, J. O. (2001). Structure, catalysis and atomic reactions on the nano-scale: A systematic approach to metal hydrides for hydrogen storage. *Applied Physics. A, Materials Science & Processing*, *72*(2), 157–165. doi:10.1007003390100783

Zhang, H., & Cooper, A. I. (2005). Synthesis and applications of emulsion-templated porous materials. *Soft Matter*, *1*(2), 107–113. doi:10.1039/b502551f PMID:32646082

Zhou, L. (2005). Progress and problems in hydrogen storage methods. *Renewable & Sustainable Energy Reviews*, *9*(4), 395–408. doi:10.1016/j.rser.2004.05.005

Zhou, W., Wu, H., & Yildirim, T. (2008). Enhanced H2 adsorption in isostructural metal– organic frameworks with open metal sites: Strong dependence of the binding strength on metal ions. *Journal of the American Chemical Society*, *130*(46), 15268–15269. doi:10.1021/ja807023q PMID:18950163

Züttel, A., Rentsch, S., Fischer, P., Wenger, P. M. C. E. P., Sudan, P. H., Mauron, P., & Emmenegger, C. (2003). Hydrogen storage properties of LiBH4. *Journal of Alloys and Compounds*, *356*, 515–520. doi:10.1016/S0925-8388(02)01253-7

Chapter 7
Hydrogen Fuel Cell Technologies for Sustainable Stationary Applications

Raluca-Andreea Felseghi
https://orcid.org/0000-0001-6310-5932
Ştefan cel Mare University of Suceava, Romania

Florin Badea
Romanian Association, Transylvania Branch for Hydrogen Energy, Romania

ABSTRACT

Science has shown that there are two sustainable alternatives to providing energy needs: renewable energy resources and fuel cells-hydrogen-based energy, which will play a complementary role in securing global energy resources. By promoting the use of hydrogen-based energy technologies, as clean energy technologies for stationary applications, at the level of local communities, industrial and commercial communities, research topics in this field will help the practical development of sustainable and clean energy systems. This chapter provides an overview of fuel cells highlighting aspects related to fuel cell short history, the main components and operating principles of fuel cells, the main constructive fuel cell types, and the main ways of powering stationary applications through the hydrogen fuel cell technologies.

Science has shown that there are two alternatives to sustaining energy needs: renewables and fuel cells - hydrogen-based energy, which will play a complementary role in securing global energy resources. By promoting the use of hydrogen-based energy technologies, as clean energy technologies for stationary applications, at the level of local communities, industrial and commercial communities, research topics in this field will help the practical development of sustainable and clean energy systems. *(Felseghi, R. A., et.al., 2019)* TN Veziroglu, stated that hydrogen is an inexhaustible source, given that it is obtained from water, and through its use is transformed into water, hydrogen production and consumption is a closed cycle that keeps the source of production constant - water and is a classic of recirculation of raw material. *(Veziroglu, T. N., 2007; Ball, M., et.al., 2015; Barbir, F., et.al, 2015; Gupta, R., et.al., 2016)*

DOI: 10.4018/978-1-7998-4945-2.ch007

SHORT HISTORY

At the beginning of the 19th century, it was the starting point in the field of fuel cells, respectively in 1801 the principle underlying the operation of the fuel cell was demonstrated by Humphry Davy, because later, in 1839, the lawyer and the man of Amateur scientist William Grove accidentally discovered the principle of the fuel cell during an electrolysis experiment, when he disconnected the battery from the electrolysis device and touched the two electrodes. He called this cell "gas battery", which consisted of platinum electrodes placed in tubes containing hydrogen gas and oxygen, respectively, tubes immersed in dilute sulfuric acid. The voltage generated was around 1 V. Something more later, Grove connected several such "gas batteries" in series and used this voltage source thus obtained to supply the electrolyzer that separates hydrogen from oxygen.

Due to the problems of electrode corrosion and the instability of the materials used, Grove's fuel cell did not give practical results. Precisely for this reason, research in the field was left for many years. Charles Langer and Ludwig Mond develop Grove's invention by observing that over time the phenomenon of platinum black reactivity in contact with electrolytes diminishes, and the life of the fuel cell is extended by keeping the electrolyte in non-conductive porous materials, using in 1889 for the first time the terminology of fuel cell - fuel cell.

A more serious approach to fuel cell research came after the 1937s, when Francis T. Bacon became concerned about the practical applicability of these components, which after many years of research produced the first practical fuel cell using an electrolyte. alkaline (KOH) instead of dilute sulfuric acid and sintered porous nickel electrodes, which allowed the diffusion of gas, developing in 1959 a 5 kW alkaline fuel cell.

Subsequently, research was deepened, making significant progress, and the 1960s and 1970s brought the first practical applications of fuel cells, which were used by NASA for spacecraft. At the same time, General Electric was developing a proton-changing membrane fuel cell, which was the basis for the construction of fuel cells used in missions of the Gemini program and the Apollo space program, for electricity generation. The oil crisis of the 1970s accelerated the development of alternative technologies, including fuel cells, for a number of fuel cell demonstration vehicle models.

In 1980, the US Navy used the fuel cell as a practical application for submarine propulsion. The major car manufacturers in Germany, USA, Japan are experiencing an increase in the power density of PEMFC batteries and are developing the first developed hydrogen storage systems.

Starting with 1990, the researches have as objectives the implementation of the fuel cell technology in stationary applications, respectively the fuel cells of different capacities have been developed for use in this field. Advances in membrane durability and improved energy performance of fuel cell assemblies have driven the use of PAFCs in cogeneration applications and their widespread use. PEMFC and SOFC technology have also been developed for this field, especially for small stationary applications. The year 2000 also makes significant contributions in the field of portable devices, when the technology of direct methanol fuel cells is adopted and widely used in this type of applications.

In the last two decades, there has been a rapid acceleration in the use of fuel cells covering a wide range of application diversity in the portable, mobile and stationary fields. These increases are due, on the one hand, to technical progress in the field of fuel cells, and on the other hand are driven by global concerns about energy security, efficiency, sustainability and energy sustainability, reduction of greenhouse gas emissions and not lastly, the reduction of dependence on the use of fossil fuels. (*Verspagen, B., 2007; Ortiz R.E.I., et.al., 2007; Akinyele, D., et.al., 2020; Felseghi, R. A., et.al., 2019)*

GENERAL ASPECTS

Fuel cells are now increasingly being researched, believing that they are revolutionizing the ways energy is produced. They use hydrogen as fuel, while ensuring the possibility of generating clean electricity, with the protection and even improvement of environmental parameters.

By definition, the fuel cell is an electric cell, which, unlike accumulator cells, can be continuously supplied with fuel, so that the electrical power at the output of this electric cell can be maintained indefinitely. Therefore, the fuel cell converts hydrogen or hydrogen-based fuels directly into electricity and heat through the electrochemical reaction of hydrogen with oxygen. The process performed at the fuel cell level is the inverse of electrolysis:

$$2H_2 + O_2 \rightarrow 2H_2O + \text{energy},$$

where, the chemical energy of the fuel (hydrogen) and the oxidant (oxygen) is converted into direct current, heat and water as reaction product.

Due to the fact that in the case of the fuel cell the hydrogen and oxygen gases are transformed by an electrochemical reaction into water, it has considerable advantages over thermal engines: higher efficiency, virtually quiet operation, no pollutant emissions if the fuel is even hydrogen. And if hydrogen is produced from renewable energy sources, the electrical power thus obtained is indeed sustainable. *(Sammes, N. (Ed.), 2006); Dicks, A., & Rand, D. A. J., 2018; Felseghi, R. A., et.al., 2019)*

Component Element

In order to analyze the importance of all the phenomena that take place in a fuel cell, the component elements must be known (see Figure 1). And especially, what are the factors that can lead to the optimization of their operation.

There are four distinct elements in a fuel cell:

Figure 1. Components of the fuel cell.
Source: Felseghi, R. A., et.al., 2019

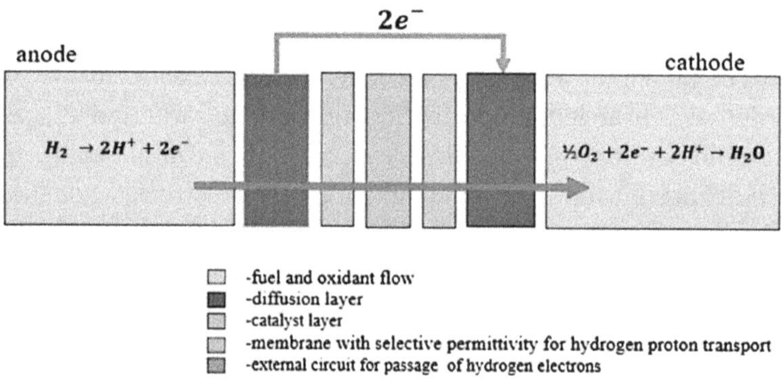

Electrolyte

In order for a substance to fulfill the role of electrolyte in a fuel cell, several conditions must be met, such as:

- high chemical stability in relation to the two electrodes, in order not to react with them;
- high melting and boiling points for cells operating at high temperatures;
- predominantly ionic conductivity and the absence of electronic conductivity.

The main types of electrolyte are:

- liquid electrolytes - the most common are basic or acidic solutions, the phenomenon of ion transport being similar to that of the electrolysis of aqueous solutions;
- solid electrolytes - ion exchange membranes and solid electrolytes with crystalline structure are used;
- electrolytes in molten state;
- liquid electrolytes with dissolved fuel.

Catalyst Layer (at Anode and Cathode)

All electrochemical reactions in the fuel cells take place on the surface of the catalyst layers. To increase the reaction rate of the cell, the electrodes (catalyst layers) contain catalyst particles. Thus, the electrode of a fuel cell is made of a porous carbon support on which the catalyst is deposited. The thickness of an electrode is usually between 5-15 μm, and the catalyst charge is between 0.1 and 0.3 mg / cm^2 .

Gas Diffusion Layer (at Anode and Cathode)

The gas diffusion layer consists of a carbon or carbon fiber texture and has the role of transporting the reactant gases to the catalyst layer. In addition, it allows the reaction product to be removed and provides a true interface for ionization and electron transfer through the solid matrix. The gas diffusion layer is mainly characterized by thickness (between 100 μm and 300 μm) and porosity, being made of a porous material and good electrical conductor.

Bipolar Plate (at the Anode and Cathode)

The bipolar plates have a double role, to guide the reactant gases towards the electrolytic exchange surface of the fuel cell and to conduct the obtained electric current. The materials used for bipolar plates must have a high conductivity and be impermeable to gases. They must also be corrosion resistant and chemically inert. Taking into account these considerations can be used: graphite or steel but also composite materials. The gas flow channels are "engraved" in the bipolar plates, which should be as thin as possible to reduce the weight and volume of the battery. The geometry of the flow channels has an influence on the flow rate of the reactants and the mass transfer, implicitly they have a decisive influence on the performance of the fuel cells, being necessary to optimize the flow surface so that the reaction surface is as large as possible. (Divisek, J., et.al., 1998; Felseghi, R. A., et.al., 2019; Song, Y., et.al, 2020; İnci, M., et.al, 2021)

Operating Principle

Although there are different types of fuel cells, they all work on the same principle:

- hydrogen or hydrogen-rich fuel is introduced to the anode, where the anode-bearing catalyst separates the electrons from the positive ions (protons);
- at the cathode, oxygen combines with electrons and, in some cases, with protons or water, resulting in water or hydrated ions;
- the electrons that form at the anode of the fuel cell cannot pass directly through the electrolyte to the cathode, but only through an electrical circuit. This movement of electrons determines the electric current.

The electrochemical conversion consists in the direct transformation into electrical energy of the chemical energy stored in various active materials. This type of conversion is called direct due to the fact that no other intermediate form is interposed between the initial and the final form of energy.

Indirect energy conversion systems contain several stages of transformation, among which the form of thermal or mechanical energy must be found. Direct energy conversion eliminates the "link" of thermal or mechanical energy by achieving superior efficiencies, which do not depend on the limited efficiency of thermal machines. The idea of obtaining electricity through the direct conversion of chemical energy arose when the problem arose of the development of the electrolysis of water (which results in its components), ie to obtain electricity from the reaction between hydrogen and oxygen.

Fuel cells are electrochemical generators of electricity that are characterized by continuous supply of reactants to the two electrodes. The qualifier fuel comes from the fact that they use as sources of chemical energy, substances with the role of fuel, natural or synthetic, which are subjected to oxidation and reduction reactions.

The anode, or fuel electrode, is the place where the oxidation of the fuel (H_2, CH_3OH, N_2H_4, hydrocarbons, that supplies the cell takes place. The cathode, or oxygen electrode (air), is the place where the reduction of molecular oxygen takes place.

Electrochemical oxidation of hydrogen is performed at an anode of a conductive material (eg activated carbon dispersed platinum) constituting the negative pole of the cell:

- for an acid electrolyte: $H_2 \rightarrow 2H^+ + 2e^-$,
- for an alkaline electrolyte: $H_2 + 2OH^- \rightarrow 2H_2 + 2e^-$,

The electrochemical reduction of oxygen occurs at a catalytic cathode constituting the positive pole of the cell::

- for an acid electrolyte: $½O_2 + 2H^+ + 2e^- \rightarrow H_2O$,
- for an alkaline electrolyte: $½O_2 + H_2O + 2e^- \rightarrow 2OH^-$,

The catalytic functions of the electrodes are very important, namely: the hydrogen electrode (anode) must ensure the adsorption of the hydrogen molecule, its activation, promoting the reaction with the hydroxyl ion; the oxygen electrode (cathode) must allow the adsorption of molecular oxygen, promoting the reaction with water.

The anode and cathode are separated by an ionic conductor, electrolyte, and / or a membrane that prevents the reactants from mixing and the electrons from traversing the heart of the cell.

In principle, the energy released in the oxidation of conventional fuels, generally used in the form of heat, can be directly converted into electricity with excellent efficiency in a fuel cell. Since in almost all oxidation reactions there is an electron transfer between fuel and oxidant, it is obvious that the chemical oxidation energy can be converted directly into electricity. An oxidation-reduction reaction occurs in which the oxidation of the fuel and the reduction of the oxidant take place with a loss on the one hand and an electron gain on the other. Any galvanic element involves oxidation at the negative pole (loss of electrons) and reduction at the positive pole (gain of electrons) and, as in all galvanic elements, fuel cells tend to separate the two partial reactions in the sense that the exchanged electrons pass through. -a circuit for outdoor use *(Ogungbemi, E., et.al., 2021; Felseghi, R. A., et.al., 2019; Akinyele, D., et.al., 2020).*

Fuel Cell Energy, Characteristic Sizes, Yield

In a fuel cell, the chemical energy stored in the fuel is converted into electricity and heat. The resulting stress can be calculated with the formulas presented above, knowing the value of the free enthalpy (- ΔG) for the overall reaction

$$-\Delta G = +\ 57{,}8 \text{ kcal/g mol} = +\ 237{,}1 \text{ kJ/g mol},$$

a voltage of 1.23 V at 25 ° C is obtained according to the standard redox potential.

$$E = (-\ \Delta G)/(z{*}F) = 237100/(2{*}96493) = 1.23 \text{ V},$$

During the overall reaction, the entropy variation is negative.

The recoverable energy in the system is:

$$\Delta G = \Delta H - T\Delta S,$$

where the notations were made: H – enthalpy; T – temperature; S – entropy.

If, the free enthalpy variation, ΔG it is negative, it is considered by convention that the reaction is spontaneous. If ΔH is negative, it is considered by convention that the cell emits heat. At a gas pressure of 1 atm. and a temperature of 25 ° C exists:

$$\Delta G^\circ = -\ 237 \text{ kJ/ moli de } H_2 \text{ consumed},$$

$$\Delta H^\circ = -\ 285 \text{kJ/ mol},$$

$$\Delta S^\circ = -\ 0{,}164 \text{ kJ/mol K},$$

$\Delta G + nFEeq = 0$ cu $\Delta G < 0$,

where n is the number of electrons exchanged, and F is the Faraday constant.

The density of the current that can be obtained (in A /m^2) varies depending on the type of cell, its operating temperature, the density of the catalyst, the stoichiometric coefficient for the supply of reactive gas. The theoretical thermodynamic potential is therefore 1.23 V. In reality, the voltage can decrease quite a lot due to the polarization phenomenon which encompasses a series of potential losses. Potential losses are, in fact, energy losses due to several causes:

- The concentration polarization (Econc) occurs due to the detachment of the active material in the immediate vicinity of the electrode, causing a change in the electrolyte concentration.
- Polarization due to the reaction mechanism (Erm) occurs because the adsorption reactions on the electrolyte surface take place through intermediate compounds that limit the voltage, in the H_2-O_2 cell the voltage is limited by the presence of hydrogen peroxide that appears as an intermediate in the reaction global.
- The activation polarization (Eact) is due to the non-uniformity of the electrochemical reactions on the electrode surface. Catalysts can reduce this type of polarization if the number of reaction centers is appropriate.
- Ohm polarization (EIR) occurs due to the internal resistance of the cell which, when the electric current passes, causes potential drops. This type of polarization can be reduced by reducing the internal resistance of the fuel cell.

As the effects of the four types of polarization on the voltage generated by a fuel cell it is found that from a current density of 10 mA / cm^2 ohmic polarization becomes very important, leading to losses that cancel the electromotive voltage of the cell type considered .

The fuel cell polarization curve is the most important feature to be analyzed to determine the overall performance, this feature being determinant in diagnosing, developing and improving certain aspects related to sizing and control both at the cell level and at the equipment level in the case of fuel cells, at the same time from this polarization curve results the character of irreversibility of certain forms of energy in the case of fuel cells.

The efficiency of fuel cells is superior to turbogenerators in current power plants because the enthalpy of reaction is converted directly into electricity, except for an entropic term.

The theoretical value of the isothermal yield (ηiz) is superior to the Carnout cycle, according to the relation:

$$\eta iz = Wmax/\Delta H = \Delta G/\Delta H = 1 - T\Delta S/\Delta H,$$

where Wmax is the maximum electricity.

The isothermal efficiency of the reactions that take place in the fuel cells can reach and theoretically exceed 80%. The electrical efficiency (ηel) of fuel cells is obtained by relating the obtained electrical energy (Wel) to the reaction enthalpy:

$$\eta el = Wel/\ \Delta H = Wel/Wmax = E/(Emax * \eta iz),$$

where: E is the voltage generated by the load cell (terminal voltage), Emax - electromotive voltage (idling voltage).

In order to make a comparison between the fuel cell and other power generation systems, respectively the internal combustion engine, an evaluation of the system efficiency is required. The overall electrical conversion efficiency of a fuel cell is higher than that of heat engine systems.

The isothermal efficiency of the fuel cells depends on the size of the reaction entropy which, when positive, implies higher yields than unity, because part of the heat of the environment is taken up. This phenomenon is found in experimental situations and is not of practical interest. The rest of the reactions show very high theoretical yields that far exceed the theoretical yield of the Carnot cycle of 30-50%.

However, practically, due to the internal polarization of the cell, voltage drops, etc. yields of 50-60% are obtained which are quite high, compared to other practical systems for converting various forms of energy *(Singh, D., et.al., 2020; Felseghi, R. A., et.al., 2019; Raychaudhuri, A., & Behera, M., 2020).*

The Main Constructive Types of Fuel Cells

The classification of fuel cells can be done according to several criteria that take into account certain common features. Thus, depending on the type of fuel (gaseous, solid, liquid), depending on the electrolyte used (liquid or solid), depending on how the fuel is consumed (directly and indirectly), but the most used classification method is that take into account the operating temperature:

- Low temperature (cold) fuel cells operating between 20 – 100 °C;
- Medium temperature (hot) fuel cells that operate between 200 – 300 °C;
- High temperature fuel cells operating between 600 – 1500 °C.

Currently in (or close to) current use are 6 types of fuel cells, at various operating temperatures, as follows:

- Alkaline fuel cells (AFCs), with operating temperatures around 70 °C.
- Direct methanol fuel cells (DMFCs), operating at temperatures between 60 - 130 °C.
- Molten carbon fuel cells (MCFCs) at temperatures up to 650 °C.
- Phosphoric acid fuel cells (PAFCs) at temperatures of 180 - 200 °C.
- Proton exchange membrane fuel cells (PEMFCs). These work at low temperatures of 100 °C, but also at temperatures of 150 °C to 200 °C.
- Solid oxide fuel cells (SOFCs). They operate at high temperatures from 800 - 1000 °C.

Each of these types of fuel cells has specific characteristics. They have obvious advantages, but also disadvantages (actually limited possibilities of use, now or in the near future). In addition to these 6 types, there are others, more or less different from the first ones, which will probably be used in the near future: ZAFC with zinc-air, which uses a zinc anode similar to a battery; PCFC with ceramic proton exchanger, a relatively new type of fuel cell; RFC regenerative fuel cell, which is based on the most attractive way to generate hydrogen and oxygen, by electrolysis, having the sun as an energy source; in the fuel cell, hydrogen and oxygen produce electricity, heat and water, which is then recycled and used for electrolysis.

With reference to the applicability of these typologies of energy conversion technologies for the stationary field, five of them have been identified in the specialized literature that are used to serve stationary type consumers, as follows: DMFC is especially suitable for the supply of electricity in small household applications, while MCFC, PAFC, PEMFC, SOFC are used to serve the market of large consumers in the stationary field. (Guaitolini, S. V. M., 2018; Felseghi, R. A., et.al., 2019; Abdelkareem, M. A., et.al., 2021)

The particularities that define and characterize these types of fuel cells are summarized as follows:

DMFC - Direct Methanol Fuel Cell

The electrolyte of this cell is generally a strong acid to repel the carbon dioxide produced by the reaction. In these cells the separating membrane acts as a solid electrolyte. Methanol is oxidized at the anode (negative electrode) after the reaction:

$CH_3OH+H_2O \rightarrow CO_2 +6H_{aq}^{+}+6e^{-}$,

Oxygen is reduced at the cathode (positive electrode) after the reaction:

$O_2+4H^{+}+4^{e-} \rightarrow 2H_2O$,

The overall reaction corresponds to the chemical combustion of methanol:

$CH_3OH+{}^{3}/_{2}O_2 \rightarrow CO_2+2H_2O$.

The catalysts used for the reactions contain quite large amounts of platinum which leads to a rather weak power. *(Alias, M. S., et.al.,2020)*

MCFC - Molten Carbon Fuel Cell

Alkaline carbonates are made up of elements that exist in abundance in nature and are not toxic. They form eutectic mixtures with relatively low melting points. It also has a good electrical conductivity in the molten state between 1.3-2.5 s / cm. The operation takes place at temperatures around 650 °C, the electrolyte in this state being relatively stable and non-volatile. *(Kouchachvili, L., et.al, 2021)*

Compared to other fuel cells, the oxidant in this case consists of a mixture of air and carbon dioxide in proportions of 70% and 30%.

The electrochemical reaction that takes place at the cathode is:

$½O_2+CO_2+2e^{-} \rightarrow CO_3^{2-\backslash}$,

This reaction always occurs, but a more detailed analysis has shown that the reduction mechanism causes O_2^{2-} si O^{2-} oxygen species to occur.

The electrochemical reaction that takes place at the anode is:

- if the fuel is hydrogen:

$H+CO_3^{2-} \rightarrow H_2O+CO_2+2e^{-}$,

- if the fuel is methane, it is previously converted into synthesis gas:

$H_2+CO+2CO_3^{2-}\rightarrow 3CO_2+H_2O+4e^-$,

The overall chemical reaction of the cell is:

$H_2+½O_2+CO_2(catod)\rightarrow(H_2O+CO_2)(anod)$,

wherein the CO_2 that forms at the anode is recycled to the cathode for consumption.

PAFC - Phosphoric Acid Fuel Cell

Cells of this type are the most developed. The electrolyte is concentrated phosphoric acid (100%) immobilized in a silicon carbon matrix. This viscous liquid has the advantages of not being volatile, of having a good ionic conductivity at high temperature and a good solubility of oxygen. It solidifies at a temperature of 42 ° C which forces the cell to remain in a warm state even when it is not working. Electrodes are composed of platinum-catalyzed carbon granules, immobilized in a sponge-shaped porous structure using polytetrafluoroethylene (which acts as a hydrophobic agent to slow the drowning of gaseous electrodes in the electrolyte). PAFC runs on natural gas which is converted to hydrogen (80%), CO_2 (20%) and CO (1%) by a reformer and a water gas reactor. The operating temperature is high enough to slow the platinum poisoning by up to 1% carbon monoxide, resulting from the reaction of the gas with water. *(Guo, X., et.al, 2021; Sadeghi, S., & Baniasad A., I.,2020).*

PEMFC - Proton Exchange Membrane Fuel Cell

Fuel cell with polymer transfer membrane is one of the most promising technologies in the field of fuel cells. These cells use a solid electrolyte, consisting of a polymeric membrane with proton conduction provided by a sulfonic function. The most commercialized and used polymer, known under the trade name NAFION, is a perfluorosulfonic compound, the acid function being of the sulfonic type, and the fluorine atoms are associated with the carbon ones. This substitution ensures this polymer close to Teflon, a higher acidity than conventional acids (sulfuric, hydrochloric, nitrogen, etc.) and a good chemical stability in oxidizing or reducing media (anodic and cathodic areas, respectively). These membranes are hydrophilic and need to be permanently hydrated to ensure proton conduction. In fact, the conduction mode is very close to conduction in a liquid.

The operating temperature is limited to a maximum of 200 °C to prevent evaporation and thus dehydration at atmospheric pressure. The catalyst used is pure platinum or alloy with transition metals. It must be resistant to the strong acidic environment and have sufficient catalytic activity to initiate oxidation and reduction reactions at the electrodes. *(Song, Y., et.al., 2020; Wu, C. W., et.al., 2020)*

SOFC - Solid Oxide Fuel Cell

Solid oxide electrolyte cells can be defined as a multilayer ceramic electrochemical system, which uses a fuel and a gaseous oxidant and operates near a temperature of 800 ÷ 1000 ° C.

The association of several elementary cells for the formation of a cell requires the use of an interconnecting material (ceramic or metal) electronic conductor. The operating characteristics of SOFCs offer them a number of advantages over other cell types, such as: no need to use catalysts, they can achieve an overall efficiency of up to 80% if used for practical cogeneration applications electricity and heat, there are no reliability problems related to electrolyte losses and electrode corrosion, they have great flexibility for electricity generation. *(Choudhury, A., et.al., 2013; Wasajja, H., et.al.,2020; Zakaria, Z., et.al., 2020)*

Stationary Fuel Cell Applications

Various pilot projects on hydrogen technology are being validated, respectively the performance of fuel cell systems operating in real conditions is analyzed and reports on technology performance, progress and new challenges are prepared. This analysis includes assemblies of fuel cells of various types, namely the fuel cell with proton exchange membrane, the cell with solid oxides, the one with phosphoric acid and the fuel cell with molten carbon, having dimensions of power generation systems with generated powers. ranging from 5 kW to 2.8 MW, and the equipment has nominal powers ranging from 0.5 kW to 400 kW. Stationary fuel cell systems are used in applications where they can be used for various purposes, namely for back-up power generation, power generation for remote locations, autonomous power plants for one or more consumers, distributed generation for buildings and cogeneration (in which excess thermal energy resulting from the production of electricity is used to provide the heating medium). *(Gahleitner, G., 2013; Niesten, E., & Alkemade, F., 2016; Taie, Z., et.al., 2021; Baumann, M., et.al., 2021; Jaunatre, M., 2021).*

The prospect of a significant increase in the number of applications of stationary fuel cells has a real potential in the coming years. On a large scale, fuel cells for the power generation sector based on hydrogen as a raw material for the production of electricity in a centralized system (power plants) have proved a real success in Korea and North America, and Europe also has a number of opportunities in this sector. A number of major plans targeting this segment are announced globally and regionally, and governments and partner organizations specializing in hydrogen and fuel cells are making a joint effort to secure infrastructure for production, storage, centralized distribution and consumer food. hydrogen, as well as a wide variety of technical, financial and management issues are being reviewed to develop a future hydrogen-based economy. (Ryghaug, M., & Skjølsvold, T. M., 2021; Tushar, W., et.al, 2021; Felseghi R.A., et.al., 2019)

The main ways of energy support of the stationary field through the technology of hydrogen and fuel cells that have been identified in the literature refer to the following aspects:

Micro – CHP With Fuel Cell in the Residential Buildings Sector

Fuel cells are suitable for micro-cogeneration because the technology inherently produces electricity and heat from a single fuel source, and the systems can also run on traditional fuels, such as natural gas. The energy efficiency of fuel cells is higher than that of conventional electricity generation systems, they operate with a efficiency of over 45% compared to traditional ones that offer maximum efficiencies of 25-35%, and if the heat is captured and used in the system, the total fuel conversion efficiency in the case of batteries can be close to 98%. This also leads to significant reductions in carbon dioxide emissions. These systems based on micro-cogeneration with fuel cells may also be eligible to contribute to lowering the unit energy tariff, as any excess electricity obtained from the system is allowed to be sold to the centralized energy grid, thus completing the list of alternative energies. Currently, micro-cogeneration fuel cell units are installed in individual houses, but such systems with low power capacities are under development, the projects having the objectives aimed at energy support of collective housing with several apartments.

In this type of application, the proton exchange membrane fuel cell (PEMFC) is currently used, which operates and ensures the energy demand both during the day when peak consumption is recorded and at night. Connectivity to the centralized energy network is necessary to ensure the start of the fuel cell or if it is implemented in autonomous operation it is necessary to be equipped with its own storage battery to ensure the start-up function for the fuel cell. *(Felseghi, R. A.,& Papp, A., 2014)*

Solid oxide fuel cells (SOFCs) can also be used in residential micro-cogeneration systems, having a relatively equal efficiency to PEMFC. Because SOFCs use higher operating temperatures than PEMFCs, they are more tolerant of carbon monoxide in the fuel, and this allows for some simplification in terms of system configuration.

The starting point for this sector is the project initiated in the 1990s by the Japanese government, which supported research to develop a city-gas-derived hydrogen system that would generate both electricity and gas. and thermal for individual private residential buildings, as a result of which the globally recognized residential micro-cogeneration system was developed under the name of Ene-Farm.

The dramatic increase in technology transfer to practical applications so far and the results expected to continue the upward trend have set a target of up to 50,000 new units for 2015. In the future, a system similar to that developed by the Ene-Farm project is planned by Japan to be installed in apartment buildings. Ene-Farm's success has inspired various demonstration projects in other parts of the world, including Korea, Denmark, Germany, the United States and the United Kingdom. (Bagherian, M. A., et.al., 2021; Ou, K., et.al., 2021; Renau, J., et.al., 2021)

Back-Up Power (Back-Up Systems) by Harnessing Renewable Energy Sources or by Converting Waste Into Energy

This type of function involves storing and increasing the degree of use by avoiding the losses associated with the excess energy produced in power plants that operate by exploiting renewable energy sources. By increasing the degree of use and capitalization of electricity obtained from renewable sources, a maximization of the profitability of these investments is implicitly obtained, but also a reduction of the costs of the energy produced. The need to store this excess can be realized by electrolytic generation of hydrogen, so the stored can be easily converted back into electricity when needed. At the same time, hydrogen can be produced from biomass or by capitalizing on different types of waste, thus making it possible to integrate this energy source into a complete energy system from renewable and alternative resources. The most efficient way to convert hydrogen back to electricity is possible through fuel cells, but this is not limited to the stationary field, in some cases, the hydrogen obtained from the excess resulting from renewable power plants can be directed for sale as fuel to power portable or transport practical applications.

Fuel cells and electrolyzers are complementary technologies. An electrolysis cell operates in the reverse mode of the fuel cell, using electricity instead of producing it. Hydrogen is obtained in its pure state in this scenario, which is why any type of fuel cell can be used to perform its conversion into electricity. (Maggio, G., et.al., 2019; Hanley, E. S., et.al., 2018; Singla, M. K., et.al, 2021; Baharuddin, N. A., et.al., 2021; Martin, A., et.al., 2020)

High Capacity Power Plants

Fuel cells are very reliable, easy to maintain, can provide a continuous supply of electricity, operate with low noise and extremely low pollutant emissions compared to fossil fuel plants. This allows fuel cell power plants to be located close to populated centers, which minimizes transmission losses. Even when fed with natural gas, the amount of carbon dioxide emitted per kilowatt-hour generated is significantly lower than conventional electricity generation. As such, fuel cells are eligible for government incentives in a large number of countries.

Several types of fuel cells find their applicability in power generation for large stationary applications. PAFC, PEMFC, SOFC and MCFC systems are used worldwide for the distributed generation of electricity for local use, with low demand. However, the capacity of a power plant must be much higher (generally over 10 MW), and in this sector molten carbon fuel cells (MCFC) are currently the optimal technology, suitable for such conditions. MFCFs operate at high temperatures, around 650 °C, are less prone to carbon monoxide poisoning than low temperature batteries, therefore they can operate safely on a wider variety of fuels, including fuel gas derived from coal, methane or natural gas.

To improve energy security and reduce greenhouse gas emissions, governments around the world are supporting initiatives to include alternative resources in the energy mix. This is complicated by the fact that the main sources of renewable energy, wind and solar, are by their nature highly variable and to some extent unpredictable. These variations in production are difficult for national grid operators to manage, as well as sudden peaks in production from renewable energy plants can overload and destabilize the grid. Fuel cells meet and solve this type of problem by providing stability, predictability and low greenhouse gas emissions. (Acar, C., & Dincer, I., 2019; El-Emam, R. S., & Özcan, H., 2019; Novosel, U., et.al., 2021)

CONCLUSION

Hydrogen and fuel cell technology has advanced considerably in the last fifteen years. Globally, this area still faces significant challenges - technical, commercial and structural - that need to be overcome before fuel cells reach their full potential, but the path today is much clearer. Policy makers have included hydrogen and the fuel cell on the map of future energy strategies and have already considered that fuel cells have great real potential and can respond successfully in meeting social, economic and environmental goals in the context of the multidisciplinary concept. sustainable development.

The industry is already engaged in exploiting these opportunities, with the fuel cell industry already well organized in Europe, North America, Japan and Korea. Globally, but also regionally, governments by allocating funds for this field are increasingly involved to further improve the competitiveness of technology. In the last five years, a large number of public-private partnerships have been formed that are involved in the development of infrastructure for hydrogen technology. It can be said that it would be necessary to meet the need for information on hydrogen and fuel cells for the general public, but in this area there are already concerns, various organizations are already well positioned to help in this endeavor.

Environmental issues including climate change that have become visible today, but also the need to switch from conventional fuels currently used to the use of alternative energy is a challenge for the entire community worldwide. In this context, new low-carbon energy technologies are needed, as well as the identification, planning and implementation of directions to capitalize on the potential of alternative energy resources.

Leveraging this potential is currently looking at finding alternatives for obtaining energy by using technologies that provide maximum efficiency and minimum pollution, such technology being that of fuel cells - one of the alternative solutions for future clean energy systems. Through their development, obtaining energy based on hydrogen becomes a reality, and hydrogen becomes a safe and sustainable energy alternative.

Hydrogen and the fuel cell are an immediate solution to the media problem of stationary power generation. Thus, fuel cells are available from manufacturers in a wide range of rated powers ranging from 0.5 kW to 400 kW, and for fuel cell power systems they can be made so as to obtain power generated with a capacity between 5 kW up to 2.8 MW. As practical applications, the fuel cells integrated in the generation systems can play various roles, either in micro- / macro-cogeneration of electricity and heat, or backup power supply (back-up) by capitalizing on renewable sources and waste, or generation of energy for remote locations, power plants of different capacities for distributed power generation.

As a typology of fuel cells, PEMFC with operation in the field of low temperatures up to 200 °C, is the first hydrogen-based technology widely marketed for practical applications for generation - cogeneration or energy reserve systems for the stationary field. Starting with 2012, when the implementation of fuel cells in large capacity power plants began, MCFC and SOFC type batteries, which operate at high temperatures up to 1000 ° C, register significant increases at the commercial level.

Fuel cells used in residential micro-cogeneration in Japan have seen rapid increases in implementation in practical applications, where this technology is expected to remain dominant in this sector to the detriment of conventional boilers.

Prospective studies are underway on the implementation of these systems at the level of the European region, and if they follow the same trajectory, it is expected that technology transfers will increase significantly when the demonstration phase is completed and marketed.

The European Union's research and innovation programs are currently structured around two pillars, transport and energy systems. The energy systems pillar will support projects in areas which cover the following aspects:

- Hydrogen production based on renewable energy sources, excess energy storage and balancing the centralized energy network, including large-scale production of green electrolytic hydrogen, with an emphasis on the crucial role of hydrogen in integrating renewable energy sources in centralized energy distribution network. Hydrogen production with a low carbon footprint from other resources - through different ways of generating hydrogen that will be developed, but also demonstrated if necessary.
- Fuel cell systems for cogeneration or power plants, which cover the technical developments necessary to reduce costs, increase service life and improve performance.
- Storage, handling, transport and distribution of hydrogen in order to deliver it to the consumer and operate the facilities safely by users.

The projects are expected to demonstrate and facilitate the regional transition to sustainable and environmentally friendly energy generation systems.

FUTURE RESEARCH DIRECTIONS

In the light of the new strategies of global policies on climate neutrality of energy systems that propose plans for the energy system of the future and place clean hydrogen in the spotlight, it is necessary to address as soon as possible the following study directions:

- carrying out a study to outline the socio-economic perception, viability and public acceptance on the use of hydrogen as an energy alternative and the regional transition to sustainable and ecological hydrogen-based energy generation systems;
- identifying and inventorying the availability of hydrogen economy implementation, estimating the potential for hydrogen production, storage, transport, distribution and use at the regional level;
- in the context of the often discussed topic of the circular economy, a series of studies can be developed on how to obtain hydrogen from various wastes, but also from wastewater by using renewable energies in production processes;
- conducting feasibility studies on the implementation of energy systems that integrate hydrogen and fuel cells in the power supply of individual consumers and at the level of local communities.

ACKNOWLEDGMENT

This work was supported by a grant from the Romanian Ministery of Research and Innovation, CCCDI-UEFISCDI, project number PN-III-P1-1.2-PCCDI-2017-0776/No. 36 PCCDI/15.03.2018, within PNCDI III.

REFERENCES

Abdelkareem, M. A., Elsaid, K., Wilberforce, T., Kamil, M., Sayed, E. T., & Olabi, A. (2021). Environmental aspects of fuel cells: A review. *The Science of the Total Environment*, *752*, 141803. doi:10.1016/j.scitotenv.2020.141803 PMID:32889267

Acar, C., & Dincer, I. (2019). Review and evaluation of hydrogen production options for better environment. *Journal of Cleaner Production*, *218*, 835–849. doi:10.1016/j.jclepro.2019.02.046

Akinyele, D., Olabode, E., & Amole, A. (2020). Review of Fuel Cell Technologies and Applications for Sustainable Microgrid Systems. *Inventions (Basel, Switzerland)*, *5*(3), 42. doi:10.3390/inventions5030042

Alias, M. S., Kamarudin, S. K., Zainoodin, A. M., & Masdar, M. S. (2020). Active direct methanol fuel cell: An overview. *International Journal of Hydrogen Energy*, *45*(38), 19620–19641. doi:10.1016/j.ijhydene.2020.04.202

Bagherian, M. A., Mehranzamir, K., Pour, A. B., Rezania, S., Taghavi, E., Nabipour-Afrouzi, H., Dalvi-Esfahani, M., & Alizadeh, S. M. (2021). Classification and Analysis of Optimization Techniques for Integrated Energy Systems Utilizing Renewable Energy Sources: A Review for CHP and CCHP Systems. *Processes (Basel, Switzerland)*, *9*(2), 339. doi:10.3390/pr9020339

Baharuddin, N. A., Yusoff, W. W., Abd Aziz, A. J., & Tahir, N. M. (2021, February). Hydrogen fuel cells for sustainable energy: Development and progress in selected developed countries. *IOP Conference Series. Materials Science and Engineering*, *1078*(1), 012011. doi:10.1088/1757-899X/1078/1/012011

Ball, M., Basile, A., & Veziroglu, T. N. (Eds.). (2015). *Compendium of hydrogen energy: hydrogen use, safety and the hydrogen economy*. Woodhead Publishing.

Barbir, F., Basile, A., & Veziroglu, T. N. (Eds.). (2015). *Compendium of Hydrogen Energy: Hydrogen Energy Conversion*. Woodhead Publishing.

Baumann, M., Domnik, T., Haase, M., Wulf, C., Emmerich, P., Rösch, C., Zapp, P., Naegler, T., & Weil, M. (2021). Comparative patent analysis for the identification of global research trends for the case of battery storage, hydrogen and bioenergy. *Technological Forecasting and Social Change*, *165*, 120505. doi:10.1016/j.techfore.2020.120505

Choudhury, A., Chandra, H., & Arora, A. (2013). Application of solid oxide fuel cell technology for power generation—A review. *Renewable & Sustainable Energy Reviews*, *20*, 430–442. doi:10.1016/j.rser.2012.11.031

Dicks, A., & Rand, D. A. J. (2018). *Fuel cell systems explained* (Vol. 479). Wiley. doi:10.1002/9781118706992

Divisek, J., Oetjen, H. F., Peinecke, V., Schmidt, V. M., & Stimming, U. (1998). Components for PEM fuel cell systems using hydrogen and CO containing fuels. *Electrochimica Acta*, *43*(24), 3811–3815. doi:10.1016/S0013-4686(98)00140-6

El-Emam, R. S., & Özcan, H. (2019). Comprehensive review on the techno-economics of sustainable large-scale clean hydrogen production. *Journal of Cleaner Production*, *220*, 593–609. doi:10.1016/j.jclepro.2019.01.309

Felseghi, R. A., Carcadea, E., Raboaca, M. S., Trufin, C. N., & Filote, C. (2019). a). Hydrogen fuel cell technology for the sustainable future of stationary applications. *Energies*, *12*(23), 4593. doi:10.3390/en12234593

Felseghi, R. A., & Papp, A. (2014). Considerations regarding hybrid systems of power generator from renewable energy sources. *Studia Universitatis Babes-Bolyai, Ambientum, 59*.

Gahleitner, G. (2013). Hydrogen from renewable electricity: An international review of power-to-gas pilot plants for stationary applications. *International Journal of Hydrogen Energy, 38*(5), 2039-2061.

Guaitolini, S. V. M., Yahyaoui, I., Fardin, J. F., Encarnação, L. F., & Tadeo, F. (2018, March). A review of fuel cell and energy cogeneration technologies. In *2018 9th International renewable energy congress (IREC)* (pp. 1-6). IEEE. 10.1109/IREC.2018.8362573

Guo, X., Zhang, H., Hu, Z., Hou, S., Ni, M., & Liao, T. (2021). Energetic, exergetic and ecological evaluations of a hybrid system based on a phosphoric acid fuel cell and an organic Rankine cycle. *Energy*, *217*, 119365. doi:10.1016/j.energy.2020.119365

Gupta, R., Basile, A., & Veziroglu, T. N. (Eds.). (2016). *Compendium of Hydrogen Energy: Hydrogen Storage, Distribution and Infrastructure*. Woodhead Publishing.

Hanley, E. S., Deane, J. P., & Gallachóir, B. Ó. (2018). The role of hydrogen in low carbon energy futures–A review of existing perspectives. *Renewable & Sustainable Energy Reviews*, *82*, 3027–3045. doi:10.1016/j.rser.2017.10.034

İnci, M., Büyük, M., Demir, M. H., & İlbey, G. (2021). A review and research on fuel cell electric vehicles: Topologies, power electronic converters, energy management methods, technical challenges, marketing and future aspects. *Renewable & Sustainable Energy Reviews*, *137*, 110648. doi:10.1016/j.rser.2020.110648

Jaunatre, M. (2021). Practical Literature Review for Establishing the Current Renewable Hydrogen Framework. *Renewable Hydrogen*, 13-28.

Kouchachvili, L., Geddis, P., & Zhuang, Q. (2021). Direct carbon fuel cell design for continuous operation. *International Journal of Hydrogen Energy*, *46*(9), 6792–6802. doi:10.1016/j.ijhydene.2020.11.179

Maggio, G., Nicita, A., & Squadrito, G. (2019). How the hydrogen production from RES could change energy and fuel markets: A review of recent literature. *International Journal of Hydrogen Energy*, *44*(23), 11371–11384. doi:10.1016/j.ijhydene.2019.03.121

Martin, A., Agnoletti, M. F., & Brangier, E. (2020). Users in the design of Hydrogen Energy Systems: A systematic review. *International Journal of Hydrogen Energy*, *45*(21), 11889–11900. doi:10.1016/j.ijhydene.2020.02.163

Niesten, E., & Alkemade, F. (2016). How is value created and captured in smart grids? A review of the literature and an analysis of pilot projects. *Renewable & Sustainable Energy Reviews*, *53*, 629–638. doi:10.1016/j.rser.2015.08.069

Novosel, U., Živić, M., & Avsec, J. (2021). The production of electricity, heat and hydrogen with the thermal power plant in combination with alternative technologies. *International Journal of Hydrogen Energy*, *46*(16), 10072–10081. doi:10.1016/j.ijhydene.2020.01.253

Ogungbemi, E., Wilberforce, T., Ijaodola, O., Thompson, J., & Olabi, A. G. (2021). Review of operating condition, design parameters and material properties for proton exchange membrane fuel cells. *International Journal of Energy Research*, *45*(2), 1227–1245. doi:10.1002/er.5810

Ortiz-Rivera, E. I., Reyes-Hernandez, A. L., & Febo, R. A. (2007, August). Understanding the history of fuel cells. In *2007 IEEE Conference on the History of Electric Power* (pp. 117-122). IEEE. 10.1109/HEP.2007.4510259

Ou, K., Yuan, W. W., & Kim, Y. B. (2021). Development of optimal energy management for a residential fuel cell hybrid power system with heat recovery. *Energy*, *219*, 119499. doi:10.1016/j.energy.2020.119499

Raychaudhuri, A., & Behera, M. (2020). Review of the process optimization in microbial fuel cell using design of experiment methodology. *Journal of Hazardous, Toxic and Radioactive Waste*, *24*(3), 04020013. doi:10.1061/(ASCE)HZ.2153-5515.0000503

Renau, J., García, V., Domenech, L., Verdejo, P., Real, A., Giménez, A., Sánchez, F., Lozano, A., & Barreras, F. (2021). Novel Use of Green Hydrogen Fuel Cell-Based Combined Heat and Power Systems to Reduce Primary Energy Intake and Greenhouse Emissions in the Building Sector. *Sustainability*, *13*(4), 1776. doi:10.3390u13041776

Ryghaug, M., & Skjølsvold, T. M. (2021). The Co-production of Pilot Projects and Society. In *Pilot Society and the Energy Transition* (pp. 23–62). Palgrave Pivot. doi:10.1007/978-3-030-61184-2_2

Sadeghi, S., & Baniasad Askari, I. (2020). Performance and economic investigation of a combined phosphoric acid fuel cell/organic Rankine cycle/electrolyzer system for sulfuric acid production; Energy-based organic fluid selection. *International Journal of Energy Research*, *44*(4), 2704–2725. doi:10.1002/er.5073

Sammes, N. (Ed.). (2006). *Fuel cell technology: reaching towards commercialization*. Springer Science & Business Media. doi:10.1007/1-84628-207-1

Singh, D., Sharma, D., Soni, S. L., Sharma, S., Sharma, P. K., & Jhalani, A. (2020). A review on feedstocks, production processes, and yield for different generations of biodiesel. *Fuel*, *262*, 116553. doi:10.1016/j.fuel.2019.116553

Singla, M. K., Nijhawan, P., & Oberoi, A. S. (2021). Hydrogen fuel and fuel cell technology for 61a cleaner future: 61a review. *Environmental Science and Pollution Research International*, 1–20.

Song, Y., Zhang, C., Ling, C. Y., Han, M., Yong, R. Y., Sun, D., & Chen, J. (2020). Review on current research of materials, fabrication and application for bipolar plate in proton exchange membrane fuel cell. *International Journal of Hydrogen Energy*, *45*(54), 29832–29847. doi:10.1016/j.ijhydene.2019.07.231

Taie, Z., Villaverde, G., Morris, J. S., Lavrich, Z., Chittum, A., White, K., & Hagen, C. (2021). Hydrogen for heat: Using underground hydrogen storage for seasonal energy shifting in northern climates. *International Journal of Hydrogen Energy*, *46*(5), 3365–3378. doi:10.1016/j.ijhydene.2020.10.236

Tushar, W., Yuen, C., Saha, T. K., Morstyn, T., Chapman, A. C., Alam, M. J. E., Hanif, S., & Poor, H. V. (2021). Peer-to-peer energy systems for connected communities: A review of recent advances and emerging challenges. *Applied Energy*, *282*, 116131. doi:10.1016/j.apenergy.2020.116131

Verspagen, B. (2007). Mapping technological trajectories as patent citation networks: A study on the history of fuel cell research. *Advances in Complex Systems*, *10*(01), 93–115. doi:10.1142/S0219525907000945

Veziroglu, T. N. (2007). 21st Century's energy: Hydrogen energy system. In *Assessment of hydrogen energy for sustainable development* (pp. 9–31). Springer. doi:10.1007/978-1-4020-6442-5_2

Wasajja, H., Lindeboom, R. E., van Lier, J. B., & Aravind, P. V. (2020). Techno-economic review of biogas cleaning technologies for small scale off-grid solid oxide fuel cell applications. *Fuel Processing Technology*, *197*, 106215. doi:10.1016/j.fuproc.2019.106215

Wu, C. W., Zhang, W., Han, X., Zhang, Y. X., & Ma, G. J. (2020). A systematic review for structure optimization and clamping load design of large proton exchange membrane fuel cell stack. *Journal of Power Sources*, *476*, 228724. doi:10.1016/j.jpowsour.2020.228724

Zakaria, Z., Awang Mat, Z., Abu Hassan, S. H., & Boon Kar, Y. (2020). A review of solid oxide fuel cell component fabrication methods toward lowering temperature. *International Journal of Energy Research*, *44*(2), 594–611. doi:10.1002/er.4907

ADDITIONAL READING

Ball, M., Basile, A., & Veziroglu, T. N. (Eds.). (2015). *Compendium of hydrogen energy: hydrogen use, safety and the hydrogen economy*. Woodhead Publishing.

Barbir, F. (2012). *PEM fuel cells: theory and practice*. Academic press.

Barbir, F., Basile, A., & Veziroglu, T. N. (Eds.). (2015). *Compendium of Hydrogen Energy: Hydrogen Energy Conversion*. Woodhead Publishing.

Graham, T. (2020). *Tomorrow's Energy, Renewables Alone? Complementary Resources for Tomorrow: Proceedings of Energy and Resources for Tomorrow 2019*. University of Windsor.

Gupta, R., Basile, A., & Veziroglu, T. N. (Eds.). (2016). *Compendium of Hydrogen Energy: Hydrogen Storage, Distribution and Infrastructure*. Woodhead Publishing.

Ito, K., Li, H., & Hao, Y. M. (2016). *Hydrogen Energy Engineering: A Japanese Perspective.*

Miranda, P. E. (Ed.). (2018). *Science and Engineering of Hydrogen-Based Energy Technologies: Hydrogen Production and Practical Applications in Energy Generation*. Academic Press.

Scipioni, A., Manzardo, A., & Ren, J. (Eds.). (2017). *Hydrogen economy: supply chain, life cycle analysis and energy transition for sustainability*. Academic Press.

Seelam, P. K., Rathnayake, B., Pitkäaho, S., Turpeinen, E., & Keiski, R. L. (2020). Overview on recent developments on hydrogen energy: Production, catalysis, and sustainability. In Current Trends and Future Developments on (Bio-) Membranes (pp. 3-32). Elsevier. doi:10.1016/B978-0-12-817110-3.00001-1

Sherif, S. A., Goswami, D. Y., Stefanakos, E. L., & Steinfeld, A. (Eds.). (2014). *Handbook of hydrogen energy*. CRC Press. doi:10.1201/b17226

Zohuri, B. (2019). Cryogenics and liquid hydrogen storage. In *Hydrogen Energy* (pp. 121–139). Springer. doi:10.1007/978-3-319-93461-7_4

KEY TERMS AND DEFINITIONS

Back-Up Power: Is defined as any device that provides instantaneous, uninterruptible power. The term UPS (uninterruptible power supply) is an often used term.

CHP With Fuel Cell in the Residential Buildings Sector: PEMFC is currently used in Combined Heat and Power for small practical applications such as the residential building sector.

Electrochemical Conversion: Consists in the direct transformation into electrical energy of the chemical energy stored in various active materials. This type of conversion is called direct due to the fact that no other intermediate form is interposed between the initial and the final form of energy.

Fuel Cell: An electric cell, which, unlike accumulator cells, can be continuously supplied with fuel, so that the electrical power at the output of this electric cell can be maintained indefinitely. Therefore, the fuel cell converts hydrogen or hydrogen-based fuels directly into electricity and heat through the electrochemical reaction of hydrogen with oxygen.

Fuel Cell Polarization Curve: Is the most important feature to be analyzed to determine the overall performance, this feature being determinant in diagnosing, developing and improving certain aspects related to sizing and control both at the cell level and at the equipment level in the case of fuel cells, at the same time from this polarization curve results the character of irreversibility of certain forms of energy in the case of fuel cells.

Power Plant: An electric utility generating station.

Stationary Application: Means any application of a Fuel Cell System in which such Fuel Cell System is designed to be attached to a fixed site, being land, a building or other immobile structure for extended use at that site, and includes stationary electrical power systems which are transportable for temporary site power supply.

Chapter 8
Three-Dimensional Numerical Study of Overheating of Two Intermediate Temperature P-AS-SOFC Geometrical Configurations

Youcef Sahli
CDER, Algeria

Bariza Zitouni
University of Batna 1, Algeria

Ben Moussa Hocine
University of Batna 2, Algeria

ABSTRACT

The purpose of this work is to perform a three-dimensional and stationary numerical study of the heat transfer phenomenon in the planar anode-supported solid oxide fuel cells operating at intermediate temperature (IT-P-AS-SOFC). With particular interest to evaluate and localize the maximum and minimum temperatures in a single cell during their stable operation according to two geometrical configuration types, repetition, and symmetry of the cell stages to determine the best configuration that minimizes and produces more homogeneous thermal stresses and logically improves their lifetime and performance. The considered heat sources are mainly due to electrical overpotentials (Ohm, activation, and concentration). The results are obtained according to a FORTRAN code based on the proposed model that is numerically modeled using the finite difference method. From the obtained result analysis, the achieved temperature values by IT-P-AS-SOFC with cell stages repetition are greater than obtained by IT-P-AS-SOFC with cell stages symmetry.

DOI: 10.4018/978-1-7998-4945-2.ch008

INTRODUCTION

Fuel cells are the devices of converting chemical energy into electrical energy electrochemically (Laribi, Mammar, Sahli, & Koussa, 2019; Monsaf, Hocine, Youcef, & Abdallah, 2017). They make it possible to produce electrical energy in a radically different way from classical and conventional production systems (Laribi, Mammar, Sahli & Koussa, 2018). Theoretically, fuel cells are non-polluting and they produce the electrical energy, water and heat at the same time (Laribi, Mammar, Hamouda, Sahli, & Koussa, 2016; Mohammedi, Sahli, & Ben-Moussa, 2020). The SOFCs are considered as the most important type of this technology due to their high performance (Sahli, Zitouni, & Ben-Moussa, 2017). In addition, these fuel cells are a promising technology because of their high overall efficiency and ability to operate through multiple fuels (Sahli, Zitouni, & Ben-Moussa, 2018b; Sahli, Ben-Moussa, & Zitouni, 2019).

The high operating temperature of SOFCs and the heat generation inside its fuel cell types produce important and inhomogeneous thermal stresses inside its cells, which logically leads to a short life, low energy production and high power cost. In this context, several researchers, in the literature, have addressed several works. Nakajo et al. (2012a, 2012b) have presented a three-dimensional analysis of the mechanical structure and their degradation under operating conditions to study the effects of operating conditions and studied designs using ABAQUS software for mechanical engineering and gPROMS for the phenomena inside the studied stack. The used model combines thermo-electrochemical models for a P-AS-SOFC with parallel channels for two agreements, counter and co-flow. They have considered that the heat transfer by conduction and convection and only the ohmic losses are responsible for the heat generation. Peksen et al. (2013) have exposed a three-dimensional thermo-mechanical analysis of a planar SOFC minor stack composed of several cells, mesh, metal frame, interconnectors and sealing materials using the CSM code to determine the mechanisms that trigger thermo-mechanical stress during heating, operation and shutdown step. Peksen (2013) has studied in a three-dimensional environment the thermo-mechanics of a minor planar SOFC stack to locate critical areas in the long-term considering that the heat generation is due solely to the Joule's effect. Kalib et al. (2019) have discussed the heat generation in SOFC, heat transfer and their impacts on the temperature and thermal stress distribution. They have concluded that the identification of almost homogeneous temperature distribution, with a minimum possible temperature difference at the SOFC, minimizes the thermal stress and helps to prolong the SOFC lifetime. Sahli et al. (2015) have analyzed the temperature distribution in a planar single cell of SOFC according to different geometric configurations; with anode supported, electrolyte supported and cathode supported in the three-dimensional environment. They have obtained the temperature fields using only the heat source due to the ohmic loss. The temperature distributions are obtained using a program in FORTRAN language that is based on the finite difference method. They have shown that the largest temperature elevations are generated by the planar SOFC with cathode supported. Conversely, the lowest temperature elevations are generated by the planar SOFC with a supported anode. Abdenebi et al. (2015) have evaluated the heat generation from the direct internal reforming in a P-AS-SOFC that feeds with methane. They have studied the influence of heat sources due to the internal reforming reactions occurring at the SOFC anode side that are the endothermic steam reforming reaction, the exothermic water-gas shift reaction, and the endothermic overall chemical reaction. The results are obtained according to a program in FORTRAN language that is developed by the authors according to a two-dimensional model that discretized using the finite difference method. They have concluded that these heat sources are a positive impact on the temperature increasing and distribution.

Thermal stress produced from the temperature gradient in SOFC leads to mechanical instability and defects, such as deformation, creep, and cracks (Huang, Lin, Liu, Ni & Zhang, 2017; Luo, Jiang, Zhang, Zhang, & Hao, 2016), where the temperature gradient is the difference between the maximal and minimal temperatures, generally between the SOFC inlet and outlet for a co-flow arrangement. The excessive developing of thermal stress in SOFC causes mechanical deformation that increases the probability of structural failure (Skrzypkiewicz, Wierzbicki, Kupecki, & Stępien, 2018).

Amedi et al. (2015) have presented a comparative study between two SOFCs that have different flow arrangements, co and counter-flow, in three-dimensional environment that counted only the heat generated from the exothermic reactions. They have concluded that the counter-flow agreement produces uniform temperature distribution to that generated using a co-flow agreement. Therefore, the counter-flow agreement permits a quasi-uniform temperature and thermal stress distributions. The SOFC power output is strongly influenced by the temperature and electrodes/electrolyte ionic conductivities, logically, the unstable and non-uniform temperature distribution leads to an unstable and non-uniform electrodes/electrolyte ionic conductivities, therefore, an unstable and non-uniform SOFC power output (Cai et al., 2018; Meng et al., 2017; Meng et al., 2018; Tucker, 2018).

The aim of the present chapter is to conduct a three-dimensional and steady state numerical study of the heat transfer phenomenon involved in the IT-P-AS-SOFC operating to evaluate and localize the maximum and minimum temperatures considering two geometrical configuration types namely repetition and symmetry of the cell stages. Moreover, this permits to knowledge, the best configuration that allows to minimizing thermal stresses and producing more homogeneous thermal stresses and logically corresponds to large lifetime, high energy production and low production cost. The used building materials are the nickel oxide–yttria-stabilized-zirconia (Ni-YSZ), lanthanum strontium manganite (LSM), yttria-stabilized-zirconia (YSZ) and lanthanum oxide-chrome (LaCrO3) for the anode, cathode, electrolyte and interconnector respectively. The heat sources are mainly due to the electrical over-potential (Ohmic, activation and concentration overpotentials). The results are obtained using a FORTRAN program that is based on the proposed model that is modelled using the finite difference method in a three-dimensional environment. The effect of the gas inlet temperatures in the anodic and cathodic channels of IT-P-AS-SOFC for the same and different gas inlet temperatures, current density and electrolyte thickness are investigated and discussed.

MATHEMATICAL MODEL

Calculation Domain

The simulation of a complete IT-P-AS-SOFC stack would have required very powerful computer hardware, High-Performance Computing Station, and excessively long simulation time. Therefore, the computational domain is limited to a single cell that consists of both cathodic and anodic gas flow channels, a cathode, an anode, an electrolyte and both anodic and cathodic interconnectors. The dimensions and the cell stages repetition and symmetry geometric configurations of the IT-P-AS-SOFC are shown in Figure 1.

Figure 1. IT-P-AS-SOFC calculation domain. a) Cell stages repetition configuration. b) Cell stages symmetry configuration. c) Cell dimensions

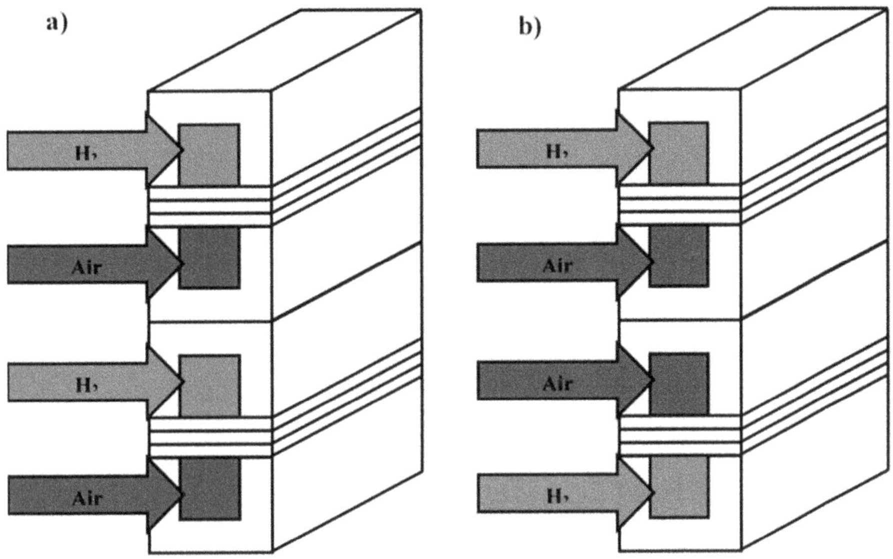

Element	*Dimension (μm)*
A	3000
B	2000
C	50
D	500
E	40
F	2000
G	1000
H	30000
K	1000

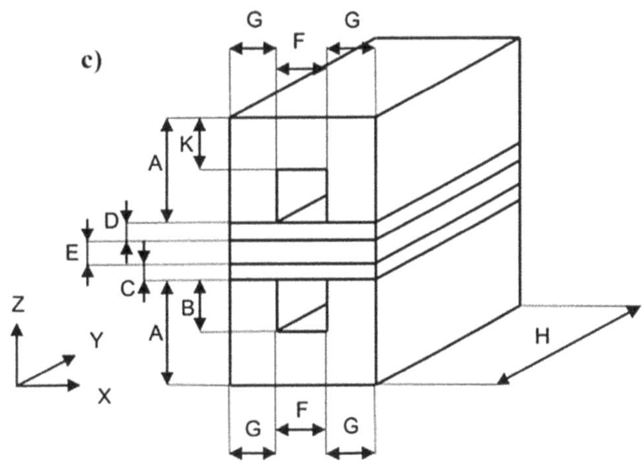

Hypotheses

The three-dimensional modeling of a single cell of IT-P-AS-SOFC taking into account all physical phenomena (real case) make this modeling very complex and the governing equations of these phenomena become very intricate, which makes the resolution of the resulting equation systems very difficult at all levels (complexity of numerical schemes, computational time, rounded off ... etc.). Therefore, it is clear tangible that the introduction of the hypotheses is necessary and imperative in order to simplify the equation system resulted from the governing equations and make this resolution operation simpler and less difficult. The used hypotheses in the present work are:

A three-dimensional and steady-state model (Sahli et al., 2015; Sahli, Zitouni, & Benmoussa, 2018a);

A continuous medium for all IT-P-AS-SOFC components (Abdenebi et al., 2015; Sahli et al., 2015; Sahli et al., 2018a);

All IT-P-AS-SOFC components are considered as homogeneous and isotropic mediums (Abdenebi et al., 2015; Sahli et al., 2015; Sahli et al., 2018a);

The fuel and oxidizer at both channel inlets are considered as brewed fluids where its temperature values are taken constants (Abdenebi et al., 2015; Sahli et al., 2015; Sahli et al., 2018a);

The reactive electrical conductivities are considered very low, hence, they are not considered (Abdenebi et al., 2015; Sahli et al., 2015; Sahli et al., 2018a);

The fuel and oxidizer velocities inside both channels are considered low (Sahli et al., 2015; Sahli et al., 2018a);

The radiative heat transfer mode inside IT-P-AS-SOFC is not considered (Abdenebi et al., 2015; Sahli et al., 2015; Sahli et al., 2018a);

A constant current density during IT-P-AS-SOFC stable operation is considered in all IT-P-AS-SOFC components (Abdenebi et al., 2015; Sahli et al., 2015; Sahli et al., 2018a);

Equations Governing the Heat Transfer Phenomenon

In this part, all equations governing the heat transfer phenomenon in all IT-P-AS-SOFC parts and the different expressions of the heat sources used inside all IT-P-AS-SOFC components taken into count in the calculation domain, considered in the present study, are presented.

Interconnectors

Thermal energy is transported by conduction in both interconnectors, the ohmic overpotential effects are taken into consideration by an additional heat source term in the energy equation. The energy conservation equation in both anodic and cathodic interconnectors is given for a three-dimensional environment by Equation (1).

$$\frac{\partial}{\partial x}\left\{\lambda_{in}\frac{\partial T}{\partial x}\right\}+\frac{\partial}{\partial y}\left\{\lambda_{in}\frac{\partial T}{\partial y}\right\}+\frac{\partial}{\partial z}\left\{\lambda_{in}\frac{\partial T}{\partial z}\right\}+S_{in}=0 \tag{1}$$

The interconnector thermal conductivity λ_{in} is given by 6 W/m.K (Sahli et al., 2015; Sahli et al., 2018a). The heat source due to the Ohmic overpotential is expressed by the ratio between the current density square and interconnector electrical conductivity. Equation (2).

$$S_{in}=S_{Ohm,in}=\frac{i^2}{\sigma_{in}} \tag{2}$$

The interconnector electrical conductivity σ_{in} is given in Table 1.

Table 1. Electrical conductivity expressions

Components	σ [Ω⁻¹ cm⁻¹]	References
Electrolyte	$\sigma_{el} = 3.34\times10^{4}\exp\left(-10300/T\right)$	(Andersson, Paradis, Yuan, & Sundén, 2013; Bao, Cai, & Croiset, 2011; Ferguson, Fiard, & Herbin, 1996; Saebea, Patcharavorachot, Assabumrungrat, & Arpornwichanop, 2013b; Yang, Chen, Wang, Peng, & Wang, 2013; Zhang, Li, Li, & Feng, 2007)
Cathode	$\sigma_{cat} = \frac{4.2\times10^{7}}{T}\exp\left(-1200/T\right)$	(Andersson et al., 2013; Ferguson et al., 1996; Saebea et al., 2013b; Yang et al., 2013; Zhang et al., 2007; Bao et al., 2011)
Anode	$\sigma_{an} = \frac{9.5\times10^{7}}{T}\exp\left(-1150/T\right)$	(Andersson et al., 2013; Bao et al., 2011;Ferguson et al., 1996; Saebea et al., 2013b; Yang et al., 2013; Zhang et al., 2007)
Interconnectors	$\sigma_{in} = 9.3\times10^{6}\exp\left(-1100/T\right)$	(Andersson et al., 2013; Bao et al., 2011; Ferguson et al., 1996; Ni, Leung, & Leung, 2007; Yang et al., 2013; Zhang et al., 2007)

Electrolyte

The thermal energy is transported by conduction through the electrolyte, the ohmic overpotential effects due to the O^{-2} ion passages through the electrolyte are taken into consideration by an additional term in the energy balance, which gives the following energy conservation equation:

$$\frac{\partial}{\partial x}\left\{\lambda_{el}\frac{\partial T}{\partial x}\right\}+\frac{\partial}{\partial y}\left\{\lambda_{el}\frac{\partial T}{\partial y}\right\}+\frac{\partial}{\partial z}\left\{\lambda_{el}\frac{\partial T}{\partial z}\right\}+S_{el}=0 \quad (3)$$

The electrolyte thermal conductivity λ_{el} is given by 2.16 $W.m^{-1}.K^{-1}$ (Kakaç, Pramuanjaroenkij, & Zhou, 2007; Sangtongkitcharoen, Vivanpatarakij, Laosiripojana, Arpornwichanop, & Assabumrungrat, 2008). The heat source existed in the electrolyte is due to the Ohmic overpotential, it is given by Equation (4).

$$S_{el}=S_{Ohm,el}=\frac{i^{2}}{\sigma_{el}} \quad (4)$$

The electrolyte electrical conductivity σ_{el} is given in Table 1.

Anode

The thermal energy is transported by conduction and convection through the porous anode. The Ohmic, concentration and activation overpotential effects are taken into account as heat sources in the energy equation by additional source terms, so the energy conservation equation in the anode solid part is given for a three-dimensional environment by the Equation (5).

$$\frac{\partial}{\partial x}\left\{\left(1-\varepsilon_{an}\right)\lambda_{an,s}\frac{\partial T}{\partial x}\right\}+\frac{\partial}{\partial y}\left\{\left(1-\varepsilon_{an}\right)\lambda_{an,s}\frac{\partial T}{\partial y}\right\}+\frac{\partial}{\partial z}\left\{\left(1-\varepsilon_{an}\right)\lambda_{an,s}\frac{\partial T}{\partial z}\right\}+S_{an,s}=0 \quad (5)$$

The existing thermal energy source in the anode solid part is due to the Ohmic overpotential; it is expressed by the following equation:

$$S_{an,s}=S_{an,Ohm}=\frac{i^2}{\sigma_{an}} \quad (6)$$

The anode electrical conductivity σ_{anl} is given in Table 1.

At the anode pores, the thermal energy is transported by conduction and convection. The gas velocity inside the anode pores is very low due to the small permeability of this electrode (10^{-12} m^2) (Kaya, Demir, Genç, & Yapici, 2014), therefore, the conductive term in the energy equation is much higher than the convective term, and this leads us to a conductive energy transport that neglects any convective transport. Therefore, the energy conservation equation in the pores of the anode is given by Equation (7).

$$\frac{\partial}{\partial x}\left\{\varepsilon_{an}\lambda_{an,f}\frac{\partial T}{\partial x}\right\}+\frac{\partial}{\partial y}\left\{\varepsilon_{an}\lambda_{an,f}\frac{\partial T}{\partial y}\right\}+\frac{\partial}{\partial z}\left\{\varepsilon_{an}\lambda_{an,f}\frac{\partial T}{\partial z}\right\}+S_{an,f}=0 \quad (7)$$

In the anode pores, the thermal energy is produced by the activation overpotential due to the electrochemical reactions inside these pores and the concentration overpotential due to the inability to maintain the initial reactant concentrations. The anodic thermal energy source due to the activation overpotential is defined by the product of the activation overpotential and current density. Equation (8).

$$S_{an,act}=\eta_{an,act}.i \quad (8)$$

The activation overpotential is described by the Butler-Volmer equation. Equation (9).

$$i=i_{0,an}\left[\exp\left(\frac{\alpha.n.F.\eta_{an,act}}{R.T}\right)-\exp\left(-\frac{\left(1-\alpha\right).n.F.\eta_{an,act}}{R.T}\right)\right] \quad (9)$$

Where α is the electronic transfer coefficient, n is the transfer electron number, which is equal to 2 for the steam water formation reaction, F is the Faraday number, R is the perfect gas constant, T is the temperature and i0,an is the exchange current density of the anode.

For an electronic transfer coefficient ($\alpha = 0.5$), the anodic activation overpotential can be simplified and written as:

$$\eta_{an,act}=\frac{R.T}{\alpha.n.F}.\sinh^{-1}\left(\frac{i}{2.i_{0,an}}\right) \quad (10)$$

The exchange current density of the anode $i_{0,an}$ is given by the following equation:

$$i_{0,an} = k_{an}.\exp\left(-\frac{E_{an}}{R.T}\right) \tag{11}$$

Where k_{an} and E_{an} represent the pre-exponential factor and the activation energy of the anode respectively, they are given in Table 2.

Table 2. Electrical conductivity expressions

Components	k [A.m^{-2}]	E [J.mol^{-1}]	References
Anode	$k_{an} = 6.54x10^{11}\frac{R.T}{2.F}$	1.4 x 10^5	(Andersson et al., 2014; Murshed, Huang, & Nandakumar, 2007; Ni et al., 2007; Saebea, Authayanun, Patcharavorachot, Paengjuntuek, & Arpornwichanop, 2013a; Tippawan, & Arpornwichanop, 2014)
Cathode	$k_{cat} = 2.35x10^{11}\frac{R.T}{2.F}$	1.37 x 10^5	

The anodic thermal energy source due to the concentration overpotential is defined by the product of the concentration overpotential and current density of the P-AS-SOFC. Equation (12).

$$S_{an,conc} = \eta_{an.conc}.i \tag{12}$$

The anodic concentration overpotential is given by:

$$\eta_{an,conc} = -\frac{R.T}{n.F}.\ln\left(1-\frac{i}{i_{an,l}}\right) \tag{13}$$

Where $i_{an,l}$ is the limit current of the anode, it is given by 2.99 x 10^4 A.m^{-2}. (Brandon, 2014; Burt, Celik, Gemmen, & Smirnov, 2004; Kang et al., 2009; Manage, Sorensen, Simons, & Brett, 2014; Yang, Zhao, Liso, & Fardadi, Mueller, & Jabbari, 2010; Zhang, Su, Chen, Zhao, & Brandon, 2011; Zhang, Wang, Guo, Shih, & Chen, 2014).

The addition of both energy conservation Equations (5 and 7), in anode solid part and anode pores, gives the following equation:

$$\left\{\begin{array}{l}\frac{\partial}{\partial x}\left\{\left(\varepsilon_{an}.\lambda_{an,f}+\left(1-\varepsilon_{an}\right).\lambda_{an.s}\right)\frac{\partial T}{\partial x}\right\}+\\ \frac{\partial}{\partial y}\left\{\left(\varepsilon_{an}.\lambda_{an,f}+\left(1-\varepsilon_{an}\right).\lambda_{an.s}\right)\frac{\partial T}{\partial y}\right\}+\\ \frac{\partial}{\partial z}\left\{\left(\varepsilon_{an}.\lambda_{an,f}+\left(1-\varepsilon_{an}\right).\lambda_{an.ss}\right)\frac{\partial T}{\partial z}\right\}\end{array}\right\}+S_{an,s}+S_{an,f}=0 \tag{14}$$

Therefore, the effective thermal conductivity $\lambda a_{n,eff}$ and total thermal source Sa_n are given by the following equations:

$$\lambda_{an,eff}=\varepsilon_{an}.\lambda_{an,f}+\left(1-\varepsilon_{an}\right).\lambda_{an,s} \tag{15}$$

$$S_{an}=S_{an,s}+S_{an,f}=S_{an,Ohm}+S_{an,act}+S_{an,conc} \tag{16}$$

Consequently, the equation governing thermal energy transport in the porous anode can be written according to the effective thermal conductivity and total thermal source as follows:

$$\frac{\partial}{\partial x}\left\{\lambda_{an,eff}\frac{\partial T}{\partial x}\right\}+\frac{\partial}{\partial y}\left\{\lambda_{an,eff}\frac{\partial T}{\partial y}\right\}+\frac{\partial}{\partial z}\left\{\lambda_{an,eff}\frac{\partial T}{\partial z}\right\}+S_{an}=0 \tag{17}$$

The effective anode thermal conductivity $\lambda_{an,eff}$ is given by 5.84 $W.m^{-1}.K^{-1}$ (Chyou, Chung, Chen, & Shie, 2005; Damm, & Fedorov, 2006; Inui, Urata, Ito, Nakajima, & Tanaka, 2006).

Cathode

At the porous cathode, the thermal energy is transported by conduction and convection. The Ohmic, concentration and activation overpotential impacts are considered as heat sources in the cathode and are accounted using additional source terms in the energy equation. The energy conservation equation in the cathode solid part is given by Equation (18).

$$\frac{\partial}{\partial x}\left\{\left(1-\varepsilon_{cat}\right)\lambda_{cat,s}\frac{\partial T}{\partial x}\right\}+\frac{\partial}{\partial y}\left\{\left(1-\varepsilon_{cat}\right)\lambda_{cat,s}\frac{\partial T}{\partial y}\right\}+\frac{\partial}{\partial z}\left\{\left(1-\varepsilon_{cat}\right)\lambda_{cat,s}\frac{\partial T}{\partial z}\right\}+S_{cat,s}=0 \tag{18}$$

The thermal energy source in the cathode solid part is due to the Ohmic overpotential; it is given by the Equation (19).

$$S_{cat,s}=S_{cat,Ohm}=\frac{i^2}{\sigma_{cat}} \tag{19}$$

The cathode electrical conductivity σ_{cat} is illustrated in Table 1.

At the cathode pores, the thermal energy is transported by conduction and convection. The oxidizer velocity inside these pores is very low due to the small permeability of this electrode (10^{-12} m^2) (Kaya et al., 2014); therefore, the convective term is neglected compared to the conductive term in the energy equation. The energy conservation equation in the cathode pores is given by the following equation:

$$\frac{\partial}{\partial x}\left\{\varepsilon_{cat}\lambda_{cat,f}\frac{\partial T}{\partial x}\right\}+\frac{\partial}{\partial y}\left\{\varepsilon_{cat}\lambda_{cat,f}\frac{\partial T}{\partial y}\right\}+\frac{\partial}{\partial z}\left\{\varepsilon_{cat}\lambda_{cat,f}\frac{\partial T}{\partial z}\right\}+S_{cat,f}=0 \quad (20)$$

In the pores of the cathode, the activation and concentration overpotentials are considered as thermal energy sources. The cathodic thermal energy source due to the activation overpotential is given by Equation (21).

$$S_{cat,act}=\eta_{cat,act}.i \quad (21)$$

Where the activation overpotential $\eta_{cat,act}$ is determined from the Butler-Volmer equation. Equation (22).

$$i=i_{0,cat}\left[\exp\left(\frac{\alpha.n.F.\eta_{cat,act}}{R.T}\right)-\exp\left(-\frac{\left(1-\alpha\right).n.F.\eta_{cat,act}}{R.T}\right)\right] \quad (22)$$

For an electronic transfer coefficient ($\alpha = 0.5$), the cathodic activation overpotential is written as:

$$\eta_{cat,act}=\frac{R.T}{\alpha.n.F}.\sinh^{-1}\left(\frac{i}{2.i_{0,cat}}\right) \quad (23)$$

The exchange current density of the cathode $i_{0,cat}$ used in Equations (22 and 23) is given by Equation (24).

$$i_{0,cat}=k_{cat}.\exp\left(-\frac{E_{cat}}{R.T}\right) \quad (24)$$

Where k_{an} and E_{an} represent the pre-exponential factor and activation energy of the cathode respectively. Table 2.

The anodic thermal energy source due to the concentration overpotential is given by the following Equation:

$$S_{cat,conc}=\eta_{cat.conc}.i \quad (25)$$

The anodic concentration overpotential is obtained using the following Equation:

$$\eta_{cat,conc} = -\frac{R.T}{n.F}.\ln\left(1 - \frac{i}{i_{cat,l}}\right) \quad (26)$$

Where $i_{cat,l}$ is the limit current density of the cathode, it is given by 2.16 x 10^4 $A.m^{-2}$. (Burt et al., 2004; Fardadi et al., 2010; Kang et al., 2009; Manage et al., 2014; Yang et al., 2014; Zhang et al., 2011; Zhang et al., 2014).

By the same way used to determine the equation governing thermal energy transport in the anode, the equation governing thermal energy transport in the cathode is given by:

$$\frac{\partial}{\partial x}\left\{\lambda_{cat,eff}\frac{\partial T}{\partial x}\right\} + \frac{\partial}{\partial y}\left\{\lambda_{cat,eff}\frac{\partial T}{\partial y}\right\} + \frac{\partial}{\partial z}\left\{\lambda_{cat,eff}\frac{\partial T}{\partial z}\right\} + S_{cat} = 0 \quad (27)$$

Where the effective thermal conductivity $\lambda c_{at,eff}$ and total thermal source Sc_{at} are given by the following equations:

$$\lambda_{cat,eff} = \varepsilon_{cat}.\lambda_{cat,f} + \left(1 - \varepsilon_{cat}\right).\lambda_{cat,s} \quad (28)$$

$$S_{cat} = S_{cat,s} + S_{cat,f} = S_{cat,Ohm} + S_{cat,act} + S_{cat,conc} \quad (29)$$

The effective cathode thermal conductivity $\lambda c_{at,eff}$ is given by 5.84 $W.m^{-1}.K^{-1}$ (Chyou et al., 2005; Damm, & Fedorov, 2006; Inui et al., 2006).

Anodic and Cathodic Channels

The thermal energy is transported by convection and conduction in the cathodic and anodic channels, the reactive velocity in both channels are assumed low, so that the convective term is very small compared to the conductive term, which leads to the negligence of the convective term and gives the energy equation in both channels by the following equation:

$$\frac{\partial}{\partial x}\left\{\lambda_f\frac{\partial T}{\partial x}\right\} + \frac{\partial}{\partial y}\left\{\lambda_f\frac{\partial T}{\partial y}\right\} + \frac{\partial}{\partial z}\left\{\lambda_f\frac{\partial T}{\partial z}\right\} = 0 \quad (30)$$

Boundary Conditions

The employed boundary conditions are Neumann type on all external surfaces and in all outer directions that get around the P-AS-SOFC, except both anodic and cathodic channel inlets where the temperature is taken constant, using Dirichlet type conditions, as shown in Figure 2 and Table 3.

Figure 2. IT-P-AS-SOFC boundary conditions

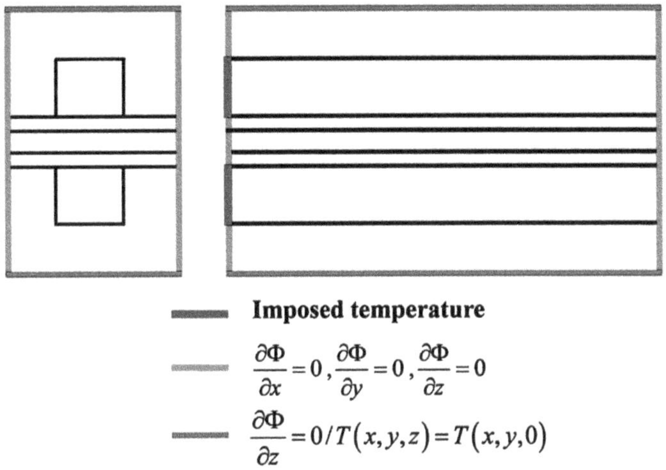

RESULTS AND DISCUSSION

In this chapter part, the obtained results, according to the developed FORTRAN program based on the presented model in the mathematical model section where the partial differential equations are numerically modeled using the finite difference method in a three-dimensional environment, are presented and discussed in three classes. In the first class, the impacts of the gas inlet temperature, current density and electrolyte thickness on the distribution and evolution of the temperature according to the cell stages symmetry configuration are presented. In the second class, the effects of the gas inlet temperature,

Table 3. IT-P-AS-SOFC boundary conditions

Components	OX Axis		OY Axis		OZ Axis	
	(0,y,z)	(Lx,y,z)	(x,0,z)	(x,Ly,z)	(x,y,0)	(x,y,Lz)
Electrolyte	$\frac{\partial \Phi}{\partial x}=0$	$\frac{\partial \Phi}{\partial x}=0$	$\frac{\partial \Phi}{\partial y}=0$	$\frac{\partial \Phi}{\partial y}=0$	$\frac{\partial \Phi}{\partial z}=0$	$\frac{\partial \Phi}{\partial z}=0$
Cathode	$\frac{\partial \Phi}{\partial x}=0$	$\frac{\partial \Phi}{\partial x}=0$	$\frac{\partial \Phi}{\partial y}=0$	$\frac{\partial \Phi}{\partial y}=0$	$\frac{\partial \Phi}{\partial z}=0$	$\frac{\partial \Phi}{\partial z}=0$
Anode	$\frac{\partial \Phi}{\partial x}=0$	$\frac{\partial \Phi}{\partial x}=0$	$\frac{\partial \Phi}{\partial y}=0$	$\frac{\partial \Phi}{\partial y}=0$	$\frac{\partial \Phi}{\partial z}=0$	$\frac{\partial \Phi}{\partial z}=0$
Anodic channel	$\frac{\partial \Phi}{\partial x}=0$	$\frac{\partial \Phi}{\partial x}=0$	$T=T_{H_2}=C^{ste}$	$\frac{\partial \Phi}{\partial y}=0$	$\frac{\partial \Phi}{\partial z}=0$	$\frac{\partial \Phi}{\partial z}=0$
Cathodic channel	$\frac{\partial \Phi}{\partial x}=0$	$\frac{\partial \Phi}{\partial x}=0$	$T=T_{air}=C^{ste}$	$\frac{\partial \Phi}{\partial y}=0$	$\frac{\partial \Phi}{\partial z}=0$	$\frac{\partial \Phi}{\partial z}=0$

current density and electrolyte thickness on the temperature fields and the extreme temperature values according to the IT-P-AS-SOFC configuration based on the cell stages repetitions are presented. Finally, a comparison between these both geometrical configuration types is conducted according to the gas inlet temperature, current density and electrolyte thickness.

IT-P-AS-SOFC With Cell Stages Symmetry

In this sub-section, the results are shown according to two different cases; the first concerns the use of the same gas inlet temperatures, and the second concerns the use of different gas inlet temperatures.

Gas Inlet Temperature Effect

The gas inlet temperature effect on the temperature fields and values is presented according to both studied cases. In the first case, the gas temperature value is the same at both anode and cathode channel inlets. For the second studied case, two different temperature values are used where the fuel temperature at the anodic channel inlet is different from the air temperature at the cathode channel inlet.

The employed electrodes and electrolyte thicknesses are given by 500 μm for the anode (Aguiar, Adjiman, & Brandon, 2004; Wang, Zhang, & Weng, 2008), 40 μm for the electrolyte (Aguiar et al., 2004; Bao et al., 2011) and 50 μm for the cathode (Aguiar et al., 2004; Ferguson et al., 1996) and an imposed current density of 2 $A.cm^{-2}$ (Aguiar et al., 2004; Wongchanapai, Iwai, Saito, & Yoshida, 2012) is used.

Same Gas Inlet Temperature

According to the several works in the literature, most authors use the same inlet temperature for both reactive and a large interval of the gas inlet temperature is obtained, therefore, three values of oxidizer and fuel inlet temperature are chosen taken in consideration in the IT-P-AS-SOFC investigation (673, 873 and 923 K). IT-P-AS-SOFC temperature fields resulted from the cell symmetry configuration according to the gas inlet temperatures of 673, 873 and 923 K are shown in Figure 3.

From Figure 3, a symmetry of IT-P-AS-SOFC temperature contours is clearly remarkable for the three gas inlet temperatures relative to the OZ axis, where the symmetry axis is the cell width center in each cross-section, which logically obtained due to the geometric symmetry of the studied calculation domain as well as the used boundary conditions. In addition, the temperature values in the anodic part (anode and anodic interconnector) are greater than the temperature values of the cathodic part (cathode and cathodic interconnector) due to the strong overpotentials produced in the anodic part, which are greater than that produced in cathodic part.

The maximum temperature is located in the electrolyte upper part at the IT-P-AS-SOFC output in the left and right wings; it is in order to 828.6400 K. The lowest temperature is that of the gases at the channel cell inlets (673 K). An increase of the temperature values is clearly remarkable from the cell input to output, it is about 155.6400 K. This increase presents a heat production case when the overpotential heat sources have a very important impact. Figure 3.a.

The lowest temperatures are located at the cell outlet, more precisely at the interconnector end of the cathodic part at the furthest places from the cathode channel. The maximum temperature is located in the electrolyte upper part in both left and right extremities at a distance of 10 mm from the gas inlet, it is in order to 877.7470 K. An increase in the temperature values is noticed in the first centimeter following the

Figure 3. IT-P-AS-SOFC temperature fields with the symmetrical configuration and the same gas inlet temperature: a) 673 K, b) 873 K and c) 923 K

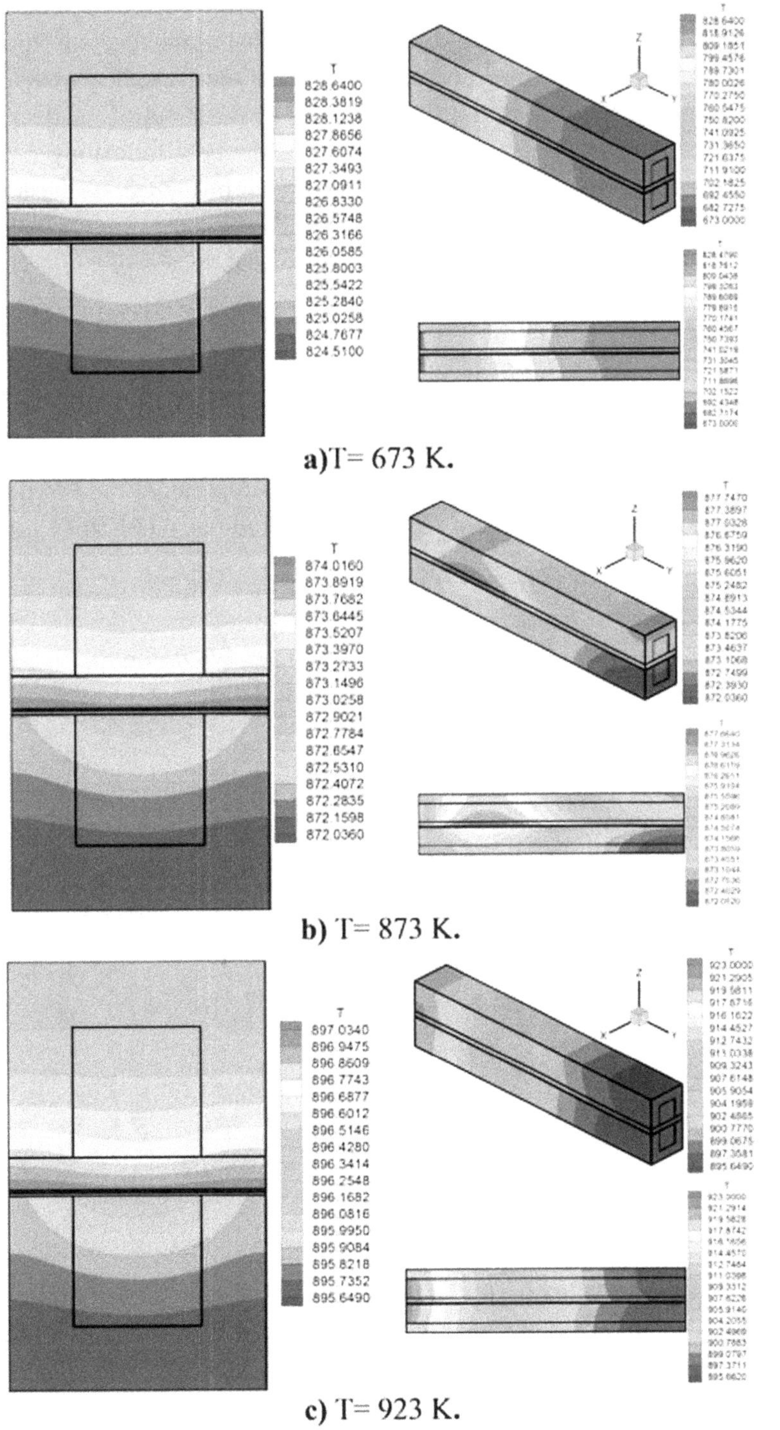

gas flow direction, and then a decrease along the P-AS-SOFC is produced. The increase in temperature values is about 4.7470 K, which presents a case of slight heat production when the overpotential heat sources have a low impact. Figure 3.b.

The maximum temperature is located at the cell inlet, it is that of the reactive (fuel and oxidizer). The minimum temperature is positioned at the interconnector end of the cathodic part at the furthest locations from the cathodic channel of the cell output. It is in order to 895.6490 K. A decrease in temperature values is clearly noticeable from the input to the output of the cell, indicating a non-productive case of the heat when the overpotential heat sources do not have an influence, because, the thermal energy produced from these heat sources are lower than the thermal energy consumed by the thermal resistance of IT-AS-SOFC components. The maximum difference between the temperature values is approximate 27.3510 K. Figure 3.c.

Different Gas Inlet Temperatures

The SOFC reactants' supply technique using two different gas inlet temperatures (for both hydrogen and oxygen) is employed by Suwanwarangkul et al. (2006). The temperature field of the IT-P-AS-SOFC at a cell stages symmetry according to different gas inlet temperatures, 673 K for hydrogen and 873 K for air, is showed in Figure 4.

Figure 4. IT-P-AS-SOFC temperature fields with the symmetrical configuration and the hydrogen and air temperatures of 673 and 873 K

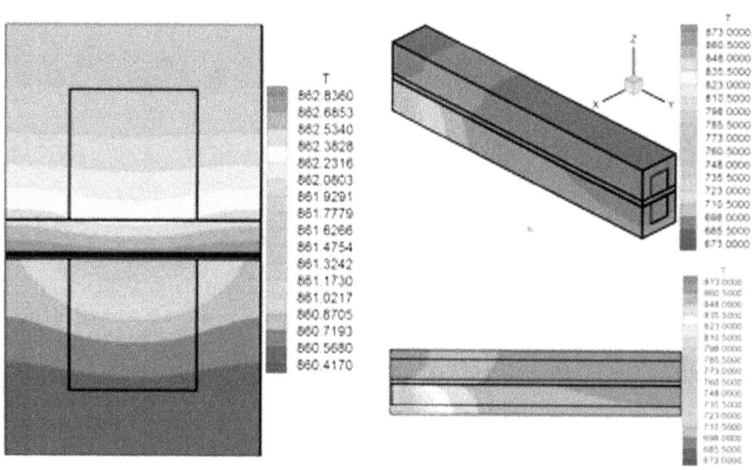

The maximum temperature value is that of hydrogen at the anode channel inlet (873 K), and the minimum temperature is located at the cathode channel inlet, it is that of air. A temperature value decrease in the cell upper part is obtained in the first centimeter, according to the gas flow direction, then an increase in temperature values is produced in the previous three millimeters, then the stability of the temperature value has recorded in the rest of the cell length following the produced thermal equilibrium, where the heat produced by the sources is identical to that consumed by thermal resistance. An increase of the temperature values at the cell lower part from the input to the output is clearly remarkable, which demonstrates the non-productivity of the heat. The maximum difference of the temperature values is that of the gas inlet temperatures. In addition, the temperature values in the anodic part are greater than the temperature values of the cathodic part. Figure 4.

Current Density Effect

In this sub-section, the current density impact on the temperature distribution is discussed for a cell stages symmetry configuration and an inlet temperature of both gases of 873 K is employed Suwanwarangkul et al. (2006). The electrodes and electrolyte thicknesses are 500 μm for the anode (Aguiar et al., 2004; Wang et al., 2008), 40 μm for the electrolyte (Aguiar et al., 2004; Bao et al., 2011) and 50 μm for the cathode (Aguiar et al., 2004; Zhang et al., 2007). Three current density values of 1.5, 2 and 2.5 $A.cm^{-2}$ are used.

The temperature fields of IT-P-AS-SOFC has a cell symmetry configuration according to the current densities of 1.5, 2 and 2.5 $A.cm^{-2}$ are shown in Figure 5.

From Figure 5, the symmetry of temperature fields is also obtained by the investigated current densities; in addition, the anodic part is hotter than the cathodic part.

The maximum temperature is located at the cell inlet; it is that of both gases and the minimum temperature is situated at the interconnector end of the cathodic part at the furthest points from the cathode channel in the cell output, it is in order to 852.2700 K. A decrease of the temperature values is clearly visible from the input to the output of the cell, which demonstrates the non- productivity of the heat. The maximum difference between the temperature values is approximate to 20.7300 K. Figure 5.a.

The lowest temperature values are located in the cell inlet, exactly at the interconnector end of the cathodic part at the most remote locations of the cathodic channel. The maximum temperature values are positioned in the electrolyte upper part at both left and right ends at a distance of one centimeter from the cell inlet, it is in order to 877.7470 K. An increase in the temperature values is produced in the first centimeter in the gas flow direction, and then a decrease in the temperature values occurs over the cell length rest. The increase in the temperature values is obtained, it is about to 4.7470 K, which represents a slight heat production case.

The maximum temperature is situated in the electrolyte upper part at the cell output in both left and right ends of the cell, it is in order to 892.8660 K. The lowest temperature is that of both gases at the cell inlet (873 K). The increase in the temperature values is clearly visible from the inlet to the outlet of the cell; this increase is about 19.8660 K, which proves the heat production in this case. Figure 5.

Electrolyte Thickness Effect

The electrode thicknesses are given by 500 μm for the anode (Aguiar et al., 2004; Wang et al., 2008) and 50 μm for the cathode (Aguiar et al., 2004; Bao et al., 2011). The temperature distribution is discussed for a cell stages symmetry configuration and an inlet temperature of both gases of 873 K is employed (Piccardo et al., 2015). The used imposed current density is 2 $A.cm^{-2}$ (Aguiar et al., 2004; Wongchanapai et al., 2012). Among several electrolyte thickness dimensions found in the bibliography, three electrolyte thickness values were chosen for the simulation namely 20, 40 and 60 μm.

The temperature distributions of IT-P-AS-SOFC with a cell stages symmetry configuration are shown in Figure 6 according to the electrolyte thicknesses of 20, 40 and 60 μm.

Figure 5. IT-P-AS-SOFC temperature fields with the symmetrical configuration and the current density: a) 1.5 A.cm^{-2}, b) 2 A.cm^{-2} and c) 2.5 A.cm^{-2}

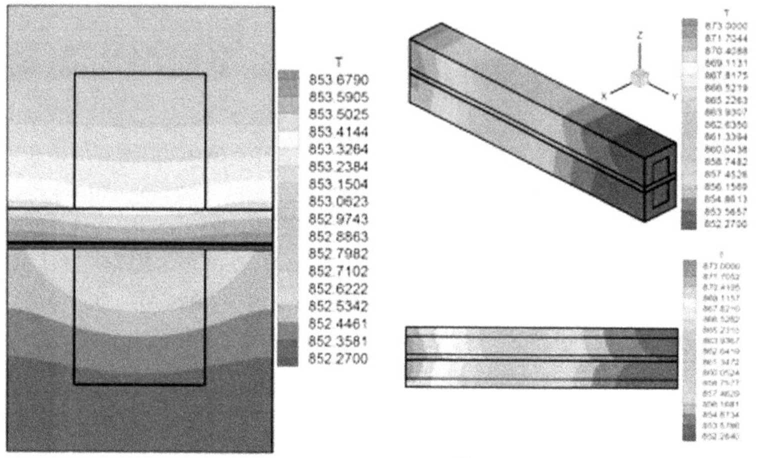

a) i= 1.5 A.cm^{-2}.

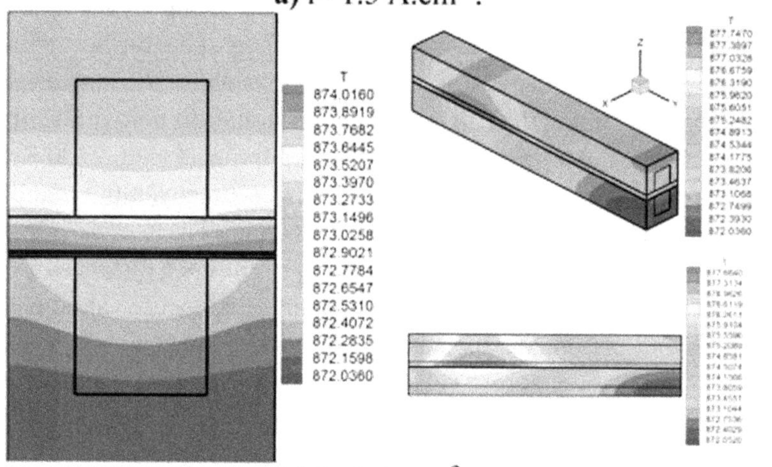

b) i= 2 A.cm^{-2}.

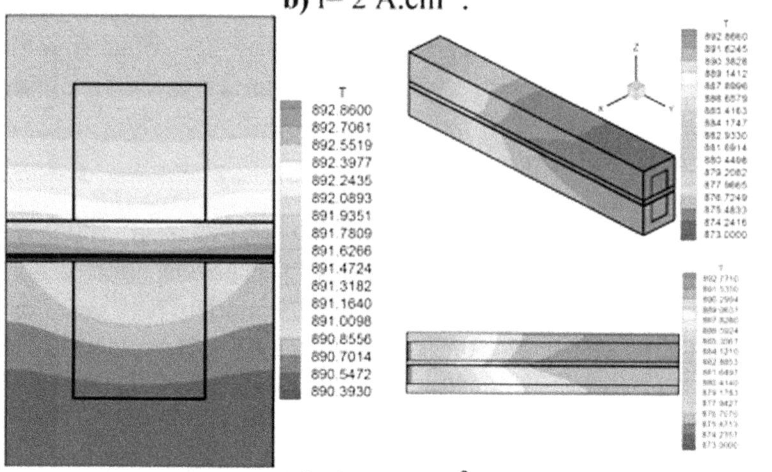

c) i= 2.5 A.cm^{-2}.

Figure 6. IT-P-AS-SOFC temperature fields with the symmetrical configuration and the electrolyte thickness: a) 20 μm, b) 40 μm and c) 60 μm

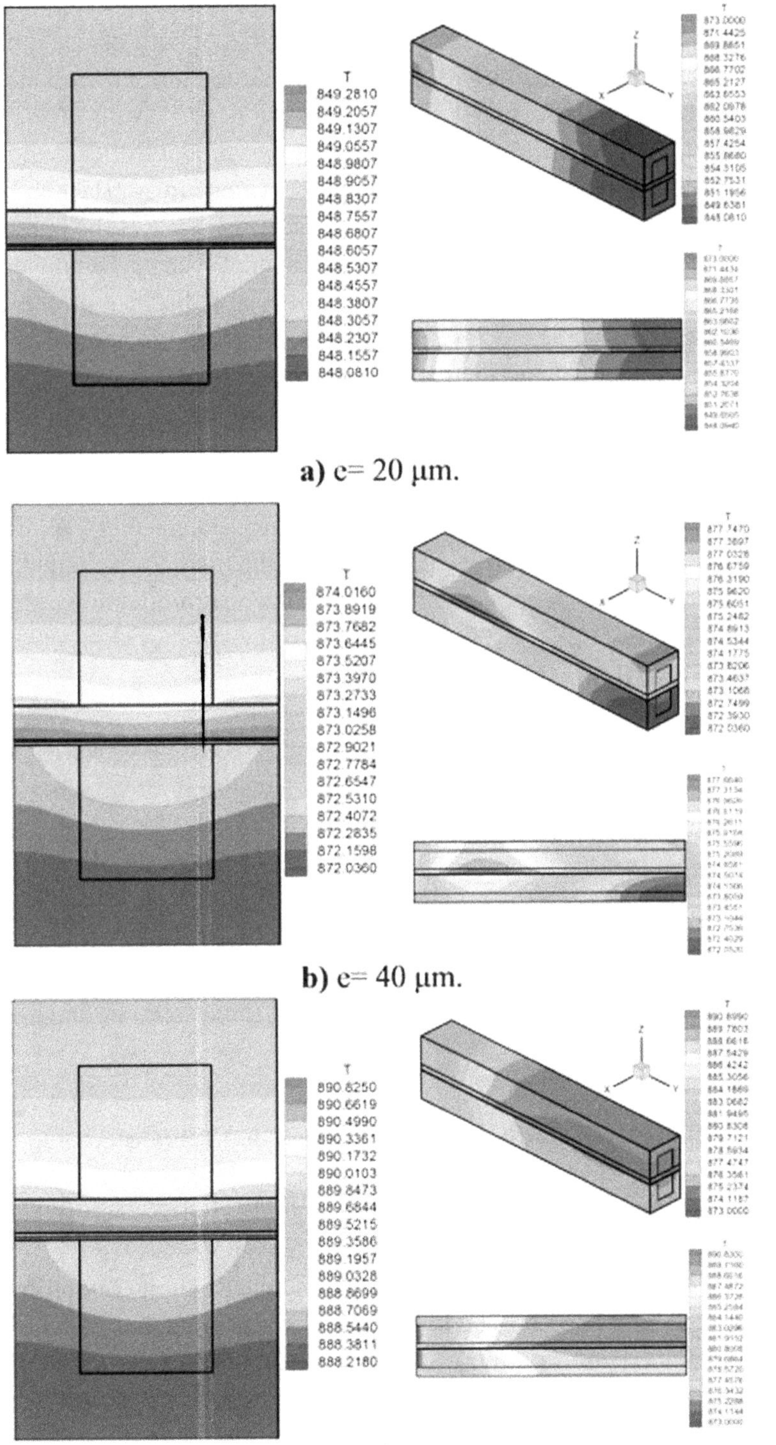

a) e= 20 μm.

b) e= 40 μm.

c) e= 60 μm.

Similarly, to the results obtained for the gas inlet temperatures and the current densities, the temperature distributions present a symmetry relative to the OZ axis at the cell width center and the high temperatures in the anodic part compared to that in the cathodic part. Figure 6.

The maximum temperature is located at the cell inlet; it is that of both gases. The minimum temperature is located at the interconnector end of the cathodic part at the most remote locations of the cathode channel of the cell output, it is in order to 848.0810 K. The temperature value decrease is clearly visible from the input to the output of the cell, which represents a non-productive case of heat. The maximum difference between the obtained temperature values is approximate to 24.9190 K. Figure 6.a.

The lowest temperatures are located at the cell outlet, more precisely at the interconnector end of the cathodic part at the most remote locations of the cathodic channel. The maximum temperature is located in the electrolyte upper part in both left and right extremities at a distance of one centimeter from the gas inlet; it is in order to 877.7470 K. A temperature value increase is noticed in the first centimeter according to the gas flow direction, and then a temperature value decrease is produced along the cell. The temperature value increase is about 4.7470 K, which represents a case of a slight heat generation. Figure 6.b.

The maximum temperature is located in the electrolyte upper part at the cell outlet in both left and right ends; it is in order to 890.8990 K. The lowest temperature obtained is that of the inlet gases (873 K). A temperature value increase is clearly remarkable from the input to the output of the cell; this temperature increase is about 13.8990 K, which demonstrates the heat production case. Figure 6.c.

IT-P-AS-SOFC With Cell Stages Repetition

To ensure a best and logical comparison between the two studied configurations, the same characteristics, dimensions and parameters used in the simulation of the cell stages symmetry configuration are used in the simulation of the cell stages repetition configuration.

Gas Inlet Temperature Effect

The gas inlet temperature impact on the temperature distributions for the IT-P-AS-SOFC cell stages repetition configuration is presented according to two supply types, same and different gas inlet temperatures.

Same Gas Inlet Temperature

The temperature fields of the IT-P-AS-SOFC with cell stages repetition according to the same gas inlet temperatures of 673, 873 and 923 K are shown in Figure 7.

The symmetry of the temperature distributions is also obtained in this sub-section, and the temperatures in the anodic part are higher than in the cathodic part. Figure 7.

The maximum temperature is located in the electrolyte upper part at the cell outlet in both left and right ends; it is in order to 832.1790 K. The lowest obtained temperature is that of the gas inlets (673 K). An increase of the temperature values is clearly remarkable from the input to the output of the cell; this increase in temperature values is about 159.1790 K, which demonstrates the heat production case. Figure 7.a.

The lowest temperatures are located at the cell inlet; they are those of the reactive gases (873 K). The maximum temperature values are located in the electrolyte upper part in both left and right cell ends at

Figure 7. IT-P-AS-SOFC temperature fields with the repetition configuration and the same gas inlet temperature: a) 673 K, b) 873 K and c) 923 K

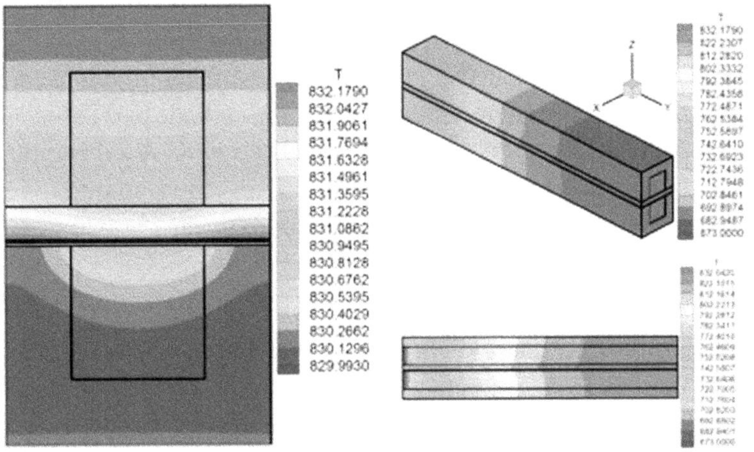

a) T= 673 K.

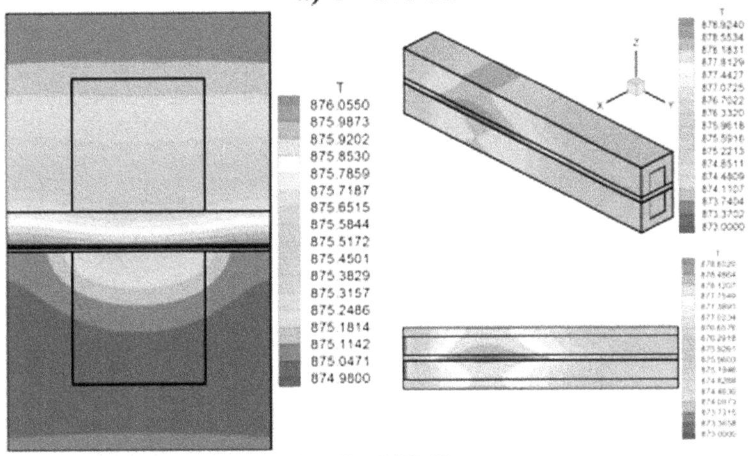

b) T= 873 K.

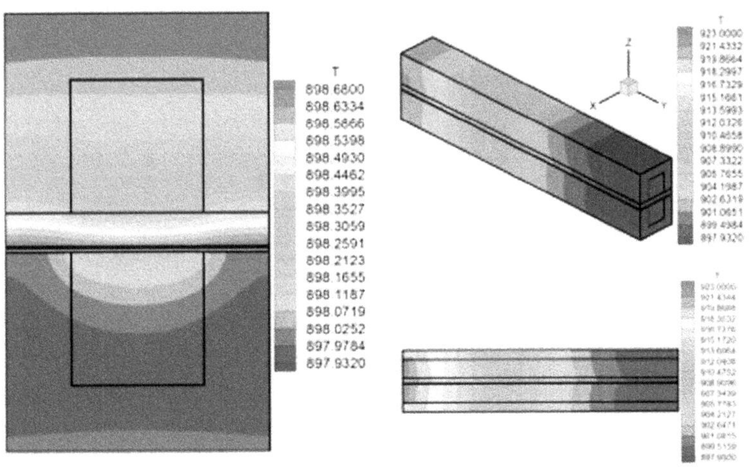

c) T= 923 K.

a distance of one centimeter from the gas inlet, they are in order to 878.9240 K. A temperature value increase is noticed in the first centimeter according to the gas flow direction, and then a temperature value decrease is produced along the cell. This increase is about 5.9240 K, which represents a case of slight heat production. Figure 7.b.

The maximum temperature is positioned at the cell inlet; it is that of the fuel and oxidizer. The minimum temperature is positioned at the interconnector end of the cathode part in the most remote locations of the cathodic channel at the cell output; it is in order to 897.9320 K. A temperature decrease is clearly remarkable from the input to the output of the cell, which gives a non-productive case of the heat. The maximum difference between the temperature values is approximate 25.0680 K. Figure.7.c.

Different Gas Inlet Temperatures

In Figure 8, it is clearly remarkable that a slight heat production along the cell length is produced. The maximum temperature value is higher than that of hydrogen at the anode channel inlet (874.0180 K). The minimum temperature is located at the cathode channel inlet; it is that of air. A temperature increase in the cell upper part is produced on gas flow direction, and then a decrease in these values is recorded in the cell length rest. A temperature value increase in the cell lower part from the input to the output of the cell is clearly remarkable. The maximum difference between the temperature values is approximate 201.0180 K. In addition, the temperature values in the anodic part are greater than the temperature values of the cathodic part.

Figure 8. IT-P-AS-SOFC temperature fields with the repetition configuration and the hydrogen and air temperatures of 673 and 873 K

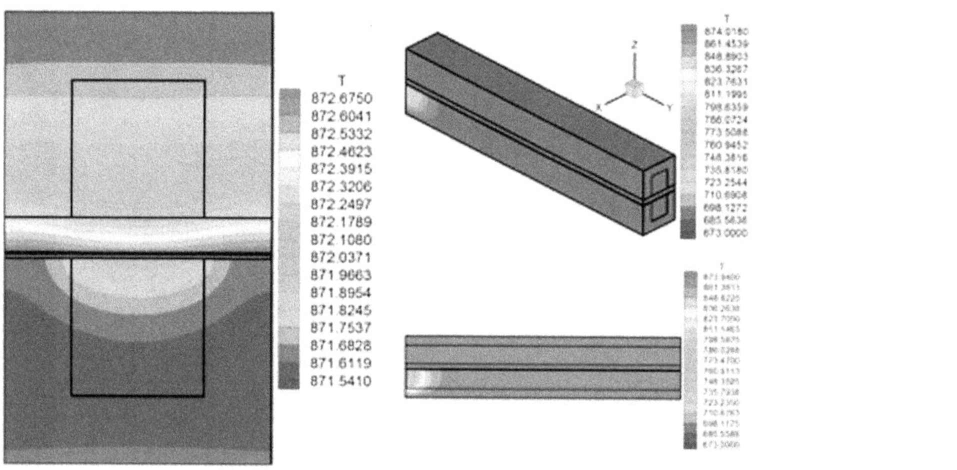

Current Density Effect

The temperature fields of the IT-P-AS-SOFC with cell stages repetitions according to the current densities of 1.5, 2 and 25 A.cm-2 are shown in Figure 9.

Figure 9. IT-P-AS-SOFC temperature fields with the repetition configuration and the current density: a) 1.5 A.cm-2, b) 2 A.cm-2 and c) 2.5 A.cm^{-2}

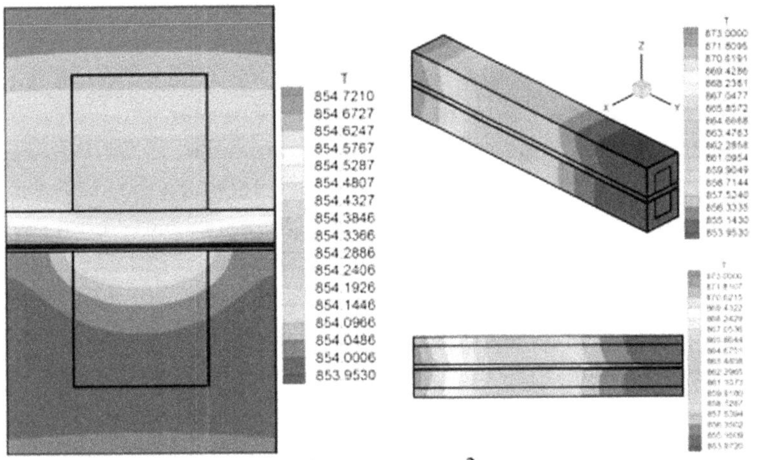

a) i= 1.5 A.cm^{-2}.

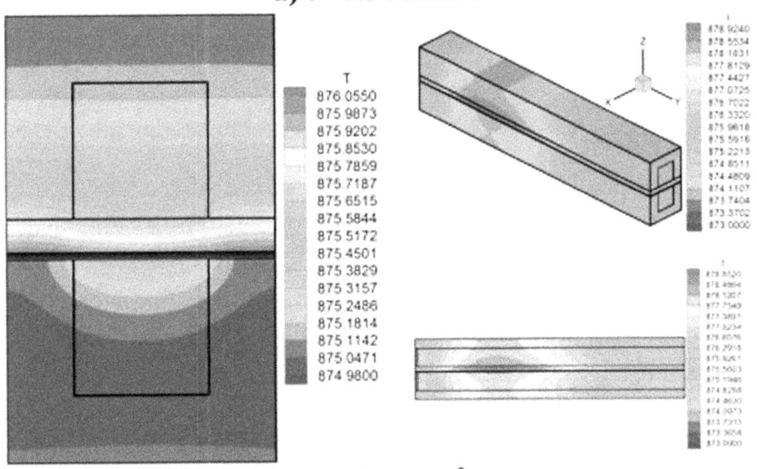

b) i= 2 A.cm^{-2}.

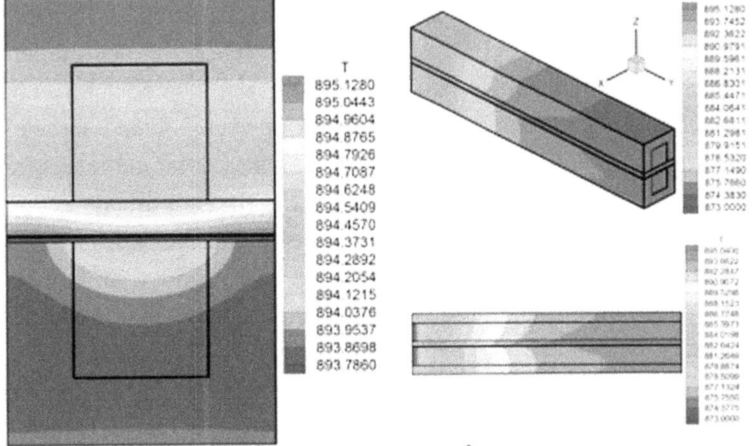

c) i= 2.5 A.cm^{-2}.

In this sub-section, the temperatures of the anodic part are also higher than of the cathodic part and the symmetry of the fields is also obtained. Figure 9.

The maximum temperature value is located at the cell input; it is that of the reactive gases (873 K). The minimum temperature is located at the interconnector end of the cathodic part at the furthest locations from the cathode channel of the cell output. It is in order to 853.9530 K. The temperature value decrease is clearly visible from the input to the output of the cell (non-productive case). The maximum difference between the temperature values in the cell is approximate 19.0470 K. Figure 9.a.

The minimum temperatures are located in the cell inlet; they are those of the two gases (873 K). The maximum temperatures are positioned at the electrolyte upper part in both left and right sides of the cell at a distance of one centimeter from the cell inlet, they are about 878.9240 K. The temperature value increase is clearly remarkable in the first centimeter according to the cell longitude, and then a temperature decrease is provoked on the cell longitude rest. The temperature increase is in order of 5.9240 K, which leads to a case of slight heat productivity. Figure 9.b.

The maximum temperature value is located in the electrolyte upper part that is close to the anode at the cell output, more exactly in both left and right sides, it is around 895.1280 K. The minimum temperature is positioned in the cell inlet; it is that of 873 K. The temperature value increase from the inlet to the outlet of the cell is clear, this increase is around 22.1280 K, and that proves the heat productivity case. Figure 9.c.

Electrolyte Thickness Effect

The electrolyte thickness effect on the temperature fields obtained by a cell stages repetition using the electrolyte thickness dimensions 20, 40 and 60 µm is presented in Figure 10.

As all other obtained results, the symmetry of the temperature distributions on the OZ axis is obtained and the anodic part is hotter than the cathodic part. Figure 10.

The maximum temperature is positioned at the cell inlet; it is that of the fuel and oxidizer. The minimum temperature is located at the interconnector end of the cathodic part at the most remote locations of the cathode channel of the cell output, it is in order to 848.9480 K. The temperature value decrease is clearly remarkable from the input to the output of the cell, which gives a non-productive case of the heat. The maximum difference between the temperature values is about 24.0520 K. Figure 10.a.

Figure 10.b shows the temperature fields for the IT-P-AS-SOFC for a cell repetition case according to the electrolyte thickness of 40 µm. The lowest temperatures are located at the cell inlet; they are those of the reactive gases (873 K). The maximum temperature values are located in the electrolyte upper part in both left and right ends of the cell at a distance of one centimeter from the gas inlet, they are in order to 878.9240 K. A temperature value increase is noticed in the first centimeter according to the gas flow direction, and then a temperature value decrease is produced along the cell. The increase in temperature values is about 5.9240 K, which represents a case of slight heat production. Figure 10.b.

In Figure 10.c, the maximum temperature is located in the electrolyte upper part of the cell outlet at both left and right sides, it is in order to 895.3410 K. The lowest temperature is that of the gases at the cell input (873 K). An increase in the temperature values is clearly remarkable from the input to the output of the cell; this increase in temperature values is about 22.3410 K, which demonstrates a heat production case. Figure 10.c.

Figure 10. IT-P-AS-SOFC temperature fields with the repetition configuration and the electrolyte thickness: a) 20 μm, b) 40 μm and c) 60 μm

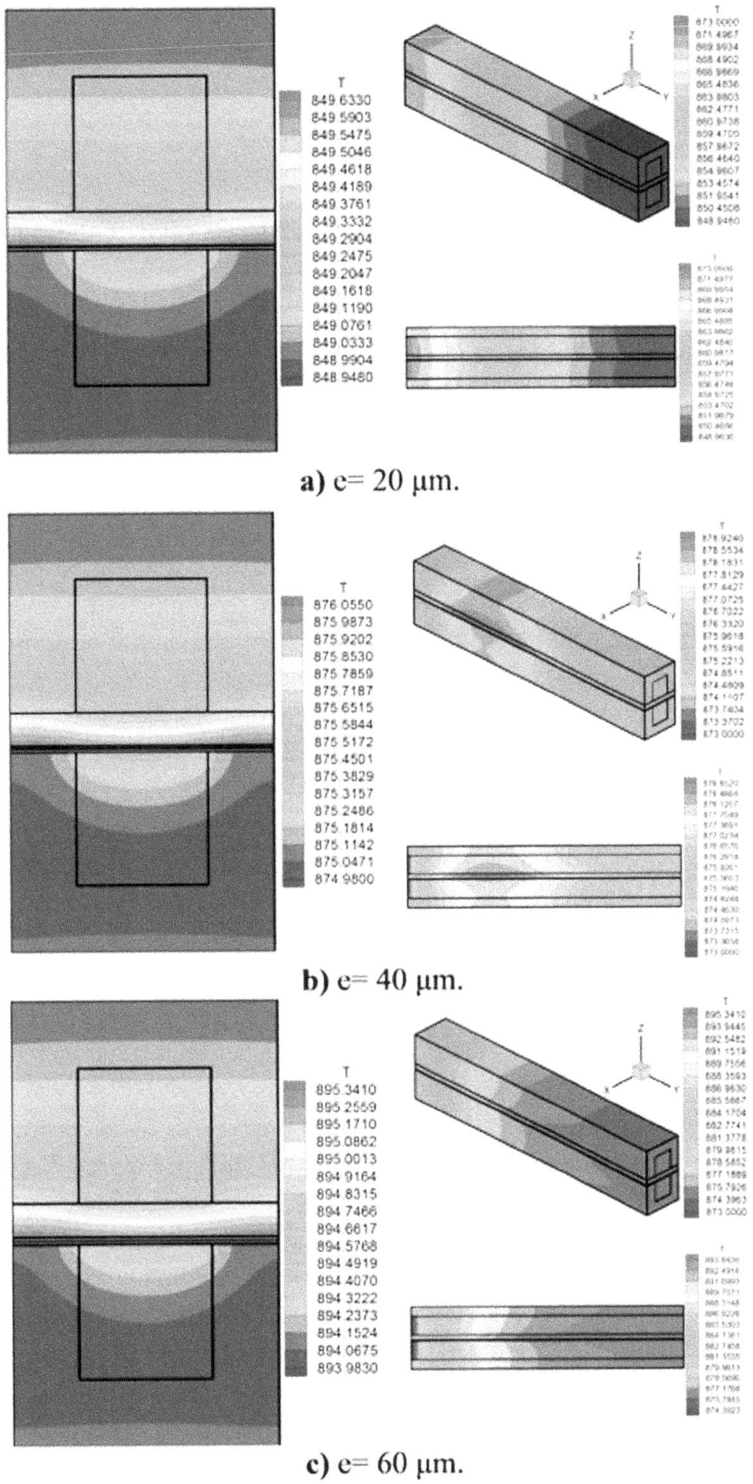

Comparison Between the Repetition and Symmetry of Cell Stages

The comparison between the results obtained by both studied geometric configurations (symmetry and repetition of the cell stages) is presented according to three parameters: gas inlet temperature, current density and electrolyte thickness.

Gas Inlet Temperature

The comparison between the two geometrical configurations according to the gas inlet temperature is presented according to the first case, where the temperature of the two gases is the same at the inlet of the anodic and cathodic channels. Because the use of the different gas inlet temperatures shows a large temperature difference in the P-AS-SOFC, which logically inducts significant thermal stresses.

Figure 11 shows the temperature value evolution of the IT-P-AS-SOFC electrolyte middle along the OY axis; for the case of the same gas inlet temperature.

For both investigated configurations, it is clearly remarkable for each studied gas inlet temperature value (673, 873 and 923 K) that the curves have almost the same form. Figure 11.

For a gas inlet temperature of 673 K, the temperature value elevation is very apparent from the inlet to the outlet of the cell, at the cell beginning, the temperature value remains constant almost one millimeter, then it increases hyperbolically. The difference between the temperature values of both geometric configurations remains the same all along the first millimeter, and then it develops proportionally according to the cell length. Figure 11.a.

For a gas inlet temperature of 873 K, a slight rise in the temperature values is obtained. At the cell inlet, the temperature value remains constant in the first millimeter, then a hyperbolically increase in 7 mm for the cell stages repetition case and 9 mm for the cell stages symmetry case, and then a parabolic decrease of the temperature values is done until the cell outlet. The difference between the temperature values of both geometric configurations remains constant in the first millimeter and then develops proportionally with the cell length. Figure 11.b.

For a gas supply temperature equal to 923 K, the decreasing of the temperature value is clearly visible in Figure 11.c. At the cell inlet, the temperature remains constant in the first millimeter, and then a parabolic decreasing of the temperature values occurs. The difference between the temperature values of both geometric configurations remains constant in the first millimeter and then develops proportionally with the cell length.

The evolution of the maximum temperature difference in IT-P-AS-SOFC according to the different input gas temperatures for both studied configurations is presented in Figure 12.

The two curves presented in Figure 12 have the same form. At the beginning of each curve, the maximum difference between the temperature values decreases quasi-linearly, from 35 K for the cell symmetry configuration and 34 K for the cell repetition configuration, until to reach a temperature differences of 5.7710 K, produced by an inlet temperature of 873 K for the cell repetition configuration, and of 5.0080 K, produced by an inlet temperature of 877 K for the cell symmetry configuration. Then increase quasi-linearly to the curve ends.

Figure 11. Evolution of the electrolyte middle temperature along the OY axis; for the same gas inlet temperature. a) T=673 K, b) T=873 K and c) T=923 K

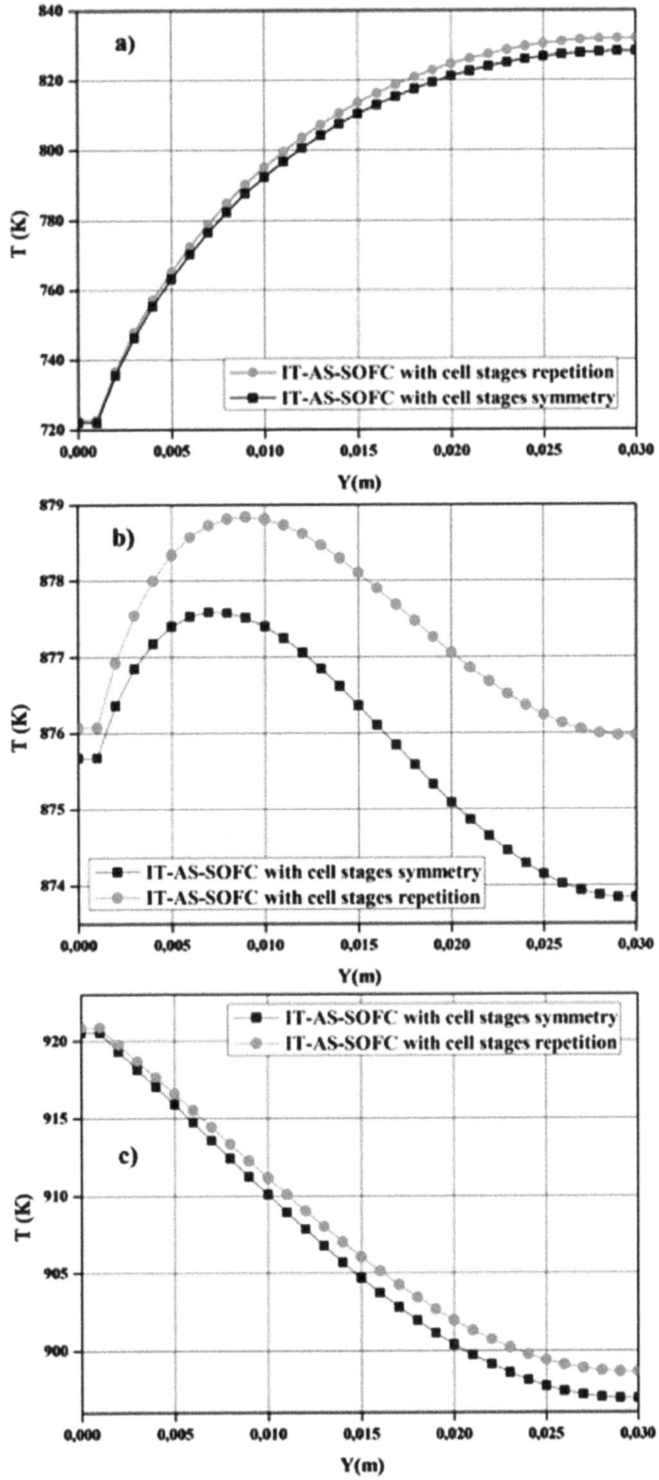

Figure 12. Evolution of the maximum temperature difference vs. the gas inlet temperature

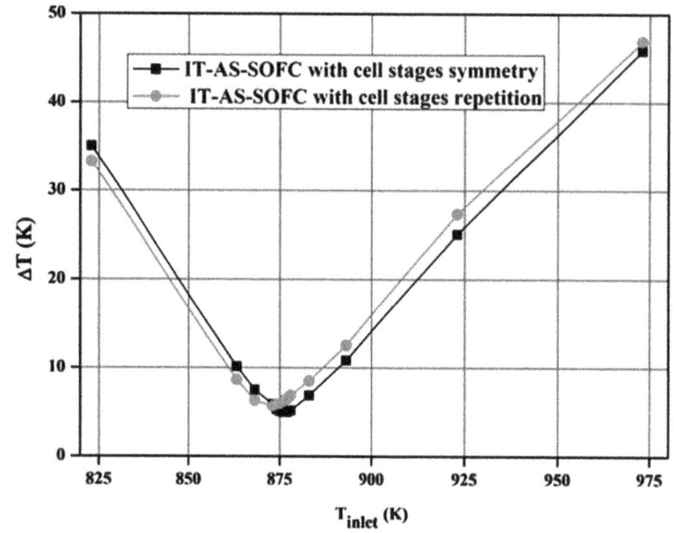

Current Density

Figure 13 presents the temperature value evolution in the electrolyte middle from the inlet to outlet of IT-P-AS-SOFC of both suited geometries according to the same gas inlet temperature for the current densities 1.5, 2 and 2.5 A.cm-2.

For each current density value, the obtained curves by both studied configurations having almost the same form. Figure 13.

For a current density of 1.5 A.cm^{-2}, the decreasing of the temperature values is clearly visible; at the cell inlet the temperature value remains constant within a distance of one millimeter, and then a parabolic decrease of the values develops. The difference between the temperature values of both configuration types remains the same in the first millimeter and then increases proportionally. Figure 13.a.

For a current density of 2 A.cm^{-2}, a slight increase in the temperature values is obtained. At the cell input, the temperature remains the same in the first millimeter, then an increase is carried out hyperbolically in 7 mm for the cell stages repetition configuration, and 9 mm for the cell stages symmetry configuration, and then a parabolic decrease of the temperature values is realized in the cell longitude. The difference between the temperature values obtained from both configurations is constant in the first millimeter and then develops proportionally. Figure 13.b.

For a current density of 2.5 A.cm^{-2}, the rise in temperature values is very clear from the IT-P-AS-SOFC input to the output; at the cell beginning, the temperature remains constant in the first millimeter, then it increases hyperbolically. The difference between the temperature values of each position of both studied configurations remains constant in the first millimeter and then increases proportionally. Figure 13.c.

The evolution of the maximum temperature difference in the AS-SOFC according to the different current density and inlet temperature values of both studied configurations is presented in Figure 14.

For each gas inlet temperature, the presented curves have the same form. At the beginning of each curve, the maximum difference between the temperature values decreases quasi-linearly until reaching the minimum difference, then it increases quasi-linearly. In addition, the deviation between both obtained

Figure 13. Evolution of the electrolyte medium temperature along the (OY) axis vs. the current density. a) i=1.5 A.cm^{-2}, b) i=2 A.cm^{-2} and c) i=2.5 A.cm^{-2}

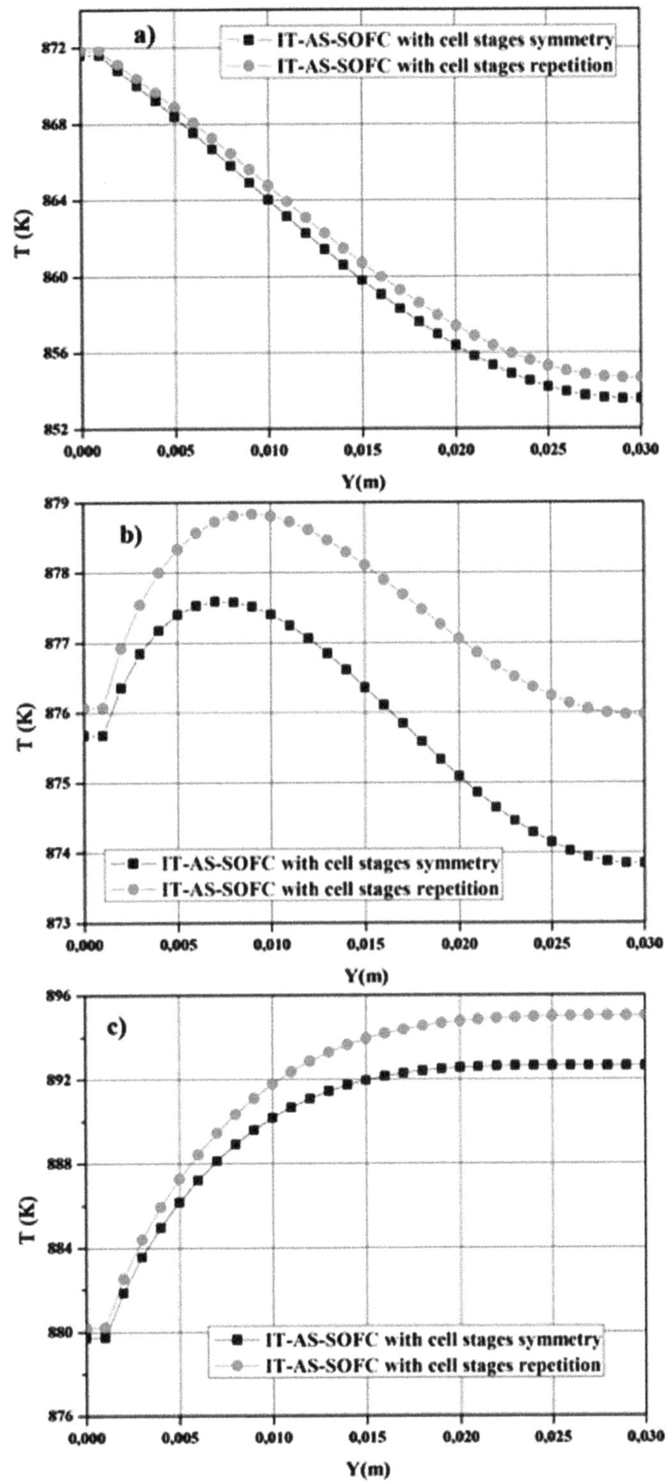

Figure 14. Evolution of the maximum temperature difference vs. the current density. Rep: with cell stages repetition and Sym: with cell stages symmetry

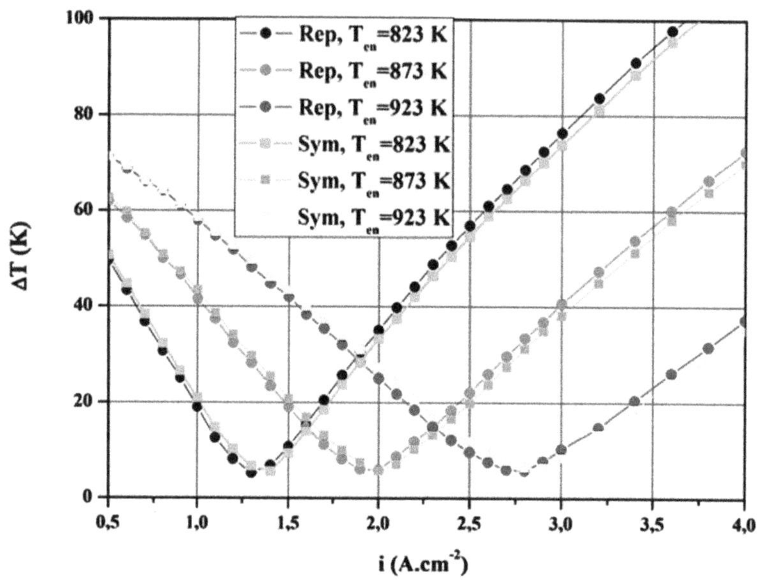

differences develops proportionally according to the current density with low influence. The minimum temperature difference is obtained by the cell stages repetition configuration. Figure 14.

Electrolyte Thickness

Figure 15 presents the temperature value evolution in the electrolyte middle from the input to the output of IT-P-AS-SOFC of both suited geometries according to the same gas inlet temperature for the electrolyte thickness 20, 40 and 60 µm.

For each electrolyte thickness, the obtained curves having the same form. Figure 15.

For an electrolyte thickness of 20 µm, the decreasing of the temperature values is clearly visible in Figure 16.a. At the cell inlet, the temperature remains constant in the first millimeter, and then a parabolic decrease is obtained. The difference between the temperature values of both studied configurations remains constant in the first millimeter and then develops proportionally with the cell length

For an electrolyte thickness of 40 µm, a slight increase in the temperature values is obtained. At the cell input, the temperature remains constant in the first millimeter, and then an increase of the temperature is realized hyperbolically in 7 mm for the cell stages repetition case and 9 mm for the cell stages symmetry case, then a parabolic decrease. The difference between the temperature values of both studied configurations remains constant in the first millimeter and then develops proportionally. Figure 15.b.

For an electrolyte thickness of 60 µm, the temperature elevation is very apparent from the input to the output of the P-AS-SOFC. At the cell inlet, the temperature remains constant in the first millimeter and then increases hyperbolically. The temperature value difference of both configurations remains in the first millimeter and then develops hyperbolically. Figure 15.c.

The evolution of the maximum temperature difference in the AS-SOFC according to the different electrolyte thickness dimensions and current densities for both studied configurations is shown in Figure 16.

Figure 15. Evolution of the electrolyte medium temperature along the (OY) axis vs. the electrolyte thickness. a) e = 20 µm, b) e = 40 µm and c) e = 60 µm

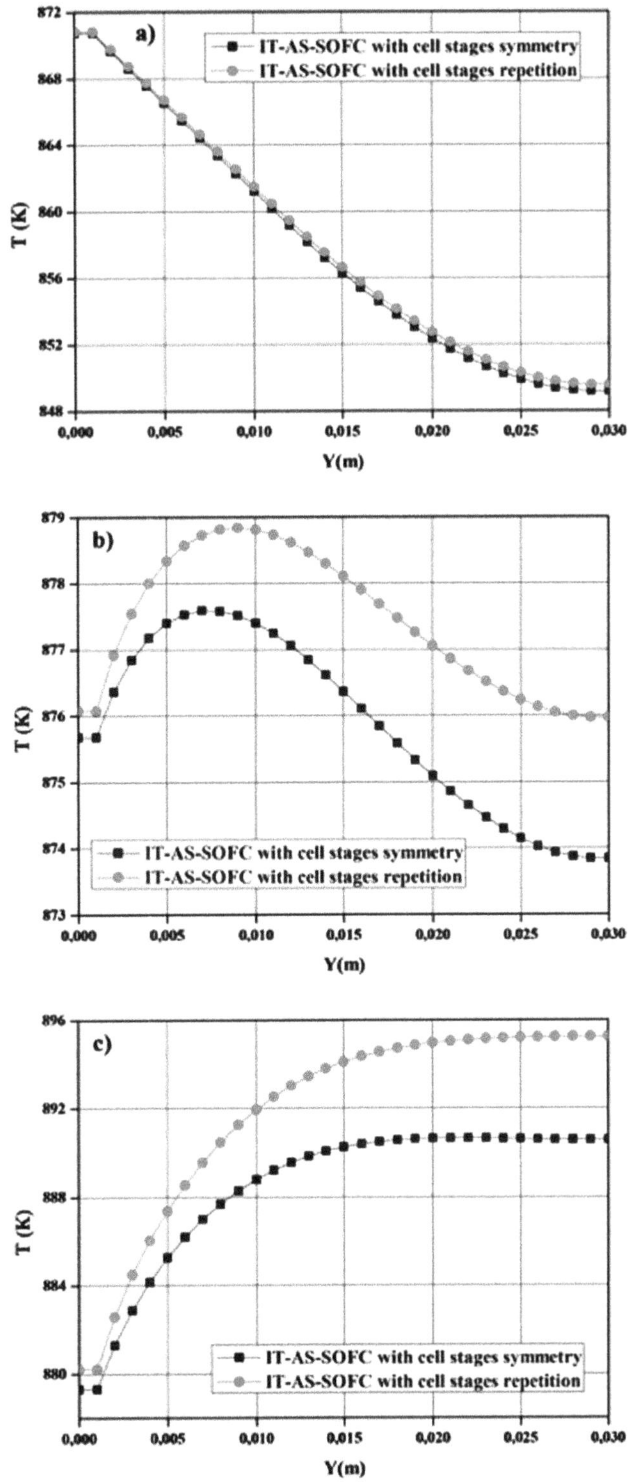

All presented curves have the same form. At the beginning of each curve, the maximum temperature difference decreases quasi-linearly until reaching the minimum temperature difference and then increases quasi-linearly. Thus, the deviation between both differences increases proportionally as a function of the electrolyte thickness whatever the gas inlet temperature. The minimum temperature difference is obtained in the cell stages repetition configuration. Figure 16.

Figure 16. Evolution of the maximum temperature difference vs. the electrolyte thickness. Rep: with cell stages repetition and Sym: with cell stages symmetry

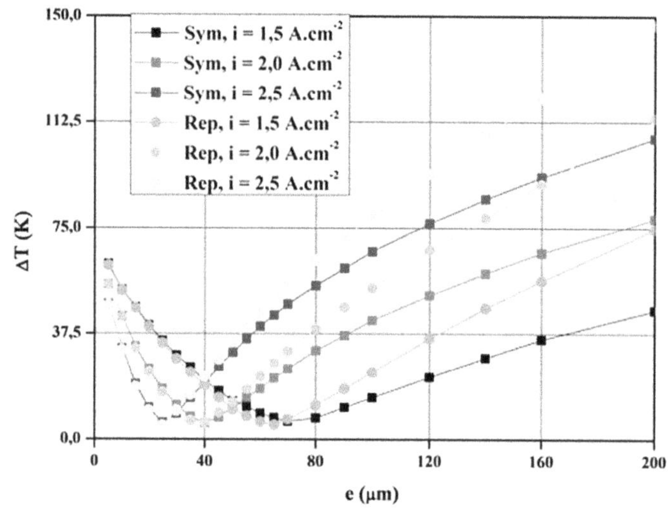

CONCLUSION

This chapter presents a study of the heat transfer phenomenon in an IT-P-AS-SOFC to evaluate and locate the maximum and minimum temperatures, in the studied elementary cell by a three-dimensional and steady state model that accounts the heat production due to the Ohmic, concentration and activation overpotentials for two geometrical configuration types; repetition and symmetry of the cell stages. The gas inlet temperature, current density and electrolyte thickness effects on the distribution and the maximum and minimum temperature values for both studied geometrical configurations are presented. Moreover, this permits the knowledge of the best configuration that has the minimal difference temperature and simultaneously allows minimizing thermal stresses and producing more homogeneous thermal stresses and logically corresponds to large lifetime, high energy production and low production cost.

From the obtained result analysis, it becomes apparent that the studied model allowed us to understand the effects of the various parameters (inlet gas temperature, current density, electrolyte thickness and geometrical configurations) on the temperature fields and values in all parts of P-AS-SOFC.

The main obtained results in this chapter are:

The use of different gas inlet temperatures allows producing a large temperature difference in the P-AS-SOFC, which logically inducts significant thermal stresses.

The temperature maximum values are located in the electrolyte upper part at the cell wings in all cases that produced heat, which means that this zone receives greater thermal stress compared to the other zones.

The minimum temperature values are positioned at the interconnector end of the cathode part in the most remote locations of the cathodic channel.

The maximum temperature difference recorded in the three directions is that of the OY axis (flow direction), which proves the importance of thermal stresses resulting in this direction compared to other directions, and the maximum temperature difference recorded is that of the Ox axis (cell width).

The cell upper part (interconnector, anode and the electrolyte upper part) is hotter than the cell lower part (interconnector, cathode and the electrolyte lower part) in all studied cases; this demonstrates that the cell lower part risk receiving more thermal stress.

The cell stages repetition configuration allows a small difference and more homogeneous distributions of temperature on the OZ axis (cell height) compared to the cell stages symmetry configuration, which logically induces more homogeneous thermal stresses.

The use of low gas inlet temperatures causes large temperature differences and logically considerable thermal stresses inside IT-P-AS-SOFC. However, the use of high gas inlet temperatures causes large temperature differences and logically considerable thermal stresses inside IT-P-AS-SOFC as well as a reduction in the net system performance due to the energy loss employed in the preheating of the reactive gases.

The production of low power densities causes large temperature differences and logically considerable thermal stresses inside IT-P-AS-SOFC as well as obtaining of low performances. However, the production of high power densities causes large temperature differences and logically considerable thermal stresses inside IT-P-AS-SOFC.

The use of thick electrolytes induces a high resistance in IT-P-AS-SOFC, which leads to the important productions of heat, large temperature differences and logically considerable and inhomogeneous thermal stresses inside IT-P- AS-SOFC. However, the use of thin electrolytes can reduce heat production, temperature differences and logically acceptable and more homogeneous thermal stresses inside IT-P-AS-SOFC.

There is a minimum temperature difference in each studied cell of about 5 K according to an optimal combination of the gas inlet temperature and current density (electrical production), which guarantees a minimum temperature difference and logically thermal constraints more homogeneous.

The obtained temperature values by IT-P-AS-SOFC with cell stages repetition configuration are smaller than of IT-P-AS-SOFC with cell stages symmetry configuration in all studied cases, and they are more homogeneous, this logically induces smaller and more homogeneous thermal stresses. Finally, the IT-P-AS-SOFC with cell stages repetition configuration is more practice compared to that with cell stages symmetry configuration.

This research received no specific grant from any funding agency in the public, commercial, or not-for-profit sectors.

REFERENCES

Abdenebi, H., Zitouni, B., Ben-Moussa, H., Haddad, D., Zitouni, H., & Sahli, Y. (2015). Inlet Methane Temperature Effect at a Planar SOFC Thermal Field under Direct Internal Reforming Condition. In I. Dincer, C. O. Colpan, O. Kizilkan & M. A. Ezan (Eds.), Progress in Clean Energy, volume I: Novel Systems and Applications (pp. 567-581). Springer. doi:10.1007/978-3-319-17031-2_41

Aguiar, P., Adjiman, C. S., & Brandon, N. P. (2004). Anode-supported intermediate temperature direct internal reforming solid oxide fuel cell. I: Model-based steady-state performance. *Journal of Power Sources*, *138*(1-2), 120–136. doi:10.1016/j.jpowsour.2004.06.040

Amedi, H. R., Bazooyar, B., & Pishvaie, M. R. (2015). Control of anode supported SOFCs (solid oxide fuel cells): Part I. mathematical modeling and state estimation within one cell. *Energy*, *90*, 605–621. doi:10.1016/j.energy.2015.07.095

Andersson, M., Nakajima, H., Kitahara, T., Shimizu, A., Koshiyama, T., Paradis, H., Yuan, J., & Sundén, B. (2014). Comparison of humidified hydrogen and partly pre-reformed natural gas as fuel for solid oxide fuel cells applying computational fluid dynamics. *International Journal of Heat and Mass Transfer*, *77*, 1008–1022. doi:10.1016/j.ijheatmasstransfer.2014.06.033

Andersson, M., Paradis, H., Yuan, J., & Sundén, B. (2013). Three dimensional modeling of an solid oxide fuel cell coupling charge transfer phenomena with transport processes and heat generation. *Electrochimica Acta*, *109*, 881–893. doi:10.1016/j.electacta.2013.08.018

Bao, C., Cai, N., & Croiset, E. (2011). A multi-level simulation platform of natural gas internal reforming solid oxide fuel cell-gas turbine hybrid generation system-Part II. Balancing units model library and system simulation. *Journal of Power Sources*, *196*(20), 8424–8434. doi:10.1016/j.jpowsour.2011.05.032

Burt, A. C., Celik, I. B., Gemmen, R. S., & Smirnov, A. V. (2004). A numerical study of cell-to-cell variations in a SOFC stack. *Journal of Power Sources*, *126*(1-2), 76–87. doi:10.1016/j.jpowsour.2003.08.034

Cai, Y., Wang, B., Wang, Y., Xia, C., Qiao, J., Aken, P. A. V., ... Lund, P. (2018). Validating the technological feasibility of yttria-stabilized zirconia-based semiconducting-ionic composite in intermediate temperature solid oxide fuel cells. *Journal of Power Sources*, *384*, 318–327. doi:10.1016/j.jpowsour.2018.03.012

Chyou, Y. P., Chung, T. D., Chen, J. S., & Shie, R. F. (2005). Integrated thermal engineering analyses with heat transfer at periphery of planar solid oxide fuel cell. *Journal of Power Sources*, *139*(1-2), 126–140. doi:10.1016/j.jpowsour.2004.07.001

Damm, D. L., & Fedorov, A. G. (2006). Reduced-order transient thermal modeling for SOFC heating and cooling. *Journal of Power Sources*, *159*(2), 956–967. doi:10.1016/j.jpowsour.2005.11.072

Fardadi, M., Mueller, F., & Jabbari, F. (2010). Feedback control of solid oxide fuel cell spatial temperature variation. *Journal of Power Sources*, *195*(13), 4222–4233. doi:10.1016/j.jpowsour.2009.12.111

Ferguson, J. R., Fiard, J. M., & Herbin, R. (1996). Three-dimensional numerical simulation for various geometries of solid oxide fuel cells. *Journal of Power Sources*, *58*(2), 109–122. doi:10.1016/0378-7753(95)02269-4

Huang, Y., Lin, Q., Liu, H., Ni, M., & Zhang, X. (2017). Evaluation of the waste heat and residual fuel from the solid oxide fuel cell and system power optimization. *International Journal of Heat and Mass Transfer, 115*, 1166–1173. doi:10.1016/j.ijheatmasstransfer.2017.07.115

Inui, Y., Urata, A., Ito, N., Nakajima, T., & Tanaka, T. (2006). Performance simulation of planar SOFC using mixed hydrogen and carbon monoxide gases as fuel. *Energy Conversion and Management, 47*(13-14), 1738–1747. doi:10.1016/j.enconman.2005.10.014

Kakaç, S., Pramuanjaroenkij, A., & Zhou, X. Y. (2007). A review of numerical modeling of solid oxide fuel cells. *International Journal of Hydrogen Energy, 32*(7), 761–786. doi:10.1016/j.ijhydene.2006.11.028

Kalib, N. S., Muchtar, A., Somalu, M. R., Ihsan, A. K. A. M., Aman, N. A. M. N., & Mah, J. C. W. (2019). Influence of Heat Transfer on Thermal Stress Development in Solid Oxide Fuel Cells: A Review. *Journal of Advanced Research in Fluid Mechanics and Thermal Sciences, 54*, 175–184.

Kang, Y. W., Li, J., Cao, G. Y., Tu, H. Y., Li, J., & Yang, J. (2009). A reduced 1D dynamic model of a planar direct internal reforming solid oxide fuel cell for system research. *Journal of Power Sources, 188*(1), 170–176. doi:10.1016/j.jpowsour.2008.11.073

Kaya, M. F., Demir, N., Genç, G., & Yapici, H. (2014). Numerically Modeling of Anode Supported Tubular SOFC. *Journal of Applied Mechanical Engineering, 3*, 1000137.

Laribi, S., Mammar, K., Hamouda, M., & Sahli, Y. (2016). Impedance model for diagnosis of water management in fuel cells using artificial neural networks methodology. *International Journal of Hydrogen Energy, 41*(38), 17093–17101. doi:10.1016/j.ijhydene.2016.07.099

Laribi, S., Mammar, K., Sahli, Y., & Koussa, K. (2018). Air supply temperature impact on the PEMFC impedance. *Journal of Energy Storage, 17*, 327–335. doi:10.1016/j.est.2018.03.020

Laribi, S., Mammar, K., Sahli, Y., & Koussa, K. (2019). Analysis and diagnosis of PEM fuel cell failure modes (flooding & drying) across the physical parameters of electrochemical impedance model: Using neural networks method. *Sustainable Energy Technologies and Assessments, 34*, 35–42. doi:10.1016/j.seta.2019.04.004

Luo, Y., Jiang, W., Zhang, Q., Zhang, W. Y., & Hao, M. (2016). Effects of anode porosity on thermal stress and failure probability of planar solid oxide fuel cell with bonded compliant seal. *International Journal of Hydrogen Energy, 41*(18), 7464–7474. doi:10.1016/j.ijhydene.2016.03.117

Manage, M. N., Sorensen, E., Simons, S., & Brett, D. J. L. (2014). A modelling approach to assessing the feasibility of the integration of power stations with steam electrolysers. *Chemical Engineering Research & Design, 92*(10), 1988–2005. doi:10.1016/j.cherd.2014.07.015

Meng, Q., Han, J., Kong, L., Liu, H., Zhang, T., & Yu, Z. (2017). Thermodynamic analysis of combined power generation system based on SOFC/GT and transcritical carbon dioxide cycle. *International Journal of Hydrogen Energy, 42*(7), 4673–4678. doi:10.1016/j.ijhydene.2016.09.067

Meng, Y., Wang, X., Xia, C., Wang, B., Dong, W., Ji, Y., & Zhu, B. (2018). High performance SOFC based on a novel semiconductor-ionic SrFeO3-δ–Ce0.8Sm0.2O2-δ membrane. *International Journal of Hydrogen Energy, 43*(28), 12697–12704. doi:10.1016/j.ijhydene.2018.01.209

Mohammedi, A., Sahli, Y., & Ben-Moussa, H. (2020). 3D investigation of the channel cross-section configuration effect on the power delivered by PEMFCs with straight channels. *Fuel*, *263*, 116713. doi:10.1016/j.fuel.2019.116713

Monsaf, T., Hocine, B. M., Youcef, S., & Abdallah, M. (2017). Unsteady three-dimensional numerical study of mass transfer in PEM fuel cell with spiral flow field. *International Journal of Hydrogen Energy*, *42*(2), 1237–5121. doi:10.1016/j.ijhydene.2016.12.084

Murshed, A. M., Huang, B., & Nandakumar, K. (2007). Control relevant modeling of planer solid oxide fuel cell system. *Journal of Power Sources*, *163*(2), 830–845. doi:10.1016/j.jpowsour.2006.09.080

Nakajo, A., Mueller, F., Brouwer, J., Van herle, J., & Favrat, D. (2012a). Mechanical reliability and durability of SOFC stacks. Part I: Modelling of the effect of operating conditions and design alternatives on the reliability. *International Journal of Hydrogen Energy*, *37*(11), 9249–9268. doi:10.1016/j.ijhydene.2012.03.043

Nakajo, A., Mueller, F., Brouwer, J., Van herle, J., & Favrat, D. (2012b). Mechanical reliability and durability of SOFC stacks. Part II: Modelling of mechanical failures during ageing and cycling. *International Journal of Hydrogen Energy*, *37*(11), 9269–9286. doi:10.1016/j.ijhydene.2012.03.023

Ni, M., Leung, M. K. H., & Leung, D. Y. C. (2007). Parametric study of solid oxide fuel cell performance. *Energy Conversion and Management*, *48*(5), 1525–1535. doi:10.1016/j.enconman.2006.11.016

Peksen, M. (2013). 3D thermomechanical behaviour of solid oxide fuel cells operating in different environments. *International Journal of Hydrogen Energy*, *38*(30), 13408–13418. doi:10.1016/j.ijhydene.2013.07.112

Peksen, M., Al-Masri, A., Blum, L., & Stolten, D. (2013). 3D transient thermomechanical behaviour of a full scale SOFC short stack. *International Journal of Hydrogen Energy*, *38*(10), 4099–4107. doi:10.1016/j.ijhydene.2013.01.072

Piccardo, P., Anelli, S., Bongiorno, V., Spotorno, R., Repetto, L., & Girardon, P. (2015). K44M ferritic stainless steel as possible interconnect material for SOFC stack operating at 600 °C: Characterization of the oxidation behavior at early working stages. *International Journal of Hydrogen Energy*, *40*(9), 3726–3738. doi:10.1016/j.ijhydene.2015.01.083

Saebea, D., Authayanun, S., Patcharavorachot, Y., Paengjuntuek, W., & Arpornwichanop, A. (2013). Use of different renewable fuels in a steam reformer integrated into a solid oxide fuel cell: Theoretical analysis and performance comparison. *Energy*, *51*, 305–313. doi:10.1016/j.energy.2012.12.014

Saebea, D., Patcharavorachot, Y., Assabumrungrat, S., & Arpornwichanop, A. (2013). Analysis of a pressurized solid oxide fuel cell-gas turbine hybrid power system with cathode gas recirculation. *International Journal of Hydrogen Energy*, *38*(11), 4748–4759. doi:10.1016/j.ijhydene.2013.01.146

Sahli, Y., Ben-Moussa, H., & Zitouni, B. (2019). Optimization Study of the Produced Electric Power by SOFCs. *International Journal of Hydrogen Energy*, *44*(39), 22445–22454. doi:10.1016/j.ijhydene.2018.08.162

Sahli, Y., Zitouni, B., & Ben-Moussa, H. (2017). Solid Oxide Fuel Cell Thermodynamic Study. *Cankaya University Journal of Science and Engineering*, *14*, 134–151.

Sahli, Y., Zitouni, B., & Ben-Moussa, H. (2018b). Thermodynamic Optimization of the Solid Oxyde Fuel Cell Electric Power. *UPB Scientific Bulletin, Series B. Chemistry and Materials Science*, *80*, 159–170.

Sahli, Y., Zitouni, B., Ben-Moussa, H., & Abdenebi, H. (2015). Three-Dimensional Numerical Study of the Heat Transfer on the Planar Solid Oxide Fuel Cell: Joules Effect. In I. Dincer, C. O. Colpan, O. Kizilkan & M. A. Ezan (Eds.), Progress in Clean Energy, volume I: Analysis and Modeling (pp. 449-461). Springer.

Sahli, Y., Zitouni, B., & Benmoussa, H. (2018a). Etude numérique tridimensionnelle de l'effet de la température d'entrée des gaz sur la production de chaleur dans une pile à combustible SOFC planaire. *Revue des Énergies Renouvelables*, *21*, 173–180.

Sangtongkitcharoen, W., Vivanpatarakij, S., Laosiripojana, N., Arpornwichanop, A., & Assabumrungrat, S. (2008). Performance analysis of methanol-fueled solid oxide fuel cell system incorporated with palladium membrane reactor. *Chemical Engineering Journal*, *138*, 436–441. doi:10.1016/j.cej.2007.06.021

Skrzypkiewicz, M., Wierzbicki, M., Kupecki, J., & Stępień, M. (2018). Selected Aspects of Design, Construction, and Operation of SOFC-Based Micro-Combined Heat and Power Systems. In J. Kupecki (Ed.), Modeling, Design, Construction, and Operation of Power Generators with Solid Oxide Fuel Cells (pp. 205-231). Springer.

Suwanwarangkul, R., Croiset, E., Pritzker, M. D., Fowler, M. W., Douglas, P. L., & Entchev, E. (2006). Mechanistic modelling of a cathode-supported tubular solid oxide fuel cell. *Journal of Power Sources*, *154*(1), 74–85. doi:10.1016/j.jpowsour.2005.03.197

Tippawan, P., & Arpornwichanop, A. (2014). Energy and exergy analysis of an ethanol reforming process for solid oxide fuel cell applications. *Bioresource Technology*, *157*, 231–239. doi:10.1016/j.biortech.2014.01.113 PMID:24561628

Tucker, M. C. (2018). Dynamic-temperature operation of metal-supported solid oxide fuel cells. *Journal of Power Sources*, *395*, 314–317. doi:10.1016/j.jpowsour.2018.05.094

Wang, L., Zhang, H., & Weng, S. (2008). Modeling and simulation of solid oxide fuel cell based on the volume–resistance characteristic modeling technique. *Journal of Power Sources*, *177*(2), 579–589. doi:10.1016/j.jpowsour.2007.10.051

Wongchanapai, S., Iwai, H., Saito, M., & Yoshida, H. (2012). Selection of suitable operating conditions for Planar Anode-supported Direct-internal-reforming Solid-Oxide Fuel Cell. *Journal of Power Sources*, *204*, 14–24. doi:10.1016/j.jpowsour.2011.12.029

Yang, S., Chen, T., Wang, Y., Peng, Z., & Wang, W. G. (2013). Electrochemical Analysis of an Anode-Supported SOFC. *International Journal of Electrochemical Science*, *8*, 2330–2344.

Yang, W., Zhao, Y., Liso, V., & Brandon, N. (2014). Optimal design and operation of a syngas-fuelled SOFC micro-CHP system for residential applications in different climate zones in China. *Energy and Building*, *80*, 613–622. doi:10.1016/j.enbuild.2014.05.015

Zhang, X., Li, G., Li, J., & Feng, Z. (2007). Numerical study on electric characteristics of solid oxide fuel cells. *Energy Conversion and Management*, *48*(3), 977–989. doi:10.1016/j.enconman.2006.08.011

Zhang, X., Su, S., Chen, J., Zhao, Y., & Brandon, N. (2011). A new analytical approach to evaluate and optimize the performance of an irreversible solid oxide fuel cell-gas turbine hybrid system. *International Journal of Hydrogen Energy*, *36*(23), 15304–15312. doi:10.1016/j.ijhydene.2011.09.004

Zhang, X., Wang, Y., Guo, J., Shih, T. M., & Chen, J. (2014). A unified model of high-temperature fuel-cell heatengine hybrid systems and analyses of its optimum performances. *International Journal of Hydrogen Energy*, *39*(4), 1811–1825. doi:10.1016/j.ijhydene.2013.11.027

Chapter 9
Hydrogen Energy Use in Comfort and Transport

Constantin Pana
University Politehnica of Bucharest, Romania

Maria Alexandra Ivan
University Politehnica of Bucharest, Romania

Alexandru Cernat
University Politehnica of Bucharest, Romania

Niculae Negurescu
University Politehnica of Bucharest, Romania

Cristian Nuțu
University Politehnica of Bucharest, Romania

Teodora Madalina Nichita
Technical University of Civil Engineering of Bucharest, Romania

Gabriel Ivan
Technical University of Civil Engineering of Bucharest, Romania

ABSTRACT

This chapter presents aspects of hydrogen use in comfort systems and thermal machines in order to improve performance and to reduce pollution emissions. Hydrogen energy is clean, and its use leads to a reduction of CO2 emissions into the atmosphere. It represents an alternative to traditional fuels, oil, coal, and gas. The use of hydrogen preserves traditional fuel resources. In the field of comfort, the energy obtained from hydrogen is applicable in the heating and air conditioning of spaces, in the production of electricity necessary to create light comfort. The use of hydrogen in boilers leads to the reduction of pollutant emissions. Hydrogen fuelling systems were designed for different experimental thermal machines, designed by authors from spark-ignition engine and compression ignition engine. The characteristic combustion parameters, like maximum pressure, the maximum rate of pressure rise, efficiency, and pollutant emissions for hydrogen fuelling are presented and analyzed.

DOI: 10.4018/978-1-7998-4945-2.ch009

INTRODUCTION

Energy Crisis and Resources

The global energy crisis has forced the finding of new resources given the fact that the traditional one's coal, oil and gas are limited. The specialists' searches in this field have led to the development of a new concept that being the sustainable development concept.

This concept involves mostly the use of green energies in order to protect the environment and to reduce the CO_2 emissions, so comfort can be achieved for both residential and daily activities buildings. With many unclear elements during the crystallization period of the concept, today's new direction of sustainable development started in Europe and it was immediately successful all over the world, from the USA to Japan.

This approach is a reaction to the chaotic development and urban agglomeration that exists nowadays in many cities around the world, with a huge waste of energy, but without clear principles that regulate the development of new and existing settlements.

The main beliefs of the concept nowadays are:

- Local environment conditions must be considered when building constructions, which must not be affected.
- In industrial applications as well as for comfort there must be used mainly energies from renewable ecological sources as solar, wind, air, earth and water.
- New architecture will be adapted to new ecological requirements, private renewable sources. The buildings shall be fitted with glazed surfaces so oriented as to allow the capture of solar energy for comfort. The architecture of the building will allow the placement of systems for the use of solar energy.
- In villages, the existing buildings must be upgraded using renewable energies for comfort, as for the waste energy sources, in local industries, there must be used for comfort but with extensive environmental protection measures.
- The transport's development must be made in new, modern directions, using clean energies, such as electricity, hydrogen.

At European level, different programs have been applied and implemented providing for energy rehabilitation measures for existing buildings and new rules to increase thermal resistance for new ones. Few examples are the EPBD 2010 (Energy Performance of Building) and ErP Directive 2016 (Energy-related-products-directive). Romania has aligned with European requirements, building a legislation framework that fosters the use of clean energies. Regarding these energies, including the use of H_2, it was drafted the Law No. 34/2017, which envisages the construction of an infrastructure for the use of alternative fuels. The law regulates the measures aimed at installing alternative fuel infrastructure, with the aim of reducing dependence on oil and thereby mitigating the environmental impact of transport. Requirements are laid down for the creation of alternative fuel infrastructure, recharging points for electric vehicles, compressed gas refuelling, liquefied gas and hydrogen (Mateescu, 2017), (Badea, 2017).

The energy used today in comfort systems is currently obtained from the combustion of fuels, fact which leads to a significant intake of CO_2 in the environment. Reducing the CO_2 emissions is a national, European, and worldwide important objective. The use of energy from renewable sources, including H_2

represents a permanent preoccupation for every specialist. In the last 70 years, research institutes and colleges from the higher education institutions in Bucharest, Cluj-Napoca, Iasi, Timisoara and Brasov, Romania have had particularly interesting results with serial production applications to the user on renewable energies (Ivan, G., 2017).

One of the methods in order to reduce these emissions is to adopt H_2 in energy production systems used in heating, air conditioning, cooling, electricity generation systems to achieve light comfort etc. So basically, adapting the comfort systems to new methods of energy production using H_2 so the comfort in people's houses/buildings would continue to be but without so much pollution. The energy obtained from hydrogen, is one of the "clean" energies, like the ones from renewable energy. It is not always the current technological level that allows us to use renewable energies in an economically and efficient way. The U.S. Department of Energy in New York in 2017 shows some economic shortcomings in renewable energy production, regarding the number of employees in each branch relative to each percentage contribution to total energy produced. In this way, by 2016, the U.S. levels were as follows.

Table 1. The U.S. Energy producer per sector in 2016

Energy producer sector	Number of employees	Energy produced per employee MWh	Percentage share of total electricity produced in the U.S.
Natural gases	398000	3812	33,80%
Coal	160000	7745	30,40%
Wind	100000	836	5,60%
Sun	374000	98	0,90%

Since wind and sun energies are only available intermittently about 15-30% of the year and must be doubled by systems based on fossil fuels, as well as the fact that the energy obtained from gas and coal is cheaper than the other two, we can find that currently obtaining energy from these sources in unprofitable. These inconveniences do not occur when using H_2.

Currently, energy generation technologies using free renewable energy sources are not of a nature to meet the requirements of sustainable development.

Considerations Regarding Innovative Design of H_2 Systems for Comfort

In general, to heat different spaces and to ensure comfort conditions in warm and cold season, there are several situations that depend on the consumer. Thus, for the household consumer, the energy needs are being covered by a low-power thermal power plant and small air conditioning units, while for large administrative buildings, office buildings or even industrial spaces, the thermal powers needed for heating, air conditioning or cooling are high and require significant fuel flows.

For small spaces, apartments, residential buildings, where there are no technical spaces for the location of the equipment, the comfort systems will be designed according to the principles of innovative design, interweaving the technical function with the aesthetic aspect. Research in the field of new materials provides composite materials with appropriate properties, which allow the production of devices with a high degree of compactness, pleasant appearance, and technical characteristics necessary to achieve comfort.

Innovation is the principle key in many industries and sectors. It has been used and applied for many years, but without calling it so. It is innovation through which things are being developed and adapted to the new requirements, requests, as well as consumers 'preferences. Due to that, systems have been adapted to new and smaller dimensions, keeping the elegance and fineness so all the needs could be fulfilled.

Comfort can be discussed and considered from different points of view, still, innovation has played an important role with technology in order to create comfort for users. During the years, more and more problems have been observed and discovered in people's houses. As part of the innovation process, research is the start of most of the processes and so it has been researched the need to increase the comfort level in people's houses. Moreover, research played an important role for the development of technology and so we reached a level of diversity that things have been created for certain types of houses, for different target groups, including the people with different diseases. Currently, through technology we can discuss about compact and customized devices, for which users can have the final decisions regarding the design of the devices and the comfort needed. Besides comfort, innovators have created beautiful scenes for the users in their homes. So, after making the decisions regarding their comfort and needs, users can play with the interfaces of the devices according to their personality and desires. That can be accomplished by a combination between a technological comfort and modern decorative design for those systems. Decorating houses, it is an important step, because through the final touches, decorations, your house can become very comfortable, so innovation took form by hiding and blending itself with its surroundings. Still, all these processes, rules and regulations have been strictly monitored and respected to achieve comfort norms and requirements. The norms refer to the parameters of comfort as in the interior temperature, relative humidity, interior light and the noise level. The systems have been designed for a specific destination, activity or environment in such a way that would suit the entire people that live and work in those specific buildings. (Ivan, M.A., 2015).

Small Consumers

For small consumers, the question arises of the use of H_2 as fuel in the wall thermal power plants, process that involves either the existence of a fuel reserve, as the bottled H_2, or the local generation of H_2 that will be immediately used in the thermal power plant.

The first solution brings to the attention, the problem of the type of space cylinder used, that is because most of the time, the user does not have a special technical space for the location of the thermal power plant of real estate or apartment. This reserve must be safe to operate, easy to handle and easy to connect to the installation. Moreover, the user must have a place from where to procure the fuel cylinders, at a price that makes this fuel attractive.

Generating locally H_2 is related to the way in which the process is made so that it can ensure the flow required by the thermal power plant. The generator system and boiler must be usable in small spaces offered by apartments or single-family houses.

For the use of H_2 by small users were obtained results such as: Viessmann and Panasonic developed a new concept of boiler based on a "cold combustion" which does not exist in conventional boilers, the CO_2 emission level and dependency of fossil felt being eliminated. These appliances eliminate CO_2 emissions, which are common for appliances that run on traditional fuel.

The "Pure Energy Centre" introduced at the All Energy Conference UK 2012 a new type of Hydrogen Boiler (installed at the Fife Innovation Centre from Methil, Fife) which uses hydrogen. The manager of the Pure Energy affirms that the boiler operates like any other natural gas boiler, being designed for radiant heating (under-floor and space heating) and the combustion inside the boiler was achieved without a flame or any dangerous emissions due to the special catalysts that operate at low temperatures, without CO_2.

The use of H_2 as fuel in cooling system encounters the same problems as those in heating systems, because H_2 combustion must be carried out in the appliances (boilers) in which the solutions are boiled.

Large Consumers

As for the large consumers, the fuel supply is large, proportional to the installed thermal power of the boilers. The reserve must ensure the operation of the thermal power plant for a period, depending on the fuel flows required for boilers. In this case, the problem of storage space and safety measures arises and must be considered.

When it comes to industrial spaces, where H_2 results from technological processes as a residue, its local use can be achieved. A situation like this is the Kawasaki Thermal Engineering Co., Ltd. Company that created the design of a conventional boiler that uses waste hydrogen with no CO_2 emissions during combustion process. This type of new boiler was used in a plant from Japan, Nankai Chemical Industry's Kozaika Plant, the necessary fuel, the waste hydrogen being provided by the factory and being used as fuel for the boiler, without being anymore released into atmosphere. Finally, 1,090.000 m^3N of natural gas was saved, fact that represents the equivalent of 2500 tons per year of CO_2 emission reduction.

For replacement of other fuels is quite difficult to achieve immediately because the volume of gaseous fuel currently in use is extremely high and the production of H_2 is not commensurate with consumption. Moreover, the conditions of the use of the new fuel are not created either. Due to that, there has been raised the possibility of creating a mixture between H_2 and natural gases.

It is known the tendency of H_2 to form sperate volumes in a mixture, process that is not stable (it does not favour a stable combustion) sometimes being even explosive. It is necessary to use technical systems that produce a homogeneous mixture of components.

Another problem is that many local natural gas usage and distribution networks are being built from HDPE pips, material that is not compatible with H_2. In this situation the use of H_2 can only be achieved by using a local system, to the consumer, in order to produce a mixture directly to the user, without using the distribution networks.

The research shows an improvement in fuel quality and a reduction in CO_2 emissions. In HyDeploy project developed at Keele University from Staffordshire, it has been tested the usage of 20% of hydrogen and 80% natural gas in blend to fuel for a boiler, this phase program being developed till 2025. All condensing gas boilers manufactured after the 1996 can be fuelled with gas blends that contain up to 23% hydrogen. During the Hy4Heat Project, the possibility of a 100% hydrogen network use is being tested. Project workers affirm that if there are already hydrogen automotive travelling on roads, then, theoretically, a hydrogen network is developed and a majority of modern boilers will be modified in order to operate with 100% hydrogen, in such a way that it limits the financial costs and trainings for operations of installing and exploitation. These aspects are related to the fact that UK must decrease its carbon emissions with 80% by year 2050, when compared to the carbon levels of the 1990s. Thus, the usage of fuels constituted from natural gas and hydrogen blends, can significantly decrease the carbon emissions. The use of H_2 as fuel in cooling systems encounters the same problems as those in heating systems, because H_2 combustion must be carried out in the appliances (boilers) in which the solutions are boiled.

Given the wide development of electricity distribution networks, it is probably not of immediate interest to create a local electricity generation system, which involves the creation of small-scale appliances with which to carry out a Rankine cycle. A direction of particular interest in the use of H_2 in the large existing thermoelectric power plants, components in national energy systems.

CONSIDERATIONS REGARDING THE USE OF H_2 IN TRANSPORT DEVELOPMENT

Environment protection preoccupations and fossil fuel reserves reduction leads to ample research to find alternative fuels suitable for thermal machines. To decrease the level of pollutant emissions and of greenhouse gases, the high carbon content fossil fuel can be replaced in burning process by alternative fuels. Between the alternative fuels, hydrogen is considered as a clean viable fuel, knowing that fossic fuels have longer energetically cycles comparative to hydrogen cycle. The use of hydrogen in the field of thermal machines can represent a real chalenge which requires solutions to many issues, but may represents a real efficient strategy used for efficiency improvement and pollution reduction. Due to its properties, table 1, hydrogen is a viable clean alternative fuel for thermal machines, with unlimited producing resources, (Nutu, 2019), (Cernat, 2016). The value of hydrogen specific volume is very high and may leads to the reduction of in-cylinder air once with the increase of hydrogen share. For example, 29% of volume is taken by hydrogen in a stoichiometric mixture with air, this issue leads to engine power decreases. Is difficult to store high quantities of hydrogen at automotive board because hydrogen has a low density and high pressure that is required for gas state store. The temperature of hydrogen liquefaction is low and the design of fuelling system represent major problems. Hydrogen diffusion velocity is very high value and may bring issues at hydrogen pass through thin jets and apertures or at mixing with other gases, (Mirica, 2015), but leads to rapid forming of homogeneous air-hydrogen mixture. Versus hydrocarbon-based fuels, hydrogen assures better combustion and engine efficiency improvement, due to hydrogen raised combustion speed, large flammability area, reduced ignition energy etc. The very wide flammability domain of hydrogen is an advantage for ICE use because the high efficiency operation at lean and very lean mixtures can be assured. Combustion velocity of hydrogen is with one order of magnitude higher comparative to other fuels, the combustion of air-fuel mixture being accelerated with hydrogen. The air-hydrogen mixture ignites very easy at contact with hot gases and surfaces. Thus, uncontrolled ignition of air-hydrogen mixture and of combustion may appear, even with phenomena of flame return into intake pipe or with the uncontrolled in-cylinder ignition. The use of lean and very lean air-hydrogen mixtures can eliminate these issues.

Hydrogen doesn't pollute, it is not toxic, and its combustion don't produce pollutant substances (HC, CO, SO_2, particulate matter-PM, ozone) and other carcinogenic substances because hydrogen carbon and Sulphur free (Nutu, 2019), (Mirica, 2015).

Higher temperatures reached at hydrogen combustion may lead to higher level of NO_x emissions, especially in the domain of stoichiometric dosage value, comparative to classic fuel engine. The use of the lean dosage mixture can eliminate this issue (Pana, 2011). In case of hydrogen use at diesel engine, (Nutu, 2019), (Cernat, 2018), the high autoignition resistance of hydrogen requires the use of an ignition source, (Negurescu, 2012), (Pana, 2011), (Sorensen, 2018). For diesel engine hydrogen can be use as

Table 2. Properties of hydrogen and diesel fuel (Negurescu, 2012), (Popa, 2003)

Property		Diesel	Gasoline	Hydrogen
Fuel formula		$C_{14.5}H_{30}$	C_8H_{15}	H_2
Molecular weight [kg·kmol^{-1}]		204	114	2.016
Density [kg/m^3] (0 [°C] and 760 [mmHg])		825-870	0.735-0.760	0.0899
Lower Heating Value, LHV [MJ/kg]		42.5	42.69	119.6
Temperature of autoignition [K]		473...493	753-823	858
R-Octane Number [-]		30	90-98	130
Cetane Number [-]		45-55	-	-
Flame speed [m/s^{-}] (a-f mixture λ=1, 20 [°C], 760 [mmHg])		0.3	0.12	2.37
Limits of flammability	$\lambda_s \ldots \lambda_i$	0.34-1.68	0.709-1.1	0.136…10.12
	$[(\%)_{vol}]$	0.7-5	1.48-2.3	4.1-75.6
A/F ratio, (kg air/kg fuel) [-]		14.76	14.5	34.32
Minimum ignition energy [mJ]		-	0.2…0.3	0.018
Diffusivity [cm^2/s]		-	-	0.63
Boiling temperature [K]		436-672	-	20.27
Adiabatic temperature of the flame [K]		2327	2411	2300

secondary fuel by applying the diesel-gas method. This fuelling method is simply to apply and assures good homogeneity of in-cylinder mixture, the economic engine operation with less smoke emission level can be assured. This method requires the injection of gaseous hydrogen into the engine inlet manifold, so high homogeneity mixture of air and hydrogen established in cylinder. In order to avoid major modifications in diesel engine design, the hydrogen can be easily added into the inlet air. The following are the most important results of hydrogen use in thermal machines.

Hydrogen Influences the Combustion Pressure

Higher combustion speed of hydrogen into a homogenized mixture (Negurescu, 2012), influences the heat release rate during combustion, the peak pressure and the rise rate of pressure starts to rise with the increase of hydrogen quantity that substitutes the classic fuel.

Depending on fuelling system and hydrogen inlet section, it can appear a zone into combustion chamber of "unused" oxygen, the oxygen distribution inside all combustion chamber being affected (Sorensen, 2018).

The peak pressure and the rise rate of pressure can be limited by decrease of diesel fuel injection timing or spark timing (Negurescu, 2012), (Cernat, 2016), (Pantile, 2012).

Hydrogen Influences the Brake Thermal Efficiency

Improved combustion properties at hydrogen use at internal combustion engines leads to brake thermal efficiency increases versus classic fuel use (Nutu, 2019), (Mirica, 2016), (Pana, 2017), (Sorensen, 2018), the LHV of H_2 being high. Comparative to classic fuel, higher combustion speed of hydrogen assures a short combustion duration and a high thermal efficiency, especially at partial engine loads operation (Neguresu, 2008).

Hydrogen Influences the Fuel Consumption

The brake thermal efficiency starts to rise at the increase of classic fuel substitution by hydrogen, the brake specific fuel consumption starts to decrease (Pana, 2005), (Mirica, 2015).

Hydrogen Influences the Pollutant Emissions

Hydrogen fuelling leads to the decrease of smoke emissions level as effect of carbon percent decrease and acceleration of combustion for classic fuel substitution (Negurescu, 2011). Only for high combustion temperature, above value of the NO_x formation, the NO_x emissions level increases at hydrogen quantity rise, at high engine loads (Negurescu, 2011), (Mirica, 2015). According to some authors results, the NO_x emissions level decreases when the engine operates at partial loads using small quantities of hydrogen injected into the inlet air (Pana, 2007), (Negurescu, 2011), (Rusu, 2011), (Mirica, 2015).

The paper presents some experimental aspects of hydrogen use in thermal machines.

EXPERIMENTAL INVESTIGATION

The purpose of the study is the use of hydrogen in order to substitute, even partially, the classic fuels, by employing as an advantage the improvements of the thermal machines' performance. Thus, this aim of research can be achieved thru experimental investigation carried on establishing the suitable hydrogen quantity in correlation with the operating regime of thermal machine-ICE, combustion parameters, efficiency and pollutant emissions level. For SIE (Spark Ignition Engine) hydrogen use study, the authors designed an experimental single cylinder engine, derived from 810.99 Dacia engine, fuelled with hydrogen, which is direct injected into the cylinder, by a special fuelling system designed by authors, figure 1. The fuelling system allows hydrogen to be used as a single fuel or as a secondary fuel in a dual fuelling mode-gasoline and hydrogen addition.

A reservoir bottle type provides hydrogen supply. The pressure reducers are used to adjust hydrogen fuelling pressure, the fuelling pressure being decreased from 150 bar to 10 bar, from H_2 reservoir to fuelling line, the fuelling pressure in front of the hydrogen valve being set up in the area of 1-10 bar (Negurescu, 2012).

For CIE (Compression Ignition Engine) hydrogen use study, the authors design a hydrogen fuelling system based on the diesel-gas method, applied for K9K engine developed for Dacia Logan MCV, figure 2.

The experimental test bed for hydrogen use in CIE is composed from the following components: HR-hydrogen reservoir, PR-pressure regulators, HF-Alicat Scientific hydrogen flowmeter, FA-flame arrestor, HI-hydrogen injector, DFR-diesel fuel reservoir, DFF-Optimass diesel fuel flowmeter, DFI- diesel fuel injectors, S-servomotor of acceleration pedal, E-engine K9K, C-coupling, D-dyno Schenck E90, DCC-dyno control cabinet, SPU-servomotor power unit, ECU-HI- electronic control unit of hydrogen injectors, C-ECU-HI- computer controller of electronic control unit of hydrogen injectors, DAS- data acquisition system, CA-AVL charge amplifier, pt-AVL pressure transducer, ae-angle encoder, TI-temperature indica-

Figure 1. Hydrogen use for fuelling a SIE: a)-hydrogen direct injection valve design; b)- time period of hydrogen fuelling: HO-hydrogen valve open, HC-hydrogen valve close (Negurescu, 2012)

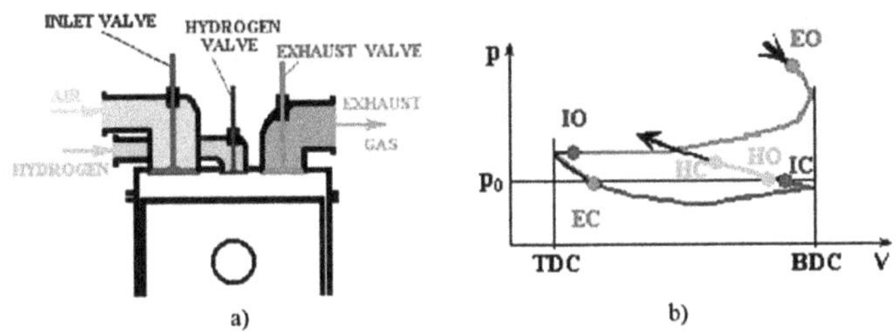

tors, GA-AVL DiCom 4000 gases analyser, AF-air flowmeter, AR-air reservoir, TC- turbo-compressor, HS- hydrogen sensor, HSU-hydrogen sensor unit.

The hydrogen injector is electronic actuated and cotrolled by a Dastek Softer trhtu a open ECU unichip type, thence the hydrogen and diesel fuel cyclic quantities are setup for similar power of classic fuelling. The hydrogen cyclic quantity is defined by xc [%] as energetic replacement percent of diesel fuel.

RESULTS

For hydrogen use as a single fuel in SIE the results at the operating regime of 3000 rev/min and WOT-full load, are presented. Spark timing was setup for maximum power. Comparative to classic engine setups, the spark ignition timing was reduced at hydrogen use in order to compensate higher combustion rate of hydrogen. At the same dosage of λ=1, the maximum pressure riches higher values for hydrogen

Figure 2. Diesel engine experimental test bed

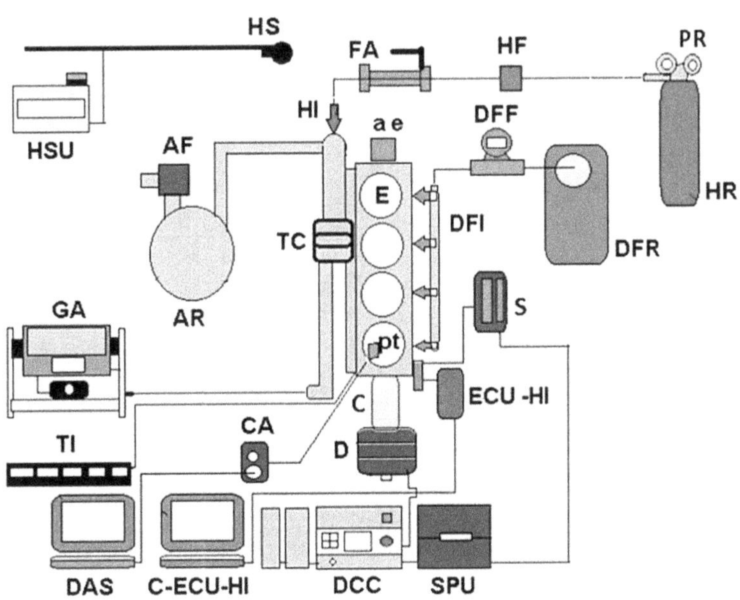

fuelling comparative to petrol engine, figure 3a, because the hydrogen combustion speed is superior to gasoline. At hydrogen fuelling the heat release rises with 25%, in correlation with a 30% increase in peak pressure (Neguresc, 2012). The molar chemical shrinking (MCS) at hydrogen combustion can represent a reduction factor of increase in maximum rate of pressure rise. At $\lambda = 1$ combustion condition, the MCS=0,85 for hydrogen and MCS=1,05 for gasoline, the molar expansion process being present. For stoichiometric dosage, at H_2 use, the rise rate of pressure is higher comparative to classic fuelling, figure 3b, due to a higher H_2 combustion speed that leads to lower combustion periods (Negurescu, 2012). Anyway, at hydrogen fuelling, the increase of peak value for rise rate of pressure, $(dp/d\alpha)_{max}$, do not significantly outweighs the normal range and spark timing adjustment can control the rise of $(dp/d\alpha)_{max}$. For lean dosages $\lambda>2$, a lower pressure rise rate during combustion is registered. To control the engine operation at hydrogen fuelling, the qualitative adjustment of the engine load was use (Negurescu, 2012). Thus, in this way, the efficiency is superior at hydrogen fuelling for $\lambda = 1$, figure 3c, due to much better combustion properties of hydrogen (Pana, 2011). The stable operation of classic engine can't be assured for $\lambda>1.3$, but for H_2 fuelling, collaborated with qualitative adjustment of engine load, the hydrogen engine can operate with dosage values till $\lambda\sim5.5$. For $\lambda=2\ldots4$ the efficiency, as brake specific fuel consumption (BSEC) is improved, figure 3c, by acceleration of the combustion process at H_2 use, (Pantile, 2012; Popa, 2003). For $\lambda > 3.2$, the BSEC is affected by the rise combustion period, figure 3c. In terms of pollution, the H_2 engine do not emits CO, CO_2, HC, PM. Reduced quantities of HC may appears in exhaust gases from the oil which may burn into cylinder. The NO_x emission may be supperior to gasoline fuelling in corelation with raised temperature of combustion at hydrogen fuelling. This issue leads to the rise of the NO_x level for $\lambda=1\text{-}2$, figure 3d.

Figure 3. Operating regime 3000 rev/min, WOT: a)- maximum pressure versus lambda; b)-maximum pressure rise rate versus lambda; c)-BSEC versus lambda; d)- NO_x emission level versus lambda

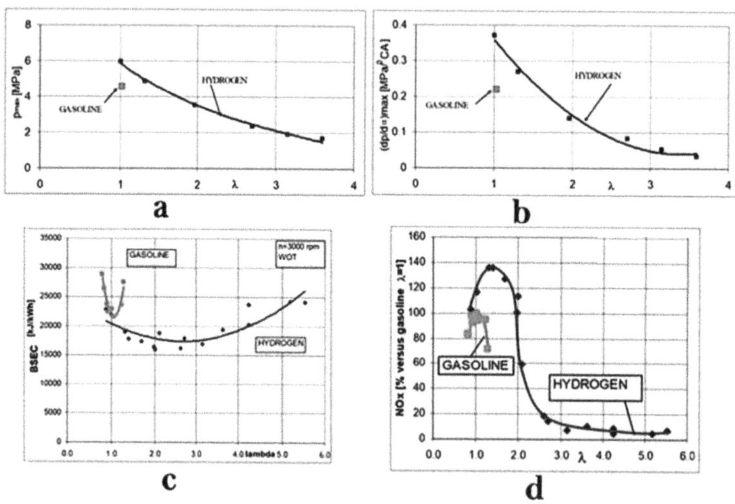

For $\lambda>2$, the NOx is significantly reduced, figure 3d. At $\lambda=1\text{-}2$, the use of catalytic converters may decrease the level of NO_x.

For hydrogen use in CIE the results for speed of 2000 rev/min and 40% load are presented. In figure 4a are presented averaged pressure diagrams for classic fuelling (*xc*=0) and for hydrogen addition (*xc*=5% and *xc*=8.7%). Values of maximum pressure and maximum pressure rise rate start to rise for hydrogen addition, figures 4b and 4c, due to the rise of fuel amount burned into the premixed stage. Comparative to classic fuelling, the maximum pressure rises with 4,4% and the maximum pressure rise rate increase with 8,4%, without exceed the normal values domain with assure the normal and reliable engine operation (Negurescu, 2012), (Sorensen, 2018), (Pantile, 2012). The raised LHV and combustion speed of H_2, the increase of thermal efficiency the brake specific energy consumption starts to decrease at partial engine loads operation.

Figure 4. Operating regime 40% engine load and 2000 rev/min speed: a)- pressure diagrams for different xc; b)-maximum pressure versus xc; c)-maximum pressure rise rate versus xc; d)-BSEC versus xc; e)- NO_x emission level versus xc; d)- smoke number versus xc

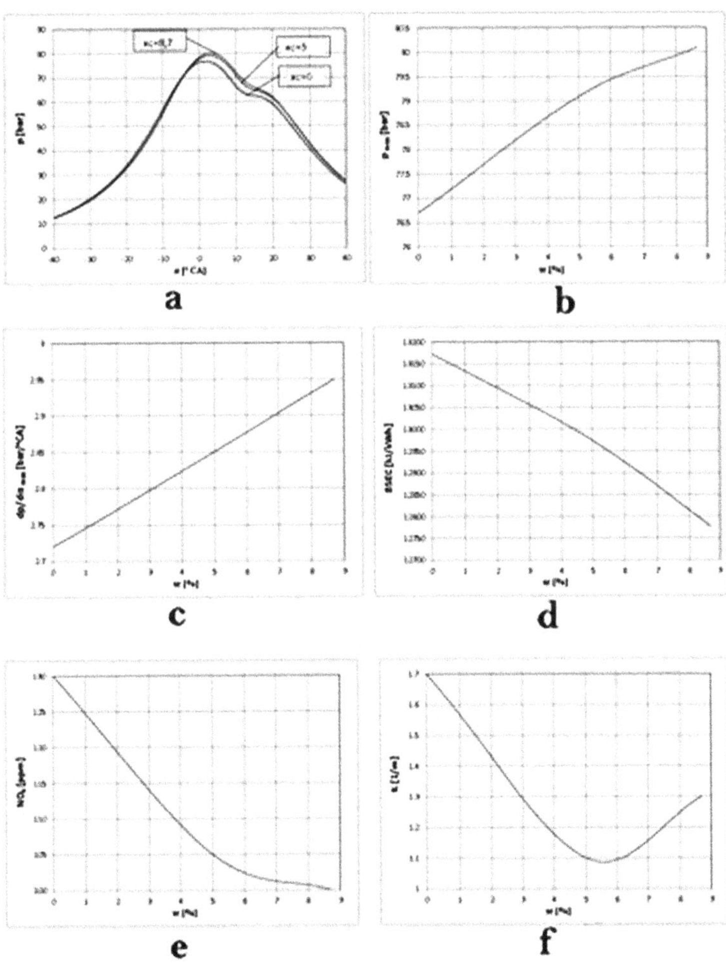

At the use of small quantities of hydrogen, the BSEC decrease with 3% comparative to classic fuel operation, figure 4d. The NO_x and smoke emission are the most important pollutant emission of diesel engine. At hydrogen fuelling the NO_x emissions level is reduced by 23%, at maximum xc, figure 4e. For small hydrogen quantities use in addition in air, hydrogen burns fast and the temperature increases, but the NO_x formation being avoided due to acceleration of the combustion ("*the combustion is so rapid that the high temperatures exist only for approx. 2 [ms]*" after Georgios Pechlivanoglou), (Pechlivanoglou, 2007). The increase of water content in reaction products at H_2 combustion may assures the decrease of NO_x emission level (Pana, 2016).

Even for low hydrogen cyclic dose the NOx emission level is decreased with 19% versus classic fuel use. The smoke emission level, defined by K smoke number, is reduced because the carbon content is decreased, and the combustion process is accelerated. The smoke number decreases with 35% and 23% for hydrogen cyclic doses of 5% and 8.7%, respectively, figure 4f.

FUTURE RESEARCH DIRECTIONS

In the field of comfort, the use of H_2 brings a new source of energy, clean, unpolluted, but whose production and use requires the creation of new technical systems, innovative, with which to be used in complete safety. As in the use of natural gas, and for the use of hydrogen, rules must be made, normative, under which comfort systems can be designed, used and operated. All this puts in front of researchers in the field from Romania and Europe some directions of work with theoretical and applicative character of great interest. In the field of transport, of great interest, our research aims to study the development of processes in different cases and to improve the procedures and technology of using H_2 in thermal engines.

CONCLUSION

Industrial gas companies, automotive players, electric utilities, materials/suppliers have existing roles when it comes to hydrogen, still these will search to adapt to new situations/opportunities in sustainable hydrogen market. It can already be discussed as in hydrogen market since this is a new way of saving energy and lots of companies are looking into new discoveries in this market.

Innovation is looking into new methods of reducing costs of production and different ways of applying faster this new type of obtaining energy.

The routes of hydrogen production are assessed, analysed through different sustainability concepts, in order to have efficiency in energy, costs, global warning potential.

The use of hydrogen in ICE can alleviate the process of combustion. For SIE, if engine load is qualitative tuned and the hydrogen is used as fuel, the engine efficiency is improved, at small engine loads. At hydrogen combustion, emissions of CO_2, CO, HC, PM are not present, but NOx emission level was increased for $\lambda=1$-2, at the investigated regime, in correlation with the rise of the combustion temperature. Thus, the NOx reduction can be accomplished by the use of catalytic converter, spark timing tune or EGR. The advantage of H_2 direct injection is reflected in the rise of engine power with 29%, the volumetric efficiency in terms of intake air being maintained. This fuelling method allows operation at $\lambda=1$, free of anomalous combustion. For CIE, hydrogen fuelling leads to the increase of maximum pressure with 4,4% and of the maximum pressure rise rate with 8,4% at the investigated regime, the decrease of BSEC with 3% due to increase engine thermal efficiency, the decrease of NO_x emissions level with 23%, the decrease of the smoke emission level, evaluated by K smoke number, with 35%. Thus, hydrogen may be defined as a viable alternative fuel which can be used to fuel ICE, or other thermal machines, with improved performance in terms of efficiency and pollution emission, without major design changes.

REFERENCES

Cernat, A., Lazaroiu, G., Pana, C., Negurescu, N., & Mirica, I. (2016, June). *Experimental Investigations of the Automotive Diesel Engine Fuelled with Hydrogen* [Poster presentation]. The 2nd South East European Conference on Sustainable Development of Energy, Water and Environment Systems, Piran, Slovenia.

Cernat, A., Pana, C., Negurescu, N., Fuiorescu, D., Nutu, C., & Mirica, I. (2018). Experimental Aspects of Hydrogen Use at Diesel Engine by Diesel Gas Method. *Thermal Science*, *22*(3), 1191–1200. doi:10.2298/TSCI170314138C

Cernat, A., Pana, C., Negurescu, N., Nutu, C., & Mirica, I. (2016). Experimental investigations of the hydrogen use at the automotive diesel engine. *Journal of Automotive Engineering*, *22*(41), 22–25.

Cernat, A., Pana, C., Negurescu, N., Nutu, C., Mirica, I., & Fuiorescu, D. (2016). Effect of Hydrogen Use on Diesel Engine Performance. In *Proceeding of 20th Innovative Manufacturing Engineering and Energy Conference (IManEE 2016), IOP Conference Series: Materials Science and Engineering* (vol. 161, no. 1). 10.1088/1757-899X/161/1/012075

Dodds, E. P., Staffell, I., Hawkes, A., Li, F., Grünewald, P., McDowall, W., & Ekins, P. (2014). Hydrogen and fuel cell technologies for heating. *RE:view*. Advance online publication. doi:10.1016/j.ijhydene.2014.11.059

Ivan, G., Badea, G., Stefanescu, I., Mateescu, I., Ivan, M. A., & Mateescu, O. (2017), From the ancient Prometheus to the modern Prometheus. *The 27th International Conference – BSEE 2017 Building Services and Energy Efficiency*

Ivan, M. A. (2015). Innovation and systems for economizing the production of energetic comfort. *World Renewable Energy Congress*.

Mateescu, I., Badea, G., Ivan, G., Aschilean, I., Mateescu, O., & Constantin, G. (2017). *Hidrogenul şi legislaţia romaneasca. Ştiinţa Modernă şi Energia.*

Mirica, I., Cernat, A., Pana, C., Negurescu, N., & Nutu, C. (2015). Performance Comparison Between Hydrogen and Diesel Fuel Fuelled Compression Ignition Engine. *Scientific Bulletin of University Politehnica of Bucharest, Series D. Mechanical Engineering*, *77*(4), 1–12.

Mirica, I., Cernat, A., Pana, C., Negurescu, N., & Nutu, C. (2016, October). *Aspects of Experimental Research on Hydrogen Fuelled Automotive Diesel Engine. The International Congress of Automotive and Transport Engineering, CONAT 2016*, Brasov, Romania.

Mirica, I., Pana, C., Negurescu, N., Cernat, A., & Nutu, C. (2015). Aspects of the Hydrogen Use at a Truck Compression Ignition Engine. In *Proceedings of the 4th International Conference of Thermal Equipment, Renewable Energy and Rural Development.* Politehnica Press.

Mirica, I., Pana, C., Negurescu, N., Cernat, A., & Nutu, C. (2016). Experimental investigations of the hydrogen addition effects on diesel engine performance. *Journal IOP Conference Series: Materials Science and Engineering, 147*(1). . doi:10.1088/1757-899X/147/1/012121

Negurescu, N., Pana, C., & Cernat, A. (2011, November). *Aspects of using hydrogen in SI engine* [Paper presentation CIEM2011-ID S4_23]. The 5th International Conference on Energy and Environment -under the patronage of World Energy Council, University Politehnica of Bucharest, Bucharest, Romania.

Negurescu, N., Pana, C., & Cernat, A. (2012). Aspects of using hydrogen in SI engine. *Scientific Bulletin of University Politehnica of Bucharest, Series D. Mechanical Engineering*, *74*(1), 11–20.

Negurescu, N., Pana, C., Popa, M., & Cernat, A. (2009). Theoretical and Experimental Researches on a Hydrogen Fueled SI Engine. In *Proceedings of the 8th International Conference, Fuel Economy, Safety and reliability of motor vehicles, ESFA 2009, under the patronage of FISITA, EAEC, SIAR* (vol. 1, pp. 39-48). Academic Press.

Negurescu, N., Pană, C., Popa, M., & Cernat, A. (2010). Experimental Researches of a Hydrogen Fuelled SI Engine. *Journal of Energy and Power Engineering*, *4*(7), 1–8.

Negurescu, N., Pana, C., Popa, M., & Cernat, A. (2011). Performance comparison between hydrogen and gasoline S.I. fuelled engine. *Journal of Thermal Science*, *15*(4), 1155–1164. doi:10.2298/TSCI110203090N

Negurescu, N., Pana, C., Popa, M., & Cernat, A. (2011). Experimental and Theoretical Researches of Hydrogen Use in S. I. Engine. *Journal of Environmental Protection and Ecology*, *12*(4A), 2275–2287.

Negurescu, N., Pană, C., Popa, M., Cernat, A., & Soare, D. (2007). Some Characteristics of the Hydrogen Combustion in SI Engine. In *Proceedings of 6th International Colloquium "FUELS 2007" (vol. 1*, pp.461-466). Technische Akademie of Esslingen.

Negurescu, N., Pana, C., Popa, M., Cernat, A., & Soare, D. (2008). Aspects Regarding the Use of Hydrogen in Spark Ignition Engine. *Scientific Bulletin of University Gh. Asachi of Iasi, 54*(2).

Negurescu, N., Pana, C., Popa, M., Cernat, A., & Soare, D. (2008, September). *Experimental Investigations Concerning the Direct Injection in Hydrogen Fuelled S.I. Engine* [Paper F2008-09-028 presentation]. FISITA2008, World Automotive Congress, Munich, Germany.

Negurescu, N., Pana, C., Popa, M., Cernat, A., & Soare, D. (2008) Experimental Researches Regarding the Direct Injection in Hydrogen Fueled S.I. Engine. In *Proceedings of Second International Congress Automotive, Safety and Environment- FISITA, EAEC, SAE International, SIAR (vol. 2*, pp. 62-70). University of Craiova Faculty of Mechanics.

Negurescu, N., Pana, C., Popa, M., Cernat, A., & Soare, D. (2008). The control of the running of the hydrogen fuelled spark ignition engine. In *Proceedings of 2th European Conference H2-Fuel-Millenium_Conference 2008, e-Hydrogenia, Academia Romana.* (vol. 1, pp. 1-10). Romanian Academy.

Nutu, C., Pana, C., Neguresc u, N., Cernat, A., Fuiorescu, D., & Mirica, I. (2019). A research on fuelling a truck diesel engine with hydrogen. *E3S Web of Conferences, 112*, 1-9. .doi:10.1051/e3sconf/201911201011

Pana, C., Negurescu, N., Cernat, A., Nutu, C., Mirica, I., & Fuiorescu, D. (2017). Experimental Aspects of the Hydrogen Use at Diesel Engine. *Procedia Engineering*, *181*, 649–657. doi:10.1016/j.proeng.2017.02.446

Pana, C., Negurescu, N., Popa, M., & Cernat, A. (2011). A control strategy of the hydrogen engine fueled load. *Termotehnica. Thermal Engineering*, *1*, 1–8.

Pana, C., Negurescu, N., Popa, M., Cernat, A., Boboc, G., & Soare, D. (2005). On the Performance of a Spark Ignition Engine fuelled with Gasoline and Hydrogen Addition. *Scientific Bulletin of University Politehnica of Bucharest, Series D. Mechanical Engineering (New York, N.Y.)*, *67*(4), 15–24.

Pana, C., Negurescu, N., Popa, M., Cernat, A., Boboc, G., & Soare, D. (2005). Influences of Hydrogen Addition in Gasoline on Spark Ignition Engine Performance. In *Proceedings of the 9th International Congress on Automotive CAR 2005, section Automotive and Environment*.University of Pitesti.

Pana, C., Negurescu, N., Popa, M., Cernat, A., & Soare, D. (2007, April). *An Investigation of the Hydrogen Addition Effects to Gasoline Fueled Spark Ignition Engine* [Paper No. 2007-01-1468 presentation]. SAE World Congress, Detroit, MI. 10.4271/2007-01-1468

Pana, C., Negurescu, N., Popa, M., Cernat, A., & Soare, D. (2007, April). *An Investigation of the Hydrogen Addition Effects to Gasoline Fueled Spark Ignition Engine* [Paper no. 2007-01-1468 presentation]. SAE World Congress 2007, Detroit, MI. 10.4271/2007-01-1468

Pantile, V., Pana, C., Negurescu, N., Radu, A., & Rusu, E. (2011, November). *The use of hydrogen as an alternative fuel in SI engines* [Paper 1142 presentation]. The FISITA 10th International Automotive Congress, Automotive Engineering and Environment CAR2011, Pitesti, Romania.

Pantile, V., Pana, C. N., & Negurescu, N. (2012, November). *An Experimental Investigation of the Hydrogen Fuelled Spark Ignition Engine* [Paper presentation]. *FISITA International Conference Motor Vehicle and Transportation MVT 2012*, Timisoara, Romania.

Pechlivanoglou, G. (2007). *Hydrogen Enhanced Combustion History. Applications and Hydrogen Supply by Plasma Reforming*. PPRE 2005-2007. University of Oldenburg.

Popa, M., Negurescu, N., & Pană, C. (2003). *Diesel Engines. Processes*. MatrixRom.

Rusu, E., Pana, C., Negurescu, N., Pantile, V., & Radu, A. (2011, November). *Analysis of energetically and emissions performance on a SI engine fueled with hydrogen in addition* [Paper 1143 presentation]. The FISITA 10th International Automotive Congress, Automotive Engineering and Environment CAR2011, Pitesti, Romania.

Sorensen, B., & Spazzafumo, G. (2018). Hydrogen, in Hydrogen and Fuel Cells (3rd ed.). Academic Press.

ADDITIONAL READING

Madhujit, D. (2015). An experimental study on combustion, performance and emission analysis of a single cylinder, 4-stroke DI-diesel engine using hydrogen in dual fuel mode of operation. *International Journal of Hydrogen Energy*, *40*(27), 8586–8598. doi:10.1016/j.ijhydene.2015.04.125

Marko, J. (2017). Impacts of low temperature combustion and diesel post injection on the in-cylinder production of hydrogen in a lean-burn compression ignition engine. *International Journal of Hydrogen Energy*, *42*(2), 1276–1286. doi:10.1016/j.ijhydene.2016.11.008

Saroj, C. D., Dhirendra, K. A., Ambica, N. T., & Mohanty, P. (2020). A review on performance of biogas and hydrogen on diesel engine in dual fuel mode. *Fuel*, *260*, ID116337. doi:10.1016/j.fuel.2019.116337

Senthil, R., Kumar, M., Loganathana, E., & Gunasekaran, J. (2015). Performance, Emission and Combustion Characteristics of CI Engine Fuelled With Hydrogen. *International Journal of Innovative Research in Science, Engineering and Technology*, *4*(2), 1–12.

Talibi, M., Hellier, P., Balachandran, R., & Ladommatos, N. (2014). Effect of hydrogen-diesel fuel co-combustion on exhaust emissions with verification using an in-cylinder gas sampling technique. *International Journal of Hydrogen Energy*, *39*(27), 1–14. doi:10.1016/j.ijhydene.2014.07.039

Tutak, W., & Jamrozik, A. (2014). Generator Gas as a Fuel to Power a Diesel Engine. *Thermal Science*, *18*(1), 205–216. doi:10.2298/TSCI130228063T

Yasin, K. (2016). Effect of hydrogen enrichment on combustion characteristics, emissions and performance of a diesel engine. *International Journal of Hydrogen Energy*, *41*(1), 656–665. doi:10.1016/j.ijhydene.2015.09.064

KEY TERMS AND DEFINITIONS:

BSEC: Break Specific Energetic Consumption, the brake specific fuel consumption that takes into consideration the lower heating value of the fuel.

CIE: Compression Ignition Engine, thermal machine that operated according to diesel cycle.

Diesel-Gas Method: A fuelling method which beside the use of classic diesel fuel, reclaims the use of o gaseous fuel that is injected into the engine inlet manifold.

ECU: Electronic Control Unit, computer that control the engine fuelling system.

EGR: Exhaust Gas Recirculation, method used for reduction of nitrogen oxides emission level.

ICE: Internal Combustion Engine, thermal machine that convert the chemic energy of the fuel into mechanical work.

Maximum Pressure Rise Rate: The maximum value registered during the combustion cycle for the increase of gases pressure during combustion until it reaches its peak value.

SIE: Spark Ignition Engine, thermal machine that operated according to Otto cycle xc- the substitute ratio that shows the percent of the energetic replacement of the diesel fuel by hydrogen, the LHV of both fuels being taken into consideration.

WOT: Wide open throttle, the operating regime at full load for SIE.

Chapter 10
Considerations Regarding the Use of Fuel Cells in Combined Heat and Power for Stationary Applications

Andrei Mircea Bolboaca
https://orcid.org/0000-0001-7750-6737
Faculty of Service Building Engineering, Technical University of Cluj-Napoca, Romania

ABSTRACT

Covering the energy demands under environmental protection and satisfying economic and social restrictions, together with decreasing polluting emissions, are impetuous necessities, considering that over half of the pollutant emissions released in the environment are the effect of the processes of electricity and heat production from the classic thermoelectric powerplant. Increasing energy efficiency and intensifying the use of alternative resources are key objectives of global policy. In this context, a range of new energy technologies has been developed, based on alternative energy conversion systems, which have recently been used more and more often for the simultaneous production of electricity and heat. An intensification of the use of combined energy production correlated with the tendency towards the use of clean energy resources can be helpful in achieving the global objectives of increasing fuel diversity and ensuring energy demand. The chapter aims at describing the fuel cell technology, in particular those of the SOFC type, used in the CHP for stationary applications.

INTRODUCTION

The present chapter aims at describing the fuel cell technology, in particular those of the solid oxide fuel cell (SOFC) type used in the cogeneration of energy for the stationary field.

The solid oxide fuel cell can be generically defined as a ceramic multilayer electrochemical energy generation system, which uses a fuel and a gas oxidant and operates near the temperature of 800÷1000 °C (*Ştefănescu I., 2010; Fuel Cell Handbook, 2004; Holtappels, P., Vogt, U., & Graule, T., 2005; Kim,*

DOI: 10.4018/978-1-7998-4945-2.ch010

Y. S., et.al. 2020). By definition, the fuel cell is an electric cell that can be continuously fed with fuel so that the electrical power from the output of this electric cell can be maintained indefinitely *(Zhu, P., et.al, 2020).* Thus, the fuel cell converts hydrogen or hydrogen-based fuels directly into electricity and heat through the electrochemical reaction of hydrogen with oxygen (*Yuksel, Y. E., et.al., 2020).*

A defining feature of the 21st century is the increasing dependence of the world economy on energy resources. The issue of the energy dimension of economic growth and development has become a very sensitive topic, both nationally and globally, and the energy paradigms will be sharpened in the future. The global economic trends of modern humans dependent on massive energy consumption have raised the problem of energy supply, respectively the issue of providing fuel in all areas (residential, commercial, industrial) *(Badea, G., et.al, 2016; Felseghi, R. A., 2015).*

In recent years, the spectrum of depletion in the not too distant future of fossil fuels, has turned the attention of modern society towards the creation of new energy sources, the higher use of materials, the diversification of products, the creation and implementation of new processes and technologies that are more efficient and efficient.

Covering the energy needs under the protection of the natural environment and satisfying economic and social restrictions, together with reducing polluting emissions, is an impetuous necessity, considering that over half of the released noises in the environment are the effect of the processes of electricity production and thermal power plants from classical thermoelectric power stations *(Badea, G., et.al, 2016).* Increasing energy efficiency, controlling energy consumption worldwide and intensifying the use of energy from alternative sources are key objectives of global policy (*Badea G., et.al, 2013).*

In this context, a range of new energy technologies have been developed, based on alternative energy conversion systems, which have recently been used more and more often for the simultaneous production of electricity and heat, falling within the concept known as cogeneration. (*Kalra, P., et.al., 2013).*

An intensification of the use of combined energy production, correlated with the tendency towards the use of clean energy resources can be helpful in achieving the global objectives of increasing fuel diversity and ensuring energy demand.

COGENERATION OF ELECTRICITY AND THERMAL ENERGY - COMBINED HEAT AND POWER (CHP)

Cogeneration, as already known, represents a cycle of transformation in which, starting from the fuel, it produces in the same plant and at the same time electricity and thermal energy, heat and power.

Performance Indicators of Combined Heat and Power Systems

In order to demonstrate the energy efficiency of cogeneration plants, the following performance indicators will be analyzed *(Mudryk, K., & Werle, S., 2018):*

1. Electric efficiency (ηE) is defined as the ratio between the electricity produced by the cogeneration cycle and the energy of the fuel introduced in the cycle, according to the relation:

$$\eta_E = \frac{E_{COG}}{E_B}, \tag{1}$$

where: E_{COG} is the electricity produced in cogeneration; E_B - the energy of the fuel introduced into the cycle.

Also, E_B is determined by a relationship of form:

$$E_B = B \, x \, Pci, \tag{2}$$

where: B is the amount of fuel introduced into the cycle; P*ci* - lower calorific value of the fuel.

2. The thermal efficiency (ηQ) is defined as the ratio between the heat energy produced by the cogeneration cycle and the energy of the fuel introduced in the cycle, according to the relation:

$$\eta_Q = \frac{E_{COG}}{E_B}, \tag{3}$$

where Q*COG* is the thermal energy produced in cogeneration.

3. The overall efficiency (ηgl) is defined as the ratio of the sum of the useful electrical and thermal energy obtained from the cogeneration cycle and the energy of the fuel introduced in the cycle, according to the relation:

$$\eta_{gl} = \frac{E_{COG} + Q_{COG}}{E_B} = \frac{E_{COG}}{E_B} + \frac{Q_{COG}}{E_B} = \eta_E + \eta_Q, \tag{4}$$

4. The degree of heat recovery (xr) represents the ratio between the useful thermal energy (recovered from the cycle) and the total amount of heat resulting from the thermodynamic process (recoverable thermal energy).
5. The heat cogeneration index (y) is defined as the ratio between the electricity produced and the heat energy actually recovered from the cogeneration cycle, according to the relation:

$$y = \frac{E_{COG}}{Q_{COG}} = \frac{\dfrac{E_{COG}}{E_B}}{\dfrac{Q_{COG}}{E_B}} = \frac{\eta_e}{\eta_Q}, \tag{5}$$

6. The cogeneration coefficient (α) is an indicator specific to the boiler where the cogeneration plant is located and is defined as the ratio between the amount of heat obtained in the cogeneration and the

amount of total heat produced in the boiler (both the cogeneration plant and the plants). according to the relationship:

$$\alpha = \frac{Q_{COG}}{Q_{COG} + Q_V}, \tag{6}$$

where: Q_{COG} is the heat produced by cogeneration; Q_V - heat obtained in the top installations. *(Badea G., et.al, 2013; Mudryk, K.,&Werle, S., 2018).*

Analyzing in detail it turns out that practically the cogeneration coefficient represents the structure of the heat generation at the level of the whole cogeneration plant *(Badea G., et.al, 2013; Mudryk, K.,&Werle, S., 2018).*

High Efficiency Cogeneration

Definitions according to the regulations in force:

- ***high efficiency cogeneration*** - the cogeneration process that realizes a useful thermal energy production and electrical and possibly mechanical energy that fulfills the certification conditions regarding the global efficiency and the primary energy economy;
- ***primary energy economy (EEP)*** - fuel economy (MWh), achieved by a configuration / unit of production in cogeneration compared to the separate production, with alternative technologies, of the same amounts of useful thermal energy and electrical and / or mechanical energy;
- ***overall efficiency*** - the sum of the useful thermal energy production and of the electrical and / or mechanical energy, related to the amount of energy contained in the fuels used to produce these energies, during the calculation period, in a cogeneration process. *(ANRE, 2015)*

Due to the advantages, materialized in particular by the fuel economy obtained by cogeneration compared to the separate production of the two forms of energy, cogeneration contributes decisively to the increase of the energy efficiency of the sources, therefore it also contributes to the efficiency of the whole system.

In Romania, Law no. 13/2007 - The law of electricity, chapter V regulates the promotion of electricity produced from renewable energy sources and in high efficiency cogeneration, being in fact the adaptation of the European Directive 2004/8 / EC to promote cogeneration based on the energy demand and thermal useful in domestic energy markets.

In order to promote the production of electricity in cogeneration, article 68 of the law defines high efficiency cogeneration as the process that fulfills one of the following criteria (*Voronca, Ş. D., Darie, G., & Pătraşcu, R., 2017*):

1. the achievement of primary energy savings when producing cogeneration electricity of at least 10% compared to the reference values established by specific regulations for the separate production of electric and thermal energy in cogeneration power plants with installed power of at least 1 MW;

2. the realization of primary energy economy when producing the cogeneration electricity compared to the separate production of the electric and thermal energy in cogeneration power stations with installed powers below 1 MW. (*Voronca, Ş. D., Darie, G., & Pătraşcu, R., 2017; Ionescu, C., et.al., 2019)*

At the same time, it regulates the conditions necessary for a cogeneration plant to be considered of high efficiency, namely the condition that the primary energy savings (fuel) at the cogeneration production of the two forms of energy (electrical and thermal) are at least 10%, compared to their separate production, ie:

$$EEP = 1 - \frac{1}{\frac{\eta_{E(COG)}}{\eta_{E(ref)}} + \frac{\eta_{Q(COG)}}{\eta_{Q(ref)}}} \geq 10\%, \tag{7}$$

where:

$\eta_{E(COG)}$ is the electrical efficiency of the cogeneration plant;
$\eta_{Q(COG)}$) - the thermal efficiency of the cogeneration plant;
$\eta_{E(ref)}$ - the electrical efficiency of the reference installation;
$\eta_{Q(ref)}$ - thermal efficiency of the reference installation.

In this context the overall yield is expressed through the relationship:

$$\eta_{gl(COG)} = \eta_{E(COG)} + \eta_{Q(COG)}, \tag{8}$$

where: $\eta_{gl(COG)}$ is the overall efficiency of the cogeneration plant.

From the relationships presented above, one can express the energy efficiency (electric and thermal) of the cogeneration depending on its overall efficiency. With their help it is possible to determine whether or not the cogeneration system complies with the requirements of European Directive 2004/8 / EC on high efficiency cogeneration, depending on the technical data that characterizes the cogeneration plant, guaranteed and certified by the equipment supplier. of cogeneration, taking into account the coefficients $\eta_{E(ref)} = 52.5\%$ and $\eta_{Q(ref)} = 90\%$, established by ANRE regulations and provisions regarding the criteria and conditions necessary to meet for a high efficiency cogeneration *(Badea G., et.al, 2013; Mudryk, K.,& Werle, S., 2018).*

Unlike conventional power plants (in condensation) where only 33% of the primary energy is transformed into electricity, in the case of high efficiency cogeneration plants, combining the 2 processes (simultaneous production of electricity and heat) results in a transformation of up to 90% of the primary energy (see Figure 1).

Figure 1. High efficiency cogeneration plant efficiency

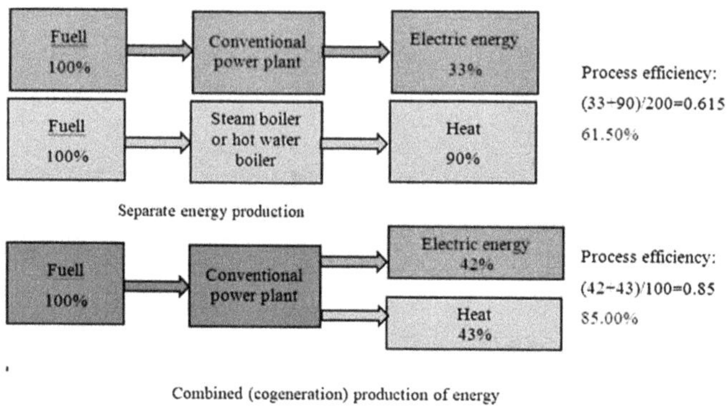

THE CURRENT STAGE OF KNOWLEDGE IN THE FIELD OF SOFC FUEL CELLS

The solid oxide fuel cell (SOFC - see Figure 2) can be defined generically as a ceramic multilayer electrochemical energy generating system, which uses a fuel and a gas oxidant and operates near the temperature of 800 ÷ 1000 °C.

Figure 2. Principle diagram of elemental cell

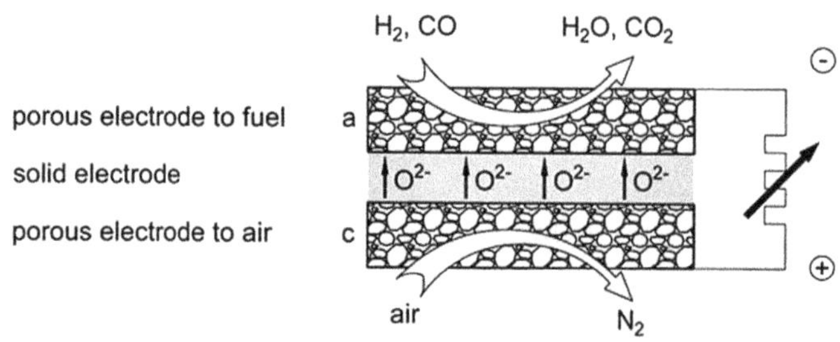

The combination of several elementary cells to form a cell requires the use of an electronic conductor (ceramic or metal) interconnection material. The operating characteristics of SOFC offer them a number of advantages over other types of fuel cells, such as: the use of catalysts is not necessary, they can reach an overall efficiency of up to 80% if used for practical applications of fuel, cogeneration of electricity and thermal energy, there are no reliability problems related to electrolyte losses and electrode corrosion, they have great flexibility for power generation *(Ştefănescu I., 2010; Hoogers G., 2003; Radenahmad, N., et.al., 2020).*

The efficiency in electricity production for a SOFC cell is about 50%, and together with the thermal energy recovery it can reach about 80-85% *(Costa, P., et.al., 2021).*

Description - Particularities of SOFC Fuel Cells

- ***operating principle***

The particular electrochemical reaction used for SOFC is illustrated in Figure 3.

Figure 3. Characteristic electrochemical reaction

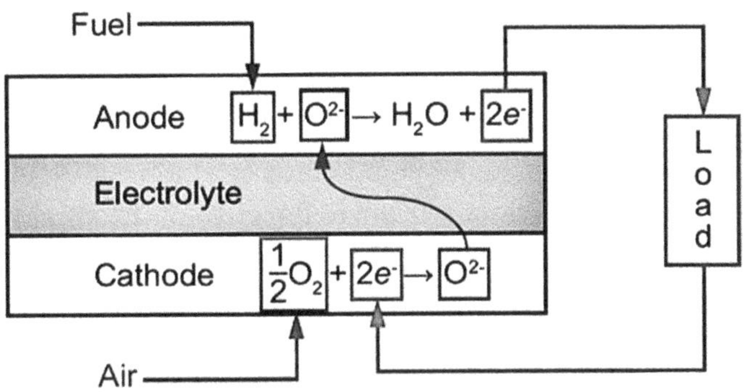

The SOFC cell is constructed of two porous electrodes permeable to the anode and cathode separated by a solid electrolyte with selective permeability for oxygen ions. The fuel and the air flow enter the cell through the anode and cathode. When the oxygen molecule reaches the contact surface at the cathode-electrolyte interface, the cathode electrons split. Oxygen ions diffuse into the electrolyte material and migrate to the other part of the cell where the anode is located. These oxygen ions meet the fuel at the electrolyte - anode interface and react catalytically, having as reaction products water, carbon dioxide, heat and electrons (electrons directed through an external circuit produce electricity) - see Figure 4. *(Kumar, V., et.al., 2020; Kalra, P., Garg, R., & Kumar, A., 2013; Chitgar, N., & Moghimi, M., 2020).*

Figure 4. SOFC Functional Scheme
(SOFC, 2003)

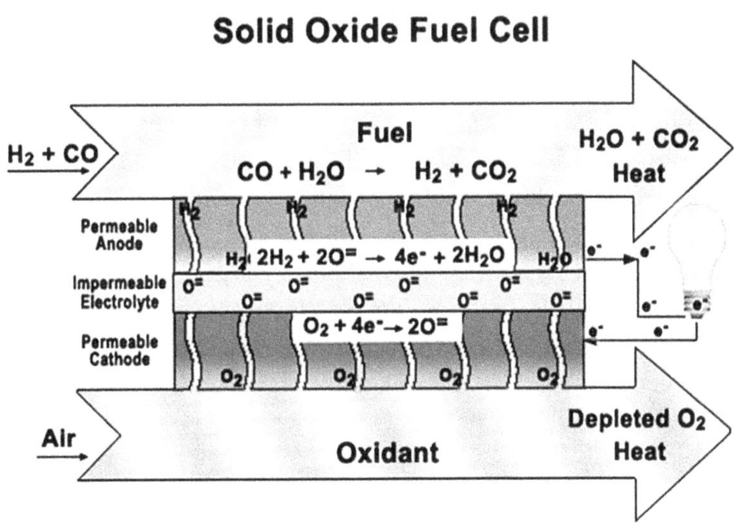

The relatively high operating temperature of the SOFC allows:

- extremely efficient conversion of fuel into electricity;
- the internal catalytic reforming of the fuel;
- heat generation usable in the high quality cogeneration process.

SOFC allows the conversion of a wide range of classic or alternative fuels, including various hydrocarbon fuels *(Ionescu, C., et.al., 2019; Ahmad, M., 2018)*. SOFC uses hydrogen as the main fuel for electricity generation, but does not require pure hydrogen as a fuel. Typically, SOFC useful fuels are hydrogen and carbon monoxide, both of which are directly consumed in specific electrochemical reactions to this type of fuel cell. These gases can be generated externally to the fuel cell by gasification of coal or steam reforming of methane. Hydrogen can also be generated by electrolysis of water, using energy from renewable sources. The particularity regarding the high operating temperature of SOFC offers the advantage that in the presence of catalysts, hydrogen can be generated by internal reforming of hydrocarbon fuels within the anodic chamber of SOFC *(Adamson K.A., 2006; Andrews J., Shabani B., 2012; Badea G., et.al., 2013)*.

However, internally for methane reforming it is necessary to introduce a large amount of steam into the fuel cell stack and use a pre-reformer for natural gas fueling (useful in removing higher hydrocarbons) to avoid carbon deposits. A number of other hydrocarbon fuels such as kerosene, gasoline, LPG, Naphtha, diesel can be used with the specification that an external fuel processing unit (reformer / cleaner) placed upstream of fuel cells is required. Direct electrochemical oxidation of natural gas at acceptable rates has not been achieved so far. SOFC can tolerate very high levels of CO and CO_2. *(Ivers-Tiffée, E., et.al., 2001; Akroot, A., Namli, L., & Ozcan, H., 2021, Minutillo, M., et.al., 2021)*.

SOFC Fuel Cell Components

A SOFC operates at temperatures between 800 ÷ 1000 °C. Due to the high operating temperature of the SOFC, there are strict requirements regarding the construction materials. In addition to electrical and electrochemical excellence, the high chemical properties and thermal compatibility between the fuel cell operating environments are especially important for long-term stability in order to achieve over 40,000-50,000 operating hours. *(Nassef, A. M., et.al., 2019; Chen, J., et.al., 2018; Choudhary, T., 2017)*

A single fuel cell produces a voltage of over 1 V in the open circuit, while the fuel cell is charging the voltage is reduced to around 0.5-0.7 V, and the current density varies between 200 ÷ 1000 mA / cm^2 (direct current) depending on the building materials, cell design and operating *conditions (Appleby A.J., Foulkes F.R., 1989; Badwal S.P.S., et.al., 2014; Hagen, A., et.al., 2020)*.

SOFCs have a series of elements as follows *(Badwal S.P.S., et.al.,2014; Zakaria, Z., et.al., 2020; Hajabdollahi, Z., & Fu, P. F., 2017):* cathode (air electrode) and anode (fuel electrode), electrolyte, interconnection materials, bipolar plates, fuel distribution system and sealing materials. These components are mostly made of ceramic, metal or metal ceramic composites. The various materials used in the construction of SOFC fuel cells and their properties are presented below *(Minh, N., et.al., 2017; Faleh, S., et.al., 2017; Chitgar, N., et.al, 2019)*.

Electrolyte

The electrolyte of a SOFC fuel cell is required to have high conductivity for oxygen ions (ie> 0.1 S / cm) over a wide range of partial oxygen pressures. For a material to prove the property of being a good electrolyte, it must demonstrate that it meets other secondary criteria. Apart from high ionic conductivity, it must also be almost completely dense, without open porosity (to avoid any transverse diffusion of the constituents of the fuel and oxidant). The electrolyte is in contact with both the air and the fuel electrodes and with the sealing materials, therefore it must be chemically inert to these materials, but it must have a coefficient of thermal expansion (CDT) in accordance with the environments with which it comes in contact. and good mechanical properties. It is exposed to extremely oxidizing media on the air electrode side and extremely reductive media on the fuel electrode side which imposes stable chemical and structural properties. It must also have high tensile strength and toughness, not only to facilitate handling, but also to withstand thermal and mechanical demands during the manufacture and operation of fuel cells. *(Jamil, S. M., et.al., 2019; Al-Attar, A. F., et.al., 2019; Yadav, A., et.al., 2020; Mojaver, P., et.al, 2020)*

Four main types of solid electrolyte systems are covered, as follows: those based on fully or partially stabilized zirconia, doped cerium, doped $LaGaO_3$ and doped Bi_2O_3 *(Fan*, L., 2020; Park, J. H., et.al., 2020; Fallah Vostakola, M., & Horri, B. A., 2021)*. Each of these systems has a number of sub-categories of materials with less or more dopant content that enhance ionic conductance or increase phase stability. The type of dopant and its concentration has a marked effect on the phase assembly, chemical stability, ionic conductivity (σ), thermal and mechanical properties. It should be emphasized that many materials are either unstable in various fuel cell operating environments, either develop low electronic conductivity under different environmental conditions or have very low ionic conductivity *(Zhang, J., et.al., 2020; Puente-Martínez, D. E., et.al., 2020; Yixiao, C., et.al., 2021)*. The electrolyte that satisfies most of the characteristics required for operation in a SOFC system and which have been used in practical applications is based on Y_2O_3-ZrO_2 with Y_2O_3 concentration of 3-10 mol%. In this system, at the operating temperature of SOFC in the immediate vicinity of 1000 ºC, 8 mol% Y_2O_3 doped with ZrO_2 (8YSZ) with cubic structure recorded the highest ionic conductivity *(Kurapova, O. Y., et.al., 2020; Fan*, L., 2020; Khajavi, P., et.al., 2021; Arun, A., et.al., 2009)*, with a mechanical resistance of about 250- 300 MPa at room temperature and 150 MPa at temperatures above 500°C *(Stolyarova, V. L., et.al., 2020)*.

Anode

A major function of the SOFC anode is to facilitate the electrochemical oxidation of the fuel and to provide free pathways for the cleaved electrons during the oxidation reaction to reach the electric collector. The characteristics of a material to be suitable as a SOFC anode include electronic conductivity, electrocatalytic activity suitable for fuel, compatibility with electrolytic collectors and microstructural stability at SOFC operating temperature, as well as thermal expansion properties suitable with other components with which it comes in direct contact. (Sarruf, B. J. M., et.al., 2020; Wang, X., et.al., 2020; Xu, J., et.al., 2017; Mo, B., et.al., 2020; Bochentyn, B., et.al., 2019)

For SOFCs fueled by hydrocarbon fuels there are additional requirements such as: catalytic activity for hydrocarbon cracking, tolerance to sulfur and resistance to coking. For the electrochemical reaction to take place at the anode - the electrolyte interface is sufficient porosity (> 30 vol%) to allow the reagent / product to be transported. A porous anode consisting of Ni and YSZ is the most commonly used anode in SOFC with YSZ type electrolytes. For cerium-based electrolytes a cerium doped Ni anode is used. In a typical SOFC, NiO is mixed with YSZ, and after manufacture of the cell the anode is subjected to reduction treatment to obtain a cermet Ni-YSZ. New materials are continually being researched and developed in order to achieve significant performance improvements for these SOFC components. (Xu, J., et.al., 2017; Venâncio, S. A., & de Miranda, P.E.V., 2017; Li, R., et.al., 2019; Silva, E. D. R., et.al., 2021)

Cathode

A SOFC fuel cell cathode is required to have high electronic conductivity, preferably high for oxygen ions, a thermal expansion coefficient (CTE) that closely matches the electrolyte, chemical stability during cell manufacture, during operation at working temperature and partial oxygenation at various pressures, good catalytic properties for oxygen molecule dissociation and reduction. The SOFC cathode is responsible for a significant portion of the fuel cell voltage losses, most of which are due to the high polarization losses associated with the oxygen reduction reaction. The most commonly used material for the cathode is strontium doped with manganese lanthanum, having compositions of the form $La1\text{-}xSrxMnO_3$ (LSM) for fuel cells operating at 800-1000 ° C. *(Ding, X., et.al., 2019; Hu, Y. Z., et.al., 2020)*

Further research efforts are being made to increase the performance of the air electrode at lower temperatures (<800 ° C), especially for low- and medium-temperature SOFC cells (IT-SOFC: ~ 800 ° C, LT-SOFC: <600 ° C). Materials that can increase the oxygen cleavage rate and possess high ionic conductivity (electronically mixed ionic conductivity - MIEC) are explored as cathodes. This would allow the oxygen reduction reaction to take place within the cathode structure or at the air / cathode interface, thus increasing the reaction yields. In this regard, alternative materials such as $La1\text{-}xSrxFeO_3$ (LSF), $La1\text{-}xSrxCoO_3$ (LSC), $La1\text{-}xSrxFe1\text{-}yCoyO_3$ (LSCF) are investigated. Another option is to use more chemically stable materials, which also offers higher ionic and electronic conductivity at lower temperatures. *(Irshad, M., et.al, 2021; Sadykov, V. A., et.al., 2020; Chick, L., et.al., 2015; Bicer, Y., & Khalid, F., 2020; Rillo, E., et.al., 2017; Beigzadeh, M., et.al., 2021)*

Interconnection

The interconnection (bipolar plate) in the SOFC aims to achieve the electrical connection between the adjacent fuel cells within a stack, separating the components of the cell destined for the fuel and the oxidant, at the same time it has a special role in the uniform distribution of the fuel and the oxidant inside the anode, respectively the cathode of each. cells. The interconnection material must have electrical conductivity and high polarization resistance, high thermal conductivity (> 20 W/m/K), CDT in accordance with other ceramic components of the cell (10-13 x 10-6 / K), stable structure adapted to cell operation for the oxidation / reduction phase and increased mechanical strength to withstand the load and thermal stresses encountered during operation, while it should be gas tight *(Yang, J., et.al., 2018; Badwal, S. P. S., et.al., 1997)*.Ceramics, metals and ceramic metal composites have been considered as suitable for achieving interconnections within the SOFC operating in the temperature range of 800-1000 ºC and below 800 ºC *(Thalmaier, G.,et.al., 2020)*. The most commonly used ceramic material is based on lanthanum chromium perovskite ($LaCrO_3$) doped with elements such as Sr and Ca *(Fergus, J. W., 2004; Balog, A. A., et.al., 2019; Bianco, M., et. al., 2020)*. Other materials that have proven to be suitable for air sintering and compatibility with YSZ (resistance to zirconate formation) are calcium and nickel doped with yttrium *chromites (Yoon, K. J., et.al., 2010; Schafföner, S., et.al, 2017)*. However, ceramic materials have low electrical and thermal conductivity. In addition, the costs of manufacturing materials for ceramic interconnections are high and demand is limited to tubular geometry that requires less interconnection material *(Tietz, F., Buchkremer, H. P., & Stöver, D., 2002).*

The interconnections through metallic materials are SOFC-friendly due to their high electrical and thermal conductivity and have lower manufacturing costs. However, metal interconnections suffer from long-term corrosion (oxidation), chromium evaporation causes cathode contamination, and CDT is asymmetrically related to other ceramic components of the cell. The main materials used are chromium alloys and ferritic stainless steels. The first chromium alloy that was considered as a substitute for ceramic materials bears the name of Plansee Ducralloy ($94Cr_5Fe_1Y_2O_3$) *(Yang, Z., 2008; Müller, M., et.al., 2002).*

It has good resistance to oxidation, mechanical strength and efficient coefficient of thermal expansion (CDT) *(Hayun, H., et.al., 2020)*, but the manufacturing costs are very high. Ferritic stainless steels offer light plasticity and suitable CDT, having low costs, but have oxidative characteristics, which leads to increased oxidation, especially in the cathode environment, which favors irreversible cell degradation *(Badwal, S. P. S., et.al., 1997; Bianco, M., et.al., 2020).*

Sealing

The sealing component of SOFC acts as a barrier to avoid the mixing of fuel and oxidant. While sealing is not a critical problem in tubular SOFCs, it is one of the key components for flat models. The necessary characteristics of the SOFC sealing elements refer to the coefficient of thermal expansion appropriate to the substrate (usually the electrolyte), the stability in the SOFC environment and the chemical inertia with respect to the cellular components in contact. *(Gross-Barsnick, S. M., 2020; Lee, H.,et.al. 2020)*

Broadly, the SOFC sealing components can be classified into two types: rigid gaskets, such as those provided by glass ceramic composites, and compression gaskets, such as those made of metal gaskets or sheets. Ceramic seals (for example, BaO-CaO-SiO_2 and BaO-Al_2O_3-SiO_2) they are cost effective options at low manufacturing costs. Also, these materials have high chemical resistance and low reactivity to other cell components. Although the glass sealing components can adequately meet the operating conditions of the typical medium-term SOFC, their long-term stability (> 20,000 operating hours) is still questionable *(Virkar, A. V., 2001; Blum, L., et.al., 2020; Cobîrzan, N., et.al., 2018).* Also, the cyclic thermal capacity of these materials is a major problem during SOFC start and stop. Rigid gasket sealing of an entire stack results in the formation of a rigid body, and if cracks occur in one of the cells the entire stack is inadequate for functioning questionable *(Virkar, A. V., 2001; Scott Weil K.,2006).*

Compression sealing components on the other hand are much more resistant to thermal shocks and stable in long-term operation *(Gross-Barsnick, S. M., 2020).* However, continuous compression is required which can cause short-circuiting if metal seals are used. Therefore, a hybrid approach, such as the use of compression sealing (eg, foil sheets) together with rigid seals (such as barium alumino-silicate) can also be used, these hybrid seals demonstrating excellent performance *(Shaur, A., et.al., 2020).*

The literature on SOFC sealing technology shows that no variant is universally accepted, and different research, development groups and companies producing such components use different *(Stolten, D., & Emonts, B., 2012; Virkar, A. V., 2001; Scott Weil K.,2006; Singh, R. N., 2007; Guski, V., et.al., 2021).*

Structure of a SOFC Fuel Cell

For practical applications, SOFC fuel cells have different structures, each with its own advantages and disadvantages, although the materials for the components in the cells of these different models are identical or very similar. Currently, there are two basic types of structure and arrangement of SOFC cells: planar and tubular. *(Su, S., et.al., 2017; Mahmud, L. S., et.al., 2017; Ashraf, M. A., et.al., 2020)*

For SOFC planes, each cell is made into a flat, square disk, or rectangular plate. The cells are serialized and connected by interconnection plates, as shown schematically in Figure 5. *(Hou, Q., et.al., 2020)*

Figure 5. SOFC flat fuel cells
(Zhao, T. S., & Ni, M., 2013).

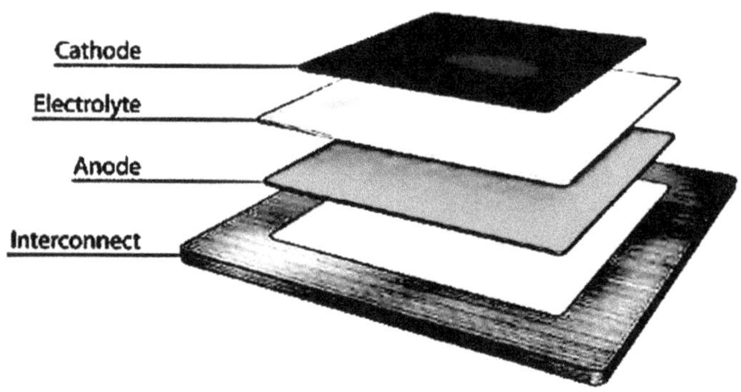

For tubular SOFCs, the electrode (either cathode or anode) is usually transformed into a long tube with a porous wall. Outside the electrode tube is the electrolyte and then the other electrode. The cells are also connected in series by interconnection, as shown schematically in Figure 6. *(Funahashi, Y., et.al., 2007; Khan, M. Z., et.al., 2020; Milcarek, R. J., & Ahn, J., 2021).*

Figure 6. Tubular SOFC fuel cells
(Brandon, N., et.al., 2016)

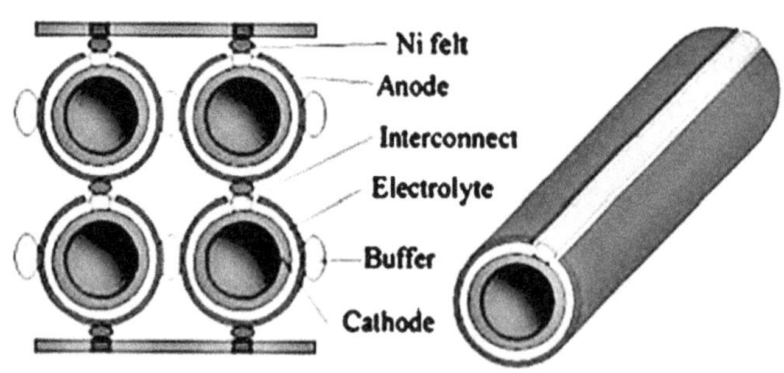

Previous studies of SOFC have focused on SOFC tubular fuel cells operating at high temperatures. At the end of the 1990s, the development of SOFC planes together with the reduction of the electrolyte layer thickness attracted great interest due to its advantages regarding the apparent power density and ease of manufacture. A simplistic comparison of SOFC fuel cells with planar and tubular structures is summarized in Table 1. *(Brandon, N., et.al., 2016; Ilbas, M., et.al., 2020; Xiong, G., et.al., 2021)*

Table 1. Comparison between SOFC with planar structures and tubular structures

	Planar	Tubular
Power per unit area	Higher	Lower
Power per unit volume	Higher	Lower
Manufacturing cost	Higher	Lower
Manufacturing ease	Easier	Difficult
Sealing ease	Difficult	Easier
Thermo-cycling stability	Fair	Good
Long-term stability	Fair	Excellent

The fuel cells are connected in series or have a parallel arrangement to form a stack that increases current/ voltage output. This arrangement forms fuel channels at the bottom and air channels at the top. The cells can be stacked in this way to an optimal stack size. A number of stacks can then be connected in series and/or in parallel to form a SOFC module (see Figure 7). *(Rosner, F., et.al, 2020; Ahmad, M., 2018; Lawlor, V., et.al.,2009)*

Figure 7. Structure of a SOFC module

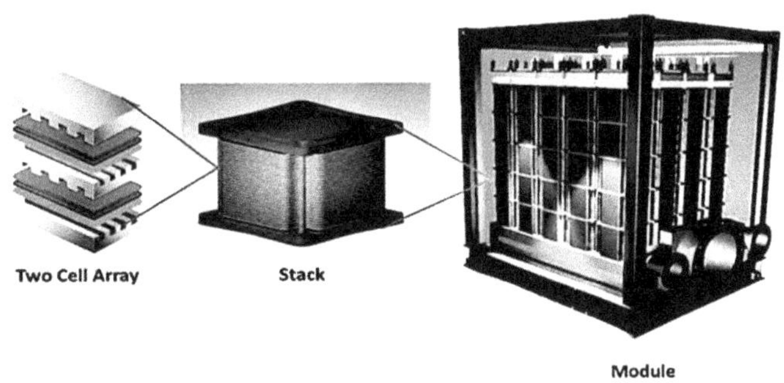

Integration of Solid Oxide Fuel Cell in CHP Systems

In order for a fuel cell to operate in the stationary field as a cogeneration system, several auxiliary systems are required to provide adequate working conditions and useful means of extracting heat and electricity from it. The fuel cell usually represents less than a quarter of the cogeneration system, both in terms of volume and cost. *(Roshandel, R., et.al., 2018; Yang, B., et. al.,2020; Baldi, F., et.al., 2020)*

Figure 8 shows schematically the main components of a fuel cell cogeneration system *(Yang, C. H., et.al., 2019; Barelli, L., et.al., 2013; Riensche, E., et.al., 1998; Wołowicz, M., et.al., 2021)*, as follows:

- fuel processor that converts natural gas (fuel) into pure hydrogen for use in the fuel cell;
- inverter to convert the DC output current into a useful alternative to the consumer;
- the SOFC fuel cell stack;
- heat exchangers to recover the heat from the stack of the battery and its use for the preparation of hot water as well as for heating;
- hot water tank to store this heat when there is no consumption;
- sensors, pumps, valves and pipes to ensure the fuel, air and coolant circuit in the fuel cell or to remove waste gas, heat and electricity from the system;
- system controller that automates and regulates fuel consumption, air intake, but also all sub-systems functioning. *(Staffell, I. 2010; Yang, C. H., et.al., 2019; Barelli, L., et.al., 2013; Riensche, E., et.al., 1998; Wołowicz, M., et.al., 2021)*

Figure 8. Functional diagram of the SOFC fuel cell cogeneration system

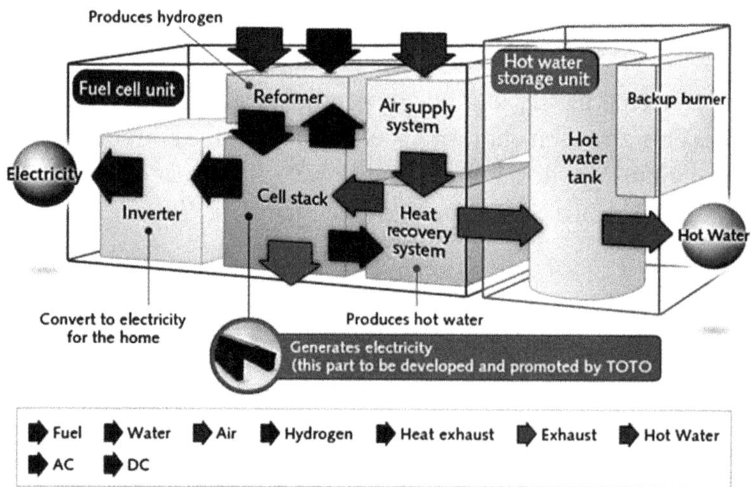

As an overview of the requirements and characteristics of these sub-systems and the principle of achieving a generic fuel cell cogeneration system, the following are relevant:

Fuel Processing

Hydrogen is the ideal fuel for fuel cells, however, it is not practical for direct use in stationary applications. Currently, there is no way to deliver hydrogen through a regional or centralized infrastructure for generating it, such applications being in the research project phase. *(Koirala, B., et.al., 2021; Ishaq, H., & Dincer, I., 2021).*

In contrast, hydrocarbons (especially natural gas) are considered as ideal fuel for cogeneration systems, because they can be reformed in hydrogen at the point of use, natural gas is abundantly available (for the moment) and has an extensive infrastructure throughout Eastern Europe. *(Sadik-Zada, E. R., & Gatto, A., 2021; Kutcherov, V., et.al., 2020).*

All cogeneration systems with commercial fuel cells are supplied with natural gas, LPG or kerosene. (*Song, Y., Han, K., & Wang, D. Y., 2020*).The literature also describes systems that operate on other hydrogen sources, for example, carbonated coal *(Homma, T., 2006; Habibi, R., & Mehrpooya, M., 2021), diesel (Sammes, N., Bove, R., & Stahl, K., 2004),* biogas from waste (*Laurencelle, F., et.al., 2001; Gummert, G., & Suttor, W., 2006; Chowdhury, T.,et.al., 2020),* biomass *(Hagiwara, A., 1999; Ramis, G., et.al., 2020; Al-Qahtani, A., et.al., 2021)* and biological hydrogen from sugar wastes *(Baeyens, J., et.al., 2020; Cieciura-Włoch, W., et.al., 2020).*

With strict reference to natural gas, their conversion to acceptable hydrogen supply requires several processing steps (see Figure 9) depending on the type of fuel cell integrated in the system *(Staffell, I., 2010; Shah, M., et. al., 2017; Schneider, S., et.al., 2020).*

The fuel processing steps required for each type of fuel cell are integrated into a single compact fuel processing unit. These units integrate equipment that performs fuel processing operations (desulfurization, steam reforming, reactors for partial preferential oxidation, etc.), plus thermal management systems and a steam generator to supply water vapor to the reforming and exchanger. *(Barrovecchio, L., 2001; Echigo, M., et.al., 2004; Cipitì, F., et.al., 2006; Welaya, Y. M., et.al., 2012)*

The difference between the fuel processors from different manufacturers is the choice of the fuel reforming method; most use steam reforming. *(Dicks, A. L., 1996; Ahmadi, M. H., et.al., 2018).*

Figure 9. Fuel reform
(Staffell, I. 2010; Ahmadi, M. H., et.al., 2018; Winkler, W., et.al, 2020)

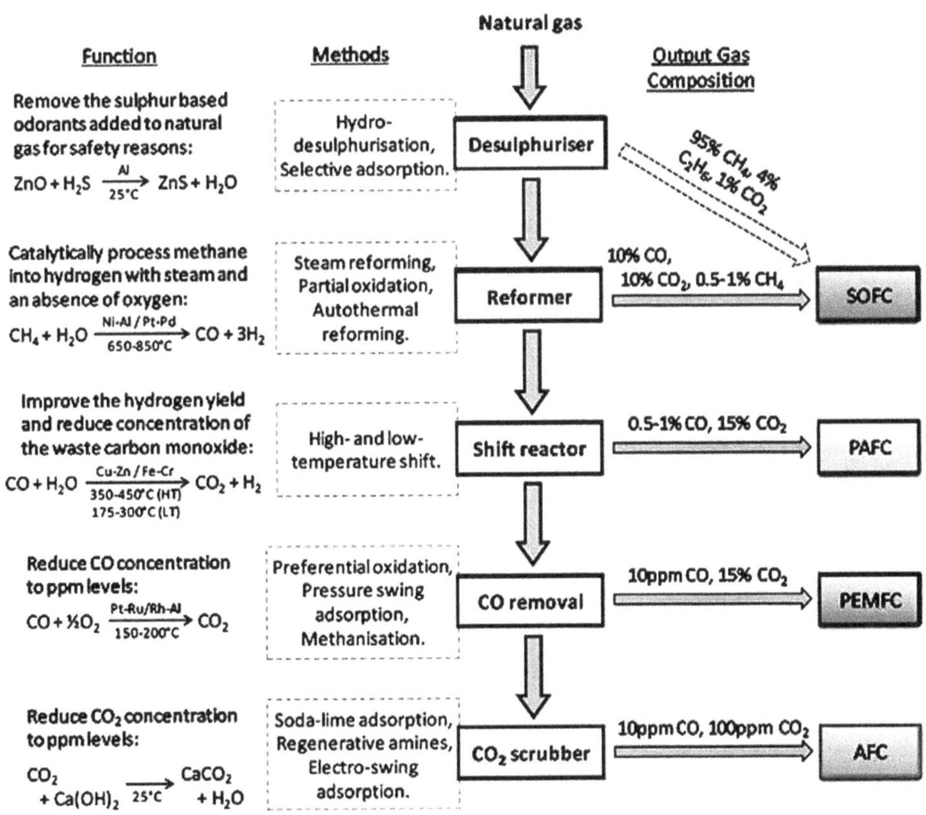

The main benefit of steam reforming is the high hydrogen concentration at the fuel output of the reformer of 70-80%, compared with 50-60% for autothermal reforming and even less for partial oxidation. Disadvantages result from the fact that the reaction is extremely endothermic (- 250 kJ / mol CH4) and is carried out at a high operating temperature (up to 800 ° C) which prevents a rapid start-up. *(Staffell, I. 2010; Ahmadi, M. H., et.al., 2018; Winkler, W., et.al, 2020)*

Heat Recovery and Storage

The heat recovery from the system is significantly different depending on the type of fuel cell used. SOFC cogeneration systems operating at high temperatures are cooled with excess air flow at the cathode level, which is then routed and "burned" into a non-consumable fuel afterburner. This heat is used to pre-heat the gas inlets and maintain the reformer temperature, and the excess is passed through a condenser heat exchanger to prepare and supply hot water to the consumer. *(Staffell, I. 2010; Zhu, P., et.al., 2020; Gandiglio, M., et.al., 2020)*

The hot water obtained is stored in a well-insulated storage tank. These heat storage tanks improve conventional hot water storage technologies by using thermal stratification with mixing valves and buffer zones. Most of the water is stored as a hot buffer (~ 45 °C) which is used as an intermediate heat exchanger between the generator and the central heating system. A smaller tank that holds ~ ¼ of the water at a higher temperature is used for direct consumption. For cogeneration systems with fuel cells, the storage of hot water and the provision of the necessary space for such a tank is a problem due to size conditions. *(Staffell, I. 2010; Yoda, M., et.al., 2017; Javidmehr, M., et.al., 2021)*

Electricity Conversion and Export

The cogeneration system with SOFC produces low voltage DC which must be converted to 50 Hz, 240 V AC for compatibility with consuming equipment. This conversion is done with a standard transformer and an inverter. Most cogeneration systems are connected to the national electricity grid, so that excess power can be exported. This allows a substantial improvement in the overall efficiency of the system use, but also economic benefits resulting from the exploitation of a fuel cell cogeneration system.

Electricity export requires the installation of a smart meter to measure both the amount of imported and exported electricity, as well as additional equipment to ensure frequency control, synchronization and other power conditioning to meet the quality demanded by the network. *(Staffell, I. 2010; Ahmadi, S., et.al., 2019; Abu-Heiba, A., et.al., 2021)*

SOFC power plant station, is composed of the main subsystems *(Fragiacomo, P., et.al., 2018; Gandiglio, M., Lanzini, A., & Santarelli, M., 2018).*

- the fuel cell stack;
- fuel processing / cleaning subsystem;
- unit for removing impurities that may be harmful to the reformer or fuel cell;
- an air supply unit for oxygen supply to the battery;
- an energy management unit that conditions the conversion power (DC / AC to meet the end user's energy requirement);
- a thermal management unit to manage the heat resulting from the fuel cell;
- an assembly of control and safety of the subsystems;

- power plant automation (see Figure 10).

A complete cogeneration product with fuel cells consists of *(Staffell, I., & Green, R., 2013):*

- The main cogeneration system
    - stack fuel cell (hydrogen conversion - electricity, heat, water, other waste products);
    - fuel processing system (conversion of natural gas or other hydrocarbons to hydrogen + CO2);
    - inverter (DC-AC conversion, with export capacity);
    - heat exchangers (with the role of transferring residual heat from the exhaust and external cooling loop of the system);
    - components to ensure the functionality of the CHP system (pumps, valves, sensors, pipes, electronic control systems, etc.);
- Auxiliary thermal management
    - Auxiliary boiler with the role of satisfying the demand for heat during peak hours;
    - High efficiency heat tank;
- Control, interaction and feedback components
    - LCD touch-screen interface;
    - Remote control system;
    - Smart meter with role in measuring consumption and production;
    - Internet for remote monitoring and control.

Figure 10. Functional diagram of the SOFC fuel cell power plant.
Source: Badwal, S. P. S., et.al. (2015)

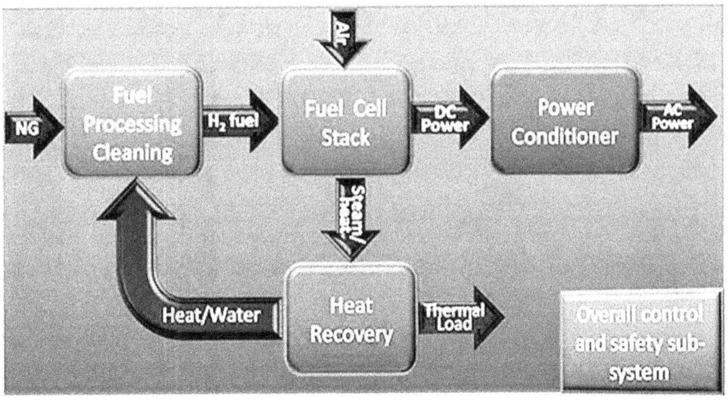

Yield of Electricity Production Systems and Electric and Thermal Cogeneration Systems

A comparison of the overall electric conversion efficiency is shown in Figure 11.

A comparison of the yields obtained by different ways of converting the chemical energy into electricity, with or without the passage through the mechanical energy, is presented in Figure 12 and 13.

Figure 11. Comparison between fuel cell conversion efficiency and other energy conversion systems
Source: Felseghi, R. A., et.al. (2019)

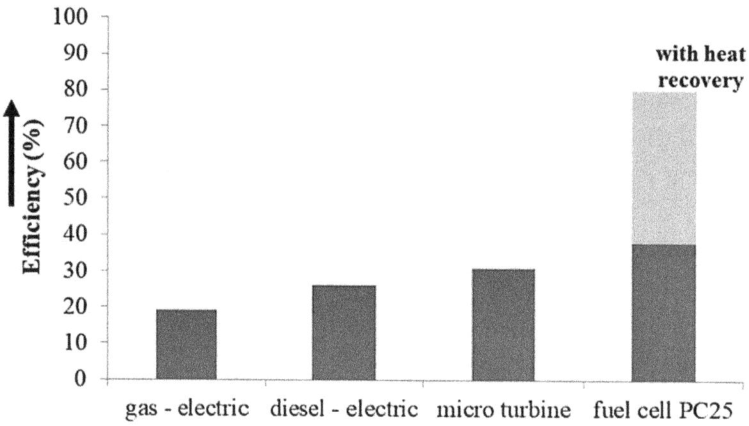

In order to make a comparison between the fuel cell and other energy production systems, namely the internal combustion engine, an evaluation of the system's efficiency is required. The overall electrical conversion efficiency of a fuel cell is higher than that of thermal motor systems.

Figure 12. Comparison of fuel cell conversion efficiency and other energy conversion systems

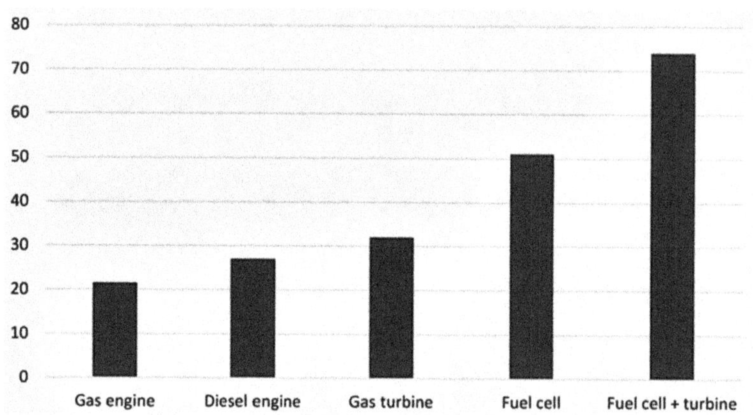

The isothermal efficiency of the fuel cells depends on the size of the reaction entropy which, when positive, implies higher yields than the unit, because some of the heat of the environment is taken up. This phenomenon is encountered in experimental situations and has no practical interest. At the rest of the reactions, very high theoretical yields can be found that far exceed the theoretical yield of the Carnot cycle of 30-50%. However, in practice, due to the internal polarization of the cell, voltage drops, etc. 50-60% yields are obtained which are quite high compared to other practical systems for converting various forms of energy. *(Badea G., et.al., 2013; Wilberforce, T., & Olabi, A. G., 2020)*

Figure 13. The efficiency of some electricity production systems

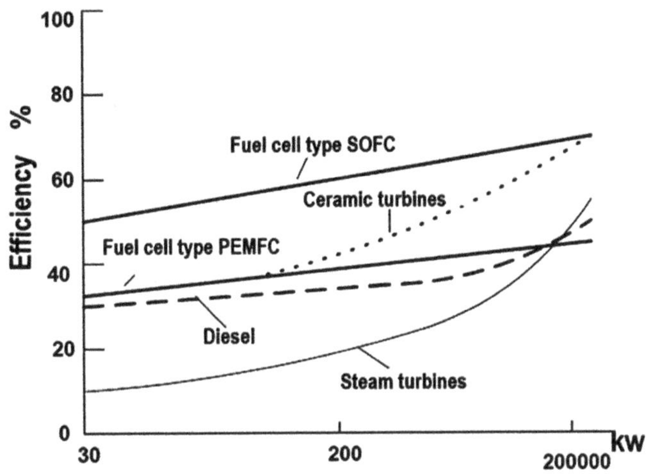

FINANCIAL ANALYSIS OF COGENERATION SYSTEMS

Specific Investment in the Case of Cogeneration Facilities

The specialized literature provides information on the characteristics and particularities typical of each category of energy cogeneration systems. In the following table this information is presented comparatively (see Table 2 and Table 3).

Table 2. Specific investment for cogeneration plants.

Type of Electrical Equipment	Typical Installation Costs (euro/kW)
Piston machine	800 ÷2,500
Gas turbine	650 ÷ 2,800
Gas turbine - combined cycle	750 ÷ 1,700
Steam turbine	550 ÷ 1,500
Geothermal system	1,850 ÷ 2,200
Fuel cell	5,500 ÷ 7,800
Wind turbine	1,200 ÷ 3,100
Hydroelectric turbine	600 ÷ 5,000
Photovoltaic module	5,200 ÷ 11,500

Source update_2020: Nissen, M. B. (2009).

Financial Considerations Regarding Fuel Cell CHP Systems

Fuel cells have been developing for the past 50 years, but are still relatively immature in commercial markets. The RDI departments, specialized entities and government agencies are still dependent on the

Table 3. Energy production systems (electrical and thermal) by cogeneration.

Characteristics	Steam Turbine	Heat Engine	Gas Turbine	Microturbines	Fuel Cell
Electric efficiency	15-40%	20-40%	20-40%	20-28%	30-65%
Overall efficiency	80%	72-88%	70-73%	60-76%	50-85%
Typical capacities MWe	0.5-300	0.01-6	0.4-300	0.01-0.30	0.004-3
Electric / Thermal report	0.1-0.3	0.5-1	0.5-2	0.4-0.7	between 1 and 2
Availability	~100%	92-97%	90-98%	90-98%	> 95%
Start time	1h-1day	15 sec	8 min-1h	55 sec	2h-2 day
Investment cost euro/KWhe	420-1,000	1,050-2,000	900-1,500	2,000-3,500	4,000-7,000
Operational cost euro/KWhe	< 0.004	0.008-0.023	0.003-0.010	0.010-0.030	0.030-0.040
Benefits	Good efficiency, high fuel range, good reliability.	Fast start-up, low cost of investment, good load scaling.	Good reliability, low emissions, high temperature heat.	Compact, few moving parts, low emissions, no cooling required.	Low noise, modular noise, constant efficiency at load variation.
Disadvantages	Slow start, poor electrical / thermal ratio.	Higher maintenance cost, noise, low temperature per thermal, generates heat even when this agent is not used	Requires high pressure on natural gas or local compressor, low efficiency on load variation, dependent on external temperature.	High cost, low mechanical efficiency, low thermal temperature.	High investment cost, relatively short lifespan, requires special fuel processing except for hydrogen.

Source update_2020: Mago, P. J., & Smith, A. D. (2015).

general and objective estimates for the total costs of a cogeneration system based on SOFC. Information on the historical and current prices of fuel cells, in particular SOFC cogeneration systems, is not widely disseminated due to trade secrecy and reduced production volumes. There is no way to determine if the projections presented in the literature and by the producers are feasible, optimistic or completely sustainable. The prices that are likely to be obtained in the near future are of great importance for government agencies and RDI companies alike, who must decide how to distribute the funding lines for research, development and grants in order to implement such projects underway incipient stage. *(Sammes, N. M., & Boersma, R., 2000; Nielsen, E. R., et.al., 2019; Lee, D., 2021).*

Cogeneration (CHP) is generally a promising application for SOFC. The governments of Japan, South Korea, Germany and the United States provide significant support to the research and development programs of these systems. *(Staffell, I., & Green, R., 2013; Hemmes, K., 2021; Baharuddin, N. A., et.al, 2021; Singla, M. K., et.al., 2021).*

When analyzing fuel cell cogeneration systems, academic and industrial estimates place the cost of a mass-produced fuel cell stack at about $ 500 / kW with an additional $ 500-1,000 / kW for the components of a complete system. of cogeneration. These estimates are related to cost targets set by government agencies. *(Nielsen, E. R., et.al., 2019; Lee, D., 2021; Singla, M. K., et.al., 2021).*

The US Department of Energy (DOE) aimed in 2007 to demonstrate the relaunch of a PEMFC system for cogeneration powered by fossil fuels at a manufacturing cost below $ 750 / kW by 2011 and under $ 450 / kW by year 2020. These targets have been revised to $ 1,200 / kW for 2015 and $ 1,000 / kW by

2020 for a 2 kW PEMFC system fueled by natural gas. Similarly, the SECA DOE program has set cost targets for stationary cogeneration systems with SOFC having a nominal capacity of between 3-10 kW, initially from \$ 800 / kW until 2005, then decreasing to \$700 / kW until in 2008 and \$ 400 / kW by 2010 [81, 82]. Their cost reduction efforts are now aimed at demonstrating fuel cell stacks manufactured at a cost of \$ 175 / kW and complete systems at a cost of \$ 700 / kW. Real sales prices are not perfectly in line with these goals. These are currently 25-50 times larger, even though some manufacturers have now started mass production. The most mature system has a commercial cost of about \$ 25,000 per kW in Japan, three years after it was launched into production [84]. Prices have dropped by more than half in just four years, however, in order to achieve the cost goal promoted by DOE, the systems require a cost reduction of about 90% over the next three years. *(Staffell, I., & Green, R., 2013; Hemmes, K., 2021; Baharuddin, N. A., et.al, 2021; Singla, M. K., et.al., 2021).*

Although the cost of manufacturing a fuel cell energy system is currently known, manufacturers do not give any indication of the costs of mass production for marketing. The transition from a low volume production to a highly specialized automatic assembly of production lines will bring huge reductions in production costs. *(Staffell, I., & Green, R., 2013; Hemmes, K., 2021; Baharuddin, N. A., et.al, 2021; Singla, M. K., et.al., 2021).*

To estimate the costs, the fuel cell energy system is broken down into components and then into individual materials, respectively production stages. The cost of each material, process and component is then estimated based on relevant information in the field or with the help of industrial cost estimation software, which contains a revised cost database for standard manufacturing goods and processes. The assumptions regarding system construction and performance are critical to the results. The fuel cell industry being close to the monopoly type, the cost of marketing the fuel cells exceeded the cost of production by a significant margin. Thus, the estimation of the costs regarding the fuel cell (cogeneration systems), presented in the specialized literature, varies greatly depending on the assumptions that are used and analyzed. *(Nielsen, E. R., et.al., 2019; Spendelow J. et al, 2012; Lee, D., 2021; Baharuddin, N. A., et.al, 2021; Singla, M. K., et.al., 2021).*

Some of the manufacturing companies, as well as the agencies involved in major demonstrations of fuel cell functionality, have published forecasts of future prices. These organizations are best placed to make such predictions because they currently have the highest experience in marketing CHP fuel cell systems.

Table 4. Cost, performance and sustainability objectives

	2015 Status	2020 Target	2030 Target
Electrical efficiency at rated power	40-45%	47%	>50%
CHP energy efficiency	68-80%	88%	90%
Equipment cost, natural gas	\$ 2,200-4,600 / kW	\$ 2,500 / kW	\$ 1,200 / kW
Installed cost, natural gas	\$ 3,400-5,500 / kW	\$ 3,500 / kW	\$ 1,800 / kW
Equipment cost, biogas	\$ 4,200-6,800 / kW	\$ 3,500 / kW	\$ 1,600 / kW
Installed cost, natural biogas	\$ 6,000-7,500 / kW	\$ 4,500 / kW	\$ 2,200 / kW
Number of planned/unplanned outages over lifetime	50	50	40
Operating lifetime	30,000-40,000 h	45,000 h	70,000 h
System availability	**90%**	**95%**	**98%**

The DOE Hydrogen and Fuel Cells Program Record has set targets and targets for performance, cost and sustainability for CHP with fuel cells running on natural gas (see Table 4.)

The report of the Assessment of the Distributed Generation Market Potential for Solid Oxide Fuel Cells by the US Department of Energy in collaboration with National Energy Technology Laboratory DOE / NETL- 342/093013, presents the main costs broken down into the two major components, respectively the cost of the cells module. fuel and cost auxiliary equipment BOP useful in the operation of the generator set.

A margin of 10% is expected to exceed the nominal design and installation costs to offset by 0.2% the degradation process of fuel cells at 1000 operating hours. This is based on the results of a detailed study highlighting the consideration of linear models of cell stack degradation in various operational scenarios, in order to optimize the additional costs, taking into account the life of the fuel cells and the frequency of replacing a stack of SOFC cell (see Table 5.) (*DOE/NETL, 2013*).

Table 5. Costs of the SOFC generator set. Source: DOE/NETL (2013)

Module Costs ($/kWe)	**$**
SOFC stack	225
Enclosure	30
Transport&placement	14
Site foundations	44
Inverter	68
Pre-reformer	29
Total Module	411
Total Module cith 10% extra installed area for 0.2%/1000 hr stack degradation	***462***
BOP Costs ($/kWe)	**$**
NQ desulfurizer	110
Cathode air blower	18
Cathode recycle gas blower	45
Cathode heat exchenger	56
Anode Recycle gas blower	17
Air combustor	51
NG Pre-Heater	2
Accessory Electric Plant	199
Instrumentation & Control	33
Total BOP	***531***
TOTAL SYSTEM	**983**

The US Department of Energy Cost Study for Manufacturing of Solid Oxide Fuel Cell Power Systems, in collaboration with Pacific Northwest National Laboratory PNNL-22732, presents a standard approach to the SOFC manufacturing process for a 270 kW capacity power system. calculated a total production cost of about $ 174,178. The highest share in total costs, excluding installation, was the BOP cost of $ 60,459, followed by the cost of the fuel cell stack of about $ 25,000 *(PNNL-22732, 2013; Chick, L., et.al., 2015).*

The study underlines the fact that currently this type of energy systems are manufactured in a small number, being analyzed production growth hypotheses developing the cost estimates at 50, 250, 1000, 4000 and 10,000 units per year.

DISCUSSIONS AND CONCLUDING REMARKS

When producing by conventional methods much of the energy contained in fuels is dissipated into the environment in the form of low heat. Cogeneration recovers this loss and simultaneously produces electricity and heat, more efficiently and at the same time, with lower pollutant emissions, compared to the conventional technologies.

Cogeneration, as already known, represents a cycle of transformation in which, starting from the fuel, it produces in the same installation and at the same time mechanical energy and thermal energy, in the vast majority of cases, the mechanical energy is transformed through a generator. electrically, in electricity.

Cogeneration, as a solution for combined and simultaneous production of electric and thermal energy, through the energetic, economic and ecological advantages it presents, falls under the category of "clean" technologies for energy production. What is interesting is the notion of cogeneration as a quality process - a cogeneration that allows a primary energy economy in relation to the production of electricity and heat. This implies a qualitative and a quantitative aspect.

From the qualitative point of view, the general rule is valid "A quality cogeneration implies that all the heat and electricity produced are effectively and efficiently used".

From a quantitative point of view, a quality cogeneration must have a high value of the global energy efficiency and a high value of the cogeneration index, which means a high quality of the fuel conversion.

In the case of using cogeneration as energy technology, currently, 85% of the energy contained in the fuel is transformed into useful energy (in the form of electricity and heat). This characteristic emphasizes that in cogeneration the fuel recovery is achieved with an energy efficiency of approximately double, compared to the efficiency in the case of electricity production by conventional processes.

In the long term, the solution can be provided by fuel cell fuel technology, an energy technology that has proven to have a significant impact on the supply of electricity and heat, making the transition to a fully renewable energy system possible.

The fuel cell technology is based on the principle of direct electrochemical energy conversion. In other words, the fuel cells are not similar to the conventional cogeneration plants because the combustion of the fuel does not take place, but the chemical energy of the fuel is converted directly into electricity, without any intermediate stage of mechanical energy production.

Cogeneration systems based on fuel cells offer extraordinary flexibility due to their modular design and lack of moving parts, which can be installed almost anywhere. Fuel cells can be installed very close to buildings allowing for easy recovery and efficient use of heat.

Fuel cells are very reliable and can provide constant supply in both centralized and independent power systems. Fuel cells can be used in decentralized power generation systems offering users autonomy in the production of electricity and heat, while fuel cells can be integrated into existing structures, as back-up or additional power and in areas that allow. metering the net consumption, having the possibility that the excess power produced by the fuel cells will be exported back to the power companies.

The fuel cell cogeneration system can become a power device using primary energy from 100% alternative sources. Also, the technology can provide renewable energy by using excess heat to generate hydrogen, this is known as cogeneration energy (electric and thermal) and hydrogen production - CHHP. This hydrogen then has the role of being reused by the cogeneration system or can be used in other applications.

Another noteworthy feature of fuel cell cogeneration systems is that they can use residual heat on refrigeration. It has been estimated that refrigeration requires almost 40% of the energy required in a supermarket. Cogeneration systems can offset this expense when combined with a chiller, which uses thermal energy to provide cooling.

The high efficiency in the production of electricity and the very low level of the pollutant emissions compared to the classic systems based on the burning of fossil fuels, make the fuel cells one of the alternative solutions, feasible, for the production of clean energy in the future. To these advantages is added the additional advantage of the possibility of using the heat generated in the process, increasing the overall efficiency of the use of fuels by cogeneration.

The combined production of electricity and heat in a co-generation system based on fuel cells is a way of increasing the energy efficiency in the clean energy production sector. Energy efficiency through primary energy savings, reduction of network losses, reduction of price and cost of electricity for consumers, reduction of environmental impact, especially of greenhouse gases, all in one place contributes to the security of energy supply to consumers.

Also, fuel cell cogeneration systems meet the criteria and conditions necessary for high efficiency cogeneration, meeting the requirements of the European Commission Directive.

The development of energy systems for cogeneration based on fuel cells is the objective of the various demonstration projects that will allow the validation of these energy production technologies as alternatives to the classical technologies.

In the long term, hydrogen fuel cell technology has the potential to put an end to dependence on oil imports into the world, drastically reduce carbon dioxide emissions and reduce the effects of global warming.

The vision of today's world energy policy corresponds to the concept of sustainable development and refers to the following important issues: consumer access to affordable and stable energy sources, sustainable development of energy production, transport and consumption, security of energy supply and emission reduction. of greenhouse gases. The concept aims to encourage smart economic growth, based on knowledge and innovation, sustainable and inclusive.

Science has shown that there are two alternatives for sustainable energy supply: renewable sources and fuel cells (hydrogen-based energy), which will play a complementary role in ensuring global energy resources.

REFERENCES

Abu-Heiba, A., Gluesenkamp, K. R., LaClair, T. J., Cheekatamarla, P., Munk, J. D., Thomas, J., & Boudreaux, P. R. (2021). Analysis of power conversion technology options for a self-powered furnace. *Applied Thermal Engineering*, *188*, 116627. doi:10.1016/j.applthermaleng.2021.116627

Adamson, K. A. (2006). *Large stationary survey*. Fuel Cell Today Ltd. www. fuelcelltoday. com

Ahmad, M. (2018). *Operation and control of renewable energy systems*. John Wiley & Sons.

Ahmadi, M. H., Sadaghiani, M. S., Pourfayaz, F., Ghazvini, M., Mahian, O., Mehrpooya, M., & Wongwises, S. (2018). Energy and exergy analyses of a solid oxide fuel cell-gas turbine-organic rankine cycle power plant with liquefied natural gas as heat sink. *Entropy (Basel, Switzerland)*, *20*(7), 484. doi:10.3390/e20070484 PMID:33265574

Ahmadi, S., Ghaebi, H., & Shokri, A. (2019). A comprehensive thermodynamic analysis of a novel CHP system based on SOFC and APC cycles. *Energy*, *186*, 115899. doi:10.1016/j.energy.2019.115899

Akroot, A., Namli, L., & Ozcan, H. (2021). Compared Thermal Modeling of Anode-and Electrolyte-Supported SOFC-Gas Turbine Hybrid Systems. *Journal of Electrochemical Energy Conversion and Storage*, *18*(1), 011001. doi:10.1115/1.4046185

Al-Attar, A. F., Farid, S. B., Hashim, F. A., & Eshraghi, M. J. (2019). Comparison in physical and mechanical properties between doped and non-doped Y-TZP electrolyte for HT-SOFC. *Energy Procedia*, *157*, 1285–1291. doi:10.1016/j.egypro.2018.11.294

Al-Qahtani, A., Parkinson, B., Hellgardt, K., Shah, N., & Guillen-Gosalbez, G. (2021). Uncovering the true cost of hydrogen production routes using life cycle monetisation. *Applied Energy*, *281*, 115958. doi:10.1016/j.apenergy.2020.115958

Andrews, J., & Shabani, B. (2012). Re-envisioning the role of hydrogen in a sustainable energy economy. *International Journal of Hydrogen Energy*, *37*(2), 1184–1203. doi:10.1016/j.ijhydene.2011.09.137

ANRE. (2015). *Metodologia de calcul pentru stabilirea cantităţilor de energie electrică produse în cogenerare de înaltă eficienţă în vederea certificării prin garanţii*. Academic Press.

Arun, A., Kumar, L., & Chowdhury, A. (2009). Structure-property relations for a phase-pure, nanograined tetragonal zirconia ceramic stabilised with minimum CaO doping. *Journal of the American Ceramic Society*.

Ashraf, M. A., Rashid, K., Rahimipetroudi, I., Kim, H. J., & Dong, S. K. (2020). Analyzing different planar biogas-fueled SOFC stack designs and their effects on the flow uniformity. *Energy*, *190*, 116450. doi:10.1016/j.energy.2019.116450

Badea, G., Felseghi, R. A., Giurca, I., Safirescu, O. C., Aşchilean, I., & Şoimoşan, T. M. (2013). Choosing the optimal fuel cell technology for application in the field of passive houses, Recent Advances in Urban Planning and Construction. *Proceedings of the 4th WSEAS International Conference USCUDAR, Energy*, 125 – 130.

Badea, G., Felseghi, R. A., Moldovan, E., & Safirescu, O. C. (2013). *Consideraţii privind utilizarea tehnologiei pilelor de combustibil în sisteme de cogenerare a energiei, Ştiinţa Modernă şi Energia XXXII*. Ed. Risoprint.

Badea, G., Felseghi, R. A., Răboaca, S., Aşchilean, I., Bolboacă, A., Mureşan, D., Moldovan, E., & Şoimoşan, T. (2016). Hybrid Solar and Wind Electric System for Romanian Nearly Zero Energy Buildings (nZEB)-Case Study. *Applied Mechanics and Materials*, *841*, 110–115. doi:10.4028/www.scientific.net/AMM.841.110

Badwal, S. P. S. (1992). Zirconia-based Solid Electrolytes: Microstructure, Stability and Ionic Conductivity. Solid State Lonics, 52, 23-32.

Badwal, S. P. S. (1994). Ceramic Superionic Conductors, Materials Science and Technology, A Comprehensive Treatment (vol. 11). VCH Verlagsgesellschaft.

Badwal, S. P. S., & Ciacchi, F. T. (2000). *Oxygen-ion Conducting Electrolyte Materials for Solid Oxide Fuel Cells. Ionics,* (6), 1–21.

Badwal, S. P. S., Deller, R., Foger, K., Ramprakash, Y., & Zhang, J. P. (1997). Interaction between chromia forming alloy interconnects and air electrode of solid oxide fuel cells. *Solid State Ionics, 99*(3-4), 297-310.

Badwal, S. P. S., Deller, R., Foger, K., Ramprakash, Y., & Zhang, J. P. (1997). Interaction between chromia forming alloy interconnects and air electrode of solid oxide fuel cells. *Solid State Ionics, 99*(3-4), 297–310. doi:10.1016/S0167-2738(97)00247-6

Badwal, S. P. S., Giddey, S., Munnings, C., & Kulkarni, A. (2015). Review of progress in high temperature solid oxide fuel cells. *ChemInform, 46*(31).

Baeyens, J., Zhang, H., Nie, J., Appels, L., Dewil, R., Ansart, R., & Deng, Y. (2020). Reviewing the potential of bio-hydrogen production by fermentation. *Renewable & Sustainable Energy Reviews, 131*, 110023. doi:10.1016/j.rser.2020.110023

Baharuddin, N. A., Yusoff, W. W., Abd Aziz, A. J., & Tahir, N. M. (2021, February). Hydrogen fuel cells for sustainable energy: Development and progress in selected developed countries. *IOP Conference Series. Materials Science and Engineering, 1078*(1), 012011. doi:10.1088/1757-899X/1078/1/012011

Baldi, F., Moret, S., Tammi, K., & Maréchal, F. (2020). The role of solid oxide fuel cells in future ship energy systems. *Energy, 194*, 116811. doi:10.1016/j.energy.2019.116811

Balog, A. A., Cobîrzan, N., Thalmaier, G., Constantinescu, H., & Voinea, M. (2019). Psyllium Seeds Used as Pore Forming Agent in Clay Bricks. *Procedia Manufacturing, 32*, 242–247. doi:10.1016/j.promfg.2019.02.209

Barelli, L., Bidini, G., Gallorini, F., & Ottaviano, P. A. (2013). Design optimization of a SOFC-based CHP system through dynamic analysis. *International Journal of Hydrogen Energy, 38*(1), 354–369. doi:10.1016/j.ijhydene.2012.09.162

Barrovecchio, L., Gehain, E., & Gibbard, F. (2001). A proton Exchange Membrane Fuel Cell in a House. *International Gas Research Conference*, 246.

Beigzadeh, M., Pourfayaz, F., Ghazvini, M., & Ahmadi, M. H. (2021). Energy and exergy analyses of solid oxide fuel cell-gas turbine hybrid systems fed by different renewable biofuels: A comparative study. *Journal of Cleaner Production, 280*, 124383. doi:10.1016/j.jclepro.2020.124383

Bianco, M., Poitel, S., Hong, J. E., Yang, S., Wang, Z. J., Willinger, M., Steinberger-Wilckens, R., & Van herle, J. (2020). Corrosion behaviour of nitrided ferritic stainless steels for use in solid oxide fuel cell devices. *Corrosion Science, 165*, 108414. doi:10.1016/j.corsci.2019.108414

Bicer, Y., & Khalid, F. (2020). Life cycle environmental impact comparison of solid oxide fuel cells fueled by natural gas, hydrogen, ammonia and methanol for combined heat and power generation. *International Journal of Hydrogen Energy*, *45*(5), 3670–3685. doi:10.1016/j.ijhydene.2018.11.122

Blum, L., Fang, Q., Gross-Barsnick, S. M., de Haart, L. B., Malzbender, J., Menzler, N. H., & Quadakkers, W. J. (2020). Long-term operation of solid oxide fuel cells and preliminary findings on accelerated testing. *International Journal of Hydrogen Energy*, *45*(15), 8955–8964. doi:10.1016/j.ijhydene.2020.01.074

Bochentyn, B., Chlipała, M., Gazda, M., Wang, S. F., & Jasiński, P. (2019). Copper and cobalt co-doped ceria as an anode catalyst for DIR-SOFCs fueled by biogas. *Solid State Ionics*, *330*, 47–53. doi:10.1016/j.ssi.2018.12.007

Brandon, N., Dincer, I., Shao, Z. P., Xia, C., Hill, J. M., Shikazono, N., ... Schüth, F. (2013). Solid oxide fuel cells: from materials to system modeling. *Royal Society of Chemistry*.

Chen, J., Chen, Y., Zhang, H., & Weng, S. (2018). Effect of different operating strategies for a SOFC-GT hybrid system equipped with anode and cathode ejectors. *Energy*, *163*, 1–14. doi:10.1016/j.energy.2018.08.032

Chick, L., Weimar, M., Whyatt, G., & Powell, M. (2015). The case for natural gas fueled solid oxide fuel cell power systems for distributed generation. *Fuel Cells (Weinheim)*, *15*(1), 49–60. doi:10.1002/fuce.201400103

Chitgar, N., Emadi, M. A., Chitsaz, A., & Rosen, M. A. (2019). Investigation of a novel multigeneration system driven by a SOFC for electricity and fresh water production. *Energy Conversion and Management*, *196*, 296–310. doi:10.1016/j.enconman.2019.06.006

Chitgar, N., & Moghimi, M. (2020). Design and evaluation of a novel multi-generation system based on SOFC-GT for electricity, fresh water and hydrogen production. *Energy*, *197*, 117162. doi:10.1016/j.energy.2020.117162

Choudhary, T., & Sanjay. (2017). Thermodynamic assessment of advanced SOFC-blade cooled gas turbine hybrid cycle. *International Journal of Hydrogen Energy*, *42*(15), 10248–10263. doi:10.1016/j.ijhydene.2017.02.178

Chowdhury, T., Chowdhury, H., Hossain, N., Ahmed, A., Hossen, M. S., Chowdhury, P., Thirugnanasambandam, M., & Saidur, R. (2020). Latest advancements on livestock waste management and biogas production: Bangladesh's perspective. *Journal of Cleaner Production*, *272*, 122818. doi:10.1016/j.jclepro.2020.122818

Cieciura-Włoch, W., Borowski, S., & Otlewska, A. (2020). Biohydrogen production from fruit and vegetable waste, sugar beet pulp and corn silage via dark fermentation. *Renewable Energy*, *153*, 1226–1237. doi:10.1016/j.renene.2020.02.085

Cipitì, F., Recupero, V., Pino, L., Vita, A., & Laganà, M. (2006). Experimental analysis of a 2 kWe LPG-based fuel processor for polymer electrolyte fuel cells. *Journal of Power Sources*, *157*(2), 914–920. doi:10.1016/j.jpowsour.2006.02.084

Cobîrzan, N., Thalmaier, G., Balog, A. A., Constantinescu, H., Timiş, I., & Streza, M. (2018). Thermophysical properties of fired clay bricks with waste ceramics and paper pulp as pore-forming agent. *Journal of Thermal Analysis and Calorimetry*, *134*(1), 843–851. doi:10.100710973-018-7092-3

Costa, P., Pinto, F., André, R. N., & Marques, P. (2021). Integration of Gasification and Solid Oxide Fuel Cells (SOFCs) for Combined Heat and Power (CHP). *Processes (Basel, Switzerland)*, *9*(2), 254. doi:10.3390/pr9020254

Dicks, A. L. (1996). Hydrogen generation from natural gas for the fuel cell systems of tomorrow. *Journal of Power Sources*, *61*(1-2), 113–124. doi:10.1016/S0378-7753(96)02347-6

Ding, X., Gao, Z., Ding, D., Zhao, X., Hou, H., Zhang, S., & Yuan, G. (2019). Cation deficiency enabled fast oxygen reduction reaction for a novel SOFC cathode with promoted CO2 tolerance. *Applied Catalysis B: Environmental*, *243*, 546–555. doi:10.1016/j.apcatb.2018.10.075

Directiva 2004/8/CE a Parlamentului European și a Consiliului din 11 februarie 2004 privind promovarea cogenerării pe baza cererii de energie termică utilă pe piața internă a energiei şi de modificare a Directivei 92/42/CEE.

Echigo, M., Shinke, N., Takami, S., & Tabata, T. (2004). Performance of a natural gas fuel processor for residential PEFC system using a novel CO preferential oxidation catalyst. *Journal of Power Sources*, *132*(1-2), 29–35. doi:10.1016/j.jpowsour.2004.01.031

Faleh, S., Khir, T., & Brahim, A. B. (2017). Energetic Performance Optimization of a SOFC–GT Hybrid Power Plant. *Arabian Journal for Science and Engineering*, *42*(4), 1505–1515. doi:10.100713369-016-2363-4

Fallah Vostakola, M., & Horri, B. A. (2021). Progress in Material Development for Low-Temperature Solid Oxide Fuel Cells: A Review. *Energies*, *14*(5), 1280. doi:10.3390/en14051280

Fan, L. (2020). Solid-State Electrolytes for SOFC. *Solid Oxide Fuel Cells: From Electrolyte-Based to Electrolyte-Free Devices*, 35-78.

Felseghi, R. A. (2015). *Contributions regarding the application of fuel cell in the passive houses domain* (Unpublished doctoral dissertation). Technical University of Cluj-Napoca, Romania.

Felseghi, R. A., Carcadea, E., Raboaca, M. S., Trufin, C. N., & Filote, C. (2019). Hydrogen fuel cell technology for the sustainable future of stationary applications. *Energies*, *12*(23), 4593.

Fergus, J. W. (2004). Lanthanum chromite-based materials for solid oxide fuel cell interconnects. *Solid State Ionics, 171*(1-2), 1-15.

Fragiacomo, P., Corigliano, O., & De Lorenzo, G. (2018). Design of an SOFC/SOE experimental station: Planning of simulation tests. *Energy Procedia*, *148*, 535–542.

Fuel Cell Handbook. (2004). *EG&G Technical Service Inc., U.S. Department of Energy* (7th ed.). Office of Fossil Energy.

Funahashi, Y., Shimamori, T., Suzuki, T., Fujishiro, Y., & Awano, M. (2007). Fabrication and characterization of components for cube shaped micro tubular SOFC bundle. *Journal of Power Sources, 163*(2), 731–736.

Gandiglio, M., Lanzini, A., & Santarelli, M. (2018). Large stationary solid oxide fuel cell (SOFC) power plants. In *Modeling, Design, Construction, and Operation of Power Generators with Solid Oxide Fuel Cells* (pp. 233–261). Springer.

Gandiglio, M., Lanzini, A., Santarelli, M., Acri, M., Hakala, T., & Rautanen, M. (2020). Results from an industrial size biogas-fed SOFC plant (the DEMOSOFC project). *International Journal of Hydrogen Energy, 45*(8), 5449–5464.

Green, R., & Staffell, I. (2012). *The cost of domestic fuel cell micro-CHP systems. Imperial College.* Imperial College Business School.

Gross-Barsnick, S. M. (2020). Interaction of glass-ceramic sealants with solid oxide fuel cell components: thermo-mechanical analysis. In *Intermediate Temperature Solid Oxide Fuel Cells* (pp. 411–426). Elsevier.

Gummert, G., & Suttor, W. (2006). *Stationary fuel cells. Technology and market; Stationaere Brennstoffzellen.* Technik und Markt.

Guski, V., Verestek, W., Rapp, D., & Schmauder, S. (2021). Microstructural Investigation of Plasma Sprayed Ceramic Coatings Focusing on the Effect of the Splat Boundary for SOFC Sealing Applications Using Peridynamics. *Theoretical and Applied Fracture Mechanics.*

Habibi, R., & Mehrpooya, M. (2021). A novel integrated Ca-Cu cycle with coal/biomass gasification unit for clean hydrogen production. *Energy Conversion and Management, 228*, 113682.

Hagen, A., Wulff, A. C., Zielke, P., Sun, X., Talic, B., Ritucci, I., ... Hendriksen, P. V. (2020). SOFC stacks for mobile applications with excellent robustness towards thermal stresses. *International Journal of Hydrogen Energy, 45*(53), 29201–29211.

Hagiwara, A. (1999, July). Fuel cell technologies in search of new directions: based on the experiences over 20 years. In *1999 IEEE Power Engineering Society Summer Meeting. Conference Proceedings* (Cat. No. 99CH36364) (Vol. 1, pp. 576-580). IEEE.

Hajabdollahi, Z., & Fu, P. F. (2017). Multi-objective based configuration optimization of SOFC-GT cogeneration plant. *Applied Thermal Engineering, 112*, 549–559.

Hayun, H., Wolf, R., Barad, C., & Gelbstein, Y. (2020). Thermal shock resistant solid oxide fuel cell ceramic composite electrolytes. *Journal of Alloys and Compounds, 821*, 153490.

Hemmes, K. (2021). A personal retrospect on three decades of high temperature fuel cell research; ideas and lessons learned. *International Journal of Hydrogen Energy.*

Hirschenhofer, J. H., Sauffer, D. B., & Engleman, R. R. (1994). *Fuel Cell: A Handbook (Revision 3).* U.S. Department of Energy, Morgantown Energy Technology Center.

Holtappels, P., Vogt, U., & Graule, T. (2005). Ceramic materials for advanced solid oxide fuel cells. *Advanced Engineering Materials, 7*(5), 292–302.

Homma, T. (2006). *The latest fuel cell news in Japan*. http://www. fcdic. com/eng/news/200808. html

Hoogers, G. (2003). *Fuel Cell Technology Handbook*. CRC Press.

Hou, Q., Wu, Y., Liu, H., Xu, X., & Xu, B. (2020, June). Numerical Simulation of Electrical Performance and Distribution of SOFC Stacks With Different Manifold Arrangement. In *ASME 2020 14th International Conference on Energy Sustainability*. American Society of Mechanical Engineers Digital Collection.

Hu, Y. Z., Su, Y. T., Li, C. X., Li, C. J., & Yang, G. J. (2020). Dense Mn1. 5Co1. 5O4 coatings with excellent long-term stability and electrical performance under the SOFC cathode environment. *Applied Surface Science*, *499*, 143726.

Ilbas, M., Kumuk, B., Alemu, M. A., & Arslan, B. (2020). Numerical investigation of a direct ammonia tubular solid oxide fuel cell in comparison with hydrogen. *International Journal of Hydrogen Energy*, *45*(60), 35108–35117.

Ionescu, C., Tuţică, D., Pătraşcu, R., Dincă, C., & Slavu, N. (2019). Evaluation of the energy efficiency of an industrial consumer in trigeneration mode. In *E3S Web of Conferences* (Vol. 85, p. 01005). EDP Sciences.

Irshad, M., Idrees, R., Siraj, K., Shakir, I., & Rafique, M., ul Ain, Q., & Raza, R. (2021). Electrochemical evaluation of mixed ionic electronic perovskite cathode LaNi1-xCoxO3-δ for IT-SOFC synthesized by high temperature decomposition. *International Journal of Hydrogen Energy*, *46*(17), 10448–10456.

Ishaq, H., & Dincer, I. (2021). Investigation and optimization of a new hybrid natural gas reforming system for cascaded hydrogen, ammonia and methanol synthesis. *Computers & Chemical Engineering*, 107234.

Ivers-Tiffée, E., Weber, A., & Herbstritt, D. (2001). Materials and technologies for SOFC-components. *Journal of the European Ceramic Society*, *21*(10-11), 1805–1811.

Jamil, S. M., Othman, M. H. D., Rahman, M. A., Jaafar, J., Ismail, A. F., Honda, S., & Iwamoto, Y. (2019). Properties and performance evaluation of dual-layer ceramic hollow fiber with modified electrolyte for MT-SOFC. *Renewable Energy*, *134*, 1423–1433.

Javidmehr, M., Pourfayaz, F., & Kasaeian, A. (2021). Optimal design and analysis of a district energy system including heat and power production for domestic applications and fuel for vehicles. *Journal of Thermal Analysis and Calorimetry*, 1–17.

Kalra, P., Garg, R., & Kumar, A. (2013). Solid oxide fuel cell-a future source of power and heat generation. *Materials Science Forum*, *757*, 217–241.

Khajavi, P., Frandsen, H. L., Gremillard, L., Chevalier, J., & Hendriksen, P. V. (2021). Strength and hydrothermal stability of NiO–stabilized zirconia solid oxide cells fuel electrode supports. *Journal of the European Ceramic Society*.

Khan, M. Z., Song, R. H., Hussain, A., Lee, S. B., Lim, T. H., & Hong, J. E. (2020). Effect of applied current density on the degradation behavior of anode-supported flat-tubular solid oxide fuel cells. *Journal of the European Ceramic Society*, *40*(4), 1407–1417.

Kharton, V. V., Marques, F. M. B., & Atkinson, A. (2004). *Transport Properties of Solid Oxide Electrolyte Ceramics: a Brief Review*. Solid State Ionics.

Kim, Y. S., Lee, Y. D., & Ahn, K. Y. (2020). System integration and proof-of-concept test results of SOFC–engine hybrid power generation system. *Applied Energy*, *277*, 115542.

Koirala, B., Hers, S., Morales-España, G., Özdemir, Ö., Sijm, J., & Weeda, M. (2021). Integrated electricity, hydrogen and methane system modelling framework: Application to the Dutch Infrastructure Outlook 2050. *Applied Energy*, *289*, 116713.

Kumar, V., Kaur, M., Kaur, G., Arya, S. K., & Pickrell, G. (2020). Stacking designs and sealing principles for IT-solid oxide fuel cell. In *Intermediate Temperature Solid Oxide Fuel Cells* (pp. 379–410). Elsevier.

Kurapova, O. Y., Shugurov, S. M., Vasil'eva, E. A., Savelev, D. A., Konakov, V. G., & Lopatin, S. I. (2020). Thermal prehistory, structure and high-temperature thermodynamic properties of Y2O3-CeO2 and Y2O3-ZrO2-CeO2 solid solutions. *Ceramics International*.

Kutcherov, V., Morgunova, M., Bessel, V., & Lopatin, A. (2020). Russian natural gas exports: An analysis of challenges and opportunities. *Energy Strategy Reviews*, *30*, 100511.

Laurencelle, F., Chahine, R., Hamelin, J., Agbossou, K., Fournier, M., Bose, T. K., & Laperriere, A. (2001). Characterization of a Ballard MK5-E Proton Exchange Membrane Fuel Cell Stack. *Fuel Cells (Weinheim)*, *1*(1), 66–71.

Lawlor, V., Griesser, S., Buchinger, G., Olabi, A. G., Cordiner, S., & Meissner, D. (2009). Review of the micro-tubular solid oxide fuel cell: Part I. Stack design issues and research activities. *Journal of Power Sources*, *193*(2), 387–399.

Lee, D. (2021). Functional material developments of fuel cells and the key factors for real commercialization of next-generation energy devices. In *Sustainable Materials for Next Generation Energy Devices* (pp. 85–100). Elsevier.

Lee, H., Kim, U. S., Kim, S. D., Woo, S. K., & Chung, W. J. (2020). SiO2–B2O3-BaO-WO3 glasses with varying Al2O3 content as a sealing material for reversible solid oxide fuel cells. *Ceramics International*, *46*(11), 18256–18261.

Li, R., Liu, F., Zhang, C., Liu, J., Zhou, J., & Xu, L. (2019). Electrical properties of Fe-doped SrTiO3 with B-site-deficient for SOFC anodes. *Ceramics International*, *45*(17), 21684–21687.

Mago, P. J., & Smith, A. D. (2015). Combined Cooling, Heating, and Power (CCHP) Systems. Handbook of Clean Energy Systems, 1-15.

Mahmud, L. S., Muchtar, A., & Somalu, M. R. (2017). Challenges in fabricating planar solid oxide fuel cells: A review. *Renewable & Sustainable Energy Reviews*, *72*, 105–116.

Midilli, A., Ay, M., Dincer, I., & Rosen, M. A. (2005). On hydrogen and hydrogen energy strategies. I: Current status and needs, J. *Renewable & Sustainable Energy Reviews*, *9*, 255–271.

Midilli, A., Ay, M., Dincer, I., & Rosen, M. A. (2005). On hydrogen and hydrogen energy strategies. II: Future projections affecting global stability and unrest, J. *Renewable & Sustainable Energy Reviews*, *9*, 273–287.

Milcarek, R. J., & Ahn, J. (2021). Micro-Tubular Solid Oxide Fuel Cell Polarization and Impedance Variation With Thin Porous Samarium-Doped Ceria and Gadolinium-Doped Ceria Buffer Layer Thickness. *Journal of Electrochemical Energy Conversion and Storage*, *18*(2), 021004.

Minh, N., Mizusaki, J., & Singhal, S. C. (2017). Advances in solid oxide fuel cells: Review of progress through three decades of the international symposia on solid oxide fuel cells. *ECS Transactions*, *78*(1), 63.

Minutillo, M., Perna, A., Di Trolio, P., Di Micco, S., & Jannelli, E. (2021). Techno-economics of novel refueling stations based on ammonia-to-hydrogen route and SOFC technology. *International Journal of Hydrogen Energy*, *46*(16), 10059–10071.

Mo, B., Rix, J., Pal, U., Basu, S., & Gopalan, S. (2020). Improving SOFC Anode Electrocatalytic Activity Using Nanoparticle Infiltration into MIEC Compositions. *Journal of the Electrochemical Society*, *167*(13), 134506.

Mojaver, P., Chitsaz, A., Sadeghi, M., & Khalilarya, S. (2020). Comprehensive comparison of SOFCs with proton-conducting electrolyte and oxygen ion-conducting electrolyte: Thermoeconomic analysis and multi-objective optimization. *Energy Conversion and Management*, *205*, 112455.

Mudryk, K., & Werle, S. (2018). *Renewable energy sources: engineering, technology, innovation*. Springer International Publishing.

Müller, M., Bouyer, E. V. Bradke, M., Branston, D. W., Heimann, R. B., Henne, R., ... Schiller, G. (2002). Thermal induction plasma processes for the synthesis of SOFC materials. *Materialwissenschaft und Werkstofftechnik: Entwicklung, Fertigung, Prüfung, Eigenschaften und Anwendungen technischer Werkstoffe*, *33*(6), 322-330.

Nassef, A. M., Fathy, A., Sayed, E. T., Abdelkareem, M. A., Rezk, H., Tanveer, W. H., & Olabi, A. G. (2019). Maximizing SOFC performance through optimal parameters identification by modern optimization algorithms. *Renewable Energy*, *138*, 458–464.

Nielsen, E. R., Prag, C. B., Bachmann, T. M., Carnicelli, F., Boyd, E., Walker, I., ... Stephens, A. (2019). Status on Demonstration of Fuel Cell Based Micro-CHP Units in Europe. *Fuel Cells (Weinheim)*, *19*(4), 340–345.

Nissen, M. B. (2009). High performance development as distributed generation. *IEEE Potentials*, *28*(6), 25–31.

Park, J. H., Kim, D., Kim, K. J., Bae, K. T., & Lee, K. T. (2020). A brief review of the bilayer electrolyte strategy to achieve high performance solid oxide fuel cells. *Ceramist*, *23*(2), 184–199.

Puente-Martínez, D. E., Díaz-Guillén, J. A., Montemayor, S. M., Díaz-Guillén, J. C., Burciaga-Díaz, O., Bazaldúa-Medellín, M. E., ... Fuentes, A. F. (2020). High ionic conductivity in CeO2 SOFC solid electrolytes; effect of Dy doping on their electrical properties. *International Journal of Hydrogen Energy*, *45*(27), 14062–14070.

Radenahmad, N., Azad, A. T., Saghir, M., Taweekun, J., Bakar, M. S. A., Reza, M. S., & Azad, A. K. (2020). A review on biomass derived syngas for SOFC based combined heat and power application. *Renewable & Sustainable Energy Reviews*, *119*, 109560.

Ramis, G., Bahadori, E., & Rossetti, I. (2020). Design of efficient photocatalytic processes for the production of hydrogen from biomass derived substrates. *International Journal of Hydrogen Energy*.

Riensche, E., Stimming, U., & Unverzagt, G. (1998). Optimization of a 200 kW SOFC cogeneration power plant: Part I: Variation of process parameters. *Journal of Power Sources*, *73*(2), 251–256.

Rillo, E., Gandiglio, M., Lanzini, A., Bobba, S., Santarelli, M., & Blengini, G. (2017). Life cycle assessment (LCA) of biogas-fed solid oxide fuel cell (SOFC) plant. *Energy*, *126*, 585–602.

Romanian Law no. 13/2007, Electricity Law; Legea nr. 13/2007, Legea energiei electrice.

Roshandel, R., Golzar, F., & Astaneh, M. (2018). Technical, economic and environmental optimization of combined heat and power systems based on solid oxide fuel cell for a greenhouse case study. *Energy Conversion and Management*, *164*, 144–156.

Rosner, F., Rao, A., & Samuelsen, S. (2020). Economics of cell design and thermal management in solid oxide fuel cells under SOFC-GT hybrid operating conditions. *Energy Conversion and Management*, *220*, 112952.

Sadik-Zada, E. R., & Gatto, A. (2021). Energy security pathways in South East Europe: Diversification of the natural gas supplies, energy transition, and energy futures. In *From Economic to Energy Transition* (pp. 491–514). Palgrave Macmillan.

Sadykov, V. A., Sadovskaya, E. M., Filonova, E. A., Eremeev, N. F., Bogdanovich, N. M., Pikalov, S. M., ... Pikalova, E. Y. (2020). Mixed ionic-electronic conductivity features of A-site deficient Nd nickelates. *Ceramics International*, *46*(16), 25527–25535.

Sammes, N., Bove, R., & Stahl, K. (2004). Phosphoric acid fuel cells: Fundamentals and applications. *Current Opinion in Solid State and Materials Science*, *8*(5), 372–378.

Sammes, N. M., & Boersma, R. (2000). Small-scale fuel cells for residential applications. *Journal of Power Sources*, *86*(1-2), 98–110.

Sarruf, B. J. M., Hong, J. E., Steinberger-Wilckens, R., & de Miranda, P. E. V. (2020). Ceria-Co-Cu-based SOFC anode for direct utilisation of methane or ethanol as fuels. *International Journal of Hydrogen Energy*, *45*(8), 5297–5308.

Schafföner, S., Fruhstorfer, J., Faßauer, C., Freitag, L., Jahn, C., & Aneziris, C. G. (2017). Influence of in situ phase formation on properties of calcium zirconate refractories. *Journal of the European Ceramic Society*, *37*(1), 305–313.

Schneider, S., Bajohr, S., Graf, F., & Kolb, T. (2020). State of the art of hydrogen production via pyrolysis of natural gas. *ChemBioEng Reviews*, *7*(5), 150–158.

Scott Weil, K. (2006). The State-Of-The-Art in Sealing Technology for Solid Oxide Fuel Cells. *J. of Materials, 58*, 37-44.

Shah, M., Mondal, P., Nayak, A. K. A. K., & Bordoloi, A. (2017). *Hydrogen from natural gas. Sustainable Utilization of Natural Resources*. CRC Press.

Shaur, A., Rehman, S. U., Kim, H. S., Song, R. H., Lim, T. H., Hong, J. E., ... Lee, S. B. (2020). Hybrid Electrochemical Deposition Route for the Facile Nanofabrication of a Cr-Poisoning-Tolerant La (Ni, Fe) O3− δ Cathode for Solid Oxide Fuel Cells. *ACS Applied Materials & Interfaces*, *12*(5), 5730–5738.

Silva, E. D. R., Curi, M., Nicolini, J. V., Furtado, J. G., Secchi, A. R., & Ferraz, H. C. (2021). Effect of doping concentration and sintering atmosphere on the microstructural and electrical characteristics of Y-doped SrTiO3 perovskite anode for SOFC. *Ceramics International*.

Singh, R. N. (2007). Sealing technology for solid oxide fuel cells (SOFC). *International Journal of Applied Ceramic Technology*, *4*(2), 134–144.

Singla, M. K., Nijhawan, P., & Oberoi, A. S. (2021). Hydrogen fuel and fuel cell technology for 61a cleaner future: 61a review. *Environmental Science and Pollution Research International*, 1–20.

Song, Y., Han, K., & Wang, D. Y. (2020). Thermodynamic analysis of fossil fuels reforming for fuel cell application. *International Journal of Hydrogen Energy*, *45*(39), 20232–20239.

Spendelow, J. (2012). DOE Hydrogen and Fuel Cells Program Record - Medium-scale CHP Fuel Cell System Targets. U.S. Department of Energy.

Staffell, I. (2010). *Fuel cells for domestic heat and power: are they worth it?* (Doctoral dissertation). University of Birmingham.

Staffell, I., & Green, R. (2013). The cost of domestic fuel cell micro-CHP systems. *International Journal of Hydrogen Energy*, *38*(2), 1088–1102.

Ştefănescu, I. (2010). *Pile de combustibil – între teorie şi practică, Editura CONPHYS, Rm*. Vâlcea.

Stokes. (2011). *Stokes, Solid Oxide Fuel Cell Applications*. Hydrogen Technical Advisory Committee.

Stolten, D., & Emonts, B. (Eds.). (2012). *Fuel Cell Science and Engineering, 2 Volume Set: Materials, Processes, Systems and Technology* (Vol. 1). John Wiley & Sons.

Stolyarova, V. L., Vorozhtcov, V. A., Masaki, K., & Costa, D. (2020). High-temperature mass spectrometric study of thermodynamic properties in the UO2–ZrO2 system. *Rapid Communications in Mass Spectrometry*, *34*(19), e8862.

Su, S., Zhang, S., Yan, C., Yang, Z., Zheng, F., & Zhang, L. (2017). Evaluations of Fuel and Air Flow Distributing Qualities within SOFC Stack Designs with Flow Manifolds Penetrated Through Cell Plane. *International Journal of Electrochemical Science*, *12*, 230–239.

Thalmaier, G., Cobîrzan, N., Balog, A. A., Constantinescu, H., Streza, M., Nasui, M., & Neamtu, B. V. (2020). Influence of sawdust particle size on fired clay brick properties. *Materiales de Construcción*, *70*(338), 215.

Tietz, F., Buchkremer, H. P., & Stöver, D. (2002). Components manufacturing for solid oxide fuel cells. *Solid State Ionics*, *152*, 373–381.

Venâncio, S. A., & de Miranda, P. E. V. (2017). Direct utilization of carbonaceous fuels in multifunctional SOFC anodes for the electrosynthesis of chemicals or the generation of electricity. *International Journal of Hydrogen Energy*, *42*(19), 13927–13938.

Virkar, A. V. (2001, February). SOFC materials and processing issues. SECA core technology program workshop.

Voronca, Ş. D., Darie, G., & Pătraşcu, R. (2017, October). Biomass ORC cogeneration—A case study. In *2017 International Conference on ENERGY and ENVIRONMENT (CIEM)* (pp. 376-380). IEEE.

Wang, X., Wei, K., Yan, S., Wu, Y., Kang, J., Feng, P., ... Ling, Y. (2020). Efficient and stable conversion of oxygen-bearing low-concentration coal mine methane by the electrochemical catalysis of SOFC anode: From pollutant to clean energy. *Applied Catalysis B: Environmental*, *268*, 118413.

Welaya, Y. M., El Gohary, M. M., & Ammar, N. R. (2012). Steam and partial oxidation reforming options for hydrogen production from fossil fuels for PEM fuel cells. *Alexandria Engineering Journal*, *51*(2), 69–75.

Wilberforce, T., & Olabi, A. G. (2020). Design of experiment (DOE) analysis of 5-cell stack fuel cell using three bipolar plate geometry designs. *Sustainability*, *12*(11), 4488.

Winkler, W., Behling, N., Behling, T., Managi, S., & Williams, M. C. (2020). Options for Natural Gas and Methane Including Fuel Cell Utilization in a Sustainable Energy Infrastructure. *ECS Transactions*, *96*(1), 81.

Wołowicz, M., Kolasiński, P., & Badyda, K. (2021). Modern Small and Microcogeneration Systems—A Review. *Energies*, *14*(3), 785.

Xiong, G., Zhang, J., Shi, D., Zhu, L., & Yuan, X. (2021). Optimal identification of solid oxide fuel cell parameters using a competitive hybrid differential evolution and Jaya algorithm. *International Journal of Hydrogen Energy*, *46*(9), 6720–6733.

Xu, J., Zhou, X., Dong, X., Pan, L., & Sun, K. (2017). Catalytic activity of infiltrated La0. 3Sr0. 7Ti0. 3Fe0. 7O3− δ–CeO2 as a composite SOFC anode material for H2 and CO oxidation. *International Journal of Hydrogen Energy*, *42*(23), 15632–15640.

Yadav, A., Pyare, R., Goodenough, J. B., & Singh, P. (2020). KTa1−x−y Ti x Ge y O3− δ: A High κ Relaxor Dielectric and Superior Oxide-Ion Electrolyte for IT-SOFC. *ACS Applied Energy Materials*, *3*(4), 3205–3211.

Yang, B., Wang, J., Zhang, M., Shu, H., Yu, T., Zhang, X., ... Sun, L. (2020). A state-of-the-art survey of solid oxide fuel cell parameter identification: Modelling, methodology, and perspectives. *Energy Conversion and Management*, *213*, 112856.

Yang, C. H., Chang, S. C., Chan, Y. H., & Chang, W. S. (2019). A dynamic analysis of the multi-stack SOFC-CHP system for power modulation. *Energies*, *12*(19), 3686.

Yang, J., Huang, W., Wang, X., Li, J., Yan, D., Pu, J., ... Li, J. (2018). Study on component interface evolution of a solid oxide fuel cell stack after long term operation. *Journal of Power Sources*, *387*, 57–63.

Yang, Z. (2008). Recent advances in metallic interconnects for solid oxide fuel cells. *International Materials Reviews, 53*(1), 39-54.

Yixiao, C., Chen, Y., Akbar, M., Jin, B., Zhengwen, T., Naveed, M., ... Huang, Y. (2021). A Bulk-Heterostructure Nanocomposite Electrolyte of Ce 0.8 Sm 0.2 O 2-δ–SrTiO 3 for Low-Temperature Solid Oxide Fuel Cells. *Nano-Micro Letters*, *13*(1).

Yoda, M., Inoue, S., Takuwa, Y., Yasuhara, K., & Suzuki, M. (2017). Development and commercialization of new residential SOFC CHP system. *ECS Transactions*, *78*(1), 125.

Yoon, K. J., Cramer, C. N., Stevenson, J. W., & Marina, O. A. (2010). Advanced ceramic interconnect material for solid oxide fuel cells: Electrical and thermal properties of calcium-and nickel-doped yttrium chromites. *Journal of Power Sources*, *195*(22), 7587–7593.

Yuksel, Y. E., Ozturk, M., & Dincer, I. (2020). Performance investigation of a combined biomass gasifier-SOFC plant for compressed hydrogen production. *International Journal of Hydrogen Energy*, *45*(60), 34679–34694.

Zakaria, Z., Awang Mat, Z., Abu Hassan, S. H., & Boon Kar, Y. (2020). A review of solid oxide fuel cell component fabrication methods toward lowering temperature. *International Journal of Energy Research*, *44*(2), 594–611.

Zhang, J., Lenser, C., Menzler, N. H., & Guillon, O. (2020). Comparison of solid oxide fuel cell (SOFC) electrolyte materials for operation at 500° C. *Solid State Ionics*, *344*, 115138.

Zhao, T. S., & Ni, M. (2013). *Solid oxide fuel cells: From materials to system modeling*. Royal Society of Chemistry.

Zhu, P., Yao, J., Qian, C., Yang, F., Porpatham, E., Zhang, Z., & Wu, Z. (2020). High-efficiency conversion of natural gas fuel to power by an integrated system of SOFC, HCCI engine, and waste heat recovery: Thermodynamic and thermo-economic analyses. *Fuel*, *275*, 117883.

KEY TERMS AND DEFINITIONS

Combined Heat and Power (CHP): Cogeneration, as already known, represents a cycle of transformation in which, starting from the fuel, it produces in the same plant and at the same time electricity and thermal energy, heat and power.

Energy: Means all forms of energy products, fuels, heat, renewable energy, electricity or any other form of energy.

Energy Efficiency: Means the ratio of the result of performance, services, goods, or energy to the energy used for that purpose.

Energy Service: Means a physical benefit, utility or good obtained from a combination of energy with an energy efficient technology or action which may include the operating, maintenance and control activities necessary for the provision of the service, which is provided on the basis of a contract and which, under normal conditions, has been shown to lead to an improvement in energy efficiency or primary energy savings, under verifiable and measurable or estimable conditions.

Improving Energy Efficiency: Means an increase in energy efficiency as a result of technological, behavioral, and/or economic changes.

Solid Oxide Fuel Cell (SOFC): Can be defined generically as a ceramic multilayer electrochemical energy generating system, which uses a fuel and a gas oxidant and operates near the temperature of 800 ÷ 1000 °C.

Compilation of References

Abanades. (2019). Metal Oxides Applied to Thermochemical Water-Splitting for Hydrogen Production Using Concentrated Solar Energy. *ChemEngineering*, *3*(3), 63. doi:10.3390/chemengineering3030063

Abbas, H. F., & Daud, W. W. (2010). Hydrogen production by methane decomposition: A review. *International Journal of Hydrogen Energy*, *35*(3), 1160–1190. doi:10.1016/j.ijhydene.2009.11.036

Abbas, S. A., Iqbal, M. I., Kim, S. H., & Jung, K. D. (2017). Catalytic activity of urchin–like Ni nanoparticles prepared by solvothermal method for hydrogen evolution reaction in alkaline solution. *Electrochimica Acta*, *227*, 382–390. doi:10.1016/j.electacta.2017.01.039

Abd, A. A., Naji, S. Z., Thian, T. C., & Othman, M. R. (2021). Evaluation of hydrogen concentration effect on the natural gas properties and flow performance. *International Journal of Hydrogen Energy*, *46*(1), 974–983. doi:10.1016/j.ijhydene.2020.09.141

Abdalla, A. M., Hossain, S., Nisfindy, O. B., Azad, A. T., Dawood, M., & Azad, A. K. (2018). Hydrogen production, storage, transportation and key challenges with applications: A review. *Energy Conversion and Management*, *165*(April), 602–627. doi:10.1016/j.enconman.2018.03.088

Abdel-Basset, M., Gamal, A., Chakrabortty, R. K., & Ryan, M. J. (2021). Evaluation of sustainable hydrogen production options using an advanced hybrid MCDM approach: A case study. *International Journal of Hydrogen Energy*, *46*(5), 4567–4591. doi:10.1016/j.ijhydene.2020.10.232

Abdelkareem, M. A., Elsaid, K., Wilberforce, T., Kamil, M., Sayed, E. T., & Olabi, A. (2021). Environmental aspects of fuel cells: A review. *The Science of the Total Environment*, *752*, 141803. doi:10.1016/j.scitotenv.2020.141803 PMID:32889267

Abdenebi, H., Zitouni, B., Ben-Moussa, H., Haddad, D., Zitouni, H., & Sahli, Y. (2015). Inlet Methane Temperature Effect at a Planar SOFC Thermal Field under Direct Internal Reforming Condition. In I. Dincer, C. O. Colpan, O. Kizilkan & M. A. Ezan (Eds.), Progress in Clean Energy, volume I: Novel Systems and Applications (pp. 567-581). Springer. doi:10.1007/978-3-319-17031-2_41

Abe, J. O., Popoola, A. P. I., Ajenifuja, E., & Popoola, O. M. (2019). Hydrogen energy, economy and storage: Review and recommendation. International Journal of Hydrogen Energy, 44(29), 15072–15086. doi:10.1016/j.ijhydene.2019.04.068

Abrell, J., Eser, P., Garrison, J. B., Savelsberg, J., & Weigt, H. (2019). Integrating economic and engineering models for future electricity market evaluation: A Swiss case study. *Energy Strategy Reviews*, *25*, 86–106. doi:10.1016/j.esr.2019.04.003

Abu-Heiba, A., Gluesenkamp, K. R., LaClair, T. J., Cheekatamarla, P., Munk, J. D., Thomas, J., & Boudreaux, P. R. (2021). Analysis of power conversion technology options for a self-powered furnace. *Applied Thermal Engineering*, *188*, 116627. doi:10.1016/j.applthermaleng.2021.116627

Acar, C., & Dincer, I. (2019). Review and evaluation of hydrogen production options for better environment. *Journal of Cleaner Production, 218*, 835–849. doi:10.1016/j.jclepro.2019.02.046

Adamson, K. A. (2006). *Large stationary survey.* Fuel Cell Today Ltd. www. fuelcelltoday. com

Adeleke, A. A., Odusote, J. K., Ikubanni, P. P., Lasode, O. A., Malathi, M., & Paswan, D. (2021). Essential basics on biomass torrefaction, densification and utilization. *International Journal of Energy Research, 45*(2), 1375–1395. doi:10.1002/er.5884

Agrawal, R., Singh, N. R., Ribeiro, F. H., & Delgass, W. N. (2007). Sustainable fuel for the transportation sector. *Proceedings of the National Academy of Sciences of the United States of America, 104*(12), 4828–4833. doi:10.1073/pnas.0609921104 PMID:17360377

Aguiar, P., Adjiman, C. S., & Brandon, N. P. (2004). Anode-supported intermediate temperature direct internal reforming solid oxide fuel cell. I: Model-based steady-state performance. *Journal of Power Sources, 138*(1-2), 120–136. doi:10.1016/j.jpowsour.2004.06.040

Ahamed, M. I., Shakeel, N., Anwar, N., Khan, A., Asiri, A. M., & Dzudzevic-Cancar, H. (2021). Graphene-based nanocomposite for hydrogen storage application. In *Nanomaterials for Hydrogen Storage Applications* (pp. 57–78). Elsevier. doi:10.1016/B978-0-12-819476-8.00006-2

Ahluwalia, R. K., Hua, T. Q., Peng, J. K., Lasher, S., McKenney, K., Sinha, J., & Gardiner, M. (2010). Technical assessment of cryo-compressed hydrogen storage tank systems for automotive applications. *International Journal of Hydrogen Energy, 35*(9), 4171–4184. doi:10.1016/j.ijhydene.2010.02.074

Ahluwalia, R. K., & Peng, J. K. (2009). Automotive hydrogen storage system using cryo-adsorption on activated carbon. *International Journal of Hydrogen Energy, 34*(13), 5476–5487. doi:10.1016/j.ijhydene.2009.05.023

Ahmadi, M. H., Sadaghiani, M. S., Pourfayaz, F., Ghazvini, M., Mahian, O., Mehrpooya, M., & Wongwises, S. (2018). Energy and exergy analyses of a solid oxide fuel cell-gas turbine-organic rankine cycle power plant with liquefied natural gas as heat sink. *Entropy (Basel, Switzerland), 20*(7), 484. doi:10.3390/e20070484 PMID:33265574

Ahmadi, S., Ghaebi, H., & Shokri, A. (2019). A comprehensive thermodynamic analysis of a novel CHP system based on SOFC and APC cycles. *Energy, 186*, 115899. doi:10.1016/j.energy.2019.115899

Ahmad, M. (2018). *Operation and control of renewable energy systems.* John Wiley & Sons.

Akhil, A. A., Huff, G., Currier, A. B., Kaun, B. C., Rastler, D. M., Chen, S. B., Cotter, A. L., Bradshaw, D. T., & Gauntlett, W. D. (2013). *DOE/EPRI 2013 Electricity Storage Handbook in Collaboration with NRECA, SANDIA REPORT, SAND2013-5131.* Sandia National Laboratories. www.sandia.gov/ess

Akinyele, D., Olabode, E., & Amole, A. (2020). Review of Fuel Cell Technologies and Applications for Sustainable Microgrid Systems. *Inventions (Basel, Switzerland), 5*(3), 42. doi:10.3390/inventions5030042

Akroot, A., Namli, L., & Ozcan, H. (2021). Compared Thermal Modeling of Anode-and Electrolyte-Supported SOFC-Gas Turbine Hybrid Systems. *Journal of Electrochemical Energy Conversion and Storage, 18*(1), 011001. doi:10.1115/1.4046185

Al-Attar, A. F., Farid, S. B., Hashim, F. A., & Eshraghi, M. J. (2019). Comparison in physical and mechanical properties between doped and non-doped Y-TZP electrolyte for HT-SOFC. *Energy Procedia, 157*, 1285–1291. doi:10.1016/j.egypro.2018.11.294

Al-Breiki, M., & Bicer, Y. (2021). Comparative life cycle assessment of sustainable energy carriers including production, storage, overseas transport and utilization. *Journal of Cleaner Production, 279*, 123481. doi:10.1016/j.jclepro.2020.123481

Alias, M. S., Kamarudin, S. K., Zainoodin, A. M., & Masdar, M. S. (2020). Active direct methanol fuel cell: An overview. *International Journal of Hydrogen Energy*, *45*(38), 19620–19641. doi:10.1016/j.ijhydene.2020.04.202

AlNouss, A., McKay, G., & Al-Ansari, T. (2020). Production of syngas via gasification using optimum blends of biomass. *Journal of Cleaner Production*, *242*, 118499. doi:10.1016/j.jclepro.2019.118499

Al-Qahtani, A., Parkinson, B., Hellgardt, K., Shah, N., & Guillen-Gosalbez, G. (2021). Uncovering the true cost of hydrogen production routes using life cycle monetisation. *Applied Energy*, *281*, 115958. doi:10.1016/j.apenergy.2020.115958

Amedi, H. R., Bazooyar, B., & Pishvaie, M. R. (2015). Control of anode supported SOFCs (solid oxide fuel cells): Part I. mathematical modeling and state estimation within one cell. *Energy*, *90*, 605–621. doi:10.1016/j.energy.2015.07.095

Anantharaj, S., Ede, S. R., Sakthikumar, K., Karthick, K., Mishra, S., & Kundu, S. (2016). Recent trends and perspectives in electrochemical water splitting with an emphasis on sulfide, selenide, and phosphide catalysts of Fe, Co, and Ni: A review. *ACS Catalysis*, *6*(12), 8069–8097. doi:10.1021/acscatal.6b02479

Andersson, M., Nakajima, H., Kitahara, T., Shimizu, A., Koshiyama, T., Paradis, H., Yuan, J., & Sundén, B. (2014). Comparison of humidified hydrogen and partly pre-reformed natural gas as fuel for solid oxide fuel cells applying computational fluid dynamics. *International Journal of Heat and Mass Transfer*, *77*, 1008–1022. doi:10.1016/j.ijheatmasstransfer.2014.06.033

Andersson, M., Paradis, H., Yuan, J., & Sundén, B. (2013). Three dimensional modeling of an solid oxide fuel cell coupling charge transfer phenomena with transport processes and heat generation. *Electrochimica Acta*, *109*, 881–893. doi:10.1016/j.electacta.2013.08.018

Andrews, J., & Shabani, B. (2012). Re-envisioning the role of hydrogen in a sustainable energy economy. *International Journal of Hydrogen Energy*, *37*(2), 1184–1203. doi:10.1016/j.ijhydene.2011.09.137

ANRE. (2015). *Metodologia de calcul pentru stabilirea cantităţilor de energie electrică produse în cogenerare de înaltă eficienţă în vederea certificării prin garanţii*. Academic Press.

Argun, H., & Kargi, F. (2011). Bio-hydrogen production by different operational modes of dark and photo-fermentation: An overview. *International Journal of Hydrogen Energy*, *36*(13), 7443–7459. doi:10.1016/j.ijhydene.2011.03.116

Armaroli, N., & Balzani, V. (2011). The hydrogen issue. *ChemSusChem*, *4*(1), 21–36. doi:10.1002/cssc.201000182 PMID:21226208

Arun, A., Kumar, L., & Chowdhury, A. (2009). Structure-property relations for a phase-pure, nanograined tetragonal zirconia ceramic stabilised with minimum CaO doping. *Journal of the American Ceramic Society*.

Ashraf, M. A., Rashid, K., Rahimipetroudi, I., Kim, H. J., & Dong, S. K. (2020). Analyzing different planar biogas-fueled SOFC stack designs and their effects on the flow uniformity. *Energy*, *190*, 116450. doi:10.1016/j.energy.2019.116450

Asif, M., & Muneer, T. (2007). Energy supply, its demand and security issues for developed and emerging economies. *Renewable & Sustainable Energy Reviews*, *11*(7), 1388–1413. doi:10.1016/j.rser.2005.12.004

Assawamongkholsiri, T., Reungsang, A., Plangkang, P., & Sittijunda, S. (2018). Repeated batch fermentation for photo-hydrogen and lipid production from wastewater of a sugar manufacturing plant. *International Journal of Hydrogen Energy*, *43*(7), 3605–3617. doi:10.1016/j.ijhydene.2017.12.119

Awate, B., Steidl, R. J., Hamlischer, T., Reguera, G., Yang, Y., Qin, M., Yang, X., He, Z., Marone, A., Ayala-Campos, O. R., & Trably, E., Carmona-Martı́nez, A. A., Moscoviz, R., Latrille, E., Steyer, J. P., Alcaraz-Gonzalez, V., Bernet, N., Montpart, N., Baeza, M., … Cheng, S. (2017). Proton transfer in microbial electrolysis cells. *Journal of Power Sources*, *356*(4), 725–736. doi:10.1016/j.ijhydene.2016.09.166

Ayers, K., Danilovic, N., Ouimet, R., Carmo, M., Pivovar, B., & Bornstein, M. (2019). Perspectives on low-temperature electrolysis and potential for renewable hydrogen at scale. *Annual Review of Chemical and Biomolecular Engineering*, *10*(1), 219–239. doi:10.1146/annurev-chembioeng-060718-030241 PMID:31173524

Bacău, C., Mateoc-Sirb, N., Ciolac, R., Mateoc, T., Teodorescu, R. I., & Tabără, V. (2021). *Study on the production and use of biomass energy in the Timis County*. Academic Press.

Badea, G., Felseghi, R. A., Aşchilean, I., Răboacă, S. M., & Şoimoşan, T. (2017, December). The role of hydrogen as a future solution to energetic and environmental problems for residential buildings. In AIP Conference Proceedings (Vol. 1917, No. 1, p. 030004). AIP Publishing LLC. doi:10.1063/1.5018277

Badea, G., Felseghi, R. A., Giurca, I., Safirescu, O. C., Aşchilean, I., & Şoimoşan, T. M. (2013). Choosing the optimal fuel cell technology for application in the field of passive houses, Recent Advances in Urban Planning and Construction. *Proceedings of the 4th WSEAS International Conference USCUDAR, Energy*, 125 – 130.

Badea, G., Felseghi, R. A., Moldovan, E., & Safirescu, O. C. (2013). *Consideraţii privind utilizarea tehnologiei pilelor de combustibil în sisteme de cogenerare a energiei, Ştiinţa Modernă şi Energia XXXII*. Ed. Risoprint.

Badea, G., Felseghi, R. A., Răboaca, S., Aşchilean, I., Bolboacă, A., Mureşan, D., Moldovan, E., & Şoimoşan, T. (2016). Hybrid Solar and Wind Electric System for Romanian Nearly Zero Energy Buildings (nZEB)-Case Study. *Applied Mechanics and Materials*, *841*, 110–115. doi:10.4028/www.scientific.net/AMM.841.110

Badea, G., Felseghi, R., Aschilean, I., Bolboaca, A., Muresan, D., Soimosan, T., ... Raboaca, S. (2015, August). Energen system for power supply of passive house: case study. In *2015 Second International Conference on Mathematics and Computers in Sciences and in Industry (MCSI)* (pp. 24-31). IEEE. 10.1109/MCSI.2015.31

Badwal, S. P. S. (1992). Zirconia-based Solid Electrolytes: Microstructure, Stability and Ionic Conductivity. Solid State Lonics, 52, 23-32.

Badwal, S. P. S. (1994). Ceramic Superionic Conductors, Materials Science and Technology, A Comprehensive Treatment (vol. 11). VCH Verlagsgesellschaft.

Badwal, S. P. S., Deller, R., Foger, K., Ramprakash, Y., & Zhang, J. P. (1997). Interaction between chromia forming alloy interconnects and air electrode of solid oxide fuel cells. *Solid State Ionics, 99*(3-4), 297-310.

Badwal, S. P. S., Giddey, S., Munnings, C., & Kulkarni, A. (2015). Review of progress in high temperature solid oxide fuel cells. *ChemInform, 46*(31).

Badwal, S. P. S., & Ciacchi, F. T. (2000). *Oxygen-ion Conducting Electrolyte Materials for Solid Oxide Fuel Cells. Ionics,* (6), 1–21.

Badwal, S. P. S., Deller, R., Foger, K., Ramprakash, Y., & Zhang, J. P. (1997). Interaction between chromia forming alloy interconnects and air electrode of solid oxide fuel cells. *Solid State Ionics*, *99*(3-4), 297–310. doi:10.1016/S0167-2738(97)00247-6

Baeyens, J., Zhang, H., Nie, J., Appels, L., Dewil, R., Ansart, R., & Deng, Y. (2020). Reviewing the potential of bio-hydrogen production by fermentation. *Renewable & Sustainable Energy Reviews*, *131*, 110023. doi:10.1016/j.rser.2020.110023

Bagherian, M. A., Mehranzamir, K., Pour, A. B., Rezania, S., Taghavi, E., Nabipour-Afrouzi, H., Dalvi-Esfahani, M., & Alizadeh, S. M. (2021). Classification and Analysis of Optimization Techniques for Integrated Energy Systems Utilizing Renewable Energy Sources: A Review for CHP and CCHP Systems. *Processes (Basel, Switzerland)*, *9*(2), 339. doi:10.3390/pr9020339

Baharuddin, N. A., Yusoff, W. W., Abd Aziz, A. J., & Tahir, N. M. (2021, February). Hydrogen fuel cells for sustainable energy: Development and progress in selected developed countries. *IOP Conference Series. Materials Science and Engineering*, *1078*(1), 012011. doi:10.1088/1757-899X/1078/1/012011

Bai, S., Tian, K., Meng, J. C., Zhao, Y., Sun, J., Zhang, K., Feng, Y., Luo, R., Li, D., & Chen, A. (2021). Reduced graphene oxide decorated SnO2/BiVO4 photoanode for photoelectrochemical water splitting. *Journal of Alloys and Compounds*, *855*, 156780. doi:10.1016/j.jallcom.2020.156780

Bajwa, D. S., Peterson, T., Sharma, N., Shojaeiarani, J., & Bajwa, S. G. (2018). A review of densified solid biomass for energy production. *Renewable & Sustainable Energy Reviews*, *96*, 296–305. doi:10.1016/j.rser.2018.07.040

Baldi, F., Moret, S., Tammi, K., & Maréchal, F. (2020). The role of solid oxide fuel cells in future ship energy systems. *Energy*, *194*, 116811. doi:10.1016/j.energy.2019.116811

Ball, M., Basile, A., & Veziroglu, T. N. (Eds.). (2015). *Compendium of hydrogen energy: hydrogen use, safety and the hydrogen economy*. Woodhead Publishing.

Balog, A. A., Cobîrzan, N., Thalmaier, G., Constantinescu, H., & Voinea, M. (2019). Psyllium Seeds Used as Pore Forming Agent in Clay Bricks. *Procedia Manufacturing*, *32*, 242–247. doi:10.1016/j.promfg.2019.02.209

Bao, C., Cai, N., & Croiset, E. (2011). A multi-level simulation platform of natural gas internal reforming solid oxide fuel cell-gas turbine hybrid generation system-Part II. Balancing units model library and system simulation. *Journal of Power Sources*, *196*(20), 8424–8434. doi:10.1016/j.jpowsour.2011.05.032

Bao, D. X., Yu, Y. Y., & Zhao, Q. M. (2020). Evaluation of the chemical composition and rheological properties of bio-asphalt from different biomass sources. *Road Materials and Pavement Design*, *21*(7), 1829–1843. doi:10.1080/14680629.2019.1568287

Barbir, F., Basile, A., & Veziroglu, T. N. (Eds.). (2015). *Compendium of Hydrogen Energy: Hydrogen Energy Conversion*. Woodhead Publishing.

Barbir, F., Veziroğlu, T. N., & Plass, H. J. Jr. (1990). Environmental damage due to fossil fuels use. *International Journal of Hydrogen Energy*, *15*(10), 739–749. doi:10.1016/0360-3199(90)90005-J

Barelli, L., Bidini, G., Gallorini, F., & Ottaviano, P. A. (2013). Design optimization of a SOFC-based CHP system through dynamic analysis. *International Journal of Hydrogen Energy*, *38*(1), 354–369. doi:10.1016/j.ijhydene.2012.09.162

Bar-On, Y. M., Phillips, R., & Milo, R. (2018). The biomass distribution on Earth. *Proceedings of the National Academy of Sciences of the United States of America*, *115*(25), 6506–6511. doi:10.1073/pnas.1711842115 PMID:29784790

Baroutaji, A., Wilberforce, T., Ramadan, M., & Olabi, A. G. (2019). Comprehensive investigation on hydrogen and fuel cell technology in the aviation and aerospace sectors. *Renewable & Sustainable Energy Reviews*, *106*, 31–40. doi:10.1016/j.rser.2019.02.022

Barreto, L., Makihira, A., & Riahi, K. (2003). The hydrogen economy in the 21st century: A sustainable development scenario. *International Journal of Hydrogen Energy*, *28*(3), 267–284. doi:10.1016/S0360-3199(02)00074-5

Barrovecchio, L., Gehain, E., & Gibbard, F. (2001). A proton Exchange Membrane Fuel Cell in a House. *International Gas Research Conference*, 246.

Barthelemy, H., Weber, M., & Barbier, F. (2017). Hydrogen storage: Recent improvements and industrial perspectives. *International Journal of Hydrogen Energy*, *42*(11), 7254–7262. doi:10.1016/j.ijhydene.2016.03.178

Batteries, R.-F. (2016). *From Metals to Organic Redox-Active Materials*. doi:10.1002/anie.201604925

Baumann, M., Domnik, T., Haase, M., Wulf, C., Emmerich, P., Rösch, C., Zapp, P., Naegler, T., & Weil, M. (2021). Comparative patent analysis for the identification of global research trends for the case of battery storage, hydrogen and bioenergy. *Technological Forecasting and Social Change*, *165*, 120505. doi:10.1016/j.techfore.2020.120505

Begum, H., Ahmed, M. S., & Jeon, S. (2019). Δ–MnO_2 nanoflowers on sulfonated graphene sheets for stable oxygen reduction and hydrogen evolution reaction. *Electrochimica Acta*, *296*, 235–242. doi:10.1016/j.electacta.2018.11.073

Beigzadeh, M., Pourfayaz, F., Ghazvini, M., & Ahmadi, M. H. (2021). Energy and exergy analyses of solid oxide fuel cell-gas turbine hybrid systems fed by different renewable biofuels: A comparative study. *Journal of Cleaner Production*, *280*, 124383. doi:10.1016/j.jclepro.2020.124383

Ben, T., Ren, H., Ma, S., Cao, D., Lan, J., Jing, X., ... Qiu, S. (2009). Targeted synthesis of a porous aromatic framework with high stability and exceptionally high surface area. *Angewandte Chemie International Edition*, *48*(50), 9457–9460. doi:10.1002/anie.200904637 PMID:19921728

Bhatia, S. K., & Myers, A. L. (2006). Optimum conditions for adsorptive storage. *Langmuir*, *22*(4), 1688–1700. doi:10.1021/la0523816 PMID:16460092

Bianco, M., Poitel, S., Hong, J. E., Yang, S., Wang, Z. J., Willinger, M., Steinberger-Wilckens, R., & Van herle, J. (2020). Corrosion behaviour of nitrided ferritic stainless steels for use in solid oxide fuel cell devices. *Corrosion Science*, *165*, 108414. doi:10.1016/j.corsci.2019.108414

Bicer, Y., & Khalid, F. (2020). Life cycle environmental impact comparison of solid oxide fuel cells fueled by natural gas, hydrogen, ammonia and methanol for combined heat and power generation. *International Journal of Hydrogen Energy*, *45*(5), 3670–3685. doi:10.1016/j.ijhydene.2018.11.122

Bluhm, M. E., Bradley, M. G., Butterick, R., Kusari, U., & Sneddon, L. G. (2006). Amineborane-based chemical hydrogen storage: Enhanced ammonia borane dehydrogenation in ionic liquids. *Journal of the American Chemical Society*, *128*(24), 7748–7749. doi:10.1021/ja062085v PMID:16771483

Blum, L., Fang, Q., Gross-Barsnick, S. M., de Haart, L. B., Malzbender, J., Menzler, N. H., & Quadakkers, W. J. (2020). Long-term operation of solid oxide fuel cells and preliminary findings on accelerated testing. *International Journal of Hydrogen Energy*, *45*(15), 8955–8964. doi:10.1016/j.ijhydene.2020.01.074

Bochentyn, B., Chlipała, M., Gazda, M., Wang, S. F., & Jasiński, P. (2019). Copper and cobalt co-doped ceria as an anode catalyst for DIR-SOFCs fueled by biogas. *Solid State Ionics*, *330*, 47–53. doi:10.1016/j.ssi.2018.12.007

Bogdanović, B., & Schwickardi, M. (1997). Ti-doped alkali metal aluminium hydrides as potential novel reversible hydrogen storage materials. *Journal of Alloys and Compounds*, *253*, 1–9. doi:10.1016/S0925-8388(96)03049-6

Borgschulte, A., Callini, E., Probst, B., Jain, A., Kato, S., Friedrichs, O., Remhof, A., Bielmann, M., Ramirez-Cuesta, A. J., & Züttel, A. (2011). Impurity gas analysis of the decomposition of complex hydrides. *The Journal of Physical Chemistry C*, *115*(34), 17220–17226. doi:10.1021/jp205566q

Borole, A. P., & Lewis, A. J. (2017). Proton transfer in microbial electrolysis cells. *Sustainable Energy & Fuels*, *1*(4), 725–736. doi:10.1039/C7SE00034K

Bossel, U., Eliasson, B., & Taylor, G. (2003). The future of the hydrogen economy: Bright or bleak? *Cogeneration and Distributed Generation Journal*, *18*(3), 29–70. doi:10.1080/15453660309509023

Boukis, N., & Stoll, I. K. (2021). Gasification of Biomass in Supercritical Water, Challenges for the Process Design—Lessons Learned from the Operation Experience of the First Dedicated Pilot Plant. *Processes (Basel, Switzerland)*, *9*(3), 455. doi:10.3390/pr9030455

Braff, W. A., Bazant, M. Z., & Buie, C. R. (2013). Membrane-less hydrogen bromine flow battery. *Nature Communications, 4*, 1–6.

Brand, M. A., Rodrigues, T. M., da Silva, J. P., & de Oliveira, J. (2021). Recovery of agricultural and wood wastes: The effect of biomass blends on the quality of pellets. *Fuel, 284*, 118881. doi:10.1016/j.fuel.2020.118881

Brandon, N., Dincer, I., Shao, Z. P., Xia, C., Hill, J. M., Shikazono, N., ... Schüth, F. (2013). Solid oxide fuel cells: from materials to system modeling. *Royal Society of Chemistry*.

Brás, I., Silva, E., & De Lemos, L. T. (2020). Feasibility of using municipal solid wastes rejected fractions as fuel in a biomass power plant. *Environment Protection Engineering, 46*(2). Advance online publication. doi:10.37190/epe200204

Burt, A. C., Celik, I. B., Gemmen, R. S., & Smirnov, A. V. (2004). A numerical study of cell-to-cell variations in a SOFC stack. *Journal of Power Sources, 126*(1-2), 76–87. doi:10.1016/j.jpowsour.2003.08.034

Cai, Y., Wang, B., Wang, Y., Xia, C., Qiao, J., Aken, P. A. V., ... Lund, P. (2018). Validating the technological feasibility of yttria-stabilized zirconia-based semiconducting-ionic composite in intermediate temperature solid oxide fuel cells. *Journal of Power Sources, 384*, 318–327. doi:10.1016/j.jpowsour.2018.03.012

Carmo, M., Fritz, D. L., Mergel, J., & Stolten, D. (2013). A comprehensive review on PEM water electrolysis. *International Journal of Hydrogen Energy, 38*, 4901–4934.

Carrillo, R. J., & Scheffe, J. R. (2017). Advances and trends in redox materials for solar thermochemical fuel production. *Solar Energy, 156*, 3–20. doi:10.1016/j.solener.2017.05.032

Casquero-Vera, J. A., Lyamani, H., Titos, G., Minguillón, M. C., Dada, L., Alastuey, A., Querol, X., Petäjä, T., Olmo, F. J., & Alados-Arboledas, L. (2021). Quantifying traffic, biomass burning and secondary source contributions to atmospheric particle number concentrations at urban and suburban sites. *The Science of the Total Environment, 768*, 145282. doi:10.1016/j.scitotenv.2021.145282

Castillo Santiago, Y., Martínez González, A., Venturini, O. J., & Yepes Maya, D. M. (2021). Assessment of the energy recovery potential of oil sludge through gasification aiming electricity generation. *Energy*, 215.

Cédric Philibert. (2017). *Producing industrial hydrogen from renewable energy*. International Energy Agency. https://www.iea.org/newsroom/news/2017/april/producing-industrial-hydrogen-from-renewable-energy.html

Cernat, A., Lazaroiu, G., Pana, C., Neguresсu, N., & Mirica, I. (2016, June). *Experimental Investigations of the Automotive Diesel Engine Fuelled with Hydrogen* [Poster presentation]. The 2nd South East European Conference on Sustainable Development of Energy, Water and Environment Systems, Piran, Slovenia.

Cernat, A., Pana, C., Negurescu, N., Fuiorescu, D., Nutu, C., & Mirica, I. (2018). Experimental Aspects of Hydrogen Use at Diesel Engine by Diesel Gas Method. *Thermal Science, 22*(3), 1191–1200. doi:10.2298/TSCI170314138C

Cernat, A., Pana, C., Negurescu, N., Nutu, C., & Mirica, I. (2016). Experimental investigations of the hydrogen use at the automotive diesel engine. *Journal of Automotive Engineering, 22*(41), 22–25.

Cernat, A., Pana, C., Negurescu, N., Nutu, C., Mirica, I., & Fuiorescu, D. (2016). Effect of Hydrogen Use on Diesel Engine Performance. In *Proceeding of 20th Innovative Manufacturing Engineering and Energy Conference (IManEE 2016), IOP Conference Series: Materials Science and Engineering* (vol. 161, no. 1). 10.1088/1757-899X/161/1/012075

Cerniauskas, S., Junco, A. J. C., Grube, T., Robinius, M., & Stolten, D. (2020). Options of natural gas pipeline reassignment for hydrogen: Cost assessment for a Germany case study. *International Journal of Hydrogen Energy, 45*(21), 12095–12107. doi:10.1016/j.ijhydene.2020.02.121

Chakraborty, D., Efthi, J. H., Khanom, M., & Mahbubul, I. M. (2020). Prospective and Challenging Issues of Biofuels. *EDU Journal of Computer and Electrical Engineering*, *1*(1), 4–10. doi:10.46603/ejcee.v1i1.5

Chalk, S. G., & Miller, J. F. (2006). Key challenges and recent progress in batteries, fuel cells, and hydrogen storage for clean energy systems. *Journal of Power Sources*, *159*(1), 73–80. doi:10.1016/j.jpowsour.2006.04.058

Chambon, C. L., Karia, T., Sandwell, P., & Hallett, J. P. (2020). Techno-economic assessment of biomass gasification-based mini-grids for productive energy applications: The case of rural India. *Renewable Energy*, *154*, 432–444. doi:10.1016/j.renene.2020.03.002

Chandra, D., Reilly, J. J., & Chellappa, R. (2006). Metal hydrides for vehicular applications: The state of the art. *JOM*, *58*(2), 26–32. doi:10.100711837-006-0005-0

Chapelle, D., Maynadier, A., Bebon, L., & Thiébaud, F. (2021). Hydrogen Storage: Different Technologies, Challenges and Stakes. Focus on TiFe Hydrides. In *Advances in Renewable Hydrogen and Other Sustainable Energy Carriers* (pp. 1–10). Springer. doi:10.1007/978-981-15-6595-3_1

Chaurasia, A. K., Goyal, H., & Mondal, P. (2019). Hydrogen gas production with Ni, Ni–Co and Ni–Co–P electrodeposits as potential cathode catalyst by microbial electrolysis cells. *International Journal of Hydrogen Energy*. Advance online publication. doi:10.1016/j.ijhydene.2019.07.175

Chaurasia, A. K., & Mondal, P. (2021). *Simultaneous Removal of Organic Load and Hydrogen Gas Production Using Electrodeposits Cathodes in MEC*. Springer. doi:10.1007/978-981-15-6595-3_34

Cheng, N., Stambula, S., Wang, D., Banis, M. N., Liu, J., Riese, A., & Botton, G. A. (2016). Platinum single–atom and cluster catalysis of the hydrogen evolution reaction. *Nature Communications*, *7*(1), 13638. doi:10.1038/ncomms13638 PMID:27901129

Chen, H. (2009). Thang Ngoc Cong, Wei Yang, Chunqing Tan, Yongliang Li, Yulong Ding Progress in electrical energy storage system: A critical review. *Progress in Natural Science*, *19*, 291–312.

Chen, J., Chen, Y., Zhang, H., & Weng, S. (2018). Effect of different operating strategies for a SOFC-GT hybrid system equipped with anode and cathode ejectors. *Energy*, *163*, 1–14. doi:10.1016/j.energy.2018.08.032

Chen, J., Kuriyama, N., Xu, Q., Takeshita, H. T., & Sakai, T. (2001). Reversible hydrogen storage via titanium-catalyzed LiAlH4 and Li3AlH6. *The Journal of Physical Chemistry B*, *105*(45), 11214–11220. doi:10.1021/jp012127w

Chen, P., Xiong, Z., Luo, J., Lin, J., & Tan, K. L. (2002). Interaction of hydrogen with metal nitrides and imides. *Nature*, *420*(6913), 302–304. doi:10.1038/nature01210 PMID:12447436

Chen, W. T., Chan, A., Waterhouse, S. D., Llorca, J., Idriss, H., & Waterhouse, G. I. N. (2018). Performance comparison of Ni/TiO_2 and Au/TiO_2 photocatalysts for H_2 production in different alcohol–water mixtures. *Journal of Catalysis*, *367*, 27–42. doi:10.1016/j.jcat.2018.08.015

Chen, W., Pei, J., He, C. T., Wan, J., Ren, H., Wang, Y., & Zheng, X. (2018). Single Tungsten Atoms Supported on MOF–Derived N–Doped Carbon for Robust Electrochemical Hydrogen Evolution. *Advanced Materials*, *30*(30), 1800396. doi:10.1002/adma.201800396 PMID:29888491

Chen, Z., Duan, X., Wei, W., Wang, S., & Ni, B. J. (2019). Recent advances in transition metal–based electrocatalysts for alkaline hydrogen evolution. *Journal of Materials Chemistry. A, Materials for Energy and Sustainability*, *7*(25), 14971–15005. doi:10.1039/C9TA03220G

Chen, Z., Fang, J., Xu, C., Xia, Z., Yan, K., & Li, X. (2020). Carbon dioxide hydrate separation from Integrated Gasification Combined Cycle (IGCC) syngas by a novel hydrate heat-mass coupling method. *Energy*, *199*, 117420. doi:10.1016/j.energy.2020.117420

Chick, L., Weimar, M., Whyatt, G., & Powell, M. (2015). The case for natural gas fueled solid oxide fuel cell power systems for distributed generation. *Fuel Cells (Weinheim)*, *15*(1), 49–60. doi:10.1002/fuce.201400103

Chi, J., & Yu, H. (2018a). Water electrolysis based on renewable energy for hydrogen production. *Cuihua Xuebao. Chinese Journal of Catalysis*, *39*(3), 390–394. doi:10.1016/S1872-2067(17)62949-8

Chitgar, N., Emadi, M. A., Chitsaz, A., & Rosen, M. A. (2019). Investigation of a novel multigeneration system driven by a SOFC for electricity and fresh water production. *Energy Conversion and Management*, *196*, 296–310. doi:10.1016/j.enconman.2019.06.006

Chitgar, N., & Moghimi, M. (2020). Design and evaluation of a novel multi-generation system based on SOFC-GT for electricity, fresh water and hydrogen production. *Energy*, *197*, 117162. doi:10.1016/j.energy.2020.117162

Choi, S. H., & Manousiouthakis, V. I. (2020). On the carbon cycle impact of combustion of harvested plant biomass vs. fossil carbon resources. *Computers & Chemical Engineering*, *140*, 106942. doi:10.1016/j.compchemeng.2020.106942

Choudhary, T., & Sanjay. (2017). Thermodynamic assessment of advanced SOFC-blade cooled gas turbine hybrid cycle. *International Journal of Hydrogen Energy*, *42*(15), 10248–10263. doi:10.1016/j.ijhydene.2017.02.178

Choudhury, A., Chandra, H., & Arora, A. (2013). Application of solid oxide fuel cell technology for power generation—A review. *Renewable & Sustainable Energy Reviews*, *20*, 430–442. doi:10.1016/j.rser.2012.11.031

Chowdhury, T., Chowdhury, H., Hossain, N., Ahmed, A., Hossen, M. S., Chowdhury, P., Thirugnanasambandam, M., & Saidur, R. (2020). Latest advancements on livestock waste management and biogas production: Bangladesh's perspective. *Journal of Cleaner Production*, *272*, 122818. doi:10.1016/j.jclepro.2020.122818

Chung, T. M., Jeong, Y., Chen, Q., Kleinhammes, A., & Wu, Y. (2008). Synthesis of microporous boron-substituted carbon (B/C) materials using polymeric precursors for hydrogen physisorption. *Journal of the American Chemical Society*, *130*(21), 6668–6669. doi:10.1021/ja800071y PMID:18454522

Chyou, Y. P., Chung, T. D., Chen, J. S., & Shie, R. F. (2005). Integrated thermal engineering analyses with heat transfer at periphery of planar solid oxide fuel cell. *Journal of Power Sources*, *139*(1-2), 126–140. doi:10.1016/j.jpowsour.2004.07.001

Cieciura-Włoch, W., Borowski, S., & Otlewska, A. (2020). Biohydrogen production from fruit and vegetable waste, sugar beet pulp and corn silage via dark fermentation. *Renewable Energy*, *153*, 1226–1237. doi:10.1016/j.renene.2020.02.085

Cipitì, F., Recupero, V., Pino, L., Vita, A., & Laganà, M. (2006). Experimental analysis of a 2 kWe LPG-based fuel processor for polymer electrolyte fuel cells. *Journal of Power Sources*, *157*(2), 914–920. doi:10.1016/j.jpowsour.2006.02.084

Cloete, S., Ruhnau, O., & Hirth, L. (2021). On capital utilization in the hydrogen economy: The quest to minimize idle capacity in renewables-rich energy systems. *International Journal of Hydrogen Energy*, *46*(1), 169–188. doi:10.1016/j.ijhydene.2020.09.197

Cobîrzan, N., Thalmaier, G., Balog, A. A., Constantinescu, H., Timiş, I., & Streza, M. (2018). Thermophysical properties of fired clay bricks with waste ceramics and paper pulp as pore-forming agent. *Journal of Thermal Analysis and Calorimetry*, *134*(1), 843–851. doi:10.100710973-018-7092-3

Colbertaldo, P., Agustin, S. B., Campanari, S., & Brouwer, J. (2019). Impact of hydrogen energy storage on California electric power system: Towards 100% renewable electricity. *International Journal of Hydrogen Energy*, *44*(19), 9558–9576. doi:10.1016/j.ijhydene.2018.11.062

Coors, W., Boxley, C., Robins, M., & Eccleston, A. (2012). *US Pat., 061823A2*. US Patent Office.

Corma, A. (1997). From microporous to mesoporous molecular sieve materials and their use in catalysis. *Chemical Reviews*, *97*(6), 2373–2420. doi:10.1021/cr960406n PMID:11848903

Cornwell, O., Radford, S. E., Ashcroft, A. E., & Ault, J. R. (2018). Comparing hydrogen deuterium exchange and fast photochemical oxidation of proteins: A structural characterisation of wild-type and ΔN6 β2-microglobulin. *Journal of the American Society for Mass Spectrometry*, *29*(12), 2413–2426. doi:10.100713361-018-2067-y PMID:30267362

Costa, P., Pinto, F., André, R. N., & Marques, P. (2021). Integration of Gasification and Solid Oxide Fuel Cells (SOFCs) for Combined Heat and Power (CHP). *Processes (Basel, Switzerland)*, *9*(2), 254. doi:10.3390/pr9020254

Crabtree, G. W., Dresselhaus, M. S., & Buchanan, M. V. (2004). The hydrogen economy. *Physics Today*, *57*(12), 39–44. doi:10.1063/1.1878333

Crabtree, R. H. (2008). Hydrogen storage in liquid organic heterocycles. *Energy & Environmental Science*, *1*(1), 134–138. doi:10.1039/b805644g

Csernica, P. M., McKone, J. R., Mulzer, C. R., Dichtel, W. R., Abruña, H. D., & DiSalvo, F. J. (2017). Electrochemical hydrogen evolution at ordered Mo_7Ni_7. *ACS Catalysis*, *7*(5), 3375–3383. doi:10.1021/acscatal.7b00344

Cunha, Á., Martins, J., Rodrigues, N. & Brito, F. P. (2014). Vanadium redox flow batteries: a technology review. *Int. J. Energy Res.* doi:10.1002/er.3260

Czagany, M., & Baumli, P. (2019). Effect of surfactants on the behavior of the Ni–P bath and on the formation of electroless Ni–P–TiC composite coatings. *Surface and Coatings Technology*, *361*, 42–49. doi:10.1016/j.surfcoat.2019.01.046

D'AprileP.NewmanJ.PinnerD. (2016). Retrieved from https://www.mckinsey.com/business-functions/sustainability/our-insights/the-new-economics-of-energy-storage

Damm, D. L., & Fedorov, A. G. (2006). Reduced-order transient thermal modeling for SOFC heating and cooling. *Journal of Power Sources*, *159*(2), 956–967. doi:10.1016/j.jpowsour.2005.11.072

Danilovic, N., Subbaraman, R., Strmcnik, D., Chang, K. C., Paulikas, A. P., Stamenkovic, V. R., & Markovic, N. M. (2012). Enhancing the alkaline hydrogen evolution reaction activity through the bifunctionality of $Ni(OH)_2$/metal catalysts. *Angewandte Chemie International Edition*, *51*(50), 12495–12498. doi:10.1002/anie.201204842 PMID:23129151

Darband, G. B., Aliofkhazraei, M., Hyun, S., Rouhaghdam, A. S., & Shanmugam, S. (2019). Electrodeposition of Ni–Co–Fe mixed sulfide ultrathin nanosheets on Ni nanocones: A low–cost, durable and high performance catalyst for electrochemical water splitting. *Nanoscale*, *11*(35), 16621–16634. doi:10.1039/C9NR04529E PMID:31460535

Das, D., Khanna, N., & Nejat Veziroğlu, T. (2008). Recent developments in biological hydrogen production processes. Chemical Industry and Chemical Engineering Quarterly, 14(2), 57–67. doi:10.2298/CICEQ0802057D

Davis, M. E. (2002). Ordered porous materials for emerging applications. *Nature*, *417*(6891), 813–821. doi:10.1038/nature00785 PMID:12075343

Dawood, F., Anda, M., & Shafiullah, G. M. (2020). Hydrogen production for energy: An overview. *International Journal of Hydrogen Energy*, *45*(7), 3847–3869. doi:10.1016/j.ijhydene.2019.12.059

De Giz, M. J., Tremiliosi-Filho, G., Gonzalez, E. R., Srinivasan, S., & Appleby, A. J. (1995). The hydrogen evolution reaction on amorphous nickel and cobalt alloys. *International Journal of Hydrogen Energy*, *20*(6), 423–427. doi:10.1016/0360-3199(94)00068-B

De Oliveira, J. G. (2011). *J.G. Power Control Systems in a Flywheel based All-Electric Driveline* (PhD thesis). Uppsala Universitet, Uppsala, Sweden.

Demirocak, D. E., Srinivasan, S. S., Ram, M. K., Kuhn, J. N., Muralidharan, R., Li, X., Goswami, D. Y., & Stefanakos, E. K. (2013). Reversible hydrogen storage in the Li–Mg–N–H system–The effects of Ru doped single walled carbon nanotubes on NH3 emission and kinetics. *International Journal of Hydrogen Energy*, *38*(24), 10039–10049. doi:10.1016/j.ijhydene.2013.05.176

De, S., Mishra, S., Poonguzhali, E., Rajesh, M., & Tamilarasan, K. (2020). Fractionation and characterization of lignin from waste rice straw: Biomass surface chemical composition analysis. *International Journal of Biological Macromolecules*, *145*, 795–803. doi:10.1016/j.ijbiomac.2019.10.068 PMID:31739019

Destek, M. A. (2017). Biomass energy consumption and economic growth: Evidence from top 10 biomass consumer countries. *Energy Sources. Part B, Economics, Planning, and Policy*, *12*(10), 853–858. doi:10.1080/15567249.2017.1314393

Dicks, A. L. (1996). Hydrogen generation from natural gas for the fuel cell systems of tomorrow. *Journal of Power Sources*, *61*(1-2), 113–124. doi:10.1016/S0378-7753(96)02347-6

Dicks, A., & Rand, D. A. J. (2018). *Fuel cell systems explained* (Vol. 479). Wiley. doi:10.1002/9781118706992

Dincer, I., & Zamfirescu, C. (2012). Sustainable hydrogen production options and the role of IAHE. *International Journal of Hydrogen Energy*, *37*(21), 16266–16286. doi:10.1016/j.ijhydene.2012.02.133

Ding, X., Gao, Z., Ding, D., Zhao, X., Hou, H., Zhang, S., & Yuan, G. (2019). Cation deficiency enabled fast oxygen reduction reaction for a novel SOFC cathode with promoted CO2 tolerance. *Applied Catalysis B: Environmental*, *243*, 546–555. doi:10.1016/j.apcatb.2018.10.075

Directiva 2004/8/CE a Parlamentului European şi a Consiliului din 11 februarie 2004 privind promovarea cogenerării pe baza cererii de energie termică utilă pe piaţa internă a energiei şi de modificare a Directivei 92/42/CEE.

Divisek, J., Oetjen, H. F., Peinecke, V., Schmidt, V. M., & Stimming, U. (1998). Components for PEM fuel cell systems using hydrogen and CO containing fuels. *Electrochimica Acta*, *43*(24), 3811–3815. doi:10.1016/S0013-4686(98)00140-6

Dodds, E. P., Staffell, I., Hawkes, A., Li, F., Grünewald, P., McDowall, W., & Ekins, P. (2014). Hydrogen and fuel cell technologies for heating. *RE:view*. Advance online publication. doi:10.1016/j.ijhydene.2014.11.059

Doman, L. E., Arora, V., Metelitsa, A., Leahy, M., Barden, J. L., Ford, M., Jones, A., Brian Murphy, E. T., Lillis, K., Slater-Thompson, N., Behjat Hojjati, B., Peterson, D., Zaretskaya, V. V, Chase, N., Morris, M., Lindstrom, P., Geagla, A., Grissom, S., Maples, J., … Smith, C. L. (2013). *World energy demand and economic outlook Macroeconomic assumptions Liquid fuels Natural gas Transportation sector Energy-related carbon dioxide emissions*. www.eia.gov

Dong, Y., Peng, W., Liu, Y., & Wang, Z. (2021). Photochemical origin of reactive radicals and halogenated organic substances in natural waters: A review. *Journal of Hazardous Materials*, *401*, 123884. doi:10.1016/j.jhazmat.2020.123884 PMID:33113752

dos Santos, R. G., & Alencar, A. C. (2020). Biomass-derived syngas production via gasification process and its catalytic conversion into fuels by Fischer Tropsch synthesis: A review. *International Journal of Hydrogen Energy*, *45*(36), 18114–18132. doi:10.1016/j.ijhydene.2019.07.133

Drake, H., Ivarsson, M., Heim, C., Snoeyenbos-West, O., Bengtson, S., Belivanova, V., & Whitehouse, M. (2021). Fossilized anaerobic and possibly methanogenesis-fueling fungi identified deep within the Siljan impact structure, Sweden. *Communications Earth & Environment*, *2*(1), 1–10. doi:10.103843247-021-00107-9

Duan, J., Chen, S., & Zhao, C. (2017). Ultrathin metal–organic framework array for efficient electrocatalytic water splitting. *Nature Communications*, *8*(1), 15341. doi:10.1038/ncomms15341 PMID:28580963

Durst, J., Siebel, A., Simon, C., Hasche, F., Herranz, J., & Gasteiger, H. A. (2014). New insights into the electrochemical hydrogen oxidation and evolution reaction mechanism. *Energy & Environmental Science*, *7*(7), 2255–2260. doi:10.1039/C4EE00440J

Dwivedi, S. K., & Vishwakarma, M. (2018). Hydrogen embrittlement in different materials: A review. *International Journal of Hydrogen Energy*, *43*(46), 21603–21616. doi:10.1016/j.ijhydene.2018.09.201

E4tech. (n.d.). *The Fuel Cell Industry Review 2018*. Available online: https://www.californiahydrogen.org/wpcontent/uploads/2019/01/TheFuelCellIndustryReview2018.pdf

EAA Annual Report 2015/2016. (n.d.). Retrieved January 6, 2020, from https://www.eaa.org.hk/en-us/Information-Centre/Publications/EAA-Annual-Report/EAA-Annual-Report-2015-2016

EASE EERA. (2017). *Storage Technology Development Roadmap 2017*. Author.

Eberle, U., Felderhoff, M., & Schueth, F. (2009). Chemical and physical solutions for hydrogen storage. *Angewandte Chemie International Edition*, *48*(36), 6608–6630. doi:10.1002/anie.200806293 PMID:19598190

Echigo, M., Shinke, N., Takami, S., & Tabata, T. (2004). Performance of a natural gas fuel processor for residential PEFC system using a novel CO preferential oxidation catalyst. *Journal of Power Sources*, *132*(1-2), 29–35. doi:10.1016/j.jpowsour.2004.01.031

Economics, B. E. (2018). BP Energy Outlook. Academic Press.

Eftekhari, A. (2017). Electrocatalysts for hydrogen evolution reaction. *International Journal of Hydrogen Energy*, *42*(16), 11053–11077. doi:10.1016/j.ijhydene.2017.02.125

Egan, D. R., Pnce de Leon, C., Wood, R. J. K., Jones, R. L., Stokes, K. R., & Walsh, F. C. (2013). Developments in Electrode Materials and Electrolytes for Aluminium-Air Batteries. *Journal of Power Sources*, *236*, 293–310.

Eiler, J. M. (2020). *The Once and Future Hydrogen Economy*. Academic Press.

Eker, S., & Sarp, M. (2017). Hydrogen gas production from waste paper by dark fermentation: Effects of initial substrate and biomass concentrations. *International Journal of Hydrogen Energy*, *42*(4), 2562–2568. doi:10.1016/j.ijhydene.2016.04.020

Electrical Energy Storage. (2011). *White paper*. The International Electrotechnical Commission.

El-Emam, R. S., & Özcan, H. (2019). Comprehensive review on the techno-economics of sustainable large-scale clean hydrogen production. *Journal of Cleaner Production*, *220*, 593–609. doi:10.1016/j.jclepro.2019.01.309

Elias, L., & Hegde, A. C. (2016). Synthesis and characterization of Ni–P–Ag composite coating as efficient electrocatalyst for alkaline hydrogen evolution reaction. *Electrochimica Acta*, *219*, 377–385. doi:10.1016/j.electacta.2016.10.024

Enescu, A., Andrei, H., Ion, V., Diaconu, E., & Angelescu, N. (2019, October). Analysis and Modeling of Biomass Plant Energy Efficieny. In *2019 6th International Symposium on Electrical and Electronics Engineering (ISEEE)* (pp. 1-6). IEEE.

European Commission. (n.d.). *Industrial policy*. Retrived from: https://ec.europa.eu/growth/industry/policy/european-battery-alliance_en

European Court of Auditorium. (2019). *EU support for energy storage*. Briefing Paper.

European Hydrogen and Fuel Cell Technology Platform. (2005). *Strategic Research Agenda*. Available online: https://www.fch.europa.eu/sites/default/files/documents/hfp-sra004_v9-2004_sra-report-final_ 22jul2005.pdf

European Market Monitor On Energy Storage (EMMES). (2019). *EMMES Edition 3.0*. Retrieved from https://www.delta-ee.com/EMMES

Faleh, S., Khir, T., & Brahim, A. B. (2017). Energetic Performance Optimization of a SOFC–GT Hybrid Power Plant. *Arabian Journal for Science and Engineering*, *42*(4), 1505–1515. doi:10.100713369-016-2363-4

Fallah Vostakola, M., & Horri, B. A. (2021). Progress in Material Development for Low-Temperature Solid Oxide Fuel Cells: A Review. *Energies*, *14*(5), 1280. doi:10.3390/en14051280

Fan, L. (2020). Solid-State Electrolytes for SOFC. *Solid Oxide Fuel Cells: From Electrolyte-Based to Electrolyte-Free Devices*, 35-78.

Fang, M., Gao, W., Dong, G., Xia, Z., Yip, S., Qin, Y., & Ho, J. C. (2016). Hierarchical NiMo–based 3D electrocatalysts for highly–efficient hydrogen evolution in alkaline conditions. *Nano Energy*, *27*, 247–254. doi:10.1016/j.nanoen.2016.07.005

Fardadi, M., Mueller, F., & Jabbari, F. (2010). Feedback control of solid oxide fuel cell spatial temperature variation. *Journal of Power Sources*, *195*(13), 4222–4233. doi:10.1016/j.jpowsour.2009.12.111

Farha, O. K., Eryazici, I., Jeong, N. C., Hauser, B. G., Wilmer, C. E., Sarjeant, A. A., Snurr, R. Q., Nguyen, S. B. T., Yazaydın, A. Ö., & Hupp, J. T. (2012). Metal–organic framework materials with ultrahigh surface areas: Is the sky the limit? *Journal of the American Chemical Society*, *134*(36), 15016–15021. doi:10.1021/ja3055639 PMID:22906112

Felderhoff, M., Weidenthaler, C., von Helmolt, R., & Eberle, U. (2007). Hydrogen storage: The remaining scientific and technological challenges. *Physical Chemistry Chemical Physics*, *9*(21), 2643–2653. doi:10.1039/b701563c PMID:17627309

Felseghi, R. A. (2015). *Contributions regarding the application of fuel cell in the passive houses domain* (Unpublished doctoral dissertation). Technical University of Cluj-Napoca, Romania.

Felseghi, R. A., & Papp, A. (2014). Considerations regarding hybrid systems of power generator from renewable energy sources. *Studia Universitatis Babes-Bolyai, Ambientum, 59*.

Felseghi, R. A., Şoimoşan, T. M., Filote, C., & Răboaca, M. S. (2019, b). Considerations Regarding the Green Retrofitting of Residential Buildings From Human Wellbeing Perspectives. In Retrofitting for Optimal Energy Performance (pp. 143-175). IGI Global.

Felseghi, R. A., Carcadea, E., Raboaca, M. S., Trufin, C. N., & Filote, C. (2019). a). Hydrogen fuel cell technology for the sustainable future of stationary applications. *Energies*, *12*(23), 4593. doi:10.3390/en12234593

Felseghi, R. A., Carcadea, E., Raboaca, M. S., Trufin, C. N., & Filote, C. (2019). Hydrogen fuel cell technology for the sustainable future of stationary applications. *Energies*, *12*(23), 4593.

Feng, Z. A., Gabaly, F. E., Ye, X., Shen, Z. X., & Chueh, W. C. (2014). Fast vacancy–mediated oxygen ion incorporation across the ceria–gas electrochemical interface. *Nature Communications*, *4374*(1), 1–9. doi:10.1038/ncomms5374 PMID:25007038

Fergus, J. W. (2004). Lanthanum chromite-based materials for solid oxide fuel cell interconnects. *Solid State Ionics, 171*(1-2), 1-15.

Ferguson, J. R., Fiard, J. M., & Herbin, R. (1996). Three-dimensional numerical simulation for various geometries of solid oxide fuel cells. *Journal of Power Sources*, *58*(2), 109–122. doi:10.1016/0378-7753(95)02269-4

Ferreira, S., Monteiro, E., Brito, P., & Vilarinho, C. (2019). A holistic review on biomass gasification modified equilibrium models. *Energies*, *12*(1), 160. doi:10.3390/en12010160

Filote, C., Felseghi, R. A., Raboaca, M. S., & Aşchilean, I. (2020). Environmental impact assessment of green energy systems for power supply of electric vehicle charging station. *International Journal of Energy Research*, *44*(13), 10471–10494. doi:10.1002/er.5678

Fiore, M., Magi, V., & Viggiano, A. (2020). Internal combustion engines powered by syngas: A review. *Applied Energy*, *276*, 115415. doi:10.1016/j.apenergy.2020.115415

Fragiacomo, P., Corigliano, O., & De Lorenzo, G. (2018). Design of an SOFC/SOE experimental station: Planning of simulation tests. *Energy Procedia*, *148*, 535–542.

Frost, H., Düren, T., & Snurr, R. Q. (2006). Effects of surface area, free volume, and heat of adsorption on hydrogen uptake in metal– organic frameworks. *The Journal of Physical Chemistry B*, *110*(19), 9565–9570. doi:10.1021/jp060433+ PMID:16686503

Fuel Cell Handbook. (2004). *EG&G Technical Service Inc., U.S. Department of Energy* (7th ed.). Office of Fossil Energy.

Fujihara, T. (1998). Development of pump turbine for seawater pumped-storage power plant. *Hitachi Rev*, *47*(5), 199–202.

Funahashi, Y., Shimamori, T., Suzuki, T., Fujishiro, Y., & Awano, M. (2007). Fabrication and characterization of components for cube shaped micro tubular SOFC bundle. *Journal of Power Sources*, *163*(2), 731–736.

Gahleitner, G. (2013). Hydrogen from renewable electricity: An international review of power-to-gas pilot plants for stationary applications. *International Journal of Hydrogen Energy, 38*(5), 2039-2061.

Gandiglio, M., Lanzini, A., & Santarelli, M. (2018). Large stationary solid oxide fuel cell (SOFC) power plants. In *Modeling, Design, Construction, and Operation of Power Generators with Solid Oxide Fuel Cells* (pp. 233–261). Springer.

Gandiglio, M., Lanzini, A., Santarelli, M., Acri, M., Hakala, T., & Rautanen, M. (2020). Results from an industrial size biogas-fed SOFC plant (the DEMOSOFC project). *International Journal of Hydrogen Energy*, *45*(8), 5449–5464.

Ganguly, G., Halder, D., Banerjee, A., Basu, S., & Paul, A. (2019). Exploring the Crucial Role of Solvation on the Viability of Sustainable Hydrogen Storage in BN-fullerene: A Combined DFT and Ab initio Molecular Dynamics investigation. *ACS Sustainable Chemistry & Engineering*, *7*(11), 9808–9821. doi:10.1021/acssuschemeng.9b00211

Gao, J., & Zhang, L. (2021). Does biomass energy consumption mitigate CO2 emissions? The role of economic growth and urbanization: Evidence from developing Asia. *Journal of the Asia Pacific Economy*, *26*(1), 96–115. doi:10.1080/13547860.2020.1717902

Gao, W., Gou, W., Zhou, X., Ho, J. C., Ma, Y., & Qu, Y. (2018). Amine–modulated/engineered interfaces of NiMo electrocatalysts for improved hydrogen evolution reaction in alkaline solutions. *ACS Applied Materials & Interfaces*, *10*(2), 1728–1733. doi:10.1021/acsami.7b16125 PMID:29282973

Garthwaite. (2011). *How ultracapacitors work (and why they fall short).* https://gigaom.com/

Gerboni, R., & Salvador, E. (2009). Hydrogen transportation systems: Elements of risk analysis. *Energy*, *34*(12), 2223–2229. doi:10.1016/j.energy.2008.12.018

Germain, J., Hradil, J., Fréchet, J. M., & Svec, F. (2006). High surface area nanoporous polymers for reversible hydrogen storage. *Chemistry of Materials*, *18*(18), 4430–4435. doi:10.1021/cm061186p

Ghiat, I., AlNouss, A., McKay, G., & Al-Ansari, T. (2020). Biomass-based integrated gasification combined cycle with post-combustion CO2 recovery by potassium carbonate: Techno-economic and environmental analysis. *Computers & Chemical Engineering*, *135*, 106758. doi:10.1016/j.compchemeng.2020.106758

Global EV Outlook 2020 – Analysis - IEA. (n.d.). Retrieved October 21, 2020, from https://www.iea.org/reports/global-ev-outlook-2020

Goddard, W. A., & Han, S. S. (2009). *U.S. Patent Application No. 12/150,046*. US Patent Office.

Gong, W., Fugetsu, B., Wang, Z., Sakata, I., Su, L., Zhang, X., Ogata, H., Li, M., Wang, C., Li, J., Medina, J., Terrones, M., & Endo, M. (2018). Carbon nanotubes and manganese oxide hybrid nanostructures as high performance fiber supercapacitors. *Communications Chemistry*, *1*(1), 161–168. doi:10.103842004-018-0017-z

Graetz, J. (2009). New approaches to hydrogen storage. *Chemical Society Reviews*, *38*(1), 73–82. doi:10.1039/B718842K PMID:19088966

Graetz, J., Reilly, J. J., Yartys, V. A., Maehlen, J. P., Bulychev, B. M., Antonov, V. E., Tarasov, B. P., & Gabis, I. E. (2011). Aluminum hydride as a hydrogen and energy storage material: Past, present and future. *Journal of Alloys and Compounds*, *509*, S517–S528. doi:10.1016/j.jallcom.2010.11.115

Green, R., & Staffell, I. (2012). *The cost of domestic fuel cell micro-CHP systems. Imperial College*. Imperial College Business School.

Grimm, A., de Jong, W. A., & Kramer, G. J. (2020). Renewable hydrogen production: A techno-economic comparison of photoelectrochemical cells and photovoltaic-electrolysis. *International Journal of Hydrogen Energy*, *45*(43), 22545–22555. doi:10.1016/j.ijhydene.2020.06.092

Gross-Barsnick, S. M. (2020). Interaction of glass-ceramic sealants with solid oxide fuel cell components: thermo-mechanical analysis. In *Intermediate Temperature Solid Oxide Fuel Cells* (pp. 411–426). Elsevier.

Guaitolini, S. V. M., Yahyaoui, I., Fardin, J. F., Encarnação, L. F., & Tadeo, F. (2018, March). A review of fuel cell and energy cogeneration technologies. In *2018 9th International renewable energy congress (IREC)* (pp. 1-6). IEEE. 10.1109/IREC.2018.8362573

Gummert, G., & Suttor, W. (2006). *Stationary fuel cells. Technology and market; Stationaere Brennstoffzellen*. Technik und Markt.

Guo, X., Zhang, H., Hu, Z., Hou, S., Ni, M., & Liao, T. (2021). Energetic, exergetic and ecological evaluations of a hybrid system based on a phosphoric acid fuel cell and an organic Rankine cycle. *Energy*, *217*, 119365. doi:10.1016/j.energy.2020.119365

Gupta, R., Basile, A., & Veziroglu, T. N. (Eds.). (2016). *Compendium of Hydrogen Energy: Hydrogen Storage, Distribution and Infrastructure*. Woodhead Publishing.

Guski, V., Verestek, W., Rapp, D., & Schmauder, S. (2021). Microstructural Investigation of Plasma Sprayed Ceramic Coatings Focusing on the Effect of the Splat Boundary for SOFC Sealing Applications Using Peridynamics. *Theoretical and Applied Fracture Mechanics*.

Gutowska, A., Li, L., Shin, Y., Wang, C. M., Li, X. S., Linehan, J. C., ... Gutowski, M. (2005). Nanoscaffold mediates hydrogen release and the reactivity of ammonia borane. *Angewandte Chemie International Edition*, *44*(23), 3578–3582. doi:10.1002/anie.200462602 PMID:15880751

Habibi, R., & Mehrpooya, M. (2021). A novel integrated Ca-Cu cycle with coal/biomass gasification unit for clean hydrogen production. *Energy Conversion and Management*, *228*, 113682.

Hagen, A., Wulff, A. C., Zielke, P., Sun, X., Talic, B., Ritucci, I., ... Hendriksen, P. V. (2020). SOFC stacks for mobile applications with excellent robustness towards thermal stresses. *International Journal of Hydrogen Energy*, *45*(53), 29201–29211.

Hagiwara, A. (1999, July). Fuel cell technologies in search of new directions: based on the experiences over 20 years. In *1999 IEEE Power Engineering Society Summer Meeting. Conference Proceedings* (Cat. No. 99CH36364) (Vol. 1, pp. 576-580). IEEE.

Hajabdollahi, Z., & Fu, P. F. (2017). Multi-objective based configuration optimization of SOFC-GT cogeneration plant. *Applied Thermal Engineering*, *112*, 549–559.

Hallenbeck, P. C., & Benemann, J. R. (2002). Biological hydrogen production; Fundamentals and limiting processes. *International Journal of Hydrogen Energy*, *27*(11–12), 1185–1193. doi:10.1016/S0360-3199(02)00131-3

Hameed, Z., Aslam, M., Khan, Z., Maqsood, K., Atabani, A. E., Ghauri, M., Khurram, M. S., Rehan, M., & Nizami, A. S. (2021). Gasification of municipal solid waste blends with biomass for energy production and resources recovery: Current status, hybrid technologies and innovative prospects. *Renewable & Sustainable Energy Reviews*, *136*, 110375. doi:10.1016/j.rser.2020.110375

Hamilton, C. W., Baker, R. T., Staubitz, A., & Manners, I. (2009). B–N compounds for chemical hydrogen storage. *Chemical Society Reviews*, *38*(1), 279–293. doi:10.1039/B800312M PMID:19088978

Hanley, E. S., Deane, J. P., & Gallachóir, B. Ó. (2018). The role of hydrogen in low carbon energy futures–A review of existing perspectives. *Renewable & Sustainable Energy Reviews*, *82*, 3027–3045. doi:10.1016/j.rser.2017.10.034

Han, S. S., Furukawa, H., Yaghi, O. M., & Goddard Iii, W. A. (2008). Covalent organic frameworks as exceptional hydrogen storage materials. *Journal of the American Chemical Society*, *130*(35), 11580–11581. doi:10.1021/ja803247y PMID:18683924

Hayun, H., Wolf, R., Barad, C., & Gelbstein, Y. (2020). Thermal shock resistant solid oxide fuel cell ceramic composite electrolytes. *Journal of Alloys and Compounds*, *821*, 153490.

Hebner, R., Beno, J., & Walls, A. (2002). Flywheel batteries come around again. *IEEE Spectrum*, *39*, 46–51.

Hemmes, K. (2021). A personal retrospect on three decades of high temperature fuel cell research; ideas and lessons learned. *International Journal of Hydrogen Energy*.

Hendriks, A. T. W. M., & Zeeman, G. (2009). Pretreatments to enhance the digestibility of lignocellulosic biomass. *Bioresource Technology*, *100*(1), 10–18. doi:10.1016/j.biortech.2008.05.027 PMID:18599291

He, S., Li, S., & Gao, L. (2021). Proposal and energy saving analysis of novel methanol–electricity polygeneration system based on staged coal gasification method. *Energy Conversion and Management*, *233*, 113931. doi:10.1016/j.enconman.2021.113931

Hirschenhofer, J. H., Sauffer, D. B., & Engleman, R. R. (1994). *Fuel Cell: A Handbook (Revision 3)*. U.S. Department of Energy, Morgantown Energy Technology Center.

Hirscher, M. (2011). Hydrogen storage by cryoadsorption in ultrahigh-porosity metal–organic frameworks. *Angewandte Chemie International Edition*, *50*(3), 581–582. doi:10.1002/anie.201006913 PMID:21226135

Hoang, T. K., & Antonelli, D. M. (2009). Exploiting the Kubas interaction in the design of hydrogen storage materials. *Advanced Materials*, *21*(18), 1787–1800. doi:10.1002/adma.200802832

Holladay, J. D., Hu, J., King, D. L., & Wang, Y. (2009). An overview of hydrogen production technologies. *Catalysis Today*, *139*(4), 244–260. doi:10.1016/j.cattod.2008.08.039

Holtappels, P., Vogt, U., & Graule, T. (2005). Ceramic materials for advanced solid oxide fuel cells. *Advanced Engineering Materials*, *7*(5), 292–302.

Homma, T. (2006). *The latest fuel cell news in Japan*. http://www. fcdic. com/eng/news/200808. html

Hoogers, G. (2003). *Fuel Cell Technology Handbook*. CRC Press.

Horcajada, P., Serre, C., Vallet-Regí, M., Sebban, M., Taulelle, F., & Férey, G. (2006). Metal–organic frameworks as efficient materials for drug delivery. *Angewandte Chemie International Edition*, *45*(36), 5974–5978. doi:10.1002/anie.200601878 PMID:16897793

Hosseini, S. E., & Wahid, M. A. (2016a). Hydrogen production from renewable and sustainable energy resources: Promising green energy carrier for clean development. *Renewable & Sustainable Energy Reviews*, *57*, 850–866. Advance online publication. doi:10.1016/j.rser.2015.12.112

Hosseini, S. E., & Wahid, M. A. (2020). Hydrogen from solar energy, a clean energy carrier from a sustainable source of energy. *International Journal of Energy Research*, *44*(6), 4110–4131. doi:10.1002/er.4930

Hou, Q., Wu, Y., Liu, H., Xu, X., & Xu, B. (2020, June). Numerical Simulation of Electrical Performance and Distribution of SOFC Stacks With Different Manifold Arrangement. In *ASME 2020 14th International Conference on Energy Sustainability*. American Society of Mechanical Engineers Digital Collection.

Houghton, J. (2005). Reports on Progress in Physics. *Global Warming, 68*, 1343-1403.

Høyer, K. G., & Holden, E. (2007). Alternative fuels and sustainable mobility: Is the future road paved by biofuels, electricity or hydrogen. *International Journal of Alternative Propulsion*, *1*(4), 352–369. doi:10.1504/IJAP.2007.013329

Huang, L., Hou, Y., Yu, Z., Peng, Z., Wang, L., Huang, J., & Li, Z. (2017). Pt/Fe–NF electrode with high double–layer capacitance for efficient hydrogen evolution reaction in alkaline media. *International Journal of Hydrogen Energy*, *42*(15), 9458–9466. doi:10.1016/j.ijhydene.2017.02.055

Huang, Y., Lin, Q., Liu, H., Ni, M., & Zhang, X. (2017). Evaluation of the waste heat and residual fuel from the solid oxide fuel cell and system power optimization. *International Journal of Heat and Mass Transfer*, *115*, 1166–1173. doi:10.1016/j.ijheatmasstransfer.2017.07.115

Huang, Z., & Autrey, T. (2012). Boron–nitrogen–hydrogen (BNH) compounds: Recent developments in hydrogen storage, applications in hydrogenation and catalysis, and new syntheses. *Energy & Environmental Science*, *5*(11), 9257–9268. doi:10.1039/c2ee23039a

Hua, T. Q., Ahluwalia, R. K., Peng, J. K., Kromer, M., Lasher, S., McKenney, K., Law, K., & Sinha, J. (2011). Technical assessment of compressed hydrogen storage tank systems for automotive applications. *International Journal of Hydrogen Energy*, *36*(4), 3037–3049. doi:10.1016/j.ijhydene.2010.11.090

Hubbert, M. K. (1956, January). Nuclear energy and the fossil fuel. In *Drilling and production practice*. American Petroleum Institute.

Huggins, R. A. (2016). *Energy Storage Fundamentals, Materials and Applications*. Springer.

Hu, H., Qiao, M., Pei, Y., Fan, K., Li, H., Zong, B., & Zhang, X. (2003). Kinetics of hydrogen evolution in alkali leaching of rapidly quenched Ni–Al alloy. *Applied Catalysis A, General*, *252*(1), 173–183. doi:10.1016/S0926-860X(03)00416-2

Hu, J., Zhang, C., Jiang, L., Lin, H., An, Y., Zhou, D., & Yang, S. (2017). Nanohybridization of MoS_2 with layered double hydroxides efficiently synergizes the hydrogen evolution in alkaline media. *Joule*, *1*(2), 383–393. doi:10.1016/j.joule.2017.07.011

Hu, X., & Gholizadeh, M. (2019). Biomass pyrolysis: A review of the process development and challenges from initial researches up to the commercialisation stage. *Journal of Energy Chemistry*, *39*, 109–143. doi:10.1016/j.jechem.2019.01.024

Hu, X., Li, Z., Zhao, Y., Sun, J., Zhao, Q., Wang, J., Tao, Z., & Chen, J. (2017). Article. *Science Advances*, *3*, e1602396.

Hu, Y. Z., Su, Y. T., Li, C. X., Li, C. J., & Yang, G. J. (2020). Dense Mn1. 5Co1. 5O4 coatings with excellent long-term stability and electrical performance under the SOFC cathode environment. *Applied Surface Science*, *499*, 143726.

Hydrogen Production by Photoelectrolysis. (n.d.). Retrieved October 21, 2020, from https://bhuang02.tripod.com/photoelectrolysis.htm

Hydrogen Production from Biomass and Organic Waste — Fuel Cell & Hydrogen Energy Association. (n.d.). Retrieved January 18, 2020, from https://www.fchea.org/in-transition/2019/7/8/hydrogen-production-from-biomass-and-organic-waste

HyUnder Project. (n.d.). Available online: www.hyunder.eu

Ichikawa, T., Hanada, N., Isobe, S., Leng, H., & Fujii, H. (2004). Mechanism of novel reaction from LiNH2 and LiH to Li2NH and H2 as a promising hydrogen storage system. *The Journal of Physical Chemistry B*, *108*(23), 7887–7892. doi:10.1021/jp049968y

IHA working paper (December 2018): "The world's water battery: Pumped hydropower storage and the clean energy transition"

Ilbas, M., Kumuk, B., Alemu, M. A., & Arslan, B. (2020). Numerical investigation of a direct ammonia tubular solid oxide fuel cell in comparison with hydrogen. *International Journal of Hydrogen Energy*, *45*(60), 35108–35117.

İnci, M., Büyük, M., Demir, M. H., & İlbey, G. (2021). A review and research on fuel cell electric vehicles: Topologies, power electronic converters, energy management methods, technical challenges, marketing and future aspects. *Renewable & Sustainable Energy Reviews*, *137*, 110648. doi:10.1016/j.rser.2020.110648

Inui, Y., Urata, A., Ito, N., Nakajima, T., & Tanaka, T. (2006). Performance simulation of planar SOFC using mixed hydrogen and carbon monoxide gases as fuel. *Energy Conversion and Management*, *47*(13-14), 1738–1747. doi:10.1016/j.enconman.2005.10.014

Ionescu, C., Tuţică, D., Pătraşcu, R., Dincă, C., & Slavu, N. (2019). Evaluation of the energy efficiency of an industrial consumer in trigeneration mode. In *E3S Web of Conferences* (Vol. 85, p. 01005). EDP Sciences.

Iordache, I., & Stefănescu, I. (2010). Tehnici Avansate De Tratare a Apei ¸Si Separare a Hidrogenului [Advanced Techniques for Water Treatment and Hydrogen Separation]. Editura AGIR.

IRENA. (2017). *Electricity Storage and Renewables: Costs and Markets to 2030*. International Renewable Energy Agency.

Irshad, M., Idrees, R., Siraj, K., Shakir, I., & Rafique, M., ul Ain, Q., & Raza, R. (2021). Electrochemical evaluation of mixed ionic electronic perovskite cathode LaNi1-xCoxO3-δ for IT-SOFC synthesized by high temperature decomposition. *International Journal of Hydrogen Energy*, *46*(17), 10448–10456.

Ishaq, H., & Dincer, I. (2019). A comparative evaluation of three CuCl cycles for hydrogen production. *International Journal of Hydrogen Energy*, *44*(16), 7958–7968. doi:10.1016/j.ijhydene.2019.01.249

Ishaq, H., & Dincer, I. (2020). A comprehensive study on using new hydrogen-natural gas and ammonia-natural gas blends for better performance. *Journal of Natural Gas Science and Engineering*, *81*, 103362. doi:10.1016/j.jngse.2020.103362

Ishaq, H., & Dincer, I. (2021). Investigation and optimization of a new hybrid natural gas reforming system for cascaded hydrogen, ammonia and methanol synthesis. *Computers & Chemical Engineering*, 107234.

Iulianelli, A., & Basile, A. (Eds.). (2014). *Advances in hydrogen production, storage and distribution*. Elsevier.

Ivan, G., Badea, G., Stefanescu, I., Mateescu, I., Ivan, M. A., & Mateescu, O. (2017), From the ancient Prometheus to the modern Prometheus. *The 27th International Conference – BSEE 2017 Building Services and Energy Efficiency*

Ivan, M. A. (2015). Innovation and systems for economizing the production of energetic comfort. *World Renewable Energy Congress.*

Ivers-Tiffée, E., Weber, A., & Herbstritt, D. (2001). Materials and technologies for SOFC-components. *Journal of the European Ceramic Society*, *21*(10-11), 1805–1811.

Jain, I. P., Jain, P., & Jain, A. (2010). Novel hydrogen storage materials: A review of lightweight complex hydrides. *Journal of Alloys and Compounds*, *503*(2), 303–339. doi:10.1016/j.jallcom.2010.04.250

Jamil, S. M., Othman, M. H. D., Rahman, M. A., Jaafar, J., Ismail, A. F., Honda, S., & Iwamoto, Y. (2019). Properties and performance evaluation of dual-layer ceramic hollow fiber with modified electrolyte for MT-SOFC. *Renewable Energy*, *134*, 1423–1433.

Janbaz, M., & Behbahaninia, A. (2021). Thermodynamic analysis of Integrated Gasification Combined Cycle power plant: Energy, Exergy. *Jixie Gongcheng Xuebao*, *50*(4), 45–49.

Jaunatre, M. (2021). Practical Literature Review for Establishing the Current Renewable Hydrogen Framework. *Renewable Hydrogen*, 13-28.

Javidmehr, M., Pourfayaz, F., & Kasaeian, A. (2021). Optimal design and analysis of a district energy system including heat and power production for domestic applications and fuel for vehicles. *Journal of Thermal Analysis and Calorimetry*, 1–17.

Jayachandran, A. (n.d.). Tata Nano could spawn electric, hybrid & air-powered variants – Report. *Indian Autos Blog*. Retrieve from: https://indianautosblog.com/tata-nano-could-spawn-electric-hybrid-air-variants-p256298

Jhi, S. H., & Ihm, J. (2011). Developing high-capacity hydrogen storage materials via quantum simulations. *MRS Bulletin*, *36*(3), 198–204. doi:10.1557/mrs.2011.32

Jiang, J. X., Su, F., Trewin, A., Wood, C. D., Niu, H., Jones, J. T., Khimyak, Y. Z., & Cooper, A. I. (2008). Synthetic control of the pore dimension and surface area in conjugated microporous polymer and copolymer networks. *Journal of the American Chemical Society*, *130*(24), 7710–7720. doi:10.1021/ja8010176 PMID:18500800

Jia, Y., Zhang, L., Gao, G., Chen, H., Wang, B., Zhou, J., Soo, M. T., Hong, M., Yan, X., Qian, G., Zou, J., Du, A., & Yao, X. (2017). A heterostructure coupling of exfoliated Ni–Fe hydroxide nanosheet and defective graphene as a bifunctional electrocatalyst for overall water splitting. *Advanced Materials*, *29*(17), 1700017. doi:10.1002/adma.201700017 PMID:28256771

Jin, H., Liu, X., Chen, S., Vasileff, A., Li, L., Jiao, Y., & Qiao, S. Z. (2019). Heteroatom–Doped Transition Metal Electrocatalysts for Hydrogen Evolution Reaction. *ACS Energy Letters*, *4*(4), 805–810. doi:10.1021/acsenergylett.9b00348

Jouzani, G. S., Aghbashlo, M., & Tabatabaei, M. (2020). Biofuels: Types, Promises, Challenges, and Role of Fungi. In *Fungi in Fuel Biotechnology* (pp. 1–14). Springer. doi:10.1007/978-3-030-44488-4_1

Kadier, A., Simayi, Y., Kalil, M. S., Abdeshahian, P., & Hamid, A. A. (2014). A review of the substrates used in microbial electrolysis cells (MECs) for producing sustainable and clean hydrogen gas. *Renewable Energy*, *71*, 466–472. doi:10.1016/j.renene.2014.05.052

Kajiura, H., Tsutsui, S., Kadono, K., Kakuta, M., Ata, M., & Murakami, Y. (2003). Hydrogen storage capacity of commercially available carbon materials at room temperature. *Applied Physics Letters*, *82*(7), 1105–1107. doi:10.1063/1.1555262

Kakaç, S., Pramuanjaroenkij, A., & Zhou, X. Y. (2007). A review of numerical modeling of solid oxide fuel cells. *International Journal of Hydrogen Energy*, *32*(7), 761–786. doi:10.1016/j.ijhydene.2006.11.028

Kakoulaki, G., Kougias, I., Taylor, N., Dolci, F., Moya, J., & Jäger-Waldau, A. (2021). Green hydrogen in Europe–A regional assessment: Substituting existing production with electrolysis powered by renewables. *Energy Conversion and Management*, *228*, 113649. doi:10.1016/j.enconman.2020.113649

Kalamaras, C. M., & Efstathiou, A. M. (2013). Hydrogen Production Technologies: Current State and Future Developments. *Conference Papers in Energy*, *2013*, 1–9. doi:10.1155/2013/690627

Kalib, N. S., Muchtar, A., Somalu, M. R., Ihsan, A. K. A. M., Aman, N. A. M. N., & Mah, J. C. W. (2019). Influence of Heat Transfer on Thermal Stress Development in Solid Oxide Fuel Cells: A Review. *Journal of Advanced Research in Fluid Mechanics and Thermal Sciences*, *54*, 175–184.

Kalra, P., Garg, R., & Kumar, A. (2013). Solid oxide fuel cell-a future source of power and heat generation. *Materials Science Forum*, *757*, 217–241.

Kang, Y. W., Li, J., Cao, G. Y., Tu, H. Y., Li, J., & Yang, J. (2009). A reduced 1D dynamic model of a planar direct internal reforming solid oxide fuel cell for system research. *Journal of Power Sources*, *188*(1), 170–176. doi:10.1016/j.jpowsour.2008.11.073

Kannah, R. Y., Kavitha, S., Karthikeyan, O. P., Kumar, G., Dai-Viet, N. V., & Banu, J. R. (2020). Techno-economic assessment of various hydrogen production methods-A review. *Bioresource Technology*, 124175. PMID:33022437

Karageorgiou, V., & Kaplan, D. (2005). Porosity of 3D biomaterial scaffolds and osteogenesis. *Biomaterials*, *26*(27), 5474–5491. doi:10.1016/j.biomaterials.2005.02.002 PMID:15860204

Kargbo, H., Harris, J. S., & Phan, A. N. (2021). "Drop-in" fuel production from biomass: Critical review on techno-economic feasibility and sustainability. *Renewable & Sustainable Energy Reviews*, *135*, 110168. doi:10.1016/j.rser.2020.110168

Karina, B. (2013). High temperature sodium batteries: Status, challenges and future trends. *Energy & Environmental Science*.

Karl, J., & Pröll, T. (2018). Steam gasification of biomass in dual fluidized bed gasifiers: A review. *Renewable & Sustainable Energy Reviews*, *98*, 64–78. doi:10.1016/j.rser.2018.09.010

Katoch, A., Goyal, N., & Gautam, S. (2019). Applications and advances in coordination cages: Metal-Organic Frameworks. *Vacuum*, *167*, 287–300. doi:10.1016/j.vacuum.2019.03.038

Kaya, M. F., Demir, N., Genç, G., & Yapici, H. (2014). Numerically Modeling of Anode Supported Tubular SOFC. *Journal of Applied Mechanical Engineering*, *3*, 1000137.

Kerea, A., Goetza, V., Pya, X., Olivesa, R., Sadikia, N., & Mercier, E. (2014). Dynamic behavior of a sensible-heat based thermal energy storage. *Energy Procedia*, *49*, 830–839.

Kerr, R. A. (2011). *Peak oil production may already be here*. Academic Press.

Khajavi, P., Frandsen, H. L., Gremillard, L., Chevalier, J., & Hendriksen, P. V. (2021). Strength and hydrothermal stability of NiO–stabilized zirconia solid oxide cells fuel electrode supports. *Journal of the European Ceramic Society*.

Khan, M. Z., Song, R. H., Hussain, A., Lee, S. B., Lim, T. H., & Hong, J. E. (2020). Effect of applied current density on the degradation behavior of anode-supported flat-tubular solid oxide fuel cells. *Journal of the European Ceramic Society*, *40*(4), 1407–1417.

Kharton, V. V., Marques, F. M. B., & Atkinson, A. (2004). *Transport Properties of Solid Oxide Electrolyte Ceramics: a Brief Review*. Solid State Ionics.

Kiani, R., Mirzaei, F., Ghanbari, F., Feizi, R., & Mehdipour, F. (2020). Real textile wastewater treatment by a sulfate radicals-Advanced Oxidation Process: Peroxydisulfate decomposition using copper oxide (CuO) supported onto activated carbon. *Journal of Water Process Engineering*, *38*, 101623. doi:10.1016/j.jwpe.2020.101623

Kilkis, B. (1989). *Energy Storage Systems* (1st ed.). Kluwer Academic Publishers.

Kim, I., Kim, K., & Song, O. (2019). Surface etching of diamond powders with electroless Ni–P coating according to phosphorus content. *Diamond and Related Materials*, *97*, 107465. doi:10.1016/j.diamond.2019.107465

Kim, Y. S., Lee, Y. D., & Ahn, K. Y. (2020). System integration and proof-of-concept test results of SOFC–engine hybrid power generation system. *Applied Energy*, *277*, 115542.

Klerke, A., Christensen, C. H., Nørskov, J. K., & Vegge, T. (2008). Ammonia for hydrogen storage: Challenges and opportunities. *Journal of Materials Chemistry*, *18*(20), 2304–2310. doi:10.1039/b720020j

Koirala, B., Hers, S., Morales-España, G., Özdemir, Ö., Sijm, J., & Weeda, M. (2021). Integrated electricity, hydrogen and methane system modelling framework: Application to the Dutch Infrastructure Outlook 2050. *Applied Energy*, *289*, 116713.

Kongjan, P., O-Thong, S., & Angelidaki, I. (2011). Performance and microbial community analysis of two-stage process with extreme thermophilic hydrogen and thermophilic methane production from hydrolysate in UASB reactors. *Bioresource Technology*, *102*(5), 4028–4035. doi:10.1016/j.biortech.2010.12.009 PMID:21216592

Konni, M., Dadhich, A. S., & Mukkamala, S. B. (2017). Impact of surface modifications on hydrogen uptake by Fe@ f-MWCNTs and Cu@ f-MWCNTs at non-cryogenic temperatures. *International Journal of Hydrogen Energy*, *42*(2), 953–959. doi:10.1016/j.ijhydene.2016.09.085

Konni, M., Dadhich, A. S., & Mukkamala, S. B. (2019). Evaluation of surface changes at the interface between TiO2 nanoparticles and COOH-MWCNTs on hydrogen adsorption capability. *Nano-Structures & Nano-Objects*, *18*, 100304. doi:10.1016/j.nanoso.2019.100304

Konni, M., Karnena, M. K., & Mukkamala, S. B. (2020). Hydrogen Storage Capacity in Ni/Pd@ f-MWCNTS Decorated Graphene Oxide/Cu-BTC Composites at Room Temperatures: A Sustainable Cleaner Energy Production. *International Journal of Surface Engineering and Interdisciplinary Materials Science*, *8*(1), 1–12. doi:10.4018/IJSEIMS.2020010101

Konni, M., Kumar Karnena, M., & Babu Mukkamala, S. (2020). Enhanced Room-Temperature Synthesis of Li@ f-MWCNTs for Hydrogen Storage Application. *ChemistrySelect*, *5*(14), 4292–4296. doi:10.1002lct.202000489

Konni, M., & Mukkamala, S. B. (2018). Hydrogen uptake behavior of Cr 2 O 3 nanoparticle decorated f-MWCNTs at non-cryogenic temperatures. *Journal of the Indian Chemical Society*, *95*, 393–397.

Kosmala, T., Coy, D. H., Komsa, H. P., Ma, Y., Krasheninnikov, A. V., Batzill, M., & Agnoli, S. (2018). Metallic twin boundaries boost the hydrogen evolution reaction on the basal plane of molybdenum selenotellurides. *Advanced Energy Materials*, *8*(20), 1–8. doi:10.1002/aenm.201800031

Kouchachvili, L., Geddis, P., & Zhuang, Q. (2021). Direct carbon fuel cell design for continuous operation. *International Journal of Hydrogen Energy*, *46*(9), 6792–6802. doi:10.1016/j.ijhydene.2020.11.179

Kou, Z., Zhang, L., Ma, Y., Liu, X., Zang, W., Zhang, J., Huang, S., Du, Y., Cheetham, A. K., & Wang, J. (2019). 2D carbide nanomeshes and their assembling into 3D microflowers for efficient water splitting. *Applied Catalysis B: Environmental*, *243*, 678–685. doi:10.1016/j.apcatb.2018.11.008

Koyuncu, I., Yilmaz, C., Alcin, M., & Tuna, M. (2020). Design and implementation of hydrogen economy using artificial neural network on field programmable gate array. *International Journal of Hydrogen Energy*, *45*(41), 20709–20720. doi:10.1016/j.ijhydene.2020.05.181

Kuang, M., Wang, Q., Han, P., & Zheng, G. (2017). Cu, Co-Embedded N-Enriched Mesoporous Carbon for Efficient Oxygen Reduction and Hydrogen Evolution Reactions. *Advanced Energy Materials*, *7*(17), 1–8. doi:10.1002/aenm.201700193

Kumar, A. (2018). *Rakesh K. Gupta: Fundamentals of Polymer Engineering* (3rd ed.). CRC Press.

Kumar, V., Kaur, M., Kaur, G., Arya, S. K., & Pickrell, G. (2020). Stacking designs and sealing principles for IT-solid oxide fuel cell. In *Intermediate Temperature Solid Oxide Fuel Cells* (pp. 379–410). Elsevier.

Kurapova, O. Y., Shugurov, S. M., Vasil'eva, E. A., Savelev, D. A., Konakov, V. G., & Lopatin, S. I. (2020). Thermal prehistory, structure and high-temperature thermodynamic properties of Y2O3-CeO2 and Y2O3-ZrO2-CeO2 solid solutions. *Ceramics International*.

Kutcherov, V., Morgunova, M., Bessel, V., & Lopatin, A. (2020). Russian natural gas exports: An analysis of challenges and opportunities. *Energy Strategy Reviews*, *30*, 100511.

LaClair, T., Gao, Z., Abdelaziz, O., & Wang, M. (2016). *Thermal Storage System for Electric Vehicle Cabin Heating - Component and System Analysis*. SAE Technical Paper 2016-01-0244.

Lai, J., Guo, H., Wang, Z., Li, X., Zhang, X., Wu, F., & Yue, P. (2012). Preparation and characterization of flake graphite/silicon/carbon spherical composite as anode materials for lithium–ion batteries. *Journal of Alloys and Compounds*, *530*, 30–35. doi:10.1016/j.jallcom.2012.03.096

Lamy, C., & Millet, P. (2020). A critical review on the definitions used to calculate the energy efficiency coefficients of water electrolysis cells working under near ambient temperature conditions. In *Journal of Power Sources* (Vol. 447). Elsevier B.V. doi:10.1016/j.jpowsour.2019.227350

Laribi, S., Mammar, K., Hamouda, M., & Sahli, Y. (2016). Impedance model for diagnosis of water management in fuel cells using artificial neural networks methodology. *International Journal of Hydrogen Energy*, *41*(38), 17093–17101. doi:10.1016/j.ijhydene.2016.07.099

Laribi, S., Mammar, K., Sahli, Y., & Koussa, K. (2018). Air supply temperature impact on the PEMFC impedance. *Journal of Energy Storage*, *17*, 327–335. doi:10.1016/j.est.2018.03.020

Laribi, S., Mammar, K., Sahli, Y., & Koussa, K. (2019). Analysis and diagnosis of PEM fuel cell failure modes (flooding & drying) across the physical parameters of electrochemical impedance model: Using neural networks method. *Sustainable Energy Technologies and Assessments*, *34*, 35–42. doi:10.1016/j.seta.2019.04.004

Lattin, W. C., & Utgikar, V. P. (2007). Transition to hydrogen economy in the United States: A 2006 status report. *International Journal of Hydrogen Energy*, *32*(15), 3230–3237. doi:10.1016/j.ijhydene.2007.02.004

Laurencelle, F., Chahine, R., Hamelin, J., Agbossou, K., Fournier, M., Bose, T. K., & Laperriere, A. (2001). Characterization of a Ballard MK5-E Proton Exchange Membrane Fuel Cell Stack. *Fuel Cells (Weinheim)*, *1*(1), 66–71.

Lawlor, V., Griesser, S., Buchinger, G., Olabi, A. G., Cordiner, S., & Meissner, D. (2009). Review of the micro-tubular solid oxide fuel cell: Part I. Stack design issues and research activities. *Journal of Power Sources*, *193*(2), 387–399.

Lazzari, E., Schena, T., Marcelo, M. C. A., Primaz, C. T., Silva, A. N., Ferrão, M. F., Bjerk, T., & Caramão, E. B. (2018). Classification of biomass through their pyrolytic bio-oil composition using FTIR and PCA analysis. *Industrial Crops and Products*, *111*, 856–864. doi:10.1016/j.indcrop.2017.11.005

Lee, D. (2021). Functional material developments of fuel cells and the key factors for real commercialization of next-generation energy devices. In *Sustainable Materials for Next Generation Energy Devices* (pp. 85–100). Elsevier.

Lee, D., Ebie, Y., Xu, K., Li, Y., & Inamori, Y. (2010). Bioresource Technology Continuous H 2 and CH 4 production from high-solid food waste in the two-stage thermophilic fermentation process with the recirculation of digester sludge. *Bioresource Technology*, *101*(1), S42–S47. doi:10.1016/j.biortech.2009.03.037 PMID:19410447

Lee, H., Kim, U. S., Kim, S. D., Woo, S. K., & Chung, W. J. (2020). SiO2–B2O3-BaO-WO3 glasses with varying Al2O3 content as a sealing material for reversible solid oxide fuel cells. *Ceramics International*, *46*(11), 18256–18261.

Lee, J. S., Tai Kim, S., Cao, R., Choi, N. S., Liu, M., Lee, K. T., & Cho, J. (2011). Metal-Air Batteries with High Energy Density: Li-Air versus Zn-Air. *Advanced Energy Materials*, *1*, 34–50.

Leng, L., Zhang, W., Leng, S., Chen, J., Yang, L., Li, H., Jiang, S., & Huang, H. (2020). Bioenergy recovery from wastewater produced by hydrothermal processing biomass: Progress, challenges, and opportunities. *The Science of the Total Environment*, *748*, 142383. doi:10.1016/j.scitotenv.2020.142383 PMID:33113702

Li, J., Wu, E., Song, J., Xiao, F., & Geng, C. (2009). Cryoadsorption of hydrogen on divalent cation-exchanged X-zeolites. *International Journal of Hydrogen Energy*, *34*(13), 5458-5465.

Liang, G., Huot, J., Boily, S., Van Neste, A., & Schulz, R. (1999). Catalytic effect of transition metals on hydrogen sorption in nanocrystalline ball milled MgH2–Tm (Tm= Ti, V, Mn, Fe and Ni) systems. *Journal of Alloys and Compounds*, *292*(1-2), 247–252. doi:10.1016/S0925-8388(99)00442-9

Liang, Z., Ahn, H. S., & Bard, A. J. (2017). A Study of the Mechanism of the Hydrogen Evolution Reaction on Nickel by Surface Interrogation Scanning Electrochemical Microscopy. *Journal of the American Chemical Society*, *139*(13), 4854–4858. doi:10.1021/jacs.7b00279 PMID:28276238

Li, F. J., Zhang, T., & Zhou, H. S. (2013). Challenges of Non-Aqueous LiO2 Batteries: Electrolytes, Catalysts, and Anodes. *Energy & Environmental Science*, *6*, 1125–1141.

Linden, D., & Reddy, T. B. (2001). *Handbook of batteries* (3rd ed.). McGraw-Hill, Inc.

Lin, J. D., & Chou, C. T. (2019). The influence of phosphorus content on the microstructure and specific capacitance of etched electroless Ni–P coatings. *Surface and Coatings Technology*, *368*, 126–137. doi:10.1016/j.surfcoat.2019.04.009

Linssen, J., & Hake, J. F. (2016). Hydrogen research, development, demonstration, and market deployment activities. *Hydrogen Science and Engineering: Materials, Processes, Systems and Technology*, 57-84.

Li, P., He, C., Yu, R., Shen, D., & Jiao, Y. (2019). Anaerobic Co-digestion of Urban Sewage Sludge with Agricultural Biomass. *Waste and Biomass Valorization*, 1–11. doi:10.100712649-019-00870-z

Li, R., Liu, F., Zhang, C., Liu, J., Zhou, J., & Xu, L. (2019). Electrical properties of Fe-doped SrTiO3 with B-site-deficient for SOFC anodes. *Ceramics International*, *45*(17), 21684–21687.

Li, T., Luo, G., Liu, K., Li, X., Sun, D., Xu, L., Li, Y., & Tang, Y. (2018). Encapsulation of Ni_3Fe Nanoparticles in N–Doped Carbon Nanotube–Grafted Carbon Nanofibers as High–Efficiency Hydrogen Evolution Electrocatalysts. *Advanced Functional Materials*, *28*(51), 1805828. doi:10.1002/adfm.201805828

Liu, H., & Jiang, J. (2007). Flywheel energy storage—An upswing technology for energy sustainability. *Energy and Building*, *39*, 599–604.

Liu, W., Sun, L., Li, Z., Fujii, M., Geng, Y., Dong, L., & Fujita, T. (2020). Trends and future challenges in hydrogen production and storage research. *Environmental Science and Pollution Research International*, *27*(25), 31092–31104. doi:10.100711356-020-09470-0 PMID:32529621

Liu, Y., Liang, X., Gu, L., Zhang, Y., Li, G. D., Zou, X., & Chen, J. S. (2018). Corrosion engineering towards efficient oxygen evolution electrodes with stable catalytic activity for over 6000 hours. *Nature Communications*, *2609*(1), 1–10. doi:10.103841467-018-05019-5 PMID:29973591

Liu, Y., Zhang, J., Li, X., Zeng, Z., Cheng, X., Wang, Y., & Pa, M. (2019). Phosphorous Modified Metal Boride as High Efficiency HER Electrocatalyst. *International Journal of Electrochemical Science*, *14*, 6123–6132. doi:10.20964/2019.07.68

Liu, Z., Sajjad, S. D., Gao, Y., Kaczur, J. J., & Masel, R. I. (2017). An alkaline water electrolyzer with Sustainion membranes: 1A/cm^2 at 1.9V with Base Metal Catalysts. *ECS Transactions*, *77*(9), 71–73. doi:10.1149/07709.0071ecst

Li, X., Liu, P. F., Zhang, L., Zu, M. Y., Yang, Y. X., & Yang, H. G. (2016). Enhancing alkaline hydrogen evolution reaction activity through Ni–Mn_3O_4 nanocomposites. *Chemical Communications*, *52*(69), 10566–10569. doi:10.1039/C6CC04141H PMID:27500290

Li, Y., & Jun, L. M. A. B. (2017). Will They Be the Future Electrochemical Energy Storage Device of Choice? *ACS Energy Letters*, *2*, 1370–1377.

Li, Y., Sun, Y., Qin, Y., Zhang, W., Wang, L., Luo, M., Yang, H., & Guo, S. (2020). Recent advances on water-splitting electrocatalysis mediated by noble-metal-based nanostructured materials. *Advanced Energy Materials*, *10*(11), 1903120. doi:10.1002/aenm.201903120

Li, Y., & Yang, R. T. (2006). Significantly enhanced hydrogen storage in metal– organic frameworks via spillover. *Journal of the American Chemical Society*, *128*(3), 726–727. doi:10.1021/ja056831s PMID:16417355

Li, Y., Zhou, L. W., & Wang, R. Z. (2017). Urban biomass and methods of estimating municipal biomass resources. *Renewable & Sustainable Energy Reviews*, *80*, 1017–1030. doi:10.1016/j.rser.2017.05.214

Li, Z., Li, Z., Tong, X., Zhang, J., Dong, L., Zheng, Y., Ma, W., Zhao, L., Wang, L., Wen, L., Dang, Z., Tuvshintogtokh, I., Liang, C., & Li, F. Y. (2020). Climatic humidity mediates the strength of the species richness–biomass relationship on the Mongolian Plateau steppe. *The Science of the Total Environment*, *718*, 137252. doi:10.1016/j.scitotenv.2020.137252 PMID:32325613

Li, Z., Xia, M., & Shao, G. (2017). The Effects of CeO_2 Nanorods and CeO_2 Nanoflakes on Ni–S Alloys in Hydrogen Evolution Reactions in Alkaline Solutions. *Catalysts*, *7*(7), 197. doi:10.3390/catal7070197

Lohse-Busch, H., Duoba, M., Rask, E., & Stutenberg, K. (2013). *Ambient Temperature (20°F, 72°F and 95°F) Impact on Fuel and Energy Consumption for Several Conventional Vehicles, Hybrid and Plug-In Hybrid Electric Vehicles and Battery Electric Vehicle.* SAE Technical Paper 2013-01-1462.

Luo, W. (2004). (LiNH2–MgH2): A viable hydrogen storage system. *Journal of Alloys and Compounds*, *381*(1-2), 284–287. doi:10.1016/j.jallcom.2004.03.119

Luo, Y., Jiang, W., Zhang, Q., Zhang, W. Y., & Hao, M. (2016). Effects of anode porosity on thermal stress and failure probability of planar solid oxide fuel cell with bonded compliant seal. *International Journal of Hydrogen Energy*, *41*(18), 7464–7474. doi:10.1016/j.ijhydene.2016.03.117

Lyu, J., Kudiiarov, V., & Lider, A. (2020). An overview of the recent progress in modifications of carbon nanotubes for hydrogen adsorption. *Nanomaterials (Basel, Switzerland)*, *10*(2), 255. doi:10.3390/nano10020255 PMID:32024092

Maggio, G., & Cacciola, G. (2012). When will oil, natural gas, and coal peak? *Fuel*, *98*, 111–123. doi:10.1016/j.fuel.2012.03.021

Maggio, G., Nicita, A., & Squadrito, G. (2019). How the hydrogen production from RES could change energy and fuel markets: A review of recent literature. *International Journal of Hydrogen Energy*, *44*(23), 11371–11384. doi:10.1016/j.ijhydene.2019.03.121

Mago, P. J., & Smith, A. D. (2015). Combined Cooling, Heating, and Power (CCHP) Systems. Handbook of Clean Energy Systems, 1-15.

Mahmood, N., Yao, Y., Zhang, J. W., Pan, L., Zhang, X., & Zou, J. J. (2018). Electrocatalysts for hydrogen evolution in alkaline electrolytes: Mechanisms, challenges, and prospective solutions. *Advancement of Science*, *5*(2), 1700464–1700487. doi:10.1002/advs.201700464 PMID:29610722

Mahmud, L. S., Muchtar, A., & Somalu, M. R. (2017). Challenges in fabricating planar solid oxide fuel cells: A review. *Renewable & Sustainable Energy Reviews*, *72*, 105–116.

Makepeace, J. W., He, T., Weidenthaler, C., Jensen, T. R., Chang, F., Vegge, T., ... & David, W. I. (2019). Reversible ammonia-based and liquid organic hydrogen carriers for high-density hydrogen storage: Recent progress. *International Journal of Hydrogen Energy, 44*(15), 7746-7767.

Manage, M. N., Sorensen, E., Simons, S., & Brett, D. J. L. (2014). A modelling approach to assessing the feasibility of the integration of power stations with steam electrolysers. *Chemical Engineering Research & Design*, *92*(10), 1988–2005. doi:10.1016/j.cherd.2014.07.015

Mandal, S., & Jana, A. K. (2020). Simulating reactive distillation of HIx (HI–H2O–I2) system in Sulphur-Iodine cycle for hydrogen production. *Nuclear Engineering and Technology*, *52*(2), 279–286. doi:10.1016/j.net.2019.07.033

Mani Kalyani, A., & VM, J. (. (2020). Mechanisms governing the "gasification to char oxidation transition" in counter-current flame propagation in packed beds: Insights from single-particle experiments. *Particulate Science and Technology*, 1–5. doi:10.1080/02726351.2020.1767245

Manla, E., Nasiri, A., Rentel, C. H., & Hughes, M. (2010). Modeling of zinc-bromide energy storage for vehicular applications. *IEEE Transactions on Industrial Electronics*, *57*(2), 624–632.

Manoharan, Y., Hosseini, S. E., Butler, B., Alzhahrani, H., Senior, B. T. F., Ashuri, T., & Krohn, J. (2019). Hydrogen Fuel Cell Vehicles; Current Status and Future Prospect. *Applied Sciences (Basel, Switzerland)*, *9*(11), 2296. doi:10.3390/app9112296

Marshall, A., Børresen, B., Hagen, H., Tsypkin, M., & Tunold, R. (2007). Hydrogen production by advanced proton exchange membrane (PEM) water electrolysers—Reduced energy consumption by improved electrocatalysis. *Energy*, *32*, 431–436.

Martin, A., Agnoletti, M. F., & Brangier, E. (2020). Users in the design of Hydrogen Energy Systems: A systematic review. *International Journal of Hydrogen Energy*, *45*(21), 11889–11900. doi:10.1016/j.ijhydene.2020.02.163

Mascaretti, L., Ferrulli, S., Mazzolini, P., Casari, C. S., Russo, V., Matarrese, R., Nova, I., Terraneo, G., Liu, N., Schmuki, P., & Li Bassi, A. (2017). Hydrogen-treated hierarchical titanium oxide nanostructures for photoelectrochemical water splitting. *Solar Energy Materials and Solar Cells*, *169*(May), 19–27. doi:10.1016/j.solmat.2017.04.045

Mateescu, I., Badea, G., Ivan, G., Aschilean, I., Mateescu, O., & Constantin, G. (2017). *Hidrogenul și legislația romaneasca. Ştiinţa Modernă şi Energia.*

Mayer, F., Bhandari, R., & Gäth, S. (2019). Critical review on life cycle assessment of conventional and innovative waste-to-energy technologies. *The Science of the Total Environment*, *672*, 708–721. doi:10.1016/j.scitotenv.2019.03.449 PMID:30974361

Ma, Z., Zhao, Q., Li, J., Tang, B., Zhang, Z., & Wang, X. (2018). Three–dimensional well–mixed/highly–densed NiS–CoS nanorod arrays: An efficient and stable bifunctional electrocatalyst for hydrogen and oxygen evolution reactions. *Electrochimica Acta*, *260*, 82–91. doi:10.1016/j.electacta.2017.11.055

McArthur, M. A., Jorge, L., Coulombe, S., & Omanovic, S. (2014). Synthesis and characterization of 3D Ni nanoparticle/carbon nanotube cathodes for hydrogen evolution in alkaline electrolyte. *Journal of Power Sources*, *266*, 365–373. doi:10.1016/j.jpowsour.2014.05.036

McCarty, R. D., Hord, J., & Roder, H. M. (1981). *Selected properties of hydrogen (engineering design data)* (Vol. 168). US Department of Commerce, National Bureau of Standards. doi:10.6028/NBS.MONO.168

McKeown, N. B., & Budd, P. M. (2006). Polymers of intrinsic microporosity (PIMs): Organic materials for membrane separations, heterogeneous catalysis and hydrogen storage. *Chemical Society Reviews*, *35*(8), 675–683. doi:10.1039/b600349d PMID:16862268

Mednikov, A. S. (2018). A review of technologies for multistage wood biomass gasification. *Thermal Engineering*, *65*(8), 531–546. doi:10.1134/S0040601518080037

Mehrpooya, M., & Habibi, R. (2020). A review on hydrogen production thermochemical water-splitting cycles. *Journal of Cleaner Production*, *275*, 123836. doi:10.1016/j.jclepro.2020.123836

Mekhilef, S., Saidur, R., & Safari, A. (2012). Comparative study of different fuel cell technologies. *Renewable & Sustainable Energy Reviews*, *16*(1), 981–989. doi:10.1016/j.rser.2011.09.020

Mendelssohn, I. A., Andersen, G. L., Baltz, D. M., Caffey, R. H., Carman, K. R., Fleeger, J. W., Joye, S. B., Lin, Q., Maltby, E., Overton, E. B., & Rozas, L. P. (2012). Oil impacts on coastal wetlands: Implications for the Mississippi River Delta ecosystem after the Deepwater Horizon oil spill. *Bioscience*, *62*(6), 562–574. doi:10.1525/bio.2012.62.6.7

Meng, Q., Han, J., Kong, L., Liu, H., Zhang, T., & Yu, Z. (2017). Thermodynamic analysis of combined power generation system based on SOFC/GT and transcritical carbon dioxide cycle. *International Journal of Hydrogen Energy*, *42*(7), 4673–4678. doi:10.1016/j.ijhydene.2016.09.067

Meng, Y., Wang, X., Xia, C., Wang, B., Dong, W., Ji, Y., & Zhu, B. (2018). High performance SOFC based on a novel semiconductor-ionic SrFeO3-δ–Ce0.8Sm0.2O2-δ membrane. *International Journal of Hydrogen Energy*, *43*(28), 12697–12704. doi:10.1016/j.ijhydene.2018.01.209

Midilli, A. D. N. A. N., Ay, M., Dincer, I., & Rosen, M. A. (2005a). On hydrogen and hydrogen energy strategies II: Future projections affecting global stability and unrest. *Renewable & Sustainable Energy Reviews*, *9*(3), 273–287. doi:10.1016/j.rser.2004.05.002

Midilli, A., Ay, M., Dincer, I., & Rosen, M. A. (2005). On hydrogen and hydrogen energy strategies. I: Current status and needs, J. *Renewable & Sustainable Energy Reviews*, *9*, 255–271.

Midilli, A., Ay, M., Dincer, I., & Rosen, M. A. (2005). On hydrogen and hydrogen energy strategies. II: Future projections affecting global stability and unrest, J. *Renewable & Sustainable Energy Reviews*, *9*, 273–287.

Midilli, A., Ay, M., Dincer, I., & Rosen, M. A. (2005b). On hydrogen and hydrogen energy strategies: I: current status and needs. *Renewable & Sustainable Energy Reviews*, *9*(3), 255–271. doi:10.1016/j.rser.2004.05.003

Milcarek, R. J., & Ahn, J. (2021). Micro-Tubular Solid Oxide Fuel Cell Polarization and Impedance Variation With Thin Porous Samarium-Doped Ceria and Gadolinium-Doped Ceria Buffer Layer Thickness. *Journal of Electrochemical Energy Conversion and Storage*, *18*(2), 021004.

Millet, P. (2013). 9 - Chlor-alkali technology: Fundamentals, processes and materials for diaphragms and membranes. *Woodhead Publishing Series in Energy*, *2*, 384–415. doi:10.1533/9780857097347.2.384

Millet, P., Mbemba, M., Grigoriev, S. A., Fateev, V. N., Aukauloo, A., & Etievant, C. (2011). Electrochemical performances of PEM water electrolysis cells and perspectives. *International Journal of Hydrogen Energy*, *36*, 4134–4142.

Millet, P., Ngameni, R., Grigoriev, S. A., Mbemba, N., Brisset, F., & Ranjbari, A. (2010). PEM water electrolyzers: From electrocatalysis to stack development. *International Journal of Hydrogen Energy*, *35*, 5043–5052.

Minh, N., Mizusaki, J., & Singhal, S. C. (2017). Advances in solid oxide fuel cells: Review of progress through three decades of the international symposia on solid oxide fuel cells. *ECS Transactions*, *78*(1), 63.

Minutillo, M., Perna, A., Di Trolio, P., Di Micco, S., & Jannelli, E. (2021). Techno-economics of novel refueling stations based on ammonia-to-hydrogen route and SOFC technology. *International Journal of Hydrogen Energy*, *46*(16), 10059–10071.

Mirica, I., Pana, C., Negurescu, N., Cernat, A., & Nutu, C. (2015). Aspects of the Hydrogen Use at a Truck Compression Ignition Engine. In *Proceedings of the 4th International Conference of Thermal Equipment, Renewable Energy and Rural Development.* Politehnica Press.

Mirica, I., Pana, C., Negurescu, N., Cernat, A., & Nutu, C. (2016). Experimental investigations of the hydrogen addition effects on diesel engine performance. *Journal IOP Conference Series: Materials Science and Engineering, 147*(1). . doi:10.1088/1757-899X/147/1/012121

Mirica, I., Cernat, A., Pana, C., Negurescu, N., & Nutu, C. (2015). Performance Comparison Between Hydrogen and Diesel Fuel Fuelled Compression Ignition Engine. *Scientific Bulletin of University Politehnica of Bucharest, Series D. Mechanical Engineering*, *77*(4), 1–12.

Mirica, I., Cernat, A., Pana, C., Negurescu, N., & Nutu, C. (2016, October). *Aspects of Experimental Research on Hydrogen Fuelled Automotive Diesel Engine. The International Congress of Automotive and Transport Engineering, CONAT 2016*, Brasov, Romania.

Moatti, A., Sachan, R., Cooper, V. R., & Narayan, J. (2019). Electrical Transition in Isostructural VO_2 Thin–Film Heterostructures. *Scientific Reports*, *9*(1), 1–10. doi:10.103841598-019-39529-z PMID:30816206

Mo, B., Rix, J., Pal, U., Basu, S., & Gopalan, S. (2020). Improving SOFC Anode Electrocatalytic Activity Using Nanoparticle Infiltration into MIEC Compositions. *Journal of the Electrochemical Society*, *167*(13), 134506.

Modi, M. A., & Patel, T. M. (2021). Syngas Assessment from Plastic Waste Using Artificial Neural Network—A Review. *Machine Learning for Predictive Analysis*, 203-208.

Mohammedi, A., Sahli, Y., & Ben-Moussa, H. (2020). 3D investigation of the channel cross-section configuration effect on the power delivered by PEMFCs with straight channels. *Fuel*, *263*, 116713. doi:10.1016/j.fuel.2019.116713

Mohan, M., Sharma, V. K., Kumar, E. A., & Gayathri, V. (2019). Hydrogen storage in carbon materials—A review. *Energy Storage*, e35.

Mojaver, P., Chitsaz, A., Sadeghi, M., & Khalilarya, S. (2020). Comprehensive comparison of SOFCs with proton-conducting electrolyte and oxygen ion-conducting electrolyte: Thermoeconomic analysis and multi-objective optimization. *Energy Conversion and Management*, *205*, 112455.

Molnarne, M., & Schroeder, V. (2019). Hazardous properties of hydrogen and hydrogen containing fuel gases. *Process Safety and Environmental Protection*, *130*, 1–5. doi:10.1016/j.psep.2019.07.012

Mondal, P., Kumari, P., Singh, J., Verma, S., Chaurasia, A. K., & Singh, R. P. (2017). Oil from Algae. In *Sustainable Utilization of Natural Resources* (pp. 213–253). CRC Press. doi:10.1201/9781315153292-8

Monsaf, T., Hocine, B. M., Youcef, S., & Abdallah, M. (2017). Unsteady three-dimensional numerical study of mass transfer in PEM fuel cell with spiral flow field. *International Journal of Hydrogen Energy*, *42*(2), 1237–5121. doi:10.1016/j.ijhydene.2016.12.084

Monte, M. C., Fuente, E., Blanco, A., & Negro, C. (2009). Waste management from pulp and paper production in the European Union. *Waste Management (New York, N.Y.)*, *29*(1), 293–308. doi:10.1016/j.wasman.2008.02.002 PMID:18406123

Morris, R. E., Wheatley, P. S., Warrender, S., & Duncan, M. (2019). *U.S. Patent No. 10,357,037*. Washington, DC: U.S. Patent and Trademark Office.

Mudryk, K., Jewiarz, M., Wróbel, M., Niemiec, M., & Dyjakon, A. (2021). Evaluation of Urban Tree Leaf Biomass-Potential, Physico-Mechanical and Chemical Parameters of Raw Material and Solid Biofuel. *Energies*, *14*(4), 818. doi:10.3390/en14040818

Mudryk, K., & Werle, S. (2018). *Renewable energy sources: engineering, technology, innovation*. Springer International Publishing.

Mularski, J., Pawlak-Kruczek, H., & Modlinski, N. (2020). A review of recent studies of the CFD modelling of coal gasification in entrained flow gasifiers, covering devolatilization, gas-phase reactions, surface reactions, models and kinetics. *Fuel*, *271*, 117620. doi:10.1016/j.fuel.2020.117620

Müller, M., Bouyer, E. V. Bradke, M., Branston, D. W., Heimann, R. B., Henne, R., ... Schiller, G. (2002). Thermal induction plasma processes for the synthesis of SOFC materials. *Materialwissenschaft und Werkstofftechnik: Entwicklung, Fertigung, Prüfung, Eigenschaften und Anwendungen technischer Werkstoffe*, *33*(6), 322-330.

MunueraL.TorilA. (2018). Retrieved from: https://www.iea.org/tcep/energyintegration/energystorage/

Murshed, A. M., Huang, B., & Nandakumar, K. (2007). Control relevant modeling of planer solid oxide fuel cell system. *Journal of Power Sources*, *163*(2), 830–845. doi:10.1016/j.jpowsour.2006.09.080

Murthy, A. P., Madhavan, J., & Murugan, K. (2018). Recent advances in hydrogen evolution reaction catalysts on carbon/carbon–based supports in acid media. *Journal of Power Sources*, *398*, 9–26. doi:10.1016/j.jpowsour.2018.07.040

Naik, S. N., Goud, V. V., Rout, P. K., & Dalai, A. K. (2010). Production of first and second generation biofuels: A comprehensive review. Renewable and Sustainable Energy Reviews, 14(2), 578–597. doi:10.1016/j.rser.2009.10.003

Nakajo, A., Mueller, F., Brouwer, J., Van herle, J., & Favrat, D. (2012a). Mechanical reliability and durability of SOFC stacks. Part I: Modelling of the effect of operating conditions and design alternatives on the reliability. *International Journal of Hydrogen Energy*, *37*(11), 9249–9268. doi:10.1016/j.ijhydene.2012.03.043

Nakajo, A., Mueller, F., Brouwer, J., Van herle, J., & Favrat, D. (2012b). Mechanical reliability and durability of SOFC stacks. Part II: Modelling of mechanical failures during ageing and cycling. *International Journal of Hydrogen Energy*, *37*(11), 9269–9286. doi:10.1016/j.ijhydene.2012.03.023

Nakamori, Y., & Orimo, S. I. (2004). Destabilization of Li-based complex hydrides. *Journal of Alloys and Compounds*, *370*(1-2), 271–275. doi:10.1016/j.jallcom.2003.08.089

Nakanishi, K., & Tanaka, N. (2007). Sol–gel with phase separation. Hierarchically porous materials optimized for high-performance liquid chromatography separations. *Accounts of Chemical Research*, *40*(9), 863–873. doi:10.1021/ar600034p PMID:17650924

Nalakath, H., Keskin, T., & Yazgin, O. (2018). ScienceDirect Biohydrogen production from autoclaved fruit and vegetable wastes by dry fermentation under thermophilic condition. *International Journal of Hydrogen Energy*. Advance online publication. doi:10.1016/j.ijhydene.2018.12.068

Narayanan, S. R., Prakash, G. K. S., Manohar, A., Yang, B., Malkhandi, S., & Kindler, A. (2012). Materials Challenges and Technical Approaches for Realizing Inexpensive and Robust Iron-Air Batteries for Large-Scale Energy Storage. *Solid State Ionics*, *216*, 105–109.

Nassef, A. M., Fathy, A., Sayed, E. T., Abdelkareem, M. A., Rezk, H., Tanveer, W. H., & Olabi, A. G. (2019). Maximizing SOFC performance through optimal parameters identification by modern optimization algorithms. *Renewable Energy*, *138*, 458–464.

Negreanu-Pirjol, B. S., Negreanu-Pirjol, T., & Paraschiv, G. M. (2020). Residual Marine Algae Biomass-An Important Raw Material for Obtaining a Soil Biostimulator-Regenerator. *European Journal of Medicine and Natural Sciences*, *4*(3), 74–85.

Negurescu, N., Pana, C., & Cernat, A. (2011, November). *Aspects of using hydrogen in SI engine* [Paper presentation CIEM2011-ID S4_23]. The 5th International Conference on Energy and Environment -under the patronage of World Energy Council, University Politehnica of Bucharest, Bucharest, Romania.

Negurescu, N., Pană, C., Popa, M., Cernat, A., & Soare, D. (2007). Some Characteristics of the Hydrogen Combustion in SI Engine. In *Proceedings of 6th International Colloquium "FUELS 2007"* (*vol. 1*, pp.461-466). Technische Akademie of Esslingen.

Negurescu, N., Pana, C., Popa, M., Cernat, A., & Soare, D. (2008) Experimental Researches Regarding the Direct Injection in Hydrogen Fueled S.I. Engine. In *Proceedings of Second International Congress Automotive, Safety and Environment- FISITA, EAEC, SAE International, SIAR* (*vol. 2*, pp. 62-70). University of Craiova Faculty of Mechanics.

Negurescu, N., Pana, C., Popa, M., Cernat, A., & Soare, D. (2008). Aspects Regarding the Use of Hydrogen in Spark Ignition Engine. *Scientific Bulletin of University Gh. Asachi of Iasi*, *54*(2).

Negurescu, N., Pana, C., Popa, M., Cernat, A., & Soare, D. (2008, September). *Experimental Investigations Concerning the Direct Injection in Hydrogen Fuelled S.I. Engine* [Paper F2008-09-028 presentation]. FISITA2008, World Automotive Congress, Munich, Germany.

Negurescu, N., Pana, C., & Cernat, A. (2012). Aspects of using hydrogen in SI engine. *Scientific Bulletin of University Politehnica of Bucharest, Series D. Mechanical Engineering*, *74*(1), 11–20.

Negurescu, N., Pana, C., Popa, M., & Cernat, A. (2009). Theoretical and Experimental Researches on a Hydrogen Fueled SI Engine. In *Proceedings of the 8th International Conference, Fuel Economy, Safety and reliability of motor vehicles, ESFA 2009, under the patronage of FISITA, EAEC, SIAR* (vol. 1, pp. 39-48). Academic Press.

Negurescu, N., Pană, C., Popa, M., & Cernat, A. (2010). Experimental Researches of a Hydrogen Fuelled SI Engine. *Journal of Energy and Power Engineering*, *4*(7), 1–8.

Negurescu, N., Pana, C., Popa, M., & Cernat, A. (2011). Experimental and Theoretical Researches of Hydrogen Use in S. I. Engine. *Journal of Environmental Protection and Ecology*, *12*(4A), 2275–2287.

Negurescu, N., Pana, C., Popa, M., & Cernat, A. (2011). Performance comparison between hydrogen and gasoline S.I. fuelled engine. *Journal of Thermal Science, 15*(4), 1155–1164. doi:10.2298/TSCI110203090N

Negurescu, N., Pana, C., Popa, M., Cernat, A., & Soare, D. (2008). The control of the running of the hydrogen fuelled spark ignition engine. In *Proceedings of 2th European Conference H2-Fuel-Millenium_Conference 2008, e-Hydrogenia, Academia Romana.* (vol. 1, pp. 1-10). Romanian Academy.

Netskina, O. V., Tayban, E. S., Ozerova, A. M., Komova, O. V., & Simagina, V. I. (2019). Solid-State NaBH4/Co Composite as Hydrogen Storage Material: Effect of the Pressing Pressure on Hydrogen Generation Rate. *Energies*, *12*(7), 1184. doi:10.3390/en12071184

Nielsen, E. R., Prag, C. B., Bachmann, T. M., Carnicelli, F., Boyd, E., Walker, I., ... Stephens, A. (2019). Status on Demonstration of Fuel Cell Based Micro-CHP Units in Europe. *Fuel Cells (Weinheim)*, *19*(4), 340–345.

Niesten, E., & Alkemade, F. (2016). How is value created and captured in smart grids? A review of the literature and an analysis of pilot projects. *Renewable & Sustainable Energy Reviews*, *53*, 629–638. doi:10.1016/j.rser.2015.08.069

Ni, M., Leung, D. Y. C., Leung, M. K. H., & Sumathy, K. (2006). An overview of hydrogen production from biomass. *Fuel Processing Technology*, *87*(5), 461–472. doi:10.1016/j.fuproc.2005.11.003

Ni, M., Leung, M. K. H., & Leung, D. Y. C. (2007). Parametric study of solid oxide fuel cell performance. *Energy Conversion and Management*, *48*(5), 1525–1535. doi:10.1016/j.enconman.2006.11.016

Nissen, M. B. (2009). High performance development as distributed generation. *IEEE Potentials*, *28*(6), 25–31.

Noritake, T., Aoki, M., Towata, S., Seno, Y., Hirose, Y., Nishibori, E., Takata, M., & Sakata, M. (2002). Chemical bonding of hydrogen in MgH 2. *Applied Physics Letters*, *81*(11), 2008–2010. doi:10.1063/1.1506007

Novosel, U., Živić, M., & Avsec, J. (2021). The production of electricity, heat and hydrogen with the thermal power plant in combination with alternative technologies. *International Journal of Hydrogen Energy*, *46*(16), 10072–10081. doi:10.1016/j.ijhydene.2020.01.253

Nunes, L. J. R., Causer, T. P., & Ciolkosz, D. (2020). Biomass for energy: A review on supply chain management models. *Renewable & Sustainable Energy Reviews*, *120*, 109658. doi:10.1016/j.rser.2019.109658

Nuttall, W. J., & Bakenne, A. T. (2020). Towards a Hydrogen Economy. In *Fossil Fuel Hydrogen* (pp. 43–52). Springer. doi:10.1007/978-3-030-30908-4_4

Nutu, C., Pana, C., Negurescu, N., Cernat, A., Fuiorescu, D., & Mirica, I. (2019). A research on fuelling a truck diesel engine with hydrogen. *E3S Web of Conferences, 112,* 1-9. . doi:10.1051/e3sconf/201911201011

Oelerich, W., Klassen, T., & Bormann, R. (2001). Metal oxides as catalysts for improved hydrogen sorption in nanocrystalline Mg-based materials. *Journal of Alloys and Compounds*, *315*(1-2), 237–242. doi:10.1016/S0925-8388(00)01284-6

Ogungbemi, E., Wilberforce, T., Ijaodola, O., Thompson, J., & Olabi, A. G. (2021). Review of operating condition, design parameters and material properties for proton exchange membrane fuel cells. *International Journal of Energy Research*, *45*(2), 1227–1245. doi:10.1002/er.5810

Oh, A., Sa, Y. J., Hwang, H., Baik, H., Kim, J., Kim, B., Joo, S. H., & Lee, K. (2016). Rational design of Pt–Ni–Co ternary alloy nanoframe crystals as highly efficient catalysts toward the alkaline hydrogen evolution reaction. *Nanoscale*, *8*(36), 16379–16386. doi:10.1039/C6NR04572C PMID:27714051

Oh, G., Ra, H. W., Yoon, S. M., Mun, T. Y., Seo, M. W., Lee, J. G., & Yoon, S. J. (2019). Syngas production through gasification of coal water mixture and power generation on dual-fuel diesel engine. *Journal of the Energy Institute*, *92*(2), 265–274. doi:10.1016/j.joei.2018.01.009

Orimo, S. I., Nakamori, Y., Kitahara, G., Miwa, K., Ohba, N., Towata, S. I., & Züttel, A. (2005). Dehydriding and rehydriding reactions of LiBH4. *Journal of Alloys and Compounds*, *404*, 427–430. doi:10.1016/j.jallcom.2004.10.091

Ortiz-Rivera, E. I., Reyes-Hernandez, A. L., & Febo, R. A. (2007, August). Understanding the history of fuel cells. In *2007 IEEE Conference on the History of Electric Power* (pp. 117-122). IEEE. 10.1109/HEP.2007.4510259

Oruc, O., & Dincer, I. (2021). Assessing the potential of thermochemical water splitting cycles: A bridge towards for clean and sustainable hydrogen generation. *Fuel*, *286*, 119325. doi:10.1016/j.fuel.2020.119325

Ou, K., Yuan, W. W., & Kim, Y. B. (2021). Development of optimal energy management for a residential fuel cell hybrid power system with heat recovery. *Energy*, *219*, 119499. doi:10.1016/j.energy.2020.119499

Pablo, G., Gomes-Dias, J. S., Rocha, C. M., Romaní, A., Garrote, G., & Domingues, L. (2020). Recent trends on seaweed fractionation for liquid biofuels production. *Bioresource Technology*, *299*, 122613. doi:10.1016/j.biortech.2019.122613 PMID:31870706

Palys, M. J., & Daoutidis, P. (2020). Using hydrogen and ammonia for renewable energy storage: A geographically comprehensive techno-economic study. *Computers & Chemical Engineering*, *136*, 106785. doi:10.1016/j.compchemeng.2020.106785

Pana, C., Negurescu, N., Popa, M., Cernat, A., & Soare, D. (2007, April). *An Investigation of the Hydrogen Addition Effects to Gasoline Fueled Spark Ignition Engine* [Paper No. 2007-01-1468 presentation]. SAE World Congress, Detroit, MI. 10.4271/2007-01-1468

Pana, C., Negurescu, N., Cernat, A., Nutu, C., Mirica, I., & Fuiorescu, D. (2017). Experimental Aspects of the Hydrogen Use at Diesel Engine. *Procedia Engineering*, *181*, 649–657. doi:10.1016/j.proeng.2017.02.446

Pana, C., Negurescu, N., Popa, M., & Cernat, A. (2011). A control strategy of the hydrogen engine fueled load. *Termotehnica. Thermal Engineering*, *1*, 1–8.

Pana, C., Negurescu, N., Popa, M., Cernat, A., Boboc, G., & Soare, D. (2005). Influences of Hydrogen Addition in Gasoline on Spark Ignition Engine Performance. In *Proceedings of the 9th International Congress on Automotive CAR 2005, section Automotive and Environment*.University of Pitesti.

Pana, C., Negurescu, N., Popa, M., Cernat, A., Boboc, G., & Soare, D. (2005). On the Performance of a Spark Ignition Engine fuelled with Gasoline and Hydrogen Addition. *Scientific Bulletin of University Politehnica of Bucharest, Series D. Mechanical Engineering (New York, N.Y.)*, *67*(4), 15–24.

Panaitescu, C., Stoicescu, M., Ponea, M. G., Nancu, D., & Nguyen, D. N. (2020). Biomass valuation in the context of sustainable agricultural development in Romania. *Ekonomika Poljoprivrede*, 3.

Pantile, V., Pana, C., Negurescu, N., Radu, A., & Rusu, E. (2011, November). *The use of hydrogen as an alternative fuel in SI engines* [Paper 1142 presentation]. The FISITA 10th International Automotive Congress, Automotive Engineering and Environment CAR2011, Pitesti, Romania.

Pantile, V., Pana, C. N., & Negurescu, N. (2012, November). *An Experimental Investigation of the Hydrogen Fuelled Spark Ignition Engine* [Paper presentation]. *FISITA International Conference Motor Vehicle and Transportation MVT 2012*, Timisoara, Romania.

Pan, X., Ichoku, C., Chin, M., Bian, H., Darmenov, A., Colarco, P., Ellison, L., Kucsera, T., da Silva, A., Wang, J., Oda, T., & Cui, G. (2020). Six global biomass burning emission datasets: Intercomparison and application in one global aerosol model. *Atmospheric Chemistry and Physics*, *20*(2), 969–994. doi:10.5194/acp-20-969-2020

Pareek, A., Dom, R., Gupta, J., Chandran, J., Adepu, V., & Borse, P. H. (2020). Insights into renewable hydrogen energy: Recent advances and prospects. *Materials Science for Energy Technologies*, *3*, 319–327. doi:10.1016/j.mset.2019.12.002

Park, J. H., Kim, D., Kim, K. J., Bae, K. T., & Lee, K. T. (2020). A brief review of the bilayer electrolyte strategy to achieve high performance solid oxide fuel cells. *Ceramist*, *23*(2), 184–199.

Park, S. J., & Seo, M. K. (2011). *Interface science and composites* (1st ed.). Academic Press.

Patrick, T. (2015). *Electrochemical Energy Storage for Renewable Sources and Grid Balancing*. Elsevier.

Patuzzi, F., Basso, D., Vakalis, S., Antolini, D., Piazzi, S., Benedetti, V., Cordioli, E., & Baratieri, M. (2021). State-of-the-art of small-scale biomass gasification systems: An extensive and unique monitoring review. *Energy*, *223*, 120039. doi:10.1016/j.energy.2021.120039

Pechlivanoglou, G. (2007). *Hydrogen Enhanced Combustion History. Applications and Hydrogen Supply by Plasma Reforming*. PPRE 2005-2007. University of Oldenburg.

Peksen, M. (2013). 3D thermomechanical behaviour of solid oxide fuel cells operating in different environments. *International Journal of Hydrogen Energy*, *38*(30), 13408–13418. doi:10.1016/j.ijhydene.2013.07.112

Peksen, M., Al-Masri, A., Blum, L., & Stolten, D. (2013). 3D transient thermomechanical behaviour of a full scale SOFC short stack. *International Journal of Hydrogen Energy*, *38*(10), 4099–4107. doi:10.1016/j.ijhydene.2013.01.072

Peng, B., & Chen, J. (2008). Ammonia borane as an efficient and lightweight hydrogen storage medium. *Energy & Environmental Science*, *1*(4), 479–483. doi:10.1039/b809243p

Peng, X., Pi, C., Zhang, X., Li, S., Huo, K., & Chu, P. K. (2019). Recent progress of transition metal nitrides for efficient electrocatalytic water splitting. *Sustainable Energy & Fuels*, *3*(2), 366–381. doi:10.1039/C8SE00525G

Piccardo, P., Anelli, S., Bongiorno, V., Spotorno, R., Repetto, L., & Girardon, P. (2015). K44M ferritic stainless steel as possible interconnect material for SOFC stack operating at 600 °C: Characterization of the oxidation behavior at early working stages. *International Journal of Hydrogen Energy*, *40*(9), 3726–3738. doi:10.1016/j.ijhydene.2015.01.083

Pinales-Márquez, C. D., Rodríguez-Jasso, R. M., Araújo, R. G., Loredo-Treviño, A., Nabarlatz, D., Gullón, B., & Ruiz, H. A. (2021). Circular bioeconomy and integrated biorefinery in the production of xylooligosaccharides from lignocellulosic biomass: A review. *Industrial Crops and Products*, *162*, 113274. doi:10.1016/j.indcrop.2021.113274

Pinaud, B. A., Benck, J. D., Seitz, L. C., Forman, A. J., Chen, Z., Deutsch, T. G., James, B. D., Baum, K. N., Baum, G. N., Ardo, S., Wang, H., Miller, E., & Jaramillo, T. F. (2013). Technical and economic feasibility of centralized facilities for solar hydrogen production via photocatalysis and photoelectrochemistry. *Energy & Environmental Science*, *6*(7), 1983–2002. doi:10.1039/c3ee40831k

Pletcher, D., & Li, X. (2011). Prospects for alkaline zero gap water electrolysers for hydrogen production. *International Journal of Hydrogen Energy*, *36*(23), 15089–15104. doi:10.1016/j.ijhydene.2011.08.080

Pokhrel, R. P., Gordon, J., Fiddler, M. N., & Bililign, S. (2021). Impact of combustion conditions on physical and morphological properties of biomass burning aerosol. *Aerosol Science and Technology*, *55*(1), 80–91. doi:10.1080/02786826.2020.1822512

Popa, M., Negurescu, N., & Pană, C. (2003). *Diesel Engines. Processes*. MatrixRom.

Popescu, A. I., & Necula, C. (2019). Determination of the Calorific Power of Densified Solid Biofuels. *Hidraulica*, (3), 68–75.

Pradecta, M. R., Majid, A. I., & Widyaparaga, A. (2021, February). Syngas Production via Supercritical Gasification of Leachate: A Review. *Journal of Physics: Conference Series*, *1772*(1), 012056.

Principi, G., Agresti, F., Maddalena, A., & Russo, S. L. (2009). The problem of solid state hydrogen storage. *Energy*, *34*(12), 2087–2091. doi:10.1016/j.energy.2008.08.027

Proskurina, S., Junginger, M., Heinimö, J., Tekinel, B., & Vakkilainen, E. (2019). Global biomass trade for energy—Part 2: Production and trade streams of wood pellets, liquid biofuels, charcoal, industrial roundwood and emerging energy biomass. *Biofuels, Bioproducts & Biorefining*, *13*(2), 371–387. doi:10.1002/bbb.1858

Puente-Martínez, D. E., Díaz-Guillén, J. A., Montemayor, S. M., Díaz-Guillén, J. C., Burciaga-Díaz, O., Bazaldúa-Medellín, M. E., ... Fuentes, A. F. (2020). High ionic conductivity in CeO2 SOFC solid electrolytes; effect of Dy doping on their electrical properties. *International Journal of Hydrogen Energy*, *45*(27), 14062–14070.

Punekar, N. S. (2018). Hallmarks of an Enzyme Catalyst. In *ENZYMES: Catalysis, Kinetics and Mechanisms* (pp. 43–51). Springer. doi:10.1007/978-981-13-0785-0_5

Qu, K., Zheng, Y., Jiao, Y., Zhang, X., Dai, S., & Qiao, S. Z. (2017). Polydopamine-Inspired, Dual Heteroatom-Doped Carbon Nanotubes for Highly Efficient Overall Water Splitting. *Advanced Energy Materials*, *7*(9), 1–8. doi:10.1002/aenm.201602068

Rabizadeh, T., Allahkaram, R. S., & Zarebidaki, A. (2010). An investigation on effects of heat treatment on corrosion properties of Ni–P electroless nano–coatings. *Materials & Design*, *7*(7), 3174–3179. doi:10.1016/j.matdes.2010.02.027

Radenahmad, N., Azad, A. T., Saghir, M., Taweekun, J., Bakar, M. S. A., Reza, M. S., & Azad, A. K. (2020). A review on biomass derived syngas for SOFC based combined heat and power application. *Renewable & Sustainable Energy Reviews*, *119*, 109560.

Rajarathnam. (2016). *The zinc/bromine flow battery: Fundamentals and novel materials for technology advancement.* Academic Press.

Ramis, G., Bahadori, E., & Rossetti, I. (2020). Design of efficient photocatalytic processes for the production of hydrogen from biomass derived substrates. *International Journal of Hydrogen Energy.*

Rana, A. R. K., & Farhat, Z. (2019). Preparation and tribological characterization of graphene incorporated electroless Ni–P composite coatings. *Surface and Coatings Technology*, *369*, 334–346. doi:10.1016/j.surfcoat.2019.04.043

Rani, M., Shanker, U., & Chaurasia, A. K. (2017). Catalytic potential of laccase immobilized on transition metal oxides nanomaterials: Degradation of alizarin red S dye. *Journal of Environmental Chemical Engineering*, *5*(3), 2730–2739. Advance online publication. doi:10.1016/j.jece.2017.05.026

Rasmussen, N. B., & Aryal, N. (2020). Syngas production using straw pellet gasification in fluidized bed allothermal reactor under different temperature conditions. *Fuel*, *263*, 116706. doi:10.1016/j.fuel.2019.116706

Raychaudhuri, A., & Behera, M. (2020). Review of the process optimization in microbial fuel cell using design of experiment methodology. *Journal of Hazardous, Toxic and Radioactive Waste*, *24*(3), 04020013. doi:10.1061/(ASCE)HZ.2153-5515.0000503

Read, C., Thomas, G., Ordaz, C., & Satyapal, S. (2007). US Department of Energy's system targets for on-board vehicular hydrogen storage. *Material Matters*, *2*(2), 3.

Renau, J., García, V., Domenech, L., Verdejo, P., Real, A., Giménez, A., Sánchez, F., Lozano, A., & Barreras, F. (2021). Novel Use of Green Hydrogen Fuel Cell-Based Combined Heat and Power Systems to Reduce Primary Energy Intake and Greenhouse Emissions in the Building Sector. *Sustainability*, *13*(4), 1776. doi:10.3390u13041776

Ren, J., Musyoka, N. M., Langmi, H. W., Mathe, M., & Liao, S. (2017). Current research trends and perspectives on materials-based hydrogen storage solutions: A critical review. *International Journal of Hydrogen Energy*, *42*(1), 289–311. doi:10.1016/j.ijhydene.2016.11.195

Ren, X., Dong, L., Xu, D., & Hu, B. (2020). Challenges towards hydrogen economy in China. *International Journal of Hydrogen Energy*, *45*(59), 34326–34345. doi:10.1016/j.ijhydene.2020.01.163

Resan, M., Hampton, M. D., Lomness, J. K., & Slattery, D. K. (2005). Effects of various catalysts on hydrogen release and uptake characteristics of LiAlH4. *International Journal of Hydrogen Energy*, *30*(13-14), 1413–1416. doi:10.1016/j.ijhydene.2004.12.009

Riensche, E., Stimming, U., & Unverzagt, G. (1998). Optimization of a 200 kW SOFC cogeneration power plant: Part I: Variation of process parameters. *Journal of Power Sources*, *73*(2), 251–256.

Rillo, E., Gandiglio, M., Lanzini, A., Bobba, S., Santarelli, M., & Blengini, G. (2017). Life cycle assessment (LCA) of biogas-fed solid oxide fuel cell (SOFC) plant. *Energy*, *126*, 585–602.

Romanian Law no. 13/2007, Electricity Law; Legea nr. 13/2007, Legea energiei electrice.

Rosalbino, F., Delsante, S., Borzone, G., & Angelini, E. (2008). Electrocatalytic behaviour of Co–Ni–R (R=Rare earth metal) crystalline alloys as electrode materials for hydrogen evolution reaction in alkaline medium. *International Journal of Hydrogen Energy*, *33*(22), 6696–6703. doi:10.1016/j.ijhydene.2008.07.125

Rosalbino, F., Maccio, D., Saccone, A., Angelini, E., & Delfino, S. (2011). Fe–Mo–R (R= rare earth metal) crystalline alloys as a cathode material for hydrogen evolution reaction in alkaline solution. *International Journal of Hydrogen Energy*, *36*(3), 1965–1973. doi:10.1016/j.ijhydene.2010.11.054

Roshandel, R., Golzar, F., & Astaneh, M. (2018). Technical, economic and environmental optimization of combined heat and power systems based on solid oxide fuel cell for a greenhouse case study. *Energy Conversion and Management*, *164*, 144–156.

Rosner, F., Rao, A., & Samuelsen, S. (2020). Economics of cell design and thermal management in solid oxide fuel cells under SOFC-GT hybrid operating conditions. *Energy Conversion and Management*, *220*, 112952.

Rowsell, J. L., & Yaghi, O. M. (2004). Metal–organic frameworks: A new class of porous materials. *Microporous and Mesoporous Materials*, *73*(1-2), 3–14. doi:10.1016/j.micromeso.2004.03.034

Rusu, E., Pana, C., Neguręscu, N., Pantile, V., & Radu, A. (2011, November). *Analysis of energetically and emissions performance on a SI engine fueled with hydrogen in addition* [Paper 1143 presentation]. The FISITA 10th International Automotive Congress, Automotive Engineering and Environment CAR2011, Pitesti, Romania.

Ryghaug, M., & Skjølsvold, T. M. (2021). The Co-production of Pilot Projects and Society. In *Pilot Society and the Energy Transition* (pp. 23–62). Palgrave Pivot. doi:10.1007/978-3-030-61184-2_2

Rzepka, M., Lamp, P., & De la Casa-Lillo, M. A. (1998). Physisorption of hydrogen on microporous carbon and carbon nanotubes. *The Journal of Physical Chemistry B*, *102*(52), 10894–10898. doi:10.1021/jp9829602

Sadeghi, S., & Baniasad Askari, I. (2020). Performance and economic investigation of a combined phosphoric acid fuel cell/organic Rankine cycle/electrolyzer system for sulfuric acid production; Energy-based organic fluid selection. *International Journal of Energy Research*, *44*(4), 2704–2725. doi:10.1002/er.5073

Sadik-Zada, E. R., & Gatto, A. (2021). Energy security pathways in South East Europe: Diversification of the natural gas supplies, energy transition, and energy futures. In *From Economic to Energy Transition* (pp. 491–514). Palgrave Macmillan.

Sadykov, V. A., Sadovskaya, E. M., Filonova, E. A., Eremeev, N. F., Bogdanovich, N. M., Pikalov, S. M., ... Pikalova, E. Y. (2020). Mixed ionic-electronic conductivity features of A-site deficient Nd nickelates. *Ceramics International, 46*(16), 25527–25535.

Saebea, D., Authayanun, S., Patcharavorachot, Y., Paengjuntuek, W., & Arpornwichanop, A. (2013). Use of different renewable fuels in a steam reformer integrated into a solid oxide fuel cell: Theoretical analysis and performance comparison. *Energy, 51*, 305–313. doi:10.1016/j.energy.2012.12.014

Saebea, D., Patcharavorachot, Y., Assabumrungrat, S., & Arpornwichanop, A. (2013). Analysis of a pressurized solid oxide fuel cell-gas turbine hybrid power system with cathode gas recirculation. *International Journal of Hydrogen Energy, 38*(11), 4748–4759. doi:10.1016/j.ijhydene.2013.01.146

Safarian, S., Unnthorsson, R., & Richter, C. (2020). Techno-economic and environmental assessment of power supply chain by using waste biomass gasification in Iceland. *BioPhysical Economics and Sustainability, 5*(2), 1–13. doi:10.100741247-020-00073-4

Safari, F., & Dincer, I. (2020). A review and comparative evaluation of thermochemical water splitting cycles for hydrogen production. *Energy Conversion and Management, 205*, 112182. doi:10.1016/j.enconman.2019.112182

Safizadeh, F., Ghali, E., & Houlachi, G. (2015). Electrocatalysis developments for hydrogen evolution reaction in alkaline solutions–a review. *International Journal of Hydrogen Energy, 40*(1), 256–274. doi:10.1016/j.ijhydene.2014.10.109

Sahasrabudhe, A., Dixit, H., Majee, R., & Bhattacharyya, S. (2018). Value added transformation of ubiquitous substrates into highly efficient and flexible electrodes for water splitting. *Nature Communications, 9*(1), 1–14. doi:10.103841467-018-04358-7 PMID:29789569

Sahli, Y., Zitouni, B., Ben-Moussa, H., & Abdenebi, H. (2015). Three-Dimensional Numerical Study of the Heat Transfer on the Planar Solid Oxide Fuel Cell: Joules Effect. In I. Dincer, C. O. Colpan, O. Kizilkan & M. A. Ezan (Eds.), Progress in Clean Energy, volume I: Analysis and Modeling (pp. 449-461). Springer.

Sahli, Y., Ben-Moussa, H., & Zitouni, B. (2019). Optimization Study of the Produced Electric Power by SOFCs. *International Journal of Hydrogen Energy, 44*(39), 22445–22454. doi:10.1016/j.ijhydene.2018.08.162

Sahli, Y., Zitouni, B., & Ben-Moussa, H. (2017). Solid Oxide Fuel Cell Thermodynamic Study. *Cankaya University Journal of Science and Engineering, 14*, 134–151.

Sahli, Y., Zitouni, B., & Benmoussa, H. (2018a). Etude numérique tridimensionnelle de l'effet de la température d'entrée des gaz sur la production de chaleur dans une pile à combustible SOFC planaire. *Revue des Énergies Renouvelables, 21*, 173–180.

Sahli, Y., Zitouni, B., & Ben-Moussa, H. (2018b). Thermodynamic Optimization of the Solid Oxyde Fuel Cell Electric Power. *UPB Scientific Bulletin, Series B. Chemistry and Materials Science, 80*, 159–170.

Sakintuna, B., Lamari-Darkrim, F., & Hirscher, M. (2007). Metal hydride materials for solid hydrogen storage: A review. *International Journal of Hydrogen Energy, 32*(9), 1121–1140. doi:10.1016/j.ijhydene.2006.11.022

Sammes, N. (Ed.). (2006). *Fuel cell technology: reaching towards commercialization*. Springer Science & Business Media. doi:10.1007/1-84628-207-1

Sammes, N. M., & Boersma, R. (2000). Small-scale fuel cells for residential applications. *Journal of Power Sources, 86*(1-2), 98–110.

Sammes, N., Bove, R., & Stahl, K. (2004). Phosphoric acid fuel cells: Fundamentals and applications. *Current Opinion in Solid State and Materials Science*, *8*(5), 372–378.

Sangtongkitcharoen, W., Vivanpatarakij, S., Laosiripojana, N., Arpornwichanop, A., & Assabumrungrat, S. (2008). Performance analysis of methanol-fueled solid oxide fuel cell system incorporated with palladium membrane reactor. *Chemical Engineering Journal*, *138*, 436–441. doi:10.1016/j.cej.2007.06.021

Sansaniwal, S. K., Pal, K., Rosen, M. A., & Tyagi, S. K. (2017). Recent advances in the development of biomass gasification technology: A comprehensive review. *Renewable & Sustainable Energy Reviews*, *72*, 363–384. doi:10.1016/j.rser.2017.01.038

Sanz, L. (2015). Study and characterization of positive electrolytes for application in the aqueous all-copper redox flow battery. *Journal of Power Sources*, *278*, 175–182.

Sapountzi, F. M., Gracia, J. M., & Weststrate, C. J. (2017). Electrocatalysts for the generation of hydrogen, oxygen and synthesis gas. In Progress in Energy and Combustion Science (Vol. 58, pp. 1–35). Elsevier Ltd. doi:10.1016/j.pecs.2016.09.001

Sarruf, B. J. M., Hong, J. E., Steinberger-Wilckens, R., & de Miranda, P. E. V. (2020). Ceria-Co-Cu-based SOFC anode for direct utilisation of methane or ethanol as fuels. *International Journal of Hydrogen Energy*, *45*(8), 5297–5308.

Sazali, N. (2020). Emerging technologies by hydrogen: A review. *International Journal of Hydrogen Energy*, *45*(38), 18753–18771. doi:10.1016/j.ijhydene.2020.05.021

Schafföner, S., Fruhstorfer, J., Faßauer, C., Freitag, L., Jahn, C., & Aneziris, C. G. (2017). Influence of in situ phase formation on properties of calcium zirconate refractories. *Journal of the European Ceramic Society*, *37*(1), 305–313.

Schill, G. P., Froyd, K. D., Bian, H., Kupc, A., Williamson, C., Brock, C. A., Ray, E., Hornbrook, R. S., Hills, A. J., Apel, E. C., Chin, M., Colarco, P. R., & Murphy, D. M. (2020). Widespread biomass burning smoke throughout the remote troposphere. *Nature Geoscience*, *13*(6), 422–427. doi:10.103841561-020-0586-1

Schmitz, B., Müller, U., Trukhan, N., Schubert, M., Férey, G., & Hirscher, M. (2008). Heat of Adsorption for Hydrogen in Microporous High-Surface-Area Materials. *ChemPhysChem*, *9*(15), 2181–2184. doi:10.1002/cphc.200800463 PMID:18821561

Schneider, S., Bajohr, S., Graf, F., & Kolb, T. (2020). State of the art of hydrogen production via pyrolysis of natural gas. *ChemBioEng Reviews*, *7*(5), 150–158.

Scott Weil, K. (2006). The State-Of-The-Art in Sealing Technology for Solid Oxide Fuel Cells. *J. of Materials, 58*, 37-44.

Sdanghi, G., Maranzana, G., Celzard, A., & Fierro, V. (2019). Review of the current technologies and performances of hydrogen compression for stationary and automotive applications. *Renewable & Sustainable Energy Reviews*, *102*, 150–170. doi:10.1016/j.rser.2018.11.028

Shahabuddin, M., Alam, M. T., Krishna, B. B., Bhaskar, T., & Perkins, G. (2020). A review of producing renewable aviation fuels from the gasification of biomass and residual wastes. *Bioresource Technology*, *312*, 123596. doi:10.1016/j.biortech.2020.123596 PMID:32507633

Shah, M., Mondal, P., Nayak, A. K. A. K., & Bordoloi, A. (2017). *Hydrogen from natural gas. Sustainable Utilization of Natural Resources*. CRC Press.

Shalom, M., Gimenez, S., Schipper, F., Herraiz-Cardona, I., Bisquert, J., & Antonietti, M. (2014). Controlled carbon nitride growth on surfaces for hydrogen evolution electrodes. *Angewandte Chemie International Edition*, *53*(14), 3654–3658. doi:10.1002/anie.201309415 PMID:24574144

Shang, X., Chi, J. Q., Lu, S. S., Dong, B., Liu, Z. Z., Yan, K. L., & Liu, C. G. (2017). Hierarchically three–level $Ni_3(VO_4)_2$@ $NiCo_2O_4$ nanostructure based on nickel foam towards highly efficient alkaline hydrogen evolution. *Electrochimica Acta*, *256*, 100–109. doi:10.1016/j.electacta.2017.10.017

Shankar, R., Pathak, N., Chaurasia, A. K., Mondal, P., & Chand, S. (2017). Energy Production through Microbial Fuel Cells. In *Sustainable Utilization of Natural Resources* (pp. 353–380). CRC Press. doi:10.1201/9781315153292-12

Shao, Q., Wang, Y., Yang, S., Lu, K., Zhang, Y., Tang, C., & Li, Y. (2018). Stabilizing and Activating Metastable Nickel Nanocrystals for Highly Efficient Hydrogen Evolution Electrocatalysis. *ACS Nano*, *12*(11), 11625–11631. doi:10.1021/acsnano.8b06896 PMID:30481970

Shao, Y., Shi, X., & Pan, H. (2017). Electronic, magnetic, and catalytic properties of thermodynamically stable two–dimensional transition–metal phosphides. *Chemistry of Materials*, *29*(20), 8892–8900. doi:10.1021/acs.chemmater.7b03832

Shaur, A., Rehman, S. U., Kim, H. S., Song, R. H., Lim, T. H., Hong, J. E., ... Lee, S. B. (2020). Hybrid Electrochemical Deposition Route for the Facile Nanofabrication of a Cr-Poisoning-Tolerant La (Ni, Fe) O3− δ Cathode for Solid Oxide Fuel Cells. *ACS Applied Materials & Interfaces*, *12*(5), 5730–5738.

Shen, Y., Zhou, Y., Wang, D., Wu, X., Li, J., & Xi, J. (2018). Nickel–copper alloy encapsulated in graphitic carbon shells as electrocatalysts for hydrogen evolution reaction. *Advanced Energy Materials*, *8*(2), 1701759–1701765. doi:10.1002/aenm.201701759

Shibli, S. M. A., Anupama, V. R., Arun, P. S., Jineesh, P., & Suji, L. (2016). Synthesis and development of nano WO_3 catalyst incorporated Ni–P coating for electrocatalytic hydrogen evolution reaction. *International Journal of Hydrogen Energy*, *41*(24), 10090–10102. doi:10.1016/j.ijhydene.2016.04.156

Shibli, S. M. A., & Dilimon, V. S. (2008). Development of TiO_2 supported nano–RuO_2–incorporated catalytic nickel coating for hydrogen evolution reaction. *International Journal of Hydrogen Energy*, *33*(4), 1104–1111. doi:10.1016/j.ijhydene.2007.12.038

Shinagawa, T., & Takanabe, K. (2017). Towards versatile and sustainable hydrogen production through electrocatalytic water splitting: Electrolyte engineering. *ChemSusChem*, *10*(7), 1318–1336. doi:10.1002/cssc.201601583 PMID:27984671

Shinnar, R. (2003). The hydrogen economy, fuel cells, and electric cars. *Technology in Society*, *25*(4), 455–476. doi:10.1016/j.techsoc.2003.09.024

Shiva Kumar, S., & Himabindu, V. (2019a). Hydrogen production by PEM water electrolysis – A review. *Materials Science for Energy Technologies*, *2*(3), 442–454. doi:10.1016/j.mset.2019.03.002

Silberberg, M. (2006). *Chemistry: The Molecular Nature of Matter and Change* (4th ed.). McGraw-Hill Education.

Silva, E. D. R., Curi, M., Nicolini, J. V., Furtado, J. G., Secchi, A. R., & Ferraz, H. C. (2021). Effect of doping concentration and sintering atmosphere on the microstructural and electrical characteristics of Y-doped SrTiO3 perovskite anode for SOFC. *Ceramics International*.

Singh, D., Sharma, D., Soni, S. L., Sharma, S., Sharma, P. K., & Jhalani, A. (2020). A review on feedstocks, production processes, and yield for different generations of biodiesel. *Fuel*, *262*, 116553. doi:10.1016/j.fuel.2019.116553

Singh, R. N. (2007). Sealing technology for solid oxide fuel cells (SOFC). *International Journal of Applied Ceramic Technology*, *4*(2), 134–144.

Singla, M. K., Nijhawan, P., & Oberoi, A. S. (2021). Hydrogen fuel and fuel cell technology for 61a cleaner future: 61a review. *Environmental Science and Pollution Research International*, 1–20.

Sittijunda, S., & Reungsang, A. (2012). Media optimization for biohydrogen production from waste glycerol by anaerobic thermophilic mixed cultures. *International Journal of Hydrogen Energy*, *37*(20), 15473–15482. doi:10.1016/j.ijhydene.2012.02.185

Sittijunda, S., & Reungsang, A. (2018). Methane Production from the Co-digestion of Algal Biomass with Crude Glycerol by Anaerobic Mixed Cultures. *Waste and Biomass Valorization*, *11*(5), 1873–1881. doi:10.100712649-018-0542-0

Situmorang, Y. A., Zhao, Z., Yoshida, A., Abudula, A., & Guan, G. (2020). Small-scale biomass gasification systems for power generation (< 200 kW class): A review. *Renewable & Sustainable Energy Reviews*, *117*, 109486. doi:10.1016/j.rser.2019.109486

Skrzypkiewicz, M., Wierzbicki, M., Kupecki, J., & Stępień, M. (2018). Selected Aspects of Design, Construction, and Operation of SOFC-Based Micro-Combined Heat and Power Systems. In J. Kupecki (Ed.), Modeling, Design, Construction, and Operation of Power Generators with Solid Oxide Fuel Cells (pp. 205-231). Springer.

Song, B., Lin, R., Lam, C. H., Wu, H., Tsui, T. H., & Yu, Y. (2021). Recent advances and challenges of inter-disciplinary biomass valorization by integrating hydrothermal and biological techniques. *Renewable & Sustainable Energy Reviews*, *135*, 110370. doi:10.1016/j.rser.2020.110370

Song, Y., & Guo, Z. X. (2006). Electronic structure, stability and bonding of the Li-NH hydrogen storage system. *Physical Review. B*, *74*(19), 195120. doi:10.1103/PhysRevB.74.195120

Song, Y., Han, K., & Wang, D. Y. (2020). Thermodynamic analysis of fossil fuels reforming for fuel cell application. *International Journal of Hydrogen Energy*, *45*(39), 20232–20239.

Song, Y., Zhang, C., Ling, C. Y., Han, M., Yong, R. Y., Sun, D., & Chen, J. (2020). Review on current research of materials, fabrication and application for bipolar plate in proton exchange membrane fuel cell. *International Journal of Hydrogen Energy*, *45*(54), 29832–29847. doi:10.1016/j.ijhydene.2019.07.231

Sonkhaskar, Y. M., Nikhade, G. R., Dharmik, S., Deshmukh, U., & Dhote, P. (2021). Development of Briquette Cum Pellet Making Machine. In *Advances in Mechanical Engineering* (pp. 391–397). Springer. doi:10.1007/978-981-15-3639-7_46

Sorensen, B., & Spazzafumo, G. (2018). Hydrogen, in Hydrogen and Fuel Cells (3rd ed.). Academic Press.

Spendelow, J. (2012). DOE Hydrogen and Fuel Cells Program Record - Medium-scale CHP Fuel Cell System Targets. U.S. Department of Energy.

Ssing, K., Everett, D., Haul, R., Moscou, L., Pierotti, R., Rouquerol, J., & Siemieniewsks, T. (1985). Reporting physisorption data for gas/solid system. *Pure and Applied Chemistry*, *57*(4), 603–619.

Staffell, I. (2010). *Fuel cells for domestic heat and power: are they worth it?* (Doctoral dissertation). University of Birmingham.

Staffell, I., & Green, R. (2013). The cost of domestic fuel cell micro-CHP systems. *International Journal of Hydrogen Energy*, *38*(2), 1088–1102.

Staffell, I., Scamman, D., Velazquez Abad, A., Balcombe, P., Dodds, P. E., Ekins, P., Shah, N., & Ward, K. R. (2019). The role of hydrogen and fuel cells in the global energy system. *Energy & Environmental Science*, *12*(2), 463–491. doi:10.1039/C8EE01157E

Staubitz, A., Robertson, A. P., & Manners, I. (2010). Ammonia-borane and related compounds as dihydrogen sources. *Chemical Reviews*, *110*(7), 4079–4124. doi:10.1021/cr100088b PMID:20672860

Ştefănescu, I. (2010). *Pile de combustibil – între teorie şi practică, Editura CONPHYS, Rm.* Vâlcea.

Stephens, F. H., Pons, V., & Baker, R. T. (2007). Ammonia–borane: The hydrogen source par excellence? *Dalton Transactions (Cambridge, England)*, (25), 2613–2626. doi:10.1039/B703053C PMID:17576485

Stokes. (2011). *Stokes, Solid Oxide Fuel Cell Applications*. Hydrogen Technical Advisory Committee.

Stolten, D., & Emonts, B. (Eds.). (2012). *Fuel Cell Science and Engineering, 2 Volume Set: Materials, Processes, Systems and Technology* (Vol. 1). John Wiley & Sons.

Stolyarova, V. L., Vorozhtcov, V. A., Masaki, K., & Costa, D. (2020). High-temperature mass spectrometric study of thermodynamic properties in the UO2–ZrO2 system. *Rapid Communications in Mass Spectrometry*, *34*(19), e8862.

Subbaraman, R., Tripkovic, D., Strmcnik, D., Chang, K. C., Uchimura, M., Paulikas, A. P., Stamenkovic, V., & Markovic, N. M. (2011). Enhancing hydrogen evolution activity in water splitting by tailoring Li^+–$Ni(OH)_2$–Pt interfaces. *Science*, *334*(6060), 1256–1260. doi:10.1126cience.1211934 PMID:22144621

Sudik, A., Yang, J., Halliday, D., & Wolverton, C. (2008). Hydrogen Storage Properties in (LiNH2) 2– LiBH4–(MgH2) X Mixtures (x= 0.0– 1.0). *The Journal of Physical Chemistry C*, *112*(11), 4384–4390. doi:10.1021/jp711393k

Su, H., Jaffer, S., & Yu, H. (2016). Article. *Energy Storage Materials*, *5*, 116–131.

Su, J., Yang, Y., Xia, G., Chen, J., Jiang, P., & Chen, Q. (2017). Ruthenium–cobalt nanoalloys encapsulated in nitrogen–doped graphene as active electrocatalysts for producing hydrogen in alkaline media. *Nature Communications*, *8*(1), 14969–14979. doi:10.1038/ncomms14969 PMID:28440269

Sunfire | Hydrogen. (n.d.). Retrieved October 21, 2020, from https://hydrogeneurope.eu/member/sunfire

Sun, Y., Shen, C., Lai, Q., Liu, W., Wang, D. W., & Aguey-Zinsou, K. F. (2018). Tailoring magnesium-based materials for hydrogen storage through synthesis: Current state of the art. *Energy Storage Materials*, *10*, 168–198. doi:10.1016/j.ensm.2017.01.010

Sun, Z., Zhang, J., Xie, J., Zheng, X., Wang, M., Li, X., & Tang, B. (2018). High–performance alkaline hydrogen evolution electrocatalyzed by a Ni_3N–CeO_2 nanohybrid. *Inorganic Chemistry Frontiers*, *5*(12), 3042–3045. doi:10.1039/C8QI00905H

Su, S., Zhang, S., Yan, C., Yang, Z., Zheng, F., & Zhang, L. (2017). Evaluations of Fuel and Air Flow Distributing Qualities within SOFC Stack Designs with Flow Manifolds Penetrated Through Cell Plane. *International Journal of Electrochemical Science*, *12*, 230–239.

Suwanwarangkul, R., Croiset, E., Pritzker, M. D., Fowler, M. W., Douglas, P. L., & Entchev, E. (2006). Mechanistic modelling of a cathode-supported tubular solid oxide fuel cell. *Journal of Power Sources*, *154*(1), 74–85. doi:10.1016/j.jpowsour.2005.03.197

Swaminathan, M. (2018). How can India's waste problem see a systemic change? *Economic and Political Weekly*. https://www.epw.in/engage/article/institutional-framework-implementing-solid-waste-management-india-macro-analysis?0=ip_login_no_cache%3D7d58c9a286a575de0525e5c5f79bc33a

Szima, S., del Pozo, C. A., Cloete, S., Chiesa, P., Alvaro, Á. J., Cormos, A. M., & Amini, S. (2021). Finding synergy between renewables and coal: Flexible power and hydrogen production from advanced IGCC plants with integrated CO2 capture. *Energy Conversion and Management*, *231*, 113866. doi:10.1016/j.enconman.2021.113866

Tabak, J. (2009). *Natural gas and hydrogen*. Infobase Publishing.

Taie, Z., Villaverde, G., Morris, J. S., Lavrich, Z., Chittum, A., White, K., & Hagen, C. (2021). Hydrogen for heat: Using underground hydrogen storage for seasonal energy shifting in northern climates. *International Journal of Hydrogen Energy*, *46*(5), 3365–3378. doi:10.1016/j.ijhydene.2020.10.236

Tang, C., Gan, L., Zhang, R., Lu, W., Jiang, X., Asiri, A. M., & Chen, L. (2016). Ternary $Fe_xCo_{1-x}P$ Nanowire Array as a Robust Hydrogen Evolution Reaction Electrocatalyst with Pt–like Activity: Experimental and Theoretical Insight. *Nano Letters*, *16*(10), 6617–6621. doi:10.1021/acs.nanolett.6b03332 PMID:27676264

Tang, S., Wang, C., & Sung, X. L. (2010). A study of corrosion resistance and formation characteristics of electroless Ni–P alloy coating on Al composite with a sample surface pretreatment. *Surface and Coatings Technology*, *205*(1), 43–49. doi:10.1016/j.surfcoat.2010.05.047

Tarasov, B. P., Lototskii, M. V., & Yartys', V. A. (2007). Problem of hydrogen storage and prospective uses of hydrides for hydrogen accumulation. *Russian Journal of General Chemistry*, *77*(4), 694–711. doi:10.1134/S1070363207040329

Teichmann, D., Arlt, W., Wasserscheid, P., & Freymann, R. (2011). A future energy supply based on Liquid Organic Hydrogen Carriers (LOHC). *Energy & Environmental Science*, *4*(8), 2767–2773. doi:10.1039/c1ee01454d

Teixeira Franco, R., Buffière, P., & Bayard, R. (2017). Cattle manure for biogas production. Does ensiling and wheat straw addition enhance preservation of biomass and methane potential? *Biofuels*, 1–12. doi:10.1080/17597269.2017.1387751

Thalmaier, G., Cobîrzan, N., Balog, A. A., Constantinescu, H., Streza, M., Nasui, M., & Neamtu, B. V. (2020). Influence of sawdust particle size on fired clay brick properties. *Materiales de Construcción*, *70*(338), 215.

Thomas, K. M. (2007). Hydrogen adsorption and storage on porous materials. *Catalysis Today*, *120*(3-4), 389–398. doi:10.1016/j.cattod.2006.09.015

Thomson, R., Kwong, P., Ahmad, E., & Nigam, K. D. P. (2020). Clean syngas from small commercial biomass gasifiers; a review of gasifier development, recent advances and performance evaluation. *International Journal of Hydrogen Energy*, *45*(41), 21087–21111. doi:10.1016/j.ijhydene.2020.05.160

Tian, X., Zhao, X., Su, Y. Q., Wang, L., & Wang, H. (2019). Engineering bunched Pt-Ni alloy nanocages for efficient oxygen reduction in practical fuel cells. *Science*.

Tie, S. F., & Tan, C. W. (2013). A review of energy sources and energy management system in electric vehicles. *Renewable & Sustainable Energy Reviews*, *20*, 82–102.

Tietz, F., Buchkremer, H. P., & Stöver, D. (2002). Components manufacturing for solid oxide fuel cells. *Solid State Ionics*, *152*, 373–381.

Tilak, B. V., Ramamurthy, A. C., & Conway, B. E. (1986) High performance electrode materials for the hydrogen evolution reaction from alkaline media. In *Proceedings of the Indian Academy of Sciences–chemical sciences* (Vol. 97, pp. 359–393). Springer.

Tippawan, P., & Arpornwichanop, A. (2014). Energy and exergy analysis of an ethanol reforming process for solid oxide fuel cell applications. *Bioresource Technology*, *157*, 231–239. doi:10.1016/j.biortech.2014.01.113 PMID:24561628

Tong, W., Forster, M., Dionigi, F., Soren, D., Roghayeh, S. E., Peter, S., Alexander, J. C., & Pau, F. (2020). Electrolysis of low-grade and saline surface water. *Nature Energy*, *5*(5), 367–377. doi:10.103841560-020-0550-8

Toro Delgado, J. A. (2020). *Biomass classification by cluster analysis based on compositional data* (Bachelor's thesis, Uniandes).

Toyoda, M., Nanbu, Y., Kito, T., Hiranob, M., & Inagaki, M. (2003). Preparation and performance of anatase-loaded porous carbons for water purification. *Desalination, 159*(3), 273–282. doi:10.1016/S0011-9164(03)90079-8

Tripa, S., & Indrie, L. (2021). Households textile waste management in the context of a circular economy in Romania. *Environmental Engineering and Management Journal, 20*(1), 81–87. doi:10.30638/eemj.2021.009

Tucker, M. C. (2018). Dynamic-temperature operation of metal-supported solid oxide fuel cells. *Journal of Power Sources, 395*, 314–317. doi:10.1016/j.jpowsour.2018.05.094

Turning organic waste into hydrogen. (2019). *C&EN Global Enterprise, 97*(14), 26–29. doi:10.1021/cen-09714-feature3

Tursi, A. (2019). A review on biomass: Importance, chemistry, classification, and conversion. *Biofuel Research Journal, 6*(2), 962–979. doi:10.18331/BRJ2019.6.2.3

Tushar, W., Yuen, C., Saha, T. K., Morstyn, T., Chapman, A. C., Alam, M. J. E., Hanif, S., & Poor, H. V. (2021). Peer-to-peer energy systems for connected communities: A review of recent advances and emerging challenges. *Applied Energy, 282*, 116131. doi:10.1016/j.apenergy.2020.116131

Tyurina, E. A., Elsukov, P. Y., & Mednikov, A. S. (2020). Synthetic liquid fuels: prospects for innovative technologies based on underground coal gasification. In *E3S Web of Conferences* (Vol. 209, p. 03026). EDP Sciences.

US DOE. (2009). *Technical System Targets: Onboard Hydrogen Storage for Light-Duty Fuel Cell Vehicles*. Author.

Valente, A., Iribarren, D., & Dufour, J. (2017). Life cycle assessment of hydrogen energy systems: A review of methodological choices. *The International Journal of Life Cycle Assessment, 22*(3), 346–363. doi:10.100711367-016-1156-z

Van de Krol, R., & Grätzel, M. (2012). *Photoelectrochemical hydrogen production* (Vol. 90). Springer. doi:10.1007/978-1-4614-1380-6

Vassilev, S. V., Baxter, D., Andersen, L. K., & Vassileva, C. G. (2010). An overview of the chemical composition of biomass. *Fuel, 89*(5), 913–933. doi:10.1016/j.fuel.2009.10.022

Vassilev, S. V., Baxter, D., Andersen, L. K., & Vassileva, C. G. (2013). An overview of the composition and application of biomass ash. Part 1. Phase–mineral and chemical composition and classification. *Fuel, 105*, 40–76. doi:10.1016/j.fuel.2012.09.041

Venâncio, S. A., & de Miranda, P. E. V. (2017). Direct utilization of carbonaceous fuels in multifunctional SOFC anodes for the electrosynthesis of chemicals or the generation of electricity. *International Journal of Hydrogen Energy, 42*(19), 13927–13938.

Verspagen, B. (2007). Mapping technological trajectories as patent citation networks: A study on the history of fuel cell research. *Advances in Complex Systems, 10*(01), 93–115. doi:10.1142/S0219525907000945

Vesborg, P. C., Seger, B., & Chorkendorff, I. B. (2015). Recent development in hydrogen evolution reaction catalysts and their practical implementation. *The Journal of Physical Chemistry Letters, 6*(6), 951–957. doi:10.1021/acs.jpclett.5b00306 PMID:26262851

Veziroğlu, T. N. (2000). Quarter century of hydrogen movement 1974–2000. *International Journal of Hydrogen Energy, 25*(12), 1143–1150. doi:10.1016/S0360-3199(00)00038-0

Veziroğlu, T. N. (2004). IJHE thrust toward meeting the needs of implementation. *International Journal of Hydrogen Energy, 1*(29), 15. doi:10.1016/j.ijhydene.2003.09.003

Veziroglu, T. N. (2007). 21st Century's energy: Hydrogen energy system. In *Assessment of hydrogen energy for sustainable development* (pp. 9–31). Springer. doi:10.1007/978-1-4020-6442-5_2

Vincent & Scrosati. (1997). *Modern Batteries*. Elsevier Science.

Virkar, A. V. (2001, February). SOFC materials and processing issues. SECA core technology program workshop.

von Helmolt, R., & Eberle, U. (2007). Fuel cell vehicles: Status 2007. *Journal of Power Sources*, *165*(2), 833–843. doi:10.1016/j.jpowsour.2006.12.073

Voronca, Ş. D., Darie, G., & Pătraşcu, R. (2017, October). Biomass ORC cogeneration—A case study. In *2017 International Conference on ENERGY and ENVIRONMENT (CIEM)* (pp. 376-380). IEEE.

Wang, G., Dai, Y., Yang, H., Xiong, Q., Wang, K., Zhou, J., Li, Y., & Wang, S. (2020). A review of recent advances in biomass pyrolysis. *Energy & Fuels*, *34*(12), 15557–15578. doi:10.1021/acs.energyfuels.0c03107

Wang, H., Kong, H., Pu, Z., Li, Y., & Hu, X. (2020). Feasibility of high efficient solar hydrogen generation system integrating photovoltaic cell/photon-enhanced thermionic emission and high-temperature electrolysis cell. *Energy Conversion and Management*, *210*, 112699. doi:10.1016/j.enconman.2020.112699

Wang, L., Husar, A., Zhou, T., & Liu, H. (2003). A parametric study of PEM fuel cell performances. *International Journal of Hydrogen Energy*, *28*(11), 1263–1272. doi:10.1016/S0360-3199(02)00284-7

Wang, L., Li, Y., Xia, M., Li, Z., Chen, Z., Ma, Z., & Shao, G. (2017). Ni nanoparticles supported on graphene layers: An excellent 3D electrode for hydrogen evolution reaction in alkaline solution. *Journal of Power Sources*, *347*, 220–228. doi:10.1016/j.jpowsour.2017.02.017

Wang, L., Zhang, H., & Weng, S. (2008). Modeling and simulation of solid oxide fuel cell based on the volume–resistance characteristic modeling technique. *Journal of Power Sources*, *177*(2), 579–589. doi:10.1016/j.jpowsour.2007.10.051

Wang, T., Tang, Z., Guo, Y., Zhang, X., Yang, Q., Xu, B., & Wang, H. (2019). Comparison of Heavy Metal Speciation of Sludge During Mesophilic and Thermophilic Anaerobic Digestion. *Waste and Biomass Valorization*, 1–10. doi:10.100712649-018-00575-9

Wang, X., Wei, K., Yan, S., Wu, Y., Kang, J., Feng, P., ... Ling, Y. (2020). Efficient and stable conversion of oxygen-bearing low-concentration coal mine methane by the electrochemical catalysis of SOFC anode: From pollutant to clean energy. *Applied Catalysis B: Environmental*, *268*, 118413.

Wang, Y., Zhuo, H., Zhang, X., Dai, X., Yu, K., Luan, C., & Gao, F. (2018). Synergistic effect between undercoordinated platinum atoms and defective nickel hydroxide on enhanced hydrogen evolution reaction in alkaline solution. *Nano Energy*, *48*, 590–599. doi:10.1016/j.nanoen.2018.03.080

Wang, Z., Bui, Q., & Zhang, B. (2020a). The relationship between biomass energy consumption and human development: Empirical evidence from BRICS countries. *Energy*, *194*, 116906. doi:10.1016/j.energy.2020.116906

Wang, Z., Bui, Q., Zhang, B., & Pham, T. L. H. (2020b). Biomass energy production and its impacts on the ecological footprint: An investigation of the G7 countries. *The Science of the Total Environment*, *743*, 140741. doi:10.1016/j.scitotenv.2020.140741 PMID:32758837

Wang, Z., Roberts, R. R., Naterer, G. F., & Gabriel, K. S. (2012). Comparison of thermochemical, electrolytic, photoelectrolytic and photochemical solar-to-hydrogen production technologies. *International Journal of Hydrogen Energy*, *37*(21), 16287–16301. doi:10.1016/j.ijhydene.2012.03.057

Wasajja, H., Lindeboom, R. E., van Lier, J. B., & Aravind, P. V. (2020). Techno-economic review of biogas cleaning technologies for small scale off-grid solid oxide fuel cell applications. *Fuel Processing Technology*, *197*, 106215. doi:10.1016/j.fuproc.2019.106215

Weber, A. Z. (2013). Hydrogen/Bromine Flow Batteries. *Proceedings of the 2013 Electrical Energy Storage Applications & Technologies (EESAT) Biennial International Conference*.

Weeda, M., & Segers, R. (2020). *The Dutch hydrogen balance, and the current and future representation of hydrogen in the energy statistics*. Academic Press.

Welaya, Y. M., El Gohary, M. M., & Ammar, N. R. (2012). Steam and partial oxidation reforming options for hydrogen production from fossil fuels for PEM fuel cells. *Alexandria Engineering Journal*, *51*(2), 69–75.

Wieczorek, A. K., Krystian, M., & Zehetbauer, M. (2006). SPD processed alloys as efficient vacancy-hydrogen systems. *Diffusion and Defect Data, Solid State Data. Part B, Solid State Phenomena*, *114*, 177–182. doi:10.4028/www.scientific.net/SSP.114.177

Wilberforce, T., & Olabi, A. G. (2020). Design of experiment (DOE) analysis of 5-cell stack fuel cell using three bipolar plate geometry designs. *Sustainability*, *12*(11), 4488.

Winkler, W., Behling, N., Behling, T., Managi, S., & Williams, M. C. (2020). Options for Natural Gas and Methane Including Fuel Cell Utilization in a Sustainable Energy Infrastructure. *ECS Transactions*, *96*(1), 81.

Wołowicz, M., Kolasiński, P., & Badyda, K. (2021). Modern Small and Microcogeneration Systems—A Review. *Energies*, *14*(3), 785.

Wongchanapai, S., Iwai, H., Saito, M., & Yoshida, H. (2012). Selection of suitable operating conditions for Planar Anode-supported Direct-internal-reforming Solid-Oxide Fuel Cell. *Journal of Power Sources*, *204*, 14–24. doi:10.1016/j.jpowsour.2011.12.029

Wu, C. W., Zhang, W., Han, X., Zhang, Y. X., & Ma, G. J. (2020). A systematic review for structure optimization and clamping load design of large proton exchange membrane fuel cell stack. *Journal of Power Sources*, *476*, 228724. doi:10.1016/j.jpowsour.2020.228724

Wulf, C., Reuß, M., Grube, T., Zapp, P., Robinius, M., Hake, J. F., & Stolten, D. (2018). Life Cycle Assessment of hydrogen transport and distribution options. *Journal of Cleaner Production*, *199*, 431–443. doi:10.1016/j.jclepro.2018.07.180

Xiang, R., Duan, Y., Peng, L., Wang, Y., Tong, C., Zhang, L., & Wei, Z. (2019). Three–dimensional Core@Shell Co@$CoMoO_4$ nanowire arrays as efficient alkaline hydrogen evolution electro–catalysts. *Applied Catalysis B: Environmental*, *246*, 41–49. doi:10.1016/j.apcatb.2019.01.035

Xiao, L., Deng, Z., Fung, K. Y., & Ng, K. M. (2013). *Biohydrogen generation from anaerobic digestion of food waste*. doi:10.1016/j.ijhydene.2013.08.072

Xing, H., Yuan, Y., Huiyuan, Z., Yong, S., Bingxi, L., & Heping, T. (2017). Solar thermochemical hydrogen production using metallic oxides. *Energy Sources. Part A, Recovery, Utilization, and Environmental Effects*, *39*(3), 257–263. doi:10.1080/15567036.2013.878769

Xing, X., Zhang, D., & Li, Y. (2015). A non-aqueous all-cobalt redox flow battery using 1,10-phenanthrolinecobalt(II) hexafluorophosphate as active species. *Journal of Power Sources*, *279*, 205–209.

Xing, Z., Han, C., Wang, D., Li, Q., & Yang, X. (2017). Ultrafine Pt nanoparticle–decorated $Co(OH)_2$ nanosheet arrays with enhanced catalytic activity toward hydrogen evolution. *ACS Catalysis*, *7*(10), 7131–7135. doi:10.1021/acscatal.7b01994

Xiong, G., Zhang, J., Shi, D., Zhu, L., & Yuan, X. (2021). Optimal identification of solid oxide fuel cell parameters using a competitive hybrid differential evolution and Jaya algorithm. *International Journal of Hydrogen Energy*, *46*(9), 6720–6733.

Xu, J., Zhou, X., Dong, X., Pan, L., & Sun, K. (2017). Catalytic activity of infiltrated La0. 3Sr0. 7Ti0. 3Fe0. 7O3−δ–CeO2 as a composite SOFC anode material for H2 and CO oxidation. *International Journal of Hydrogen Energy*, *42*(23), 15632–15640.

Yadav, A., Pyare, R., Goodenough, J. B., & Singh, P. (2020). KTa1−x−y Ti x Ge y O3− δ: A High κ Relaxor Dielectric and Superior Oxide-Ion Electrolyte for IT-SOFC. *ACS Applied Energy Materials*, *3*(4), 3205–3211.

Yadav, D., & Banerjee, R. (2018). Economic assessment of hydrogen production from solar driven high-temperature steam electrolysis process. *Journal of Cleaner Production*, *183*, 1131–1155. doi:10.1016/j.jclepro.2018.01.074

Yang, Z. (2008). Recent advances in metallic interconnects for solid oxide fuel cells. *International Materials Reviews*, *53*(1), 39-54.

Yang, B., Wang, J., Zhang, M., Shu, H., Yu, T., Zhang, X., ... Sun, L. (2020). A state-of-the-art survey of solid oxide fuel cell parameter identification: Modelling, methodology, and perspectives. *Energy Conversion and Management*, *213*, 112856.

Yang, C. H., Chang, S. C., Chan, Y. H., & Chang, W. S. (2019). A dynamic analysis of the multi-stack SOFC-CHP system for power modulation. *Energies*, *12*(19), 3686.

Yang, J., Huang, W., Wang, X., Li, J., Yan, D., Pu, J., ... Li, J. (2018). Study on component interface evolution of a solid oxide fuel cell stack after long term operation. *Journal of Power Sources*, *387*, 57–63.

Yang, L., Zhou, H., Qin, X., Guo, X., Cui, G., Asiri, A. M., & Sun, X. (2018). Cathodic electrochemical activation of Co_3O_4 nanoarrays: A smart strategy to significantly boost the hydrogen evolution activity. *Chemical Communications*, *54*(17), 2150–2153. doi:10.1039/C7CC09416G PMID:29424845

Yang, S., Chen, T., Wang, Y., Peng, Z., & Wang, W. G. (2013). Electrochemical Analysis of an Anode-Supported SOFC. *International Journal of Electrochemical Science*, *8*, 2330–2344.

Yang, W., Zhang, J., Ma, Q., Zhao, Y., Liu, Y., & He, H. (2017). Heterogeneous Re action of SO_2 on Manganese Oxides: The Effect of Crystal Structure and Relative Humidity. *Scientific Reports*, *4550*, 1–14.

Yang, W., Zhao, Y., Liso, V., & Brandon, N. (2014). Optimal design and operation of a syngas-fuelled SOFC micro-CHP system for residential applications in different climate zones in China. *Energy and Building*, *80*, 613–622. doi:10.1016/j.enbuild.2014.05.015

Yang, Y., Liu, J., Guo, S., Liu, Y., & Kang, Z. (2015). A nickel nanoparticle/carbon quantum dot hybrid as an efficient electrocatalyst for hydrogen evolution under alkaline conditions. *Journal of Materials Chemistry. A, Materials for Energy and Sustainability*, *36*(36), 18598–18604. doi:10.1039/C5TA04867B

Yang, Z., Xia, Y., & Mokaya, R. (2007). Enhanced hydrogen storage capacity of high surface area zeolite-like carbon materials. *Journal of the American Chemical Society*, *129*(6), 1673–1679. doi:10.1021/ja067149g PMID:17243684

Yi, H. S. (1996). *Climate change 1995: Economic and social dimensions of climate change: Contribution of Working Group III to the second assessment report of the Intergovernmental Panel on Climate Change* (Vol. 2). Cambridge University Press.

Ying, Z., Wang, Y., Zheng, X., Geng, Z., Dou, B., & Cui, G. (2020). Experimental study and development of an improved sulfur–iodine cycle integrated with HI electrolysis for hydrogen production. *International Journal of Hydrogen Energy*, *45*(24), 13176–13188. doi:10.1016/j.ijhydene.2020.03.037

Yin, H., Zhao, S., Zhao, K., Muqsit, A., Tang, H., Chang, L., & Tang, Z. (2015). Ultrathin platinum nanowires grown on single–layered nickel hydroxide with high hydrogen evolution activity. *Nature Communications*, *6*(1), 6430. doi:10.1038/ncomms7430 PMID:25728293

Yixiao, C., Chen, Y., Akbar, M., Jin, B., Zhengwen, T., Naveed, M., ... Huang, Y. (2021). A Bulk-Heterostructure Nanocomposite Electrolyte of Ce 0.8 Sm 0.2 O 2-δ–SrTiO 3 for Low-Temperature Solid Oxide Fuel Cells. *Nano-Micro Letters*, *13*(1).

Yoda, M., Inoue, S., Takuwa, Y., Yasuhara, K., & Suzuki, M. (2017). Development and commercialization of new residential SOFC CHP system. *ECS Transactions*, *78*(1), 125.

Yoon, K. J., Cramer, C. N., Stevenson, J. W., & Marina, O. A. (2010). Advanced ceramic interconnect material for solid oxide fuel cells: Electrical and thermal properties of calcium-and nickel-doped yttrium chromites. *Journal of Power Sources*, *195*(22), 7587–7593.

You, J. (2016). Waste and Wastewater Clean-up using Microbial. *Fuel Cells (Weinheim)*, *208*. http://eprints.uwe.ac.uk/26953

Yu, F., Zhou, H., Huang, Y., Sun, J., Qin, F., Bao, J., & Ren, Z. (2018). High–performance bifunctional porous non–noble metal phosphide catalyst for overall water splitting. *Nature Communications*, *9*(1), 2551. doi:10.103841467-018-04746-z PMID:29959325

Yu, H., Ren, Y., Xiao, D., Guo, S., Zhu, Y., Qian, Y., Gu, L., & Zhou, H. (2015). *Angewandte Chemie International Edition in English*. Academic Press.

Yuksel, Y. E., Ozturk, M., & Dincer, I. (2020). Performance investigation of a combined biomass gasifier-SOFC plant for compressed hydrogen production. *International Journal of Hydrogen Energy*, *45*(60), 34679–34694.

Yuzer, B., Selcuk, H., Chehade, G., Demir, M. E., & Dincer, I. (2020). Evaluation of hydrogen production via electrolysis with ion exchange membranes. *Energy*, *190*, 116420. doi:10.1016/j.energy.2019.116420

Zakaria, Z., Awang Mat, Z., Abu Hassan, S. H., & Boon Kar, Y. (2020). A review of solid oxide fuel cell component fabrication methods toward lowering temperature. *International Journal of Energy Research*, *44*(2), 594–611. doi:10.1002/er.4907

Zaluska, A., Zaluski, L., & Ström-Olsen, J. O. (2001). Structure, catalysis and atomic reactions on the nano-scale: A systematic approach to metal hydrides for hydrogen storage. *Applied Physics. A, Materials Science & Processing*, *72*(2), 157–165. doi:10.1007003390100783

Zhang, B., Wang, H. H., Su, H., Lv, L. B., Zhao, T. J., Ge, J. M., & Chen, J. S. (2016). Nitrogen–doped graphene microtubes with opened inner voids: Highly efficient metal–free electrocatalysts for alkaline hydrogen evolution reaction. *Nano Research*, *9*(9), 2606–2615. doi:10.100712274-016-1147-1

Zhang, B., Zhang, J., Shi, J., Tan, D., Liu, L., Zhang, F., & Han, B. (2019). Manganese acting as a high–performance heterogeneous electrocatalyst in carbon dioxide reduction. *Nature Communications*, *2980*(1), 1–8. doi:10.103841467-019-10854-1 PMID:31278257

Zhang, H., & Cooper, A. I. (2005). Synthesis and applications of emulsion-templated porous materials. *Soft Matter*, *1*(2), 107–113. doi:10.1039/b502551f PMID:32646082

Zhang, J., Lenser, C., Menzler, N. H., & Guillon, O. (2020). Comparison of solid oxide fuel cell (SOFC) electrolyte materials for operation at 500° C. *Solid State Ionics*, *344*, 115138.

Zhang, R., Ren, X., Hao, S., Ge, R., Liu, Z., Asiri, A. M., & Sun, X. (2018). Selective phosphidation: An effective strategy toward CoP/CeO_2 interface engineering for superior alkaline hydrogen evolution electrocatalysis. *Journal of Materials Chemistry. A, Materials for Energy and Sustainability*, *5*(5), 1985–1990. doi:10.1039/C7TA10237B

Zhang, T., Wu, M. Y., Yan, D. Y., Mao, J., Liu, H., Hu, W. B., & Qiao, S. Z. (2018). Engineering oxygen vacancy on NiO nanorod arrays for alkaline hydrogen evolution. *Nano Energy*, *43*, 103–109. doi:10.1016/j.nanoen.2017.11.015

Zhang, X., Gu, W., & Wang, E. (2017). Wire–on–flake heterostructured ternary $Co_{0.5}Ni_{0.5}$ P/CC: An efficient hydrogen evolution electrocatalyst. *Journal of Materials Chemistry. A, Materials for Energy and Sustainability*, *5*(3), 982–987. doi:10.1039/C6TA08094D

Zhang, X., Li, G., Li, J., & Feng, Z. (2007). Numerical study on electric characteristics of solid oxide fuel cells. *Energy Conversion and Management*, *48*(3), 977–989. doi:10.1016/j.enconman.2006.08.011

Zhang, X., Su, S., Chen, J., Zhao, Y., & Brandon, N. (2011). A new analytical approach to evaluate and optimize the performance of an irreversible solid oxide fuel cell-gas turbine hybrid system. *International Journal of Hydrogen Energy*, *36*(23), 15304–15312. doi:10.1016/j.ijhydene.2011.09.004

Zhang, X., Wang, Y., Guo, J., Shih, T. M., & Chen, J. (2014). A unified model of high-temperature fuel-cell heatengine hybrid systems and analyses of its optimum performances. *International Journal of Hydrogen Energy*, *39*(4), 1811–1825. doi:10.1016/j.ijhydene.2013.11.027

Zhan, X., Daniil, M., Liu, X., Willard, M., Akolkar, R., & Ernst, F. (2019). Effect of tungsten alloying on magnetic properties of amorphous Ni–P. *Journal of Alloys and Compounds*, *786*, 742–749. doi:10.1016/j.jallcom.2019.01.109

Zhao, T. S., & Ni, M. (2013). *Solid oxide fuel cells: From materials to system modeling*. Royal Society of Chemistry.

Zhao, X., Babu Joseph, J. K., & Ozcan, S. (2020). Biogas Reforming to Syngas: A Review. *iScience*, *23*(5), 101082. doi:10.1016/j.isci.2020.101082 PMID:32380422

Zheng, L., Ma, S., Wang, Z., Zhang, Y. Q., Xu, X., & Chen, X. (2019). Ni–P nanostructures on flexible paper for morphology effect of nonenzymatic electrocatalysis for urea. *Electrochimica Acta*, *320*, 134586. doi:10.1016/j.electacta.2019.134586

Zheng, Y., Jiao, Y., Zhu, Y., Li, L. H., Han, Y., Chen, Y., & Qiao, S. Z. (2016). High electrocatalytic hydrogen evolution activity of an anomalous ruthenium catalyst. *Journal of the American Chemical Society*, *138*(49), 16174–16181. doi:10.1021/jacs.6b11291 PMID:27960327

Zhiznin, S. Z., Vassilev, S., & Gusev, A. L. (2019). Economics of secondary renewable energy sources with hydrogen generation. *International Journal of Hydrogen Energy*, *44*(23), 11385–11393. doi:10.1016/j.ijhydene.2019.03.072

Zhou, H., Meng, W., Wang, D., Li, G., Li, H., Liu, Z., & Yang, S. (2021). A novel coal chemical looping gasification scheme for synthetic natural gas with low energy consumption for CO2 capture: Modelling, parameters optimization, and performance analysis. *Energy*, •••, 120249. doi:10.1016/j.energy.2021.120249

Zhou, L. (2005). Progress and problems in hydrogen storage methods. *Renewable & Sustainable Energy Reviews*, *9*(4), 395–408. doi:10.1016/j.rser.2004.05.005

Zhou, Q., Li, T. T., Qian, J., Hu, Y., Guo, F., & Zheng, Y. Q. (2018). Self–supported hierarchical CuO_x@Co_3O_4 heterostructures as efficient bifunctional electrocatalysts for water splitting. *Journal of Materials Chemistry. A, Materials for Energy and Sustainability*, *29*(29), 14431–14439. doi:10.1039/C8TA03120G

Zhou, W., Lu, X. F., Chen, J. J., Zhou, T., Liao, P. Q., Wu, M., & Li, G. R. (2018). Hierarchical Porous Prism Arrays Composed of Hybrid Ni–NiO–Carbon as Highly Efficient Electrocatalysts for Overall Water Splitting. *ACS Applied Materials & Interfaces*, *45*(45), 38906–38914. doi:10.1021/acsami.8b13542 PMID:30360101

Zhou, W., Wu, H., & Yildirim, T. (2008). Enhanced H2 adsorption in isostructural metal– organic frameworks with open metal sites: Strong dependence of the binding strength on metal ions. *Journal of the American Chemical Society, 130*(46), 15268–15269. doi:10.1021/ja807023q PMID:18950163

Zhu, P., Yao, J., Qian, C., Yang, F., Porpatham, E., Zhang, Z., & Wu, Z. (2020). High-efficiency conversion of natural gas fuel to power by an integrated system of SOFC, HCCI engine, and waste heat recovery: Thermodynamic and thermo-economic analyses. *Fuel, 275*, 117883.

Züttel, A., Rentsch, S., Fischer, P., Wenger, P. M. C. E. P., Sudan, P. H., Mauron, P., & Emmenegger, C. (2003). Hydrogen storage properties of LiBH4. *Journal of Alloys and Compounds, 356*, 515–520. doi:10.1016/S0925-8388(02)01253-7

About the Contributors

Gheorghe Badea hold the title of Emeritus professor and Doctor in Engineering within Technical University of Cluj-Napoca, Romania, Doctor in Engineering of the National Institute of Nuclear Physics of Bucharest, Romania and PhD supervisor in the fundamental field of Engineering Sciences. He has been a dean of the Faculty of Installations within Technical University of Cluj-Napoca and, for over two decades, a member of the Council of Faculties of Civil Engineering and Installations and the Senate of Technical University of Cluj-Napoca, Romania. In addition, he has been a technical expert within the Romanian Council for Research in Higher Education (CNCSIS), as well as a member of the Specialty Commission- Architecture, Urban Planning, and Civil Engineering of the National Council for University Titles, Diploma, Certificates (CNATDCU), Romania. President of the international conference „Modern Science and Energy-SME" from the 1st Edition (1981)- Edition XL (2021), Cluj-Napoca, Romania. Editor-in-Chief of Building Services Engineering Journal, Technical University of Cluj-Napoca, U.T.PRESS Publishing Press, 2011, Cluj-Napoca, Romania. Member in several national and international professional associations: Member of the International Association for Energy Economics (ATEE) -France; Member of the American Society of Heating, Refrigerating and Air-Conditioning Engineers (ASHRE) in the USA; Member of the Romanian Electric Installations and Automation Society (SIEAR); Member of the Romanian Society for Atmospheric Protection (SOROPA); Member of the Romanian Water Association (ARA); Founding member of the Romanian Association of Installations Engineers (AIIR) and President of Transylvania Branch; Founding member of the Romanian Association of Construction Entrepreneurs (ARACO); President and Founding Member of the Romanian Association of Pipeline Rehabilitation Technologies, without Drilling (ROSTT); Founding member and vice-president of the Romanian Association for Tritium and Torus Energy (T & T ENERGY); Founding member and vice-president of the Association for Hydrogen Energy in Romania (AEHR). During his academic career, he received numberless acknowledgements for his contributions in the field of engineering: Diploma and Medal - 200 OUTSTANDING PEOPLE OF THE 20 TH CENTURY - International Biographical Center Cambridge, England, 1998; Diploma and medal - 20th century CONSTRUCTOR - from the Romanian Association of Construction Entrepreneurs (ARACO), 2001; Honorary member of the Energy Association of Bordeaux - France; Diploma of Excellence from the Faculty of Installations at the Technical University of Civil Engineering Bucharest, Romania; Diploma of Excellence from the Romanian Water Association (ARA); Diploma of Excellence and Honorary Member of the Romanian Association of Installations Engineers (AIIR); The General Engineering Association of Romania (AGIR) Award for the „2002 Installation Manual" (co-author); Order "Merit for Education" as Officer degree granted by Presidential Decree – Romania.

Raluca-Andreea Felseghi is a Civil Engineer and Independent Researcher, she graduated the Technical University of Cluj-Napoca, Faculty of Building Services Engineering. Starting in 2011, she attended the postgraduate work (Ph.D.) within Technical University of Cluj-Napoca and obtained the Doctoral Diploma in a fundamental field: Engineering Sciences, Civil Engineering. During the doctoral studies, the main research direction was the concern related to the implementation and use of hydrogen fuel cell technologies in energy-efficient buildings. She is currently working at the Faculty of Electrical Engineering and Computer Science, "Ştefan cel Mare" University of Suceava, Romania, being involved in a research project that aims to integrate hydrogen energy into the electricity supply systems of the electric vehicles charging stations (Smart conductive charging station, fixed and mobile, for electric propulsion transportation: SMiLE-EV). She is also an associate professor at the Technical University of Cluj-Napoca, Faculty of Civil Engineering where she conducts teaching activities for students and research activities in the field of sustainable development in Civil Engineering. In order to deepen the knowledge in the field of hydrogen energy, she carries out practical research activity within the center with a tradition in hydrogen research, National Research and Development Institute for Cryogenics and Isotopic Technologies Rm. Valcea, Romania. Also, she is the head of the research, development and innovation department of ACI S.A., one of the main economic operators in Romania, with experience in the field of civil and industrial constructions and concerns in the direction of eco-responsibility, consumption efficiency and implementation of clean technologies in construction (residential, educational, sports, industrial, commercial buildings). Her main fields of interest are energy efficiency of buildings, stationary application of renewable energies, fuel cell hydrogen technology, hybrid power generation systems, the impact of alternative energies on the environment and climate change. She is president of the Scientific, Innovation and Technical Education Commission and secretary of the Romanian Association of Building Services Engineering (AIIR) - Transylvania Branch, member of Romanian Association for Hydrogen Energy (AEHR) - secretary of the Cluj-Napoca Branch and member of Women in Green Hydrogen community (https://women-in-green-hydrogen.net/raluca-andreea-felseghi/).

Ioan Aşchilean graduated from the Technical University of Cluj-Napoca, Faculty of Civil Engineering, Department Installations for Civil Engineering, in 1992. In 1998, he was awarded the post-academic diploma in the field of architectural lighting installations. Starting in 2002, he attended the training program at Doctoral School, within Technical University of Cluj-Napoca and obtained the Doctoral Diploma in a fundamental field: Engineering Sciences, Civil Engineering. During 2004-2007, he has been an associated teaching staff of the Faculty of Civil Engineering and Installations, while during 2007 until 2014 he has been an associated teaching staff of the Faculty of Installations, within Technical University of Cluj-Napoca. Starting in 2018, he holds the academic position of associate professor at the Department of Civil Engineering and Management within the Faculty of Civil Engineering of Technical University, Cluj-Napoca, being habilitated as a doctoral supervisor starting with 2019 in the fundamental field: Engineering Sciences, Civil Engineering. Since 1993, he has been the head engineer responsible for installations at AICI CLUJ SA. Since 2001, he has been the general manager of Grup 4 Instalații SA company and, starting in 2004, he has been the Chairman of the Board of Directors of ACI CLUJ SA Company. Author of numberless scientific papers, published in ISI-indexed journals, and ISI-indexed journals with high impact factor, books, guides, invention patents. Due to the experience and expertise in the fundamental field of engineering sciences- civil engineering- he has been certified as a technical verifier and expert for installation works, technologic equipment and industrial installations, verifier and expert in electric installations and expert in natural gas transport. Currently, he is a vice-president of ARACO (Romanian Association of Construction Entrepreneurs), president of AIIR - Transylvania Branch and member of Hydrogen Association.

* * *

Florin Badea holds a BA in German and English Language from the Babes-Bolyai University of Cluj-Napoca, a European Master in Media, Communication and Cultural Studies from the University of London and the University of Kassel. He attended the International Performance and Media Studies PhD Program at the University of Mainz and holds a PhD in Sociology from the Vest University of Timisoara. He lectured in journalism at the University of Mainz and at the Babes Bolyai University of Cluj-Napoca. His research interests are commercial communication and their integration in the program flow, journalistic forms, intercultural communication and renewable energy.

Houcine Ben Moussa received their Ph.D. in Energy from the University of Poitiers- France in 1989. In 2007 he has received their promotion of Professor at the University of Batna-Algeria. He writes and presents widely contributions in the fuel cell and other domain.

Andrei Mircea Bolboaca is a Service Building Engineer and assistant professor. He graduated the Technical University of Cluj-Napoca, Faculty of Building Services. Starting with 2009, he attended the postgraduate work (PhD) within The Technical University of Cluj-Napoca and obtained the PhD title in a fundamental field: Engineering Sciences, Civil Engineering. He is currently working at the Faculty of Service Building Engineering, Technical University of Cluj-Napoca, Romania, being involved in a research project that aims to improve energy efficiency in buildings and indoor comfort qulity. His main fields of interests are energy efficiency of buildings, renewable energies and building comfort. He is vice-president of the Romanian Association of Building Services Engineering (AIIR). He is member of Romanian Association for Hydrogen Energy (AEHR) and member of Technical Committee CEN/CENELEC TC 6 Hydrogen in energy systems of Romanian Standards Association.

Alexandru Cernat is a lecturer, doctor engineer in internal combustion engines, specialist in alternative fuel use, in-cylinder process modeling and analysis, supercharging technology.

Amit Kumar Chaurasia is a research scholar IIT Roorkee. He has done b tech, m tech and pursuing a Ph.D. on the topic of hydrogen production on fermentation technologies. he has more than 5 publications in highly referred journals with 20 seminars and conferences across the globe.

Alin Corbu works as a researcher at the National Research and Development Institute for Cryogenics and Isotopic Technologies - ICSI Rm. Vâlcea, Hydrogen and Fuel Cell Department. During his career, he was responsible for the utilization of proton-exchange membrane fuel cells (PEMFC) in automotive applications. In his research he published, two articles on international databases and he participated at four national and international conferences.

Gabriel Ivan is a PhD Professor at the Technical University of Civil Engineering of Bucharest. He is working in the field of thermal comfort in buildings, thermal systems and plants using green energy. He is a member of the Hydrogen Energy Association from Romanian through which they are researching and promoting clean energies. The association is part of the National Institutes of Romania that studies renewable sources, including H2 energy.

Manoj Kumar Karnena received the B.S. degree in Environmental Science and the M.S. degree in Environmental science from GITAM (Deemed to be) University, India. Later, he pursued his Ph.D. in Water and Waste Water treatment at the same institution. He was the university topper and gold medalist in bachelors and post-graduation. He holds an Environmental Law degree from National Law University, Delhi, and Industrial Safety Degree from Annamalai University, Tamilnadu. Since 2019, he has been working as a teaching assistant with the Department of Environmental Science, GITAM Institute of Science, GITAM (Deemed to be) University, India. His greatest strengths are his research and communication skills. He started teaching and working in technical work in 2019 in various backgrounds. Manoj Kumar Karnena has written more than 50 papers since he started writing in 2017. He has written several book chapters and books on the topic within his research interest and holds two patents. His research interest is in Water and Wastewater Treatment, Natural Coagulants, Nanocatalysis, Storage Technologies, and membrane technology. Manoj Kumar Karnena had won several engineering designs and environmental protection awards in water and wastewater treatment.

Madhavi Konni is an Associate Professor in the Department of Basic Science and Humanities with 7 years of teaching and research experience. She has published her research papers in reputed international and national journals. her research interests include Hydrogen Storage Technologies, Metal-Organic Frame Works., Nano Catalysis, etc.

Ciprian Lupu is IFAC and IEEE member and is with the Faculty of Automatic Control and Computer Science, "Politehnica" University of Bucharest, Romania. He received his M.S. and PhD degree in Automatic Control from the "Politehnica" University of Bucharest, Romania, in 1996 and 2003 respectively. In 1995 he joined the same university. Since 2016 he has been the head of the Research Laboratory for Advanced Real-Time Control Systems. He was invited professor at Savoie Polytechnic School. He has co-authored more than 100 research papers and 6 books. His research interests include adaptive control, embedded systems and real time control.

Ameen Sha M. is working as Assistant Professor and Head, Department of Science and Humanities Muslim Association College of Engineering, Venjaramoodu, Trivandrum, Kerala, India. He received his Ph. D. degree from University of Kerala, Kerala, India on June 2020. He completed his M. Phil. Chemistry degree (2015) and M. Sc. Chemistry (2013) from the same University. He has authored 17 international publications. His research interests include electrochemistry, surface chemistry, photo and electrochemical hydrogen production and material and corrosion science.

Nichita Madalina is an installation engineer for constructions and I graduated from the Technical University of Civil Engineering Bucharest, Faculty of Building Services. Since 2000, I have attended postgraduate courses: master - specialization in thermal engineering and Doctoral Engineering - doctoral degree in mechanical engineering at the Technical University of Civil Engineering Bucharest. My main areas of pedagogical activity and research are: heat and mass transfer, heating equipment and installations, cold and air conditioning, heat pumps, renewable energy, energy efficiency of buildings. I am a member of the Romanian Association of Construction Services Engineering (AIIR) - Valahia Bucharest Branch, the Association of Romanian Refrigeration Technicians and Cryogenists (AFCR) and the American Society of Heating, Refrigeration and Air Conditioning Engineers (ASHRAE).

Prasenjit Mondal is presently working as Associate Professor in the Department of Chemical Engineering, Indian Institute of Technology Roorkee, India. His area of research is Energy and Environmental Engineering (Water /wastewater treatment through adsorption, electrocoagulation and biological processes including phytoremediation, microbial fuel cells, oil from algae, energy from coal, biomass and wastes, life cycle assessment).

Saratchandra Babu Mukkamala has been working in GITAM since 1991 and currently heading the GITAM Institute of Science since January 2019. He received his Master's degree in Chemistry in 1988 and Ph.D. in 1992 from Andhra University, Visakhapatnam. He has been a Visiting Scientist at the University of Durban-Westville, South Africa (2001), a DFG Postdoctoral Fellow at Institute of Inorganic Chemistry, University of Karlsruhe, Germany (2001-2004) and a NSF Postdoctoral Fellow at National Tsing Hua University, Taiwan (2005-2007). He visited several universities at USA, Canada, Japan, Australia and Singapore on academic assignment. He is a recipient of the Best Researcher award twice from GITAM in 2010 and 2012. His research interests include Synthesis of Metal-Organic Frameworks and Carbon nanotubes for gas storage applications, Luminescent nanomaterials for Technological & biomedical applications.

Ana-Maria Nasture is a Automation Engineer and Research National Institute for Cryogenics and Isotopic Technologies-ICSI-Rm. Valcea, she graduated from the Faculty of Automatic Control and Computers, Politehnica University of Bucharest, Bucharest, Romania. Starting in 2019, she attended the postgraduate work (Ph.D.) within the Faculty of Automatic Control and Computers, Politehnica University of Bucharest. Her main areas of interest are control algorithms for fuel cell hydrogen technology; mobile and stationary application of renewable energies; hybrid power generation systems; the impact of alternative energies on the environment and climate change.

Niculae Negurescu is professor, doctor engineer in internal combustion engine, specialist in alternative fuel use, in-cylinder process modeling and analysis, supercharging technology.

Cristian Nutu is a lecturer, doctor engineer in internal combustion engines, specialist in alternative fuel use, in-cylinder process modeling and analysis, fuelling systems management, pollutant emissions management.

Meenu P. C. is a research scholar at BITS Pilani, Hyderabad Campus. She completed his M. Phil. Chemistry degree (2018) and M. Sc. Chemistry (2016) from University of Kerala, India. She has authored 3 international publications. Her research interests include electrochemistry, fuel cells and material chemistry.

Constantin Pana is a professor, doctor engineer in internal combustion engines, a specialist in alternative fuel use, in-cylinder process modeling and analysis, supercharging technology, dynamic calculus and reduction of noise and vibrations of internal combustion engines, variable valves systems design and calculus, calculus and design of internal combustion engines.

Laurentiu Patularu is an Electrical Engineer, member of ICSI Energy, an Research Department of National Institute for Cryogenics and Isotopic Technologies-ICSI-Rm. Valcea. He graduated University Politehnica of Bucharest, Faculty of Electrical Engineering, with a Doctoral Diploma from 2014, that investigates a particular aspect of the PEM hydrogen fuel cell, linked to operating parameters, gas channels configuration, in order to obtain maximum power under high efficiency for cathode supply. Main field of interest is hydrogen: production (PEM electrolyzes), storage (metal hydride and pressurized systems), conversion (PEM fuel cells) and ultra purification (palladium membranes and Pressure Swing Adsorption Dryers). He is involved in the design, realization and test of different installations based on increased efficiency for electrochemical conversion of energy in Power to Gas and Gas to Power concepts. He is a Member of the Romanian Association for Hydrogen Energy (AEHR)- Ramnicu Valcea Division, Member of Romanian National Standardization Body – ASRO, with 6 national pattents from Romanian State Office for Inventions and Trademarks.

Maria Simona Raboaca is working as a Researcher at to National Research and Development Institute for Cryogenics and Isotopic Technologies ICSI Rm. Valcea, Hydrogen and Fuel Cell Department. Her Ph.D. is "Theoretical and practical Contribution regarding to sustain with hybrid energy a Passive House" in Faculty of Building Services Engineering in Technical University of Cluj-Napoca, Romania. Now, she is a project manager at ICSI to project "Smart conductive charging station, fixed and mobile, for electric propulsion transportation (SMiLE-EV)" proposes the deployment of fixed and mobile EV & PHEV charging stations to meet the mobility needs of tomorrow's society and to prepare active/potential industrial partners for knowledge/technology transfer at the component or system level in prepare launching new products. She has been contributing to the field of Renewable Energy, Green Buildings, Passive House Concept, Hydrogen Energy and Stationary and Mobile applications. She is the author and co-author of more technical papers in scientific conference proceedings and ISI journals.

Youcef Sahli received their Ph.D. in Energetic Mechanics from the University of Batna 2-Algeria in 2019. In 2019 he has received their promotion of Senior researcher B at the Renewable Energy Research Unit in the Saharan medium -Adrar-Algeria. He writes and presents several contributions in the fuel cell and other domains.

Sudeep Tanwar is an Associate Professor in the Computer Science and Engineering Department at the Institute of Technology, Nirma University, Ahmedabad, Gujarat, India. He is visiting Professor in Jan Wyzykowski University in Polkowice, Poland and University of Pitesti in Pitesti, Romania. He received B.Tech in 2002 from Kurukshetra University, India, M.Tech (Honor's) in 2009 from Guru Gobind Singh Indraprastha University, Delhi, India and Ph.D. in 2016 with specialization in Wireless Sensor Network. He has authored or coauthored more than 200 technical research papers published in leading journals and conferences from the IEEE, Elsevier, Springer, Wiley, etc. Some of his research findings are published in top cited journals such as IEEE TNSE, IEEE TVT, IEEE TII, Transactions on Emerging, Telecommunications Technologies, IEEE WCM, IEEE Networks, IEEE Systems Journal, IEEE Access, IET Software, IET Networks, JISA, Computer Communication, Applied Soft Computing, JPDC, JNCA, PMC, SUSCOM, CEE, IJCS, Software: Practice and Experience, MTAP, Telecommunication System. He has also edited/authored more than 20 books with International/National Publishers like IET, Springer. One of the edited text-books entitled, "Multimedia Big Data Computing for IoT Applications: Concepts, Paradigms, and Solutions" published in Springer in 2019 is having 3.7 million downloads till 10 September 2020. It attracts the attention of the researchers across the globe. (https://link.springer.com/book/10.1007/978-981-13-8759-3). He has guided many students leading to M.E./M.Tech and guiding students leading to Ph.D. He is Associate Editor of IJCS, Wiley and Security and Privacy Journal, Wiley. His current interest includes Wireless Sensor Networks, Fog Computing, Smart Grid, IoT, and Blockchain Technology. He was invited as Guest Editors/Editorial Board Members of many International Journals, invited for keynote Speaker in many International Conferences held in Asia and invited as Program Chair, Publications Chair, Publicity Chair, and Session Chair in many International Conferences held in North America, Europe, Asia and Africa. He has been awarded best research paper awards from IEEE GLOBECOM 2018, IEEE ICC 2019, and Springer ICRIC-2019.

Bariza Zitouni received her Ph.D. in Energetic Mechanics from the University of Batna-Algeria in 2007. In 2012 she has received her promotion of Professor at the University of Batna 1-Algeria. She writes and presents widely contributions in the fuel cell domain.

Index

A

B

C

D

E

F

G

H

I

L

M

N

O

P

R

S

T

U

W

www.ingramcontent.com/pod-product-compliance
Lightning Source LLC
LaVergne TN
LVHW060417300926
849162LV00006B/1457